INTERACTION OF TRANSLATIONAL AND TRANSCRIPTIONAL CONTROLS IN THE REGULATION OF GENE EXPRESSION

DEVELOPMENTS IN BIOCHEMISTRY

INTERACTION OF TRANSLATIONAL AND TRANSCRIPTIONAL CONTROLS IN THE REGULATION OF GENE EXPRESSION

Proceedings of the Fogarty International Conference on Translational/Transcriptional Regulation of Gene Expression, held at the National Institutes of Health, Bethesda, Maryland, U.S.A., on April 7–9, 1982.

Editors:

MARIANNE GRUNBERG-MANAGO, Ph.D.
Fogarty Scholar in Residence, Research Director, Institut de Biologie Physico-chimique, Paris, France

and

BRIAN SAFER, M.D., Ph.D.
Section on Protein Biosynthesis, Laboratory of Molecular Hematology, National Heart, Lung and Blood Institute, National Institutes of Health, Bethesda, Maryland, U.S.A.

ELSEVIER BIOMEDICAL
New York • Amsterdam • Oxford

Published by:

Elsevier Science Publishing Co., Inc.
52 Vanderbilt Avenue, New York, New York 10017

Sole distributors outside the USA and Canada:

Elsevier Science Publishers B.V.
P.O. Box 211, 1000 AE, Amsterdam, The Netherlands

Library of Congress Cataloging in Publication Data

Fogarty International Conference on Translational/Transcriptional Regulation of Gene
    Expression (1982: National Institutes of Health)
    Interaction of translational and transcriptional controls in the regulation of gene
    expression.

    (Developments in biochemistry, ISSN 0165-1714; v. 24)

    Sponsored by the Fogarty International Center.
    Includes bibliographical references and index.
    1. Gene expression—Congresses. 2. Genetic regulation—Congresses.
    3. Genetic translation—Congresses. 4. Genetic transcription—Congresses.
    I. Grunberg-Manago, Marianne, 1921-   II. Safer, Brian. III. John E.
    Fogarty International Center for Advanced Study in the Health Sciences.
    IV. Title. V. Series.
QH450.F63 1982   574.87'322   82-16283
ISBN 0-444-00760-1

Manufactured in the United States of America

Contents

Preface

A detailed understanding of the molecular events which occur during protein synthesis is fundamental to increasing our knowledge of both normal and pathologic processes. Since the isolation and purification of most translational components required for in vitro assembly of initiation complexes was achieved, a major shift in emphasis has occurred from the study of the mechanism of protein synthesis to the study of how this process is regulated. Recently, it has become increasingly apparent that there exists a close interaction of translational components and products with the transcriptional machinery of the cell. A greater appreciation has also emerged for the details of the molecular structure of translational components which specify their complex interactions with one another, as well as with the ultrastructure of the cell.

This Conference focuses on the molecular strategies employed during the modulation of gene expression subsequent to transcriptional initiation. The intention is to survey recent developments in several key areas in which transcriptional and translational components specifically interact. Both prokaryotic and eukaryotic systems are explored, and whenever possible structure-function correlations are considered. Because of the broad area covered, it has not been possible to cover all aspects and present all viewpoints. It is our hope, however, that this Conference

will stimulate productive interactions and provide an opportunity to
exchange new concepts between the diverse areas represented.

This Conference was planned during the tenure at the National
Institutes of Health of one of us (M. G-M.) as a Fogarty International
Scholar. This program provides the opportunity for Fogarty Scholars
to broaden their outlook from their own specialized fields, as well
as contribute their own areas of expertise to the NIH community.

The editors would like to express their gratitude to the Fogarty
International Center* for sponsoring this meeting, to Dr. E. Stadtman,
Dr. T. Stadtman, and Dr. B. Williams whose help in the organization of
this meeting was invaluable, and to Mrs. E. Church for her excellent
editing of the manuscripts. In addition, appreciation is extended to
all the speakers and discussants who were responsible for making this
meeting a highly successful one and the completion of this volume an
enjoyable and rewarding experience.

Marianne Grunberg-Manago

Brian Safer

* Fogarty International Center Director   — Dr. Claude Lenfant
  Scholar-in-Residence Branch Chief   — Dr. Peter G. Condliffe
  Conference and Seminar Program Branch Chief — Dr. Earl C. Chamberlayne
  Conference Coordinator   — Mrs. Nancy E. Shapiro

SECTION I: TRANSLATIONAL / TRANSCRIPTIONAL CONTROLS IN PROKARYOTES

Published 1982 by Elsevier Science Publishing Co., Inc.
Marianne Grunberg-Manago and Brian Safer, editors
INTERACTION OF TRANSLATIONAL AND TRANSCRIPTIONAL
CONTROLS IN THE REGULATION OF GENE EXPRESSION

REGULATION OF GENE EXPRESSION BY TRANSCRIPTION TERMINATION AND RNA PROCESSING

MARTIN ROSENBERG AND URSULA SCHMEISSNER*
Laboratory of Biochemistry, National Cancer Institute, National Institutes of
Health, Bethesda, Maryland 20205 (301-496-5226)
*Present address: BIOGEN, S.A. 3, Route de Troinex, 1227 Carouge/Geneva,
Switzerland

INTRODUCTION

Lysogenic development by phage λ requires the insertion of the phage genome into the E. coli chromosome. This site-specific integration event requires several host proteins and one phage gene product, integrase.[1-3] The gene for integrase (int) is part of the major leftward transcription unit of lambda and is positioned immediately preceding the site at which the integrative recombination event occurs (att) (Fig. 1). The att regulatory region extends for 200 bp beyond int and contains a variety of regulatory sequences involved in the recombination events.[4,5]

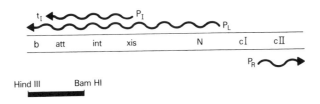

Fig. 1. Schematic genetic map of phage λ showing the two transcripts which traverse the int gene and initiate at the P_I and P_L promoters, respectively. Also indicated is the major rightward mRNA which initiates at the P_R promoter and the BamHI-HindIII restriction fragment used as a hybridization probe for the t_I region (see text for details). This figure, as well as figures 2-8 and 10, are from Schmeissner et al., in preparation.

During a normal phage infection, int is transcribed at different times from
two promoter signals.[6,7] Early after infection, int transcription derives
from the major leftward phage promoter, P_L, positioned approximately 8 kb
upstream of int. This high molecular weight polycistronic mRNA expresses
integrase very poorly.[8,9] In contrast, efficient int expression occurs later
in phage development from another transcript. This monocistronic mRNA derives
from the positively regulated promoter, P_I, positioned only 137 bp upstream
of int.[10-13] The dramatically different levels of int expression obtained
from these two transcripts does not appear to result simply from differential
message translation (Schmeissner, U., McKenney, K., Court, D. and Rosenberg,
M. in preparation). Instead, the phage has evolved a rather remarkable
regulatory mechanism which utilizes overlapping and alternative signals for
transcription termination and RNA processing to control int expression from
the two different mRNAs. This article summarizes the work which has helped to
elucidate the molecular features of this regulatory phenomenon and in doing
so, has demonstrated a new and important role for transcription termination
in gene expression.

P_I transcription terminates at t_I. The first indication of the unusual
nature of the regulation of int expression was the finding by Guaneros and
coworkers of mutations located beyond the int coding sequence which allowed
efficient int expression from the P_L transcription unit.[14,15] It was
suggested that these mutations defined a regulatory site responsible for
selectively inhibiting int expression from the P_L transcript (i.e., sib, site
of inhibition in the λ b region). In an effort to characterize the function
of this regulatory signal, we examined int specific transcription from both
the P_I and P_L promoters specifically in the region distal to the int coding
sequence. For these analyses, a 492 bp λ DNA fragment which spans the end of
the int gene, the entire att regulatory region, and extends 250 bp beyond att
into the b region, was used as a hybridization probe (see Fig. 1). The DNA
sequence of this region has been defined (Fig. 2).[4,16]

We first monitored int transcription directed by the P_I promoter. Cells
were pulse-labeled with ^{32}P between 10 and 11.5 minutes after infection, a
time when int is being transcribed from P_I. RNA was prepared, hybridized to
the DNA fragment probe, and the hybridized RNA characterized by standard two-
dimensional fingerprinting procedures.[17] Each oligonucleotide was identified
and unambiguously positioned within the DNA sequence of the region (see Fig.
2).

3

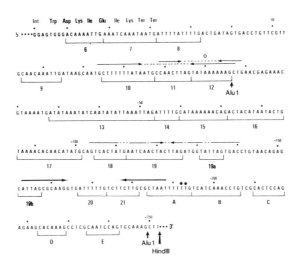

Fig. 2. Partial DNA sequence of the coding strand of the BamHI-HindIII fragment shown in Fig. 1.[4,16] The orientation of the sequence is inverted relative to the λ map shown in Fig. 1. Numbering starts at the center of att with positive numbers extending toward int and negative numbers extending toward b. T_1 oligonucleotides are indicated by either numbered or lettered brackets. Major dyad symmetries are designated by arrows. Asterisks indicate the exact positions of transcription termination at t_I.

The results demonstrate that P_I transcription traverses the entire att regulatory region and extends some 180 nucleotides beyond att. The detection of oligonucleotide #21, but the absence of oligonucleotides A-E indicates that the P_I transcript must terminate beyond residue position -182 but before residue -194. Consistent with this analysis is the fact that the DNA sequence of this region exhibits those features usually associated with transcription termination signals (Fig. 3): 1) an extensive dyad symmetry rich in G-C pairs which gives rise to a potential base paired stem and loop structure in the corresponding RNA transcript of the region and 2) a run of consecutive thymidylate residues immediately following the symmetry element.[18] By analogy with other terminators the P_I int mRNA should stop within or just beyond this T-rich sequence. We have designated this region t_I, the terminator signal for the P_I directed int transcript.

4

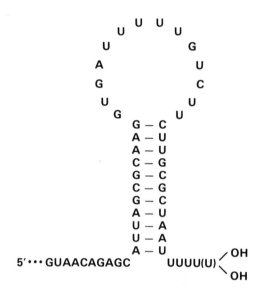

Fig. 3. Potential secondary structure of the 3'-end of the P_I directed <u>int</u> mRNA. The 3'-terminal heterogeneity of the transcript is indicated by parentheses.

In order to better characterize the t_I signal, we cloned a 242 bp λ DNA fragment carrying t_I into a plasmid vector designed specifically for studying transcription termination signals (Fig. 4).[19] This vector, pKG1800, carries the <u>E. coli</u> galactokinase gene (<u>galK</u>) such that <u>galK</u> expression is controlled by the <u>gal</u> promoter (Pgal). Insertion of a DNA fragment carrying a terminator between Pgal and <u>galK</u> results in a reduction in <u>galK</u> expression. The extent of the reduction is a direct measure of the <u>in vivo</u> termination efficiency of the signal. In pKG1800, the t_I fragment terminates transcription from Pgal with an efficiency of 98%. This result is consistent with the results obtained on the phage: no read-through transcription at the t_I site was detected from the P_I promoter.

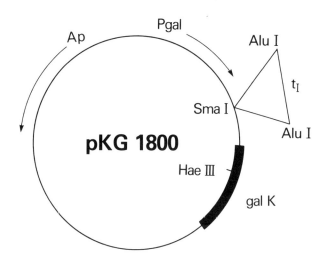

Fig. 4. Construction of the plasmid pKG1800sib. The 242 bp AluI restriction fragment (Fig. 2), which carries the t_I terminator region was cloned into the SmaI site of the vector pKG1800 between the gal promoter (Pgal) and the galactokinase gene (galK).[19] Ap indicates the β-lactamase gene from pBR322. Arrows indicate the direction of transcription.

The pKG1800 vector also allows in vitro analysis of terminator function. The vector containing t_I was linearized and used as a template for in vitro transcription. The RNA products were resolved by polyacrylamide gel electrophoresis (Fig. 5) and the major transcripts characterized by fingerprint analysis. The prominent 630 nucleotide transcript (T) was identified as a Pgal-initiated RNA which terminated precisely at the t_I signal. Two 3'-terminal oligonucleotides were identified indicating that the RNA terminated with some heterogeneity within the T-rich sequence at residue positions -192 and -193 (Fig. 2 and 3). Apparently, the t_I signal also functions efficiently in vitro and termination at this site does not require any ancillary factors such as rho protein (i.e. t_I is an independent termination signal).

P_L transcription read-through t_I. We next monitored int gene transcription directed by the P_L promoter. Although this transcription also traverses the int gene coding region, little if any, integrase is expressed.[8,9] Our method of analysis was identical to that used for P_I transcription, except that the cells were pulse-labeled very early after infection using a phage which carried an inactive P_I promoter. Under these conditions int transcription is from P_L.[20]

6

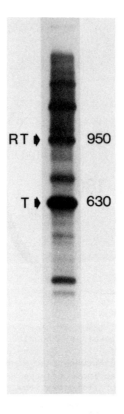

RT ▶ 950

T ▶ 630

Fig. 5. Polyacrylamide gel analysis of the products resulting from in vitro transcription of plasmid pKG1800sib cleaved with restriction endonuclease HaeIII (see text for details).

The P_L transcription unit (unlike P_I) is known to be subject to regulation by the phage antitermination function N.[21,22] In the presence of N, P_L-directed transcription will read-through termination signals. Indeed, the oligonucleotide fingerprint analysis of P_L transcription beyond int indicated efficient read-through transcription at the t_I site. Oligonucleotides characteristic of the region beyond t_I (e.g. oligonucleotides B, C, D, E, Fig. 2 and 6) were detected in mole quantities equivalent to those which precede t_I. Most unexpectedly however, we found several oligonucleotides characteristic of the t_I signal either absent from the fingerprint (e.g. oligonucleotides 20, 21, and A) or in dramatically reduced amounts (e.g. 19a, 19b). It seems that although the P_L transcript reads-through t_I into the b region of the phage, this RNA is altered such that oligonucleotides specific for t_I are not detected by the hybridization-fingerprint analysis.

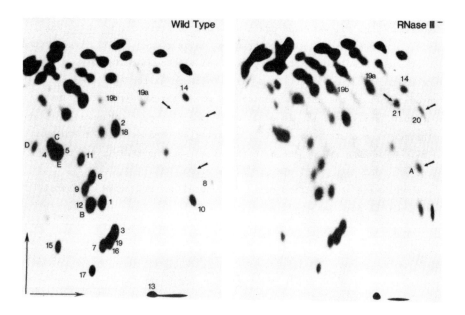

Fig. 6. Two-dimensional T_1 oligonucleotide fingerprint analysis of the int-att region of the λ P_L mRNA synthesized in vivo either wild type cells (left) or in cells deficient in RNase III activity (right). Oligonucleotide numbers correspond to those shown in Fig. 2. Arrows indicate those oligonucleotides found present in the RNase III⁻ host which are absent in the wild type host (see text for details).

One possible explanation for the missing oligonucleotides is that the transcript is being processed in the t_I region. Experiments by Belfort (1981) suggested that RNA processing plays an important role in the regulation of int expression.[23] She demonstrated that the int gene could be expressed efficiently from the P_L transcript in cells deficient in the RNA processing enzyme, ribonuclease III (RNase III). For this reason, we repeated our analysis of the P_L transcript in an RNase III deficient host. Again, RNA from the t_I region was selectively hybridized and characterized by fingerprint analysis (Fig. 6). In contrast to the results obtained using the wild type host, we now detected full mole amounts of the t_I specific oligonucleotides, in addition to the oligonucleotides preceding and distal to t_I. It appears that RNase III is responsible for the selective loss of the t_I specific RNA sequence and moreover, that the loss of this region correlates precisely with the ability of the transcript to express integrase.

Transcripts reading through t_I are processed. We next sought to define in detail the effect of RNase III on transcripts passing through or terminating at the t_I signal. The plasmid vector pKG1800, into which the t_I region had been inserted (Fig. 4), was used again as a DNA template for in vitro RNA synthesis. The DNA was cut with restriction endonuclease HaeIII, transcribed, and the products resolved by polyacryamide gel electrophoresis (Fig. 5). Two discrete transcripts were identified which initiate at the Pgal promoter. The major product, T, is the 630 nucleotide RNA which terminates at t_I. The other product, designated RT, is a 950 nucleotide "run-off" transcript which reads-through t_I and stops at a HaeIII restriction site positioned 320 bp beyond t_I. Both RNAs are produced in the same reaction because the t_I signal functions with about only 80% efficiency in vitro.

The two RNAs were gel-purified (Fig. 7, lane a) and subjected separately to digestion with pure RNase III.

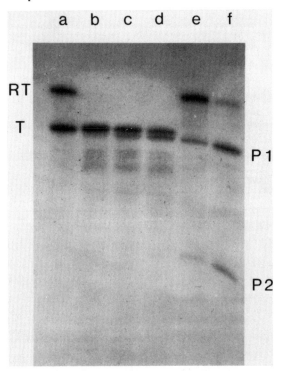

Fig. 7. Polyacryamide gel analysis of the products of RNase III processing of the T and RT in vitro synthesized transcripts (see Fig. 5 and text for details). lane a: unprocessed gel-purified T and RT substrates; lanes b, c and d: T RNA digested with increasing concentrations of RNase III (b = 1x, c = 2.5x, d = 5x); lanes e and f: RT RNA digested with RNase III concentrations identical to those used in lanes b and c, respectively. P1 and P2 are the major products of RNase III digestion (see text for details).

Analysis of the products on polyacrylamide gels indicates that the RT RNA substrate is cleaved into three discrete fragments (lanes e and f) of approximately 595 nucleotides (Pl), 330 nucleotides (P2), and 25 nucleotides (not seen in Fig. 7 but resolved on a higher percentage gel). RT processing was more than 90% efficient using digestion conditions similar to those known · to process the RNase III sites in the phage T7 early region mRNA.[24,25] Surprisingly, the terminated RNA (T) also was cleaved by RNase III. In this case two products resulted identical to the 595 nucleotide (Pl) and 25 nucleotide long products obtained with the RT transcript. Most importantly however, the T RNA substrate was processed much less efficiently than the RT RNA. Using the same digestion conditions, only 10-20% of the T RNA was cleaved (compare lanes b and e, c and f). Even at 5x higher RNase III concentrations less than 50% of the T RNA was processed (lane d). Presumably, the markedly increased susceptability of the read-through RNA is due to additional sequence information encoded selectively in this transcript beyond the t_I signal.

Examination of the DNA sequence in the region beyond t_I indicates a region which exhibits complementarity to a sequence immediately preceding the stem and loop structure of the t_I signal. Base-pairing of these two sequences gives rise to an extended region of potential secondary structure (Fig. 8). This structure occurs only in transcripts which read-through t_I (i.e. the phage λ P_L transcript and the plasmid RT RNA) and exhibits features similar to those associated with other known sites of RNase III cleavage.[26]

The exact sites of RNase III processing in the t_I region were positioned unambiguously by fingerprint analysis of each of the cleavage products. Two cuts are made 24 nucleotides apart on opposite sides of the terminator stem and loop structure (Fig. 8). The cuts are staggered two base-pairs apart, exactly analogous to the relative position of the cuts made by RNase III both in the E. coli ribosomal RNA precursor transcript[27] and at the end of the T7 gene 1.1 message.[28] In the case of the t_I site, cleavage releases a 24 nucleotide RNA fragment from the transcript. It is precisely this 24 nucleotide region which was missing from our analysis of the P_L read-through transcript synthesized in wild type cells (Fig. 6). Apparently, RNase III makes the same cuts at t_I on the P_L mRNA as it does in vitro on the plasmid transcript.

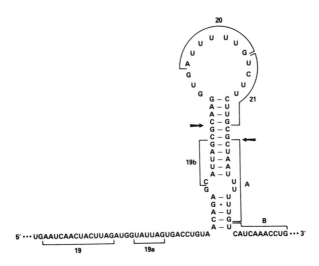

Fig. 8. Potential secondary structure of the RNA transcripts which read-through t_I. T_1 oligonucleotides are bracketed and are numbered or lettered as in Figs. 2 and 6. Arrows designate the defined sites of RNase III cleavage.

Thus, we see again an exact correlation between the transcriptional events which occur in the t_I region and the ability of the λ P_I and P_L mRNAs to express int. As summarized in Fig. 9, transcripts which terminate at t_I express int, whereas transcripts which read-through t_I are processed and inactive for int expression. If the processing step is blocked, then int is expressed from the read-through RNA. Apparently, the t_I region positioned more than 250 bp downstream of the int gene is the regulatory region con-trolling int expression (i.e. sib). Regulation is achieved by the mutually exclusive function of two structurally overlapping transcriptional regulatory elements, one specifying termination and the other RNA processing. Signal function is dependent upon the nature of the transcription within this region.

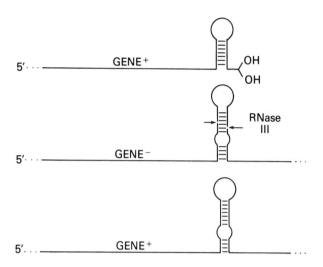

Fig. 9. Schematic representation showing the correspondence between the transcriptional events which occur in the t_I region and the expression of the *int* gene positioned upstream. Arrows designate the cleavage sites of RNase III.

Mechanism of sib regulation. How does the RNase III processing event at t_I lead to *int* gene inactivation? A number of models have been proposed.[14,15,23,29] Most invoke some form of RNA folding based on secondary structural interactions between sequences in the *sib* regulatory region and the *int* coding region. For example, sequences in *sib* might base-pair with sequences in the translation initiation region of *int*, thereby inhibiting *int* translation. Alternatively, base-pairing between sequences in *sib* and *int* might create yet another RNase III processing site, resulting in cleavage of the *int* coding sequence. Of course, in either case, the mechanism requires that the potential interactions occur selectively in the processed P_L transcript, since both the P_I mRNA and the unprocessed P_L mRNA express *int* efficiently.

We find these models unlikely for several reasons. It is difficult to rationalize any of these possibilities with the positioning of the RNase III cleavage sites in the t_I region. Presumably, these cleavages are responsible for making the appropriate sequences available for the proposed structural interactions within *int*. Yet, sequences within and beyond t_I do not exhibit extensive complemtarity, either to sequences in the translation initiation region or coding region of *int*. Moreover, our examination of transcription within the int gene coding region does not support the existence of additional RNase III processing events. Instead, our data suggest another mechanism by which the t_I region regulates int expression.

We have demonstrated that the primary event leading to inactivation of int expression from the P_L mRNA is the occurrence of two specific RNase III directed cuts in the t_I site. Moreover, we have shown that this processing occurs independently of int region sequences and even occurs when the t_I site is contained within an entirely different transcription unit (i.e. the pKG1800 plasmid construction). The most obvious result of the processing event is the selective destruction and removal of the terminator stem and loop structure from the read-through transcript. Both transcripts which express int, the P_I mRNA and the unprocessed P_L mRNA, retain this stem and loop structure. We propose that the selective destruction of this structure results in destabilization of the processed message and thus, int inactivation results from rapid turnover of the processed mRNA. This model implies that the terminator structure at the end of the P_I mRNA is also required to stabilize the message in order to insure its proper translation. Perhaps terminator structures in general have this dual function, and thus play an important role in gene expression by controlling mRNA stability. The experiments described in the following section provide strong support for these contentions.

Sib regulation functions downstream of another gene. One important prediction of our model is that the sib regulatory system might function independently of the λ int gene. If sib is controlling only mRNA stability, then it should function when placed downstream of other genes and thereby regulate their expression. In order to test this possibility the sib region (same fragment used to construct pKG1800sib, Fig. 4) was placed downstream of the E. coli galactokinase gene on a plasmid vector, pAK1 (Fig. 10, A. Shatzman and M.R., unpublished), in which galK expression is controlled by the P_L promoter. Constructions were obtained which had the insert in both possible orientations. The vector with t_I positioned correctly was called pAK1sib, whereas the control plasmid, carrying t_I inverted, was called pAK1bis. In pAK1sib, the t_I site is positioned 330 bp beyond the galK coding sequence, a distance similar to that which occurs between t_I and int on phage λ.

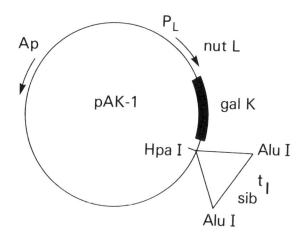

Fig. 10. Construction of the plasmid pAKlsib. The 242 bp AluI restriction fragment (Fig. 2), which carries the t_I region (sib) was inserted into a HpaI site positioned beyond the galK gene in the vector pAKl (A. Shatzman and M.R., unpublished result). galK in this vector is transcribed from the phage λ P_L promoter. Ap indicates the β-lactamase gene of pBR322; NutL indicates the N utilization site on the P_L transcription unit[22] (see text for details). Arrows indicate the direction of the transcription.

The ability of pAKl, pAKlbis, and pAKlsib to express galK was examined. Expression was monitored in a galK deficient host lysogen carrying a temperature sensitive (ts) mutation in the phage repressor gene (λN99cI857).[30] The use of the ts lysogen allows us to 1) grow cells at low temperature (32°C) such that galK expression from the plasmid is repressed, 2) selectively induce P_L transcription of galK by simply raising the temperature of the culture (to 42°C), and 3) concomitantly provide the antitermination function N which is required for read-through transcription at the t_I signal. Expression was monitored initially by examining the gal phenotype of each cell on an indicator media MacConkey galactose.[19] Temperature induction of cells containing the plasmids pAKl and pAKlbis resulted in red colonies, indicating that these vectors expressed sufficient galK levels to complement the galK deficiency of the host. In contrast, pAKlsib gave white colonies even after prolonged periods of induction. Apparently, the sib regulatory element was functioning to inactivate galK expression on the plasmid identically to its mode of action on the phage λ int gene.

14

The extent of the sib regulatory effect was examined by determining the levels of galK activity produced by each vector after induction. Consistent with their phenotypes the pAK1 and pAK1bis constructions produced high levels of galK activity (> 125 units), whereas the pAK1sib vector produced very low levels (< 10 units). In order to demonstrate that this effect was dependent on processing, each plasmid was moved to a lysogenic host deficient in RNase III. In this background the pAK1sib vector now expressed galK activity at levels equivalent to the control vectors pAK1 and pAK1bis. Clearly, galK expression was under sib control and thus, t_I functions as an independent regulatory element controlling the expression of other upstream positioned genes. Gene inactivation requires only the removal of the t_I stem and loop structure by RNase III processing. Presumably, this removal leads to rapid message degradation.

Recombinant Selection For Mutants

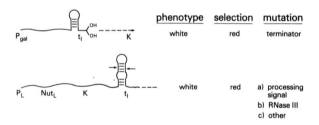

Fig. 11. Schematic representation of the galK transcription units constructed in the plasmids pKG1800sib (top) and pAK1sib (bottom). Also indicated is the phenotype of those plasmids on MacConkey galactose indicator plates, the phenotypic selection which is used to generate mutations, and the type of mutations obtained (see text for details). All other designations are those used in Figures 4 and 10.

The ability to remove the sib regulatory element from phage λ and establish its function as both a termination signal (pKG1800sib, Fig. 4) and as a processing site (pAK1sib, Fig. 10) allows an entirely new approach to the study of this regulation. These vectors provide both genetic selection and functional assessment of mutations which effect <u>sib</u> regulation. As shown in Fig. 11, mutations are being obtained and characterized which effect both t_I terminator function and RNase III processing at t_I. The system allows selective mutagenesis of either the plasmid or the bacterial host depending upon the mutagenic procedure. Plasmid mutations affect the regulatory signals (e.g. terminator or RNase processing site) whereas cellular mutations affect the host components required for regulatory site function. Of particular interest are those cellular mutations other than those in RNase III, which allow <u>gal</u>K expression from the pAK1sib vector. These mutations may be in important cellular components involved in mRNA degradation. It is likely that these recombinant systems will provide us with a new level of understanding of the processes of transcription termination, RNA processing, and message degradation in <u>E</u>. <u>coli</u>.

ACKNOWLEDGEMENTS

We wish to thank Charles Mock for photographic work and Gail Taff for typing and editing this manuscript.

REFERENCES

1. Nash, H.A. (1978) Curr. Top. Microbiol. Immunol. 78, 171-199.
2. Miller, H.I., Abraham, J., Benedik, M., Campbell, A., Court, D., Echols, H., Fischer, R., Galindo, J.M., Guarneros, G., Hernandez, T., Mascarenhas, D., Montanez, C., Schindler, D., Schmeissner, U. and Sosa, L. (1980) Cold Spring Harbor Symp. Quant. Biol. 45, 439-445.
3. Herskowitz, I. and Hagen, D. (1980) Annu. Rev. Genet. 14, 339-445.
4. Hsu, P.L., Ross, W. and Landy, A. (1980) Nature (London) 285, 85-91.
5. Mizuuchi, M. and Mizuuchi, K. (1980) Proc. Natl. Acad. Sci. 77, 3220-3224.
6. Shimada, K. and Campbell, A. (1974) Proc. Natl. Acad. Sci. 71, 237-241.
7. Singer, E. (1970) Virology 40, 624-633.
8. Court, D., Adhya, S., Nash, H. and Enquist, L. (1977) in DNA Insertion Elements, Plasmids and Episomes, Bukhari, A.I., Shapior, J.A. and Adhya, S.L. ed., Cold Spring Harbor Laboratory, Cold Spring Harbor, NY, pp. 389-394.
9. Chung, S. and Echols, H (1977) Virology 79, 312-319.
10. Davis, R.W. (1980) Nucleic Acid Res. 8, 1765-1782.
11. Hoess, R.H., Foeller, C., Bidwell, K. and Landy, A. (1980) Proc. Natl. Acad. Sci. 77, 2482-2486.
12. Abraham, J., Mascarenhas, D., Fischer, R., Benedik, M., Campbell, A. and Echols, H. (1980) Proc. Natl. Acad. Sci. 77, 2477-2481.
13. Schmeissner, U., Court, D., McKenney, K. and Rosenberg, M. (1981) Nature (London) 292, 173-175.
14. Guarneros, G. and Galindo, J.M. (1979) Virology 95, 119-126.
15. Guarneros, G., Montanez, C., Hernandez, T. and Court, D. (1982) Proc. Natl. Acad. Sci. 79, 238-242.
16. Hoess, R.H., Foeller, C., Bidwell, K. and Landy, A. (1980) Proc. Natl. Acad. Sci. 77, 2482-2486.
17. Barrell, B.G. (1971) in Procedures in Nucleic Acids Research Vol. 2, Cantoni, G. and Davies, D. ed., Harper and Row, New York, pp. 751-779.
18. Rosenberg, M. and Court, D. (1979) Annu. Rev. Genet. 13, 319-353.
19. McKenney, K., Shimatake, H., Court, D., Schmeissner, U., Brady, C. and Rosenberg, M. (1981) in Gene Amplification and Analysis, Vol II: Analysis of Nucleic Acids by Enzymatic Methods, Chirikjian, J.C. and Papas, T.S., ed., Elsevier, North Holland, pp. 383-415.
20. Bovre, K. and Szybalski, W. (1969) Virology 38, 614-626.
21. Franklin, N.C. (1974) J. Mol. Biol. 89, 33-48.
22. Salstrom, J.S. and Szybalski, W. (1978) Virology 88, 252-260.
23. Belfort, M. (1980) Gene 11, 149-155.
24. Rosenberg, M., Kramer, R.A. and Steitz, J.A. (1974) J. Mol. Biol. 89, 777.
25. Rosenberg, M. and Kramer, R. (1977) Proc. Natl. Acad. Sci. 74, 984-988.
26. Gegenheimer, P. and Apirion, D. (1981) Microbiological Review Vol 45, 502-541.
27. Bram, R.J., Young, R.A. and Steitz, J.A. (1980) Cell 19, 393-401.
28. Dunn, J.J. (1976) J. Biol. Chem. 251, 3807-3814.
29. Schindler, D. and Echols, H. (1981) Proc. Natl. Acad. Sci. 78, 4475-4479.
30. Sussman, R. and Jacob, F. (1962) C.R. Seances Acad. Sci., Paris 254, 1517-1519.

Published 1982 by Elsevier Science Publishing Co., Inc.
Marianne Grunberg-Manago and Brian Safer, editors
INTERACTION OF TRANSLATIONAL AND TRANSCRIPTIONAL
CONTROLS IN THE REGULATION OF GENE EXPRESSION

ATTENUATION IN THE CONTROL OF TRYPTOPHAN OPERON EXPRESSION

CHARLES YANOFSKY

Department of Biological Sciences, Stanford University, Stanford, California
94305 USA

Several bacterial operons encoding enzymes that participate in amino acid biosynthesis are regulated by attenuation[1]. This regulatory mechanism involves controlled transcription termination at a site, the attenuator, located in the leader region of the operon, the segment between the transcription start site and the first structural gene. Comparison of the sequences of the leader regions of the 6 amino acid biosynthetic operons known to be regulated by attenuation[2-10] reveals considerable similarities (Fig. 1). In each there is a short coding region specifying a peptide rich in the regulatory amino acid. Slightly beyond this coding region there is a site of transcription termination. This site consists of an A+T rich region preceded by a G+C rich region exhibiting dyad symmetry. The transcript segment corresponding to the G+C rich region can form a stable hydrogen-bonded stem and loop structure[11,12]. This RNA structure, which is called the terminator, is thought to be the termination signal that is recognized by the transcribing RNA polymerase molecule. In addition, each leader transcript can potentially form an alternate RNA secondary structure involving an earlier segment of the transcript and one segment of the GC-rich stem[11,12]. We have proposed that formation of this alternate secondary structure, which we call the antiterminator, prevents formation of the terminator, thereby eliminating termination at the attenuator[11]. The choice between alternate RNA secondary structures was hypothesized to be regulated in vivo by ribosome movement along the transcript during synthesis of the leader peptide[11]. Specifically in the case of the trp operon, ribosome stalling over the tandem Trp codons in the leader transcript coding segment -- a consequence of the unavailability of charged $tRNA^{trp}$ -- was thought to mask a segment of the transcript, thereby favoring formation of the antiterminator RNA structure that is mutually exclusive with formation of the terminator[11-13]. This model[1] can explain attenuation control in each of the six amino acid biosynthetic operons regulated by attenuation.

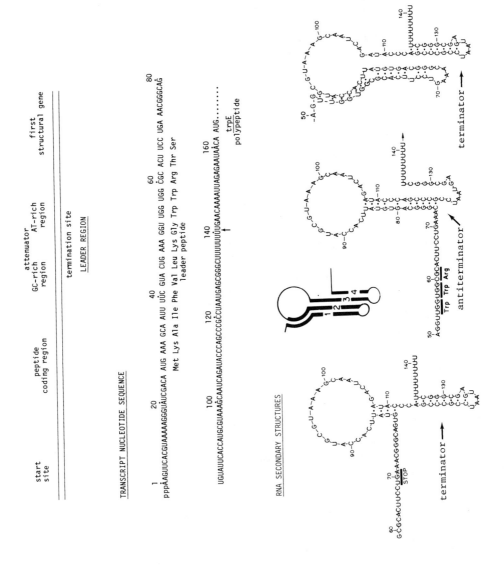

start site · peptide coding region · attenuator GC-rich region AT-rich region · first structural gene

termination site

LEADER REGION

TRANSCRIPT NUCLEOTIDE SEQUENCE

1   20   40   60   80
pppAAGUUCACGUAAAAAGGGUAUUCGACA AUG AAA GCA AUU UUC GUA CUG AAA GGU UGG CGC ACU UCC UGA AACGGGCAG
                                Met Lys Ala Ile Phe Val Leu Lys Gly Trp Trp Arg Thr Ser
                                leader peptide

100   120   140   160
UGUAUUCACCAUGCUAAAGCUAAAGCAAUCAGAUACCCAGCCCGCCUAAUGAGCGGGCGUUUUUUUUGAACAAAAUUAGAGAAUAACA AUG........
                                      trpE polypeptide

RNA SECONDARY STRUCTURES

We and others have examined the predictions of the model in in vivo and in vitro studies with the trp operons of Escherichia coli and Serratia marcescens. The evidence obtained with E. coli indicates that (1) the peptide coding region of the leader transcript is used as an efficient site of translation initiation[14,15]; (2) defective translation of Trp codons rather than a Trp deficiency as such is responsible for relief of termination[16]; (3) both in vivo and in vitro, transcription can be terminated in the leader region at about bp 140[17,18]; (4) the terminated transcript has secondary structure[11,12], the segment from nucleotides 114 to 134 forms an RNase Tl resistant stem and loop -- this structure is believed to be the terminator; the RNA segment from bp 52 to 69 pairs with segment 75 to 94, also forming an RNase Tl resistant structure. In addition, segment 74 to 85 can theoretically pair with segment 108 to 121 to form the presumed antiterminator. This structure is not demonstrable experimentally, presumably because formation of the other two structures precludes formation of this alternate RNA structure; (5) single bp mutations in the leader region that relieve transcription termination at the attenuator in vivo and in vitro occur in the G-C rich region that forms the RNA terminator [with one exceptional change in the AT rich region] and reduce the predicted stability of the terminator[19,20]; (6) single bp mutations that prevent formation of the antiterminator have no effect on termination in vitro but prevent transcription termination relief associated with tryptophan starvation in vivo[21]; (7) mutations altering the translation start codon similarly prevent the tryptophan starvation response, presumably by preventing ribosome movement on the transcript[21]; (8) starvation for Arg as well as Trp, but not other amino acids in the leader peptide, relieves transcription termination in vivo[21]. The single Arg codon in the leader transcript is the codon immediately following the tandem Trp codons. If we assume that a translating ribosome masks 10 nucleotides on the 3' side of the codon being read, then stalling over only the Trp and Arg codons could permit formation of the antiterminator secondary structure that prevents formation of the terminator; (9) mutations in the gene for the β-subunit of RNA polymerase may either increase or decrease transcrip-

---

Fig. 1. The trp leader region and the primary and secondary structures of the trp leader transcript. In the schematic insert the pairing segments are numbered. Structure 3:4, the terminator, is believed to be the termination signal that the transcribing RNA polymerase molecule recognizes (left structure). Structure 2:3, the antiterminator, is thought to form when the translating ribosome stalls over the Trp or Arg codons (center structure). Structure 1:2 presumably forms when translation initiation does not occur (right structure). The leader regions and leader transcripts of all amino acid biosynthetic operons known to be regulated by attenuation have similar structures.

tion termination at the attenuator[22]. These mutations probably alter the recognition site for the RNA terminator. These mutations have no effect on termination in vivo under conditions where the terminator does not form; (10) in vitro studies with base analogs indicate that RNA secondary structures rather than DNA secondary structures constitute the termination signal that RNA polymerase recognizes[11,23]; (11) when RNA polymerase transcribes the leader region of the operon it pauses at bp 90 after synthesizing the first RNA secondary structure[24,25]; it is thought that this pause serves to synchronize transcription with translation, thereby ensuring that the ribosome translating the leader transcript will reach the Trp codons or the leader peptide stop codon at an appropriate time to determine which of the alternate RNA secondary structures will form[1].

The findings summarized, although providing information on many aspects of attenuation, leave unanswered questions on key features of this regulatory process. In our continuing studies on attenuation we have obtained additional insight into this regulatory mechanism. We have recently completed an analysis of the in vivo and in vitro effects of deletions in the leader region of the trp operon of Serratia marcescens[26,27]. In this study, performed by Iwona Stroynowski, we used Bal 31 exonuclease to produce deletions in vitro of virtually every segment of the leader region. The deletion operons, contained in plasmids, were transferred to bacteriophage lambda and subsequently into the bacterial chromosome in single copy form, where their effects were analyzed in gene expression studies. The plasmid intermediates were used to determine the sequence deleted from the leader region. In addition, a small restriction fragment containing the trp promoter/operator, leader region and initial segment of trpE, was isolated from each plasmid and used in in vitro transcription termination/readthrough experiments with pure RNA polymerase (I. Stroynowski & C. Yanofsky, in preparation). The trp leader region of Serratia[28] was selected for this project because (1) the entire promoter-operator-leader region is on a single 250 bp EcoRl fragment, (2) the second EcoRl site is 25 bp before trpE, allowing convenient fusion of any manipulated EcoRl fragment to the trpE-containing segment, and (3) there are many convenient restriction sites in the trp leader region of Serratia at which the DNA may be cleaved and subsequently digested with Bal 31. Using this Serratia system we planned to test one of the basic assumptions of our attenuation model. We reasoned that if the role of the ribosome were simply to mask a specific segment of the transcript and prevent it from participating in the formation of a particular RNA secondary structure, then a deletion removing the segment that is presumed to be masked should have the same effect as ribosome stalling. Accordingly, we isolated deletions that removed crucial portions of the leader region (and hence corresponding segments of the leader transcript) and examined their effects in vivo and in vitro[26,27].

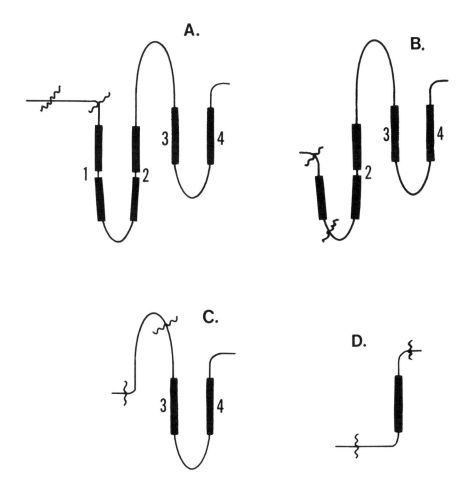

Fig. 2. Schematic representation of the leader transcripts of mutants with deletions in the leader region of the trp operon of S. marcescens. The wiggly lines indicate the segments fused as a consequence of deletions. The black bars mark the transcript segments that can participate in the formation of base-paired stems. 2A: deletions terminating just before or after the translation start codon. 2B: deletions terminating within pairing segment 1 or between segments 1 and 2. 2C: deletions removing segments 1 and 2. 2D: deletions removing segments 1, 2 and 3 or 1, 2, 3 and 4.

Four classes of deletions warrant discussion:

(1)   Deletions that alter the nucleotide sequence of the Shine/Dalgarno region or remove the leader peptide start codon (Fig. 2A).

Deletions that remove RNA segments necessary for translation of the leader peptide coding segment, as expected, eliminate the relief of termination that normally accompanies tryptophan starvation. These deletions have a second important effect, however. They _increase_ transcription termination at the attenuator _in vivo_ about 8-fold, i.e., expression of the _trp_ operon in cells growing in the presence of tryptophan is only 15% that observed in cells with an unaltered leader sequence. We call this phenomenon _superattenuation_ and we presume that its purpose is to turn off _trp_ operon expression maximally when the bacterium has a limited supply of any cell component needed for translation initiation. We in fact reported in 1972 that during a nutritional shift-up in the presence of excess tryptophan, an E. coli _trpR_ culture exhibited a 3-fold drop in _trp_ mRNA synthesis. We labeled this phenomenon "metabolic regulation"[29]. It now seems possible that the observed reduction in _trp_ mRNA synthesis was due to superattenuation and reflects a deficiency in the capacity to initiate translation immediately following a nutritional shift-up.

(2)   Deletions that remove the start codon and RNA segment 1 (Fig. 2B).

Deletions of this type relieve transcription termination at the attenuator _in vivo_ and _in vitro_; the further the location of the deletion endpoint in segment 1, and the greater the concomitant reduction in stability of structure 1:2, the greater the relief of termination. We interpret these results as indicating that if the antiterminator (structure 2:3) can form, transcription termination at the attenuator will be prevented.

(3)   Deletions that remove the start codon and RNA segments 1 and 2 (Fig. 2C).

Deletions of this type show superattenuation _in vivo_ and _in vitro_. This important result indicates that structure 3:4 alone is recognized by the transcribing RNA polymerase as the transcription termination signal _in vivo_ and _in vitro_. In view of this finding we believe that the basis of attenuation is formation of the RNA terminator; if this structure forms, termination is the

result, if it does not, RNA polymerase continues transcription into the structural genes of the operon.

(4)   Deletions that remove the start codon and RNA segments 1, 2 and 3, or
        the start codon and RNA segments 1, 2, 3 and 4 (Fig. 2D).

Deletions of this type eliminate transcription termination in vivo and
in vitro, reinforcing the conclusion that structure 3:4, the terminator, is the
transcription termination signal.

The findings summarized above provide strong support for our model of attenuation. Most importantly they show that deletions of different specific segments of the leader region have precisely the effect predicted by our hypothesis of steric masking of RNA segments by a stalled ribosome.

In other studies[30] we obtained additional evidence supporting a role for
RNA secondary structure in the transcription termination event. A synthetic
single stranded deoxy 15-mer complementary to RNA segment 1 of the E. coli trp
leader transcript (Fig. 3) was added to an in vitro transcription reaction

15-mer

C-C-A-A-C-C-A-C-C-G-C-G-T-G-A
  · · · · · · · · · · · · · · ·
-A-G-G-U-U-G-G-U-G-G-C-G-C-A-C-U-U-C-C-U-G-
50   60   70

leader transcript segment

Fig. 3. Nucleotides 50-70 of the E. coli trp leader transcript and the synthetic oligonucleotide (15-mer) that is complementary to this segment.

mixture containing an E. coli trp leader template. The objective of the
experiment was to determine if, by preventing pairing of RNA segment 1 with RNA
segment 2, we would allow structure 2:3 to form, thereby relieving termination. The exciting result was that addition of the 15-mer increased read-
through transcription 4-fold. Analyses with mutant templates confirmed that
pairing of segment 2 with segment 3 must occur for the 15-mer to influence
termination. These studies support the view that RNA polymerase recognizes RNA
secondary structures as signals and indicate that transcript segments in the
transcription complex are accessible for pairing with added oligonucleotides.

Our continuing studies therefore provide support for the basic features of
our model of attenuation, and reveal new regulatory aspects such as the role of
pausing[24,25], and superattenuation[28], that were not previously appreciated.

24

ACKNOWLEDGEMENTS

The studies summarized in this article represent the efforts of many coworkers. Their contributions to the development of our understanding of attenuation have been invaluable. These studies were supported by grants from the National Science Foundation, the United States Public Health Service and the American Heart Association. The author is a Career Investigator of the American Heart Association.

REFERENCES
1. Yanofsky, C. (1981) Nature, 289, 751-758.
2. Di Nocera, P.P., Blasi, F., DiLauro, R., Funzio, R. & Bruni, C.B. (1978) Proc. Nat. Acad. Sci., 75, 4276-4280.
3. Barnes, W.M. (1978) Proc. Nat. Acad. Sci., 75, 4281-4285.
4. Johnston, H.M., Barnes, W.M., Chumley, F.G., Bossi, L., and Roth, J. (1980) Proc. Nat. Acad. Sci., 77, 508-512.
5. Gardner, J.F. (1978) Proc. Nat. Acad. Sci., 76, 1706-1710.
6. Lawther, R.P. and Hatfield, G.W. (1980) Proc. Nat. Acad. Sci., 77, 1862-1866.
7. Nargang, F.E., Subrahmanyam, C.S., and Umbarger, H.E. (1980) Proc. Nat. Acad. Sci., 77, 1823-1827.
8. Gemmill, R.M., Wessler, S.R., Keller, E.B., and Calvo, J.M. (1979) Proc. Nat. Acad. Sci., 76, 4941-4945.
9. Zurawski, G., Brown, K., Killingly, D., and Yanofsky, C. (1978) Proc. Nat. Acad. Sci., 75, 4271-4275.
10. Lee, F., Bertrand, K., Bennett, G., and Yanofsky, C. (1978) J. Mol. Biol., 121, 193-217.
11. Lee, F. and Yanofsky, C. (1977) Proc. Nat. Acad. Sci., 74, 4365-4369.
12. Oxender, D., Zurawski, G., and Yanofsky, C. (1979) Proc. Nat. Acad. Sci., 76, 5524-5528.
13. Eisenberg, S.P., Soll, L., and Yarus, M. (1980) in Transfer RNA: Biological Aspects, Soll, D., Schimmel, P. and Abelson, J. ed., Cold Spring Harbor, New York, pp. 469-479.
14. Miozzari, G.F. and Yanofsky, C. (1978) J. Bact., 133, 1457-1466.
15. Schmeissner, U., Ganem, D., and Miller, J.H. (1977) J. Mol. Biol., 109, 303-326.
16. Yanofsky, C. and Soll, L. (1977) J. Mol. Biol., 113, 663-677.
17. Lee, F., Squires, C.L., Squires, C., and Yanofsky, C. (1976) J. Mol. Biol., 103, 383-393.
18. Bertrand, K., Korn, L.J., Lee, F., and Yanofsky, C. (1977) J. Mol. Biol., 117, 227-247.
19. Stauffer, G.V., Zurawski, G., and Yanofsky, C. (1978) Proc. Nat. Acad. Sci., 75, 4833-4837.
20. Zurawski, G. and Yanofsky, C. (1980) J. Mol. Biol., 142, 123-129.
21. Zurawski, G., Elseviers, D., Stauffer, G.V., and Yanofsky, C. (1978) Proc. Nat. Acad. Sci., 75, 5988-5992.
22. Yanofsky, C. and Horn, V. (1981) J. Bacteriol., 145, 1334-1341.
23. Farnham, P.J. and Platt, T. (1982) Proc. Nat. Acad. Sci., 79, 998-1002.
24. Farnham, P.J. and Platt, T., (1980) Cell, 20, 739-748.
25. Winkler, M.E. and Yanofsky, C. (1981) Biochemistry, 20, 3738-3444.
26. Stroynowski, I., VanCleemput, M., and Yanofsky, C., Nature, (submitted).
27. Stroynowski, I. and Yanofsky, C., Nature, (submitted).
28. Miozzari, G.F. and Yanofsky, C. (1978) Nature, 276, 684-689.
29. Rose, J.K. and Yanofsky, C. (1972) J. Mol. Biol., 69, 103-118.
30. Winkler, M., Mullis, K., Barnett, J., Stroynowski, I., and Yanofsky, C., Proc. Nat. Acad. Sci., (in press).

ORGANISATION AND EXPRESSION OF THE E.COLI PHENYLALANYL-TRANSFER RNA SYNTHETASE OPERON

MATHIAS SPRINGER, JACQUELINE PLUMBRIDGE, MARIE TRUDEL, MARIANNE GRUNBERG-MANAGO
Institut de Biologie Physico-Chimique, 13, rue Pierre et Marie Curie,
75005 Paris, France

and

GUY FAYAT, JEAN-FRANCOIS MAYAUX, CHRISTINE SACERDOT, PHILIPPE DESSEN,
MICHEL FROMANT, SYLVAIN BLANQUET
Laboratoire de Biochimie, Ecole Polytechnique, 91128 Palaiseau Cedex, France

INTRODUCTION

Aminoacyl-tRNA synthetases, as well as having indispensable roles in protein synthesis, are also involved in several other cellular processes, especially the regulation of the expression of amino acid biosynthetic pathways[1,2]. The regulation of the expression of aminoacyl-tRNA synthetases themselves has been extensively studied. Somewhat over the half of these enzymes have been examined for the effect of amino acid deprivation on the synthesis of the cognate synthetase. Among the subset of synthetases studied only arginyl-tRNA synthetase, isoleucyl-tRNA synthetase and phenylalanyl-tRNA synthetase were shown to exhibit long term derepression, with seven other aminoacyl-tRNA synthetases the derepression was shown to be only transient[1,2]. Many aminoacyl-tRNA synthetases have been shown to be metabolically regulated, i.e. their cellular concentration increases with bacterial growth rates[2,3]. Studies on the stringent response of several synthetases indicate the effect of amino acid deprivation on non-cognate synthetases is different for different synthetases[4].

Mutants have been isolated in almost every synthetase and their genetic loci are generally scattered on the E.coli chromosome. Several cis-acting (promoter type) mutants have been isolated and a few trans-acting regulatory genes have been identified[2,5,6]. Until recently no attempts were made to explain this mass of still descriptive features of synthetase regulation in precise molecular terms. The only exception being alanyl-tRNA synthetase which was shown to be

Abbreviations : PheRS, phenylalanyl-tRNA synthetase (E.C.6.1.20) ; pheS and pheT, the structural genes for the small and large subunits of PheRS ; thrS, structural gene for threonyl-tRNA synthetase ; infC, structural gene for translation initiation factor IF3 ; Kb, kilobase pairs ; bp, base pairs ; Kd, kilodaltons ; SDS, sodium dodecyl sulphate ; bla, structural gene for β-lactamase.

autoregulated in vitro at the transcriptional level[7]. We focused on phenylalanyl-tRNA synthetase for two principle reasons : (1) The two genes corresponding to the two subunits of phenylalanyl-tRNA synthetase (which is an $\alpha_2\beta_2$ type enzyme) are clustered on the E.coli chromosome at 38 min with the structural genes for threonyl-tRNA synthetase and translation initiation factor IF3[8,9,10]. As all these macromolecules are involved in translation and as clustered genes involved in translation are often expressed as long transcription units, we originally wondered if these genes were cotranscribed. (2) Phenylalanyl-tRNA synthetase is one of the most studied and most interestingly regulated synthetases ; it is one of the three synthetases to show long term derepression upon cognate amino acid starvation and it has been shown to be metabolically regulated. Moreover mutants are available in the two subunits of phenylalanyl-tRNA synthetase[11]. We have ourselves isolated mutants in both subunits of this enzyme and constructed the necessary genetic tools for their detailed analysis (Springer et al., in preparation).

The present paper (1) describes the mode of expression of phenylalanyl-tRNA synthetase and the surrounding genes ; (2) describes experiments which shed some light on how the synthesis of the enzyme is regulated ; (3) presents the nucleotide sequence of part of the pheS,T operon and (4) describes in vitro transcription experiments relevant to the expression of PheRS.

RESULTS

Mode of expression of phenylalanyl-tRNA synthetase and surrounding genes

The genes thrS, infC, pheS and pheT coding respectively for threonyl-tRNA synthetase (ThrRS), translation initiation factor IF3, the small and large subunit of phenylalanyl-tRNA synthetase (PheRS) were originally cloned on a single λ transducing phage (λp2)[8,9,10]. An EcoRI to HindIII fragment of E.coli DNA from this transducing phage was recloned in pBR322 to give the plasmid pB1 (Fig. 1). The genes were physically localised on the pB1 DNA by different methods : (1) Deletion mapping with a set of λ transducing phages carrying different fragments of this region of the chromosome[12] ; (2) Classical subcloning techniques[13] ; (3) thrS which was not precisely located by the two first methods, was localised with the aid of DNA sequencing data[14] and by the existence of hybrid proteins between ThrRS and IF3[15] or PheRS. For instance, the recombinant plasmid pB5-54 (Fig. 1) synthesizes a hybrid (of 78 Kd molecular weight) between the amino-terminal end of ThrRS and the carboxy-terminal end of the large subunit of PheRS[15]. As the carboxy-terminal end of the large

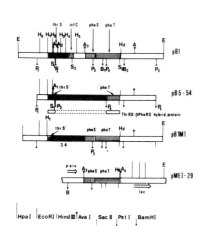

Fig. 1. Physical map of plasmids pB1, pB5-54, pB1M1 and pMEI-29. pB1 is a 10.5 Kb fragment of E.coli DNA cloned between the EcoRI and HindIII sites of pBR322[13]. The positions of the four genes thrS indicated by ▨▨▨ infC ▦▦▦ , pheS ▥▥▥ and pheT ▤▤▤ are taken from data in ref. 13, 14, 15. The position of the restriction sites are indicated as E = EcoRI, H = HpaI, A = AvaI, P = PstI, S = SacII, B = BamHI, Hd = HindIII. Sites are numbered from left to right in pB1 and are consistent throughout the figures. pB5-54 is the PstI$_1$-PstI$_2$ and PstI$_4$-PstI$_5$ fragments of pB1 joined together. α and β are respectively the small and large subunits of PheRS. pB1M1 is an in vivo derived deletion of 3.4 Kb. ▬▬ Deleted DNA. Serrated edges indicate that the exact end of in vivo deletions is not known. Primes on the gene names indicate that the gene is incomplete on that side. Only the relevant part of pMEI-29 is shown. The AvaI$_4$ site is not present on pB1 but derived from λp2 (Fig. 2).

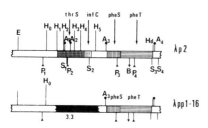

Fig. 2. Physical map of the E.coli DNA insert of λp2 and λpp1-16. Genes, restriction enzyme sites and symbols are as described in Fig. 1. The EcoRI to HindIII fragment of λp2 was cloned in pBR322 giving pB1[13]. Additional E.coli DNA is present on either side of this fragment. λpp1-16 carries an in vivo derived deletion of about 3.3 Kb.

subunit of PheRS is well localised on pB1 DNA we could easily deduce the location of the amino-terminal end of the hybrid on pB5-54 which corresponds to the location of the amino-terminal end of ThrRS on pB1. Knowing the length of the thrS gene, the carboxy-terminal end of the gene is easily localized on pB1 DNA.

All characterized genes of pB1 are transcribed in the same direction from thrS to pheT. This was shown by the fact that pulse labelled mRNA extracted from pB1 carrying strains hybridizes only to one strand of λp2 (λp2 carries all the E.coli genes of pB1)[16]. These four genes are transcribed as three separate transcription units : the first unit covering thrS, the second infC and the third both pheS and pheT[15,16]. The two genes for PheRS are thus expressed as a multicistronic operon. These results were obtained using two different types of strategies : (1) Selected restriction fragments were cloned into plasmid vectors able to reveal promoter activities. For instance the fragment PstI$_2$-PstI$_3$ was shown to carry a promoter which permits transcription from left to right (Fig. 1) over the PstI$_3$ site[16]. This promoter can only be the pheS promoter. However the fragment $\underline{PstI}_3$-$\underline{PstI}_4$ which carries the end of $\underline{pheS}$ and the start of $\underline{pheT}$ was shown not to carry any promoter. This indicates that $\underline{pheT}$ is expressed by a promoter in front of $\underline{PstI}_3$, i.e. either within $\underline{pheS}$ or in front of $\underline{pheS}$. The existence of $\lambda pp1$-16 (Fig. 2) deleted in front of $\underline{pheS}$ and $\underline{pheT}$ (see later), which expresses neither $\underline{pheS}$ nor $\underline{pheT}$[12] shows that $\underline{pheT}$ is indeed expressed from a promoter in front of $\underline{pheS}$. (2) Selected fragments were also cloned into λ bacteriophage. Cloning in λ is convenient for promoter loca-lisation. Expression of a gene from a fragment cloned in λ under lysogenic conditions, when λ promoters are repressed, indicates the existence of a promo-ter within the cloned fragment. A recombinant carrying the $\underline{SacII}_2$-$\underline{SacII}_3$ (Fig. 1) fragment was shown to express $\underline{pheS}$ under such conditions where the expression can only be from a promoter inside the $\underline{SacII}$ fragment[15]. As this fragment does not carry the promoter proximal part of $\underline{infC}$, $\underline{pheS}$ expression from that recombinant cannot be from $\underline{infC}$'s promoter. This proves that the $\underline{pheS,T}$ operon is expressed from a promoter located after $\underline{infC}$.

As it is relevant for data presented later, we shall now give some evidence that the regulatory regions of the $\underline{pheS,T}$ operon are located to the left hand side and the structural regions to the right hand side of the $\underline{AvaI}_3$ site of pB1 (Fig. 1).

In two cases we were able to replace the region to the left of $\underline{AvaI}_3$ by foreign DNA and obtain overproduction of PheRS with exactly the same molecular weight subunits as wild type PheRS. In plasmid pME1-29 (Fig. 1) the left hand side of $\underline{AvaI}_3$ was replaced by the arabinose promoter. This plasmid was obtained by cloning $\underline{AvaI}_3$-$\underline{AvaI}_4$, i.e. the structural genes of PheRS (both $\underline{AvaI}_3$ and $\underline{AvaI}_4$ sites are also $\underline{XmaI}$ sites) of $\lambda p2$ (Fig. 2) in the $\underline{XmaI}$ site located between the arabinose promoter and the lac genes of pMC306[17] (Fig. 4). The plasmid pME1-29 overproduces in the presence of arabinose both the large and small subunits of PheRS at the same molecular weight as pB1 (data not shown). Secondly, the plasmid pB1M1 (Fig. 1) derived from pB1 $\underline{in\ vivo}$ carries a 3.4 Kb deletion which fuses $\underline{thrS}$ to the $\underline{pheS,T}$ operon[18]. The deletion eliminates the $\underline{AvaI}_3$ site, however PheRS subunits are overproduced at their normal mole-cular weights from that plasmid[18]. The overproduction of apparently wild type PheRS from these two plasmids where the left hand side of $\underline{AvaI}_3$ has been repla-ced by foreign DNA clearly indicates that the structural parts of the operon are located to the right hand side of $\underline{AvaI}_3$. This is confirmed by the nucleotide sequence analysis.

The bacteriophage $\lambda pp1$-16 (Fig. 2) carries a deletion located to the left hand side of $\underline{AvaI}_3$ which eliminates $\underline{pheS}$ and $\underline{pheT}$ expression[12]. This proves that regions located to the left of $\underline{AvaI}_3$ are necessary for in vivo expression of PheRS. The proof that the left hand side of $\underline{AvaI}_3$ carries a functional promoter comes from the fact that $\underline{AvaI}_2$-$\underline{AvaI}_3$ cloned in the XmaI site of pMC306 (Fig. 4) expresses β-galactosidase from a promoter within the cloned fragment (see later).

Studies on the regulation of the pheS,T operon

Transcription studies of wild type and mutated operons. Strains carrying the plasmid pB1 overproduce PheRS about 100 fold[18] (Ducruix et al., in preparation). This overproduction is somehow detrimental to the cell and many non-overproducing mutants can be found by simple screening[18]. The non-overproduction was always associated with a mutational alteration in the plasmid. The fact that the mutations are isolated on plasmids has several advantages : (1) Mutants can be isolated that would be lethal on the chromosome ; (2) The analysis of the mutant plasmid DNA makes it easy to classify the mutations as deletions, insertions or point mutations ; (3) As wild type pheS,T mRNA is strongly expressed from pB1, significant measurements can be made even with "down" mutations.

A set of deletions, insertions and point mutations of the pheS,T operon was thus isolated on multicopy plasmids which, because of their mutational alteration, no longer caused the overproduction of PheRS[18]. The properties of some mutant plasmids (Fig. 3) belonging to this set are particularly informative.

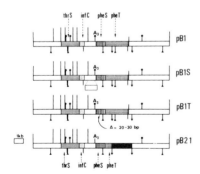

Fig. 3. Physical map of pB1, pB1S, pB1T and pB21. Genes, restriction enzyme sites and symbols are as described in Fig. 1. pB1S and pB1T are spontaneous in vivo mutants of pB1 isolated because they no longer cause host strains to overproduce PheRS. pB1S carries a 1.3 Kb insertion between sites $\underline{HpaI}_5$ and $\underline{AvaI}_3$. pB1T carries a 20-30 bp deletion between sites $\underline{AvaI}_3$ and $\underline{PstI}_3$. pB21 is derived from pB1 and was constructed by deletion between the BamHI sites in pheT and pBR322 DNA.

The plasmid pB1S carries a 1.3 Kb insertion in the regulatory region of the pheS,T operon (to the left hand side of $\underline{AvaI}_3$). This insertion was shown, using a cis-trans test, to decrease PheRS expression by interruption of a cis acting sequence rather than a trans-acting regulatory gene. The plasmid pB1T carries a 20 to 30 bp deletion within pheS and causes the synthesis of a smaller than wild type small subunit of PheRS which is less stable than the wild type subunit. This was shown by SDS polyacrylamide gel analysis of anti-PheRS immunoprecipitates of pulse labelled and chased extracts from pB1 and pB1T carrying strains[18]. The decreased stability of one subunit presumably destabilises the whole molecule and eliminates the overproduction. The plasmid pB21 was isolated by in vitro deletion of the BamHI fragment which carries the end of pheT and the beginning of the genes conferring resistance to tetracycline in pBR322.

Cells carrying pB1 and the three mutant plasmids were pulse labelled with tritiated uridine and the extracted mRNAs were hybridized to probes specific for thrS, pheS and pheT. The probes are fragments of the different genes cloned into λ. Liquid hybridization was carried out using the separated strands of these λ's. The thrS probe carries the $AvaI_1$-$AvaI_2$ fragment (Fig. 1), the pheS probe the $AvaI_3$-$PstI_3$ fragment and the pheT probe the $BamHI_1$-$SacII_3$ fragment[19]. The result of these hybridization experiments are shown in Table 1. The values were calculated as c.p.m. hybridized per Kb of DNA inserted into the probes and then normalized to the thrS value. The gene thrS is wild type in all these mutant plasmids and is thus an internal control to correct for possible differences in copy number of the different plasmids.

TABLE 1

RELATIVE VALUES OF pheS, pheT AND thrS mRNA SYNTHESIZED FROM pB1 AND MUTANT PLASMIDS

Plasmid	Genetic alteration	pheT	pheS	thrS
pB1	wild type	1.39	1.76	1
pB1S	insertion with in the regulatory regions	0.1	0.14	1
pB1T	small deletion within pheS	0.38	0.49	1
pB21	deletion at the end of pheT	-	0.4	1

Cells were grown in MOPS minimal medium[25] supplemented with auxotrophic amino acids required by the host bacteria arginine, histidine and proline at 50 μg/ml and uridine and cytidine at 50 μg/ml. At A_{650} = 0.4 the cells were washed free of pyrimidines, pulse labelled with ^{3}H-uridine, lysed by the boiling SDS method and mRNA purified by phenol-chloroform extraction and ethanol precipitation. The relative amounts of pheS, pheT and thrS mRNA were measured by hybridization to λ's carrying DNA specific to each of the three genes. Further details are given in ref. 19.

The data for pB1 indicates that transcription is slightly stronger for pheS than for pheT, this could mean that there is some natural polarity within the pheS,T operon. The plasmid pB1S synthesizes about ten times less pheS and

pheT mRNA than pB1. A decrease in mRNA is expected since the insertion inter-
rupts a cis-acting regulatory region of the pheS,T operon. The decreased level
of mRNA in pB1S carrying strains could be due to the insertion interrupting
either the promoter of the pheS,T operon or some sequence between the promoter
and the structural genes.

The remaining two plasmids pB1T and pB21 differ from pB1S by the fact that
they carry mutational alterations in the structural part of the pheS,T operon
which should a priori not alter mRNA levels of the operon unless the mutations
are polar. The pheS and pheT mRNA levels synthesized from pB1T are down 3 to 4
fold when compared to pB1. This result cannot be explained by polarity since
the 20 to 30 bp deletion within pheS makes an in phase fusion within the gene.
The molecular weight of the small subunit of PheRS synthesized from pB1T is
1 Kd less than that of wild type ; 1 Kd corresponds to about 10 amino acids .
which correlates with the 20 to 30 bp deletion measured on the DNA of pB1T. If
the 20 to 30 bp deletion created an out-of-phase fusion within pheS, a premature
termination could occur at a normally out-of-phase termination codon. Inspec-
tion of the nucleotide sequence shows that pheS finishes 216 bp after the $PstI_3$
site and that an out-of-phase fusion within $AvaI_3$-$PstI_3$ fragment must result in
the loss of at lease 4 Kd in one reading frame and even more in the third reading frame. Hence we are sure that the deletion within pheS of pB1T produces
an in phase fusion and hence no polar effect on mRNA level is possible.

We are thus forced to conclude that a defect in the small subunit of PheRS
has an effect on the transcription of the pheS,T operon. Analysis of the mRNA
from pB21 shows that a defect in the large subunit has a similar effect. The
in vitro deletion of $BamHI_1$-$BamHI_2$ from pB1 to give pB21 removes the carboxy-
terminal part of the pheT gene (including the DNA corresponding to the pheT
probe). Measurement of mRNA levels in pB21 carrying strains shows that pheS
mRNA level is reduced to a quarter its value in pB1 (Table 1). For this strain,
a carboxy-terminal deletion of the promoter distal gene of the operon affecting
the mRNA level of a promoter proximal gene, no polarity argument is possible.

Thus a mutational alteration in either subunit provokes a decrease in the
mRNA levels to about a third that of a wild type plasmid. Plasmids which
carry only one of the genes of the pheS,T operon do not overproduce the single
subunit but synthesis of the individual subunits at appreciable levels does
take place (as seen by gel electrophoresis of (^{35}S) methionine pulse labelled
cultures). For pB1T the small mutated subunit was less stable than the wild
type large subunit which should result in an excess of the large subunit. Thus
two models can be proposed for the effect of PheRS on its own mRNA level. Either

the wild type tetrameric PheRS protein positively regulates its own mRNA level
or the excess synthesis of one subunit negatively regulates PheRS mRNA level.
At what level this regulation is exerted is unknown - the most likely possibilities are either the rate of synthesis or the stability of the mRNA.

Effect of phenylalanine starvation on the expression of β-galactosidase
synthesized from the pheS,T promoter. Nass and Neidhardt[20] showed that phenylalanine starvation in a phenylalanine bradytroph (leaky auxotroph) caused a
long term derepression of PheRS. As the amplitude of the derepression is low
(about 2.5 fold) the phenomenon is not easy to study using activity or even
immunological methods. We have used fusions between pheS,T promoter and lac
structural genes to investigate this observation. The structure of the fusions
used are shown in Fig. 4. The $AvaI_2$-$AvaI_3$ fragment of pB1 was cloned in the
right orientation in the XmaI site of pMC306[17] between the arabinose promoter
and the lac genes. The resulting plasmid pMF3 does not synthesize more
β-galactosidase when arabinose is added to the growth medium[21]. This means
that the transcription initiated in the presence of arabinose at the ara
promoter stops somewhere in the insert, most probably at the end of thrS.
However pMF3 synthesizes a constitutive level of β-galactosidase which, as
transcription from the arabinose promoter does not proceed into lac, comes
from the pheS,T promoter located to the left hand side of $AvaI_3$.

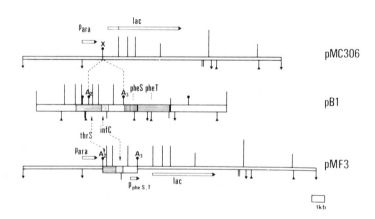

Fig. 4. Physical map of pMC306 and pMF3. The structure of pMC306 is taken from
ref. 17. Genes and restriction sites are indicated as in Fig. 1. X is the
unique XmaI site of pMC306. P_{ara} indicates the arabinose BAD promoter. lac
indicates β-galactosidase structural gene. $P_{pheS,T}$ indicates the promoter for
the pheS,T operon. Genes, restriction enzyme sites and symbols are as described
in Fig. 1.

We first investigated the effect of starvation for phenylalanine in a pheA bradytroph on β-galactosidase synthesis from pMF3. The gene pheA codes for the principal enzyme of the phenylalanine biosynthetic pathway. Phenylalanine starvation clearly induces β-galactosidase synthesis in a pheA strain carrying pMF3 (Fig. 5), β-galactosidase levels in a non-starved strain is shown in the figure by the horizontal line corresponding to 180 units per A_{650} of bacteria. The β-galactosidase induction is specific for phenylalanine starvation, since starvation for histidine induced by 3 amino-1,2,4-triazole[22] has no effect on β-galactosidase synthesis (Fig. 5). Starvation for phenylalanine or histidine in the same pheA strain carrying pMC306 (the parental plasmid, shown in Fig. 4, which expresses β-galactosidase from the arabinose promoter) has no effect on β-galactosidase synthesis (data not shown).

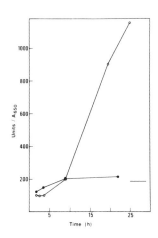

Fig. 5. Effect of phenylalanine starvation on β-galactosidase synthesis from a pheA strain carrying pMF3. β-galactosidase activity is indicated in units/A_{650} [24]. One unit of β-galactosidase liberates 1 nmole of o-nitrophenol at 28°C pH 7, per min, per ml. Time in the figure indicates time after the start of the starvation. MA3 (araD136, Δ(ara,leu)7697, ΔlacX74, galU, galK, strA, pheA97) with pMF3 was grown at 37°C in MOPS medium[25] supplemented with 50 μg/ml of leucine, isoleucine, phenylalanine and ampicillin to 0.2 A_{650}. The bacteria were then centrifuged, washed twice and suspended in the same medium, in the same medium minus phenylalanine and in the same medium + 20 mM 3-amino-1,2-4-triazole. Aliquots for enzymatic tests were taken from the cultures at about 0.2, 0.4, 0.6 and 0.8 A_{650}. Ampicillin was added at 50 μg/ml every 5 to 8 hours. Doubling times are : in the complete medium,, 55 min ; complete medium minus phenylalanine, 8 h 55 min ; complete medium with aminotriazole, 7 h 35 min. ●——● starvation for histidine ; o——o starvation for phenylalanine ; —— level of β-galactosidase activity in unstarved strain.

This result indicates that the starvation response is specific for the pheS,T promoter which synthesizes β-galactosidase from pMF3. The induction of β-galactosidase from pMF3 during phenylalanine starvation does probably not reflect an increase in plasmid copy number since β-lactamase levels are the same in histidine and phenylalanine starved strains (data not shown). The effect of phenylalanine starvation on β-galactosidase synthesis from pMF3 could be explained by a decreased aminoacylation level of tRNA[Phe]. To test this hypothesis we measured the effect of phenylalanine starvation in a pheS strain on β-galactosidase synthesis from pMF3. The pheS mutant used was an auxotroph (also quite leaky) for phenylalanine due to a mutation in the small subunit of PheRS. This mutant is likely to be a mutant with a modified K_M for phenylalanine. Phenylalanine starvation clearly induces β-galactosidase synthesis from pMF3 (Table 2). As for the effect with pheA, this induction is specific for phenylalanine starvation since histidine starvation has no effect ; the induction is probably not explained by copy number effects since β-lactamase levels are similar in unstarved, phenylalanine starved, or histidine starved strains (Table 2) and no effect of phenylalanine starvation is seen with pMC306 instead of pMF3 which indicates that this effect is specific for the pheS,T promoter.

TABLE 2

EFFECT OF PHENYLALANINE STARVATION ON β-GALACTOSIDASE SYNTHESIS FROM A pheS STRAIN CARRYING pMF3

	Non starved	Starved for Phe	Starved for His
β-galactosidase (u/A$_{650}$)	98	1033	78
β-lactamase (u/mg prot)	3.27	2.97	2.19

β-galactosidase activity is given in units/A$_{650}$[24]. β-lactamase is given in units/mg of protein[26]. One unit of β-lactamase hydrolyses 1 μmole of ampicillin per min at 28°C. IBPC5041 (thi-1, argE3, his-4, proA2, lacY1, galK2, mtl-1, xyl-5, tsx-29, supE44,pheS76) with pMF3 was grown at 37°C in MOPS medium[25] supplemented with 50 μg/ml of arginine, histidine, proline, phenylalanine and ampicillin to 0.3 A$_{650}$. The bacteria were then centrifuged, washed twice and suspended in the same medium, in the same medium minus Phe and in the same medium minus His. Aliquots were taken from cultures at regular intervals for β-galactosidase and β-lactamase tests. Ampicillin was added at 50 μg/ml every 5 to 8 hours. Doubling times are : in complete medium, 80 min ; complete minus Phe, 6 h 20 min ; complete minus His, only very low residual growth. All the values indicated in the table are averages over at least six determinations. The minus phenylalanine determinations are started 8 hours after starvation. The precise conditions for β-lactamase activity measurements are going to be published elsewhere (M. Trudel et al., in preparation).

Phenylalanine starvation in a pheS or pheA mutant is expected to decrease aminoacylation of tRNA^{Phe}. These experiments thus indicate that aminoacylation could modulate β-galactosidase levels synthesized from the pheS,T promoter. However aminoacylation levels were never checked directly under phenylalanine starvation conditions. We have thus no direct proof that deacylated tRNA^{Phe}

is involved in β-galactosidase induction. Moreover β-lactamase activity measurements might not really give a true value of the copy number of pMF3 under starvation conditions. The β-lactamase activity level would not correlate with bla gene copy number if histidine and phenylalanine starvation differentially affected β-lactamase and β-galactosidase synthesis. A better proof of phenylalanine starvation inducing β-galactosidase synthesis from pMF3 would be if the same fusion behaved in a similar manner in a single copy vector such as λ. With these provisos, under our experimental conditions phenylalanine starvation probably derepresses the pheS,T promoter as shown by induction of β-galactosidase synthesis. The derepression almost certainly occurs at the transcriptional level since pMF3 is an operon fusion : β-galactosidase is expressed from the pheS,T promoter but from its own translation initiation site. These results mean that expression of the pheS,T operon seems to be modulated at the transcriptional level by tRNA^{Phe} aminoacylation levels. This situation is analogous to many amino acid biosynthetic operons which are regulated by attenuation[23]. The regulatory regions of these operons show many features recognizable on a nucleotide sequence. The next part of this paper describes the features of the nucleotide sequence of pheS,T operon relevant to the regulation of PheRS expression.

Nucleotide sequence analysis of pheS-pheT

A DNA fragment of 3.3 Kb of plasmid pB1 (Fig. 1), carrying infC, pheS and the beginning of pheT has been sequenced[14] (Fayat et al., in preparation). The location and orientation of these genes agrees with the genetic studies[15,16]. In particular the open reading frame corresponding to the structural part of pheS starts 87 nucleotides downstream from the center of the AvaI restriction site (Fig. 6) which the genetic studies of Springer et al. have shown to separate pheS from its promoter region. This open reading frame is preceded 6 bases upstream by a GAGG sequence capable of pairing with the sequence CCUC at the 3'-end of 16S rRNA[27].

The aminoacid sequence of the small subunit of PheRS deduced from the DNA sequence (Fig. 7), accounts for a protein of 36.577 Kd in very good agreement with 37 Kd, the measured molecular weight of the subunit[28]. The DNA predicted amino-terminal of this sequence is identical to that determined for the first seven residues of the isolated subunit (Ducruix et al., in preparation), with the exception that the protein starts with serine, indicating that the DNA-encoded amino-terminal methionine has been removed in vivo. The carboxy-terminal analysis of the isolated subunit shows the lysine and penultimate

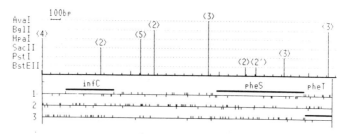

Fig. 6. Restriction map of the HpaI₄-BglI₅ portion of pB1 plasmid DNA with the
location of the open reading frames corresponding to the structural parts of
infC, pheS and pheT. Initiation and termination codons in each of the three
phase frames (1,2,3) are shown on the sense-strand of DNA. Vertical bars above
the lines indicate termination codons and bars under the lines initiation
codons. The heavy horizontal lines indicate the identified open reading frames.
The part of the open reading frame corresponding to pheT is capable of encoding
the first 100 N-terminal aminoacid residues of the large subunit of phenylalanyl-
tRNA synthetase.

phenylalanine residues predicted by the DNA sequence. Finally, the aminoacid
composition deduced from DNA analysis is in agreement with that determined for
the isolated subunit (Ducruix et al., in preparation). Examination of the aminoacid composition shows an abnormally high level of phenylalanine residues :
there are 20 phenylalanine out of a total of 327 aminoacids. The majority of
these phenylalanine residues are clustered in the carboxy-terminal half of the
polypeptide chain (Fig. 7). There seems to be a similar clustering of alanine
residues within the amino-terminal half of the molecule. The sequence of Fig. 7
was also examined for the possible occurrence of repeating sequences. The only
significant repeated peptide is a tetrapeptide of composition Ala-Leu-Asn-Ala.

Downstream from pheS, on the same coding DNA strand but in a different
coding phase, we find another open reading frame which ends outside the sequenced DNA (Fig. 6). This coding sequence is shown to correspond to the beginning
of the structural part of pheT by comparison of the DNA-encoded amino-terminal
end with the first four amino-terminal residues of the isolated large subunit
of PheRS (Ducruix et al., in preparation). There are 14 nucleotides between
the stop codon of pheS (UAA) and the start codon of pheT :

```
        Phe Lys              Met Lys Phe Ser Glu
        TTT AAA TAA GGCAGGAATAGATT ATG AAA TTC AGT GAA
```

The initiator codon for pheT is preceded by an AGGA sequence, capable of pairing

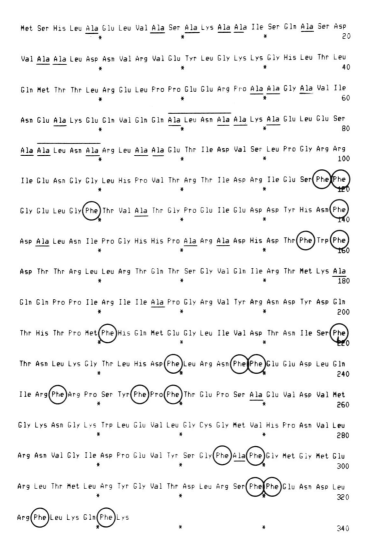

Fig. 7. Aminoacid sequence of the small subunit (36.577 Kd) of E.coli phenylalanyl-tRNA synthetase as deduced from DNA sequencing. The locations of the
20 phenylalanine residues present in the protein are indicated by circles and
the alanine residues by underlining. A line above the sequence shows the location of the repeated peptide Ala-Leu-Asn-Ala.

with the sequence UCCU of 16S rRNA[27]. On the other hand, we do not find any
RNA structure resembling a transcription terminator[29] in the part of the
sequenced DNA downstream from the pheS structural gene. This supports the conclusion of Plumbridge and Springer[16] that pheS and pheT are expressed as two
adjacent cistrons from the same transcription unit. However, 72 nucleotides
upstream from the carboxy-end of pheS, within the structural part of pheS, we
note a TATGTTG sequence which resembles others RNA-polymerase binding sites[30].

Upstream to the AvaI restriction site which separates the structural part of
pheS-pheT from its promoter region, we have searched for homologies to the
- 35 and Pribnow box regions, using the consensus sequences TTGACA and TATAATG,
respectively[29]. The best combinations were located more than 400 bp away from
the initiator codon of pheS (Fig. 8).

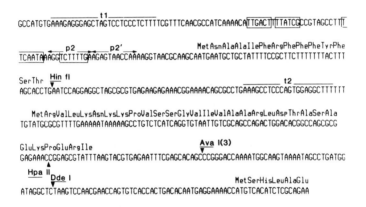

Fig. 8. Nucleotide sequence of the regulatory region of the pheS-pheT operon.
The positions of the restriction sites HinfI, HpaII, AvaI₃, DdeI are shown.
The dyad symmetries corresponding to transcription terminators t1 and t2 are
indicated by lines above the sequence, p2 and p2' corresponding to the mapped
start points of transcription are shown by horizontal arrows the length of
which indicates the uncertainty in mapping. The Pribnow box-like sequences
TTCAATA and TCTTTTC and their corresponding -35 regions, TTGACT and TTATCG,
respectively are shown by boxes around the sequences. The aminoacid composition
of the 14 residue phenylalanine-rich peptide is shown above the DNA sequence,
as is the aminoacid composition of the 31 residue peptide devoid of phenylalanine located after the proposed attenuator DNA sequence, t2. The N-terminal
sequence of the pheS structural gene is shown at the 3' end of the DNA sequence.

Two possible promoter-like sequences (TTCATA and TCTTTTG), separated by only
four nucleotides, each with its corresponding -35 region (TTGACT and TTATCG)
precede a short open reading frame coding for a 14-residue peptide. Interestingly, this coding sequence contains 5 phenylalanine residues, 3 of which are

adjacent. The initiator codon is preceded by several possible Shine and Dalgarno
type sequences capable of pairing with 16S rRNA. In addition, downstream from
the stop codon (UGA) of this open reading frame, the DNA sequence indicates an
RNA which can be folded in several alternatively base paired structures (Fig. 9).
Such a combination of features strongly recalls the mechanism of transcription
attenuation reported in the cases of several aminoacid biosynthetic operons
(reviewed in Yanofsky[23]). Finally, immediately downstream from these alternative secondary structures, the DNA sequence indicates an open reading frame
capable of encoding a 31-aminoacid peptide characterized by the absence of
phenylalanine residues (Fig. 8). This open reading frame is preceded by a strong
signal for translation initiation (GGAGG).

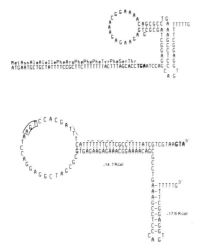

Fig. 9. Possible alternate secondary structures of the RNA in the region of the
transcription terminator, t2. In the top
scheme, the DNA sequence comprising the
leader peptide coding sequence and the two
alternative base paired stems of the attenuator like arrangement are indicated. The
bottom scheme shows another secondary structure of mRNA in this region involving
possible base pairs between mRNA of the
leader peptide coding sequence and the
large loop of the upper structure (see text
for details). The free energies associated
with this structure are given. The phenylalanine codons of the leader peptide are indicated by parentheses and the termination
codon by a box.

In vitro transcription of the pheS-pheT regulatory region

The mechanisms of attenuation implies that (1) there is at least one functional origin of transcription upstream of the initiation codon of the leader
peptide and (2) transcription starting from this origin can either proceed
through the leader peptide into the.structural genes of the operon or terminate
at a site located after the leader peptide and in front of the structural genes.

We have searched for the existence of functional promoters in the $HpaI_5$-
$AvaI_3$ region of pB1 (Figs. 1 and 10) using in vitro transcription of purified
restriction DNA fragments. The first DNA fragment studied was the 881 bp HpaII
fragment, located as shown on the restriction map of Fig. 10. It begins 153 bp
to the left of the $HpaI_5$ site and ends 39 bp to the left of the $AvaI_3$ site.
In vitro transcription of this DNA fragment (HpaII-HpaII) by purified E.coli
RNA polymerase produces 5 major low-molecular weight RNAs (Fig. 10), with

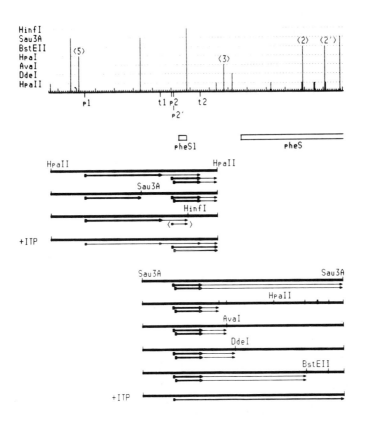

Fig. 10. Promoter mapping and organisation of transcription units in the
regulatory region of the pheS-pheT operon. In vitro transcription was carried
out with purified DNA restriction fragments prepared by HpaII or Sau3A restriction enzyme digestion of plasmid DNA pB1. The in vitro assay of transcription was performed in 20 mM Tris-HCl (pH 7.8), 6.5 % glycerol, 150 mM KCl,
10 mM MgCl$_2$, 0.1 mM Na EDTA, 1 mM 1-4 dithioerythritol. Restriction fragment
DNA template (0.3 to 1 μg) was incubated at 37°C for 15 minutes with 0.2 mM
ATP, CTP, GTP (or ITP), 50 μM (^{32}P)-UTP (NEN, 10 Ci/mmole) and 3.5 units of
RNA polymerase (NENzymes, 1200 units/mg). For transcription products analysis,
the reaction mixture was phenol-extracted, ether-extracted and ethanol-
precipitated ; then, the pellet was dissolved in 50 mM Tris-borate buffer
(pH 8.3) containing 1 mM Na$_2$EDTA and 80 % formamide, heated to 90°C for
3 minutes and electrophoresed on a 6 % polyacrylamide gel containing 8.3 M
urea. The molecular weight of RNA transcripts was estimated by comparison with
the migration of RNAs of known molecular weight and 5'-end labelled DNA restriction fragments. The fine restriction map of the 1.55 Kb HpaII-Sau3A DNA
fragment carrying 581 bp of the pheS structural gene is shown. The DNA restriction fragments(HpaII-HpaII and Sau3A-Sau3A, respectively) are represented
by heavy lines. The upper line for the two fragments shows the transcripts
obtained when the whole fragment was used as template. Prior to their use in
the in vitro transcription assay, these purified DNA fragments were digested

by the restriction enzyme marked on the figure. The RNA transcripts obtained are represented by arrows, the relative thickness of which roughly reflects the relative intensity of transcription in the in vitro assay. The major observed transcripts RNA_1 (starting at p1), RNA_2, $\overline{RNA_3}$, $\overline{RNA_4}$ and RNA_5 (starting at p2 and p2') are defined in the text. A readthrough RNA product is symbolized by a fainter arrow extending from a stronger arrow. When indicated (last line for each DNA fragment) transcription of the DNA fragment was carried out with ITP instead of GTP. The mapped start and end points of transcription (p1, p2, p2' and t1, t2, respectively) are positioned on the DNA restriction map of this region with respect to the location of the open reading frames corresponding to pheS and to pheS1 (leader peptide).

lengths of 403 ± 5 (RNA_1), 149 ± 1 (RNA_2), 147 ± 1 (RNA_3), 142 ± 1 (RNA_4) and 139 ± 1 (RNA_5) nucleotides respectively. A minor longer transcript of 600 ± 10 bases was also distinguishable. These RNA species were also observed with in vitro transcription experiments using intact plasmids carrying this DNA region as templates. The RNA_1 species was strongly expressed from intact DNA plasmids as well as from the HpaII-HpaII DNA restriction fragment.

The start point of RNA_1 (called p1) (Fig. 10) was determined by 5'-end sequencing. By comparison with the known DNA sequence, we localised it 33 bp downstream from the center of the $HpaI_5$ site. It is preceded by a TAGAATA and a TTAACG sequences which resemble a Pribnow box and its -35 region. In addition an open reading frame starts at the 15th base downstream from the 5'-end of RNA_1. This coding sequence is preceded by a good Shine and Dalgarno sequence (AGGAG).

On the DNA sequence, 400 bp downstream from the beginning of RNA_1, there is a G + C - rich region of dyad symmetry, which can result in an RNA hairpin (ΔG = -18.4 kcal/Mole), immediately followed by a stretch of U. The position of this structure indicating a transcription terminator (called t1)(Figs. 8 and 10) fully agrees with the length of RNA_1 (403 ± 5 nucleotides).

In addition, sequence of the minor 600 ± 10 bases RNA shows it to start with the same 5'-end sequence as RNA_1. This indicates that in vitro, RNA polymerase reads through terminator t1 and stops at the level of the attenuator (t2) (see later). The readthrough frequency is 30 ± 5 %.

The DNA template, HpaII-HpaII fragment, was digested by different restriction enzymes prior to in vitro transcription. Cleavage with Sau3A did not affect the synthesis of RNA_2, RNA_3, RNA_4 or RNA_5 but RNA_1 was eliminated and a new shorter RNA observed. HinfI digestion left RNA_1 intact, but RNA_2 to RNA_5 disappeared on the 6 % polyacrylamide gel, 8.3 M urea. This implies that the origins of transcription of these RNAs are less than 90 bases from the HinfI site (Fig. 10).

Using ITP instead of GTP in the in vitro transcription assay, all the RNAs described above decreased in intensity. RNA_1 could still be detected, but not RNA_2 to RNA_5. The main products were two new RNAs of 225 and 232 ± 1 bases, respectively. These transcripts are likely to be read through products of RNA_2, RNA_3, RNA_4 and RNA_5[31].

Another DNA restriction fragment obtained by Sau3A digestion was used as template in the in vitro transcription experiments. This DNA fragment of 1052 bp (Sau3A-Sau3A in Fig. 10) overlaps 398 bp of the above HpaII-HpaII DNA fragment and 581 bp of the pheS structural gene. The transcripts RNA_2, RNA_3, RNA_4 and RNA_5 are strongly expressed but no RNA_1 is synthesized. In addition one long minor transcript of 900 ± 40 bases is distinguishable. The origin of this long transcript was mapped by digestion of the Sau3A-Sau3A DNA fragment with various restriction enzymes before using it as a template for in vitro transcription (Fig. 10). The decrease in length observed for a transcript after restriction enzyme digestion of the template allows the positioning of the start site for transcription relative to the restriction site. In addition, in vitro transcription of the intact Sau3A-Sau3A DNA fragment was performed using ITP instead of GTP. RNA_2 to RNA_5 disappeared resulting in an enhancement of the expression of the 900 nucleotide RNA transcript. This indicates that the 900 nucleotide RNA corresponds to a read-through product of all RNA_2 to RNA_5. The in vitro read-through frequency is low, however (10 ± 5 %).

From this set of results we conclude that the short RNAs (RNA_2 to RNA_5) originate from at least two promoters (p2 and p2') corresponding to transcription start sites located, respectively at 77 ± 5 and 66 ± 5 bases upstream to the centre of the HinfI restriction site. The lengths of these short RNAs precisely locate a transcription terminator at 69 ± 5 bases downstream from HinfI. This location closely corresponds to that of the stretch of Us following the attenuator (t2), as positioned by DNA sequencing. In addition all the in vitro transcription products RNA_2-RNA_5 are 5' end labelled with $(\gamma^{32}P)ATP$ showing that they all start with an A residue.

The location by in vitro transcription of two RNA startpoints, p2 and p2', is consistent with the positions of the two promoter-like regions on the DNA sequence shown in Fig. 8. The presence of four distinguishable transcripts instead of two may be the consequence of a degeneracy of either the start and/or the end points of transcription.

DISCUSSION

The genetic and structural studies described here serve to show that expression of the pheS-pheT operon is a regulated process with controls apparently
acting at several levels.

The in vivo studies suggest at least two types of regulation : (1) an effect
of the gene products on their own expression ; (2) an induction of pheS-pheT
promoter activity under the condition of phenylalanine starvation. This effect
was seen whether a pheA or pheS bradytrophic strain was used. It suggests
a role for the level of aminoacylation of tRNA^{Phe} in the control of phenylalanyl-tRNA synthetase expression. This phenomenon resembles the attenuation
mechanism established for several aminoacid biosynthetic operons[23]. Thus it
is gratifying to find that the DNA sequence analysis of the pheS-pheT operon
gives strong evidence for this type of control mechanism. The nucleotide
sequence of pheS-pheT indicates a putative phenylalanine-rich "leader peptide"
followed by a DNA sequence indicating an RNA capable of being folded in several
alternative secondary structures (Fig. 9).

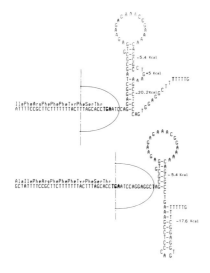

Fig. 11. Possible alternate RNA secondary
structures thought to regulate
transcription termination at the
attenuator-like structure of the pheS-pheT
operon. The top scheme represents the situation where cells are starved of phenylalanine.
The bottom scheme shows the case of non-
starvation. According to Yanofsky[23], it is
proposed that a ribosome stalled over one of
the codons of the transcript covers about
10 more nucleotides on the downstream region
of the transcript, indicated in the figure
by the half oval structure.

One may speculate on the possible relation existing between the translation
of this "leader peptide" and the secondary structure adopted by the mRNA
(Fig. 11). Under normal growth conditions, translation of the "leader peptide"
follows transcription. Ribosomes would stop on the UGA terminator codon.
Supposing that one ribosome is able to cover downstream at least 10 more
nucleotides of the mRNA, this position of the ribosome would prevent the
formation of the first base paired stem in the mRNA (Fig. 9 upper part)

and result in the structure represented in Fig. 11, lower part . As transcription continues, the second base paired mRNA structure, corresponding to a transcription terminator, is formed and premature termination of transcription can occur. Consequently the rate of transcription of the pheS-pheT structural genes will be decreased. An identical effect, i.e. transcription termination, is obtained if one supposes that the ribosome dissociates from mRNA at the position of the UGA codon. In this case, the part of the mRNA encoding the leader peptide is capable of base pairing with the RNA downstream forming the large loop of the alternative RNA structure lower scheme (Fig. 9). This pairing also would prevent the formation of the first base paired structure of the mRNA and favour the formation of the second structure, i.e. the attenuator. In addition, the latter secondary structure of the mRNA would explain that attenuated transcripts are mainly observed in the in vitro transcription experiments.

In the case of cells starved of phenylalanine, i.e. of lack of aminoacylated tRNA^{Phe}, a ribosome translating the "leader peptide" would stall at one of the 5 Phe codons. This position of the ribosome on the transcript will not affect the formation of the first mRNA structure. The establishment of a termination structure is thus prevented and the transcribing polymerase does not stop at the attenuator. Consequently, the structural genes of the pheS-pheT operon will be transcribed.

This mechanism is supported by the results of the in vitro transcription experiments. The regulatory region of the pheS-pheT operon is the source of several transcripts of variable lengths and intensities (Fig. 10). The origins of these transcripts have been mapped by 5' end sequencing or by using DNA restriction fragments of variable lengths derived from different parts of this regulatory region. The main conclusion is that the DNA region encoding the putative "leader peptide" is covered by several transcripts originating from several start points located upstream to the leader peptide's initiation codon Moreover, in vitro, these transcripts are capable of extending into the pheS structural region. This read-through process is enhanced when ITP is used instead of GTP. This has been shown to promote read-through of other attenuators due to the weak base pairing between I and C.

The in vivo studies using fusions between the promoter for pheS-pheT and the structural gene of β-galactosidase have indicated that regulation of phenyl-alanyl-tRNA synthetase depends on the level of aminoacylation of tRNA^{Phe}. Maximal 10-fold derepression of β-galactosidase activity was measured. This factor is of the same order of derepression observed in aminoacid biosynthetic

operons controlled by attenuation[23]. On the other hand, phenylalanine deprivation results in only a slight derepression of phenylalanyl-tRNA synthetase
intracellular levels giving only a maximal two-and-half-fold elevation of the
enzyme activity[20,21].

A possible explanation for such a small derepression could be the presence
of the large number of phenylalanine codons within the pheS structural gene.
Thus, extreme phenylalanine deprivation, although stimulating transcription
at the level of the attenuator structure, might induce ribosome stalling
within the pheS structural gene and hence transcription termination due to
polarity. In this case, premature termination of transcription within pheS
would counter-react with the positive effect of phenyalalnine starvation on RNA
read-through at the level of the attenuator. So, it is possible that the
10-fold stimulation of β-galactosidase expression from the pheS-pheT promoter
reflects the normal tuning of the attenuator in the absence of polarity effects
at the level of the structural gene of pheS.

One may ask if there is any advantage in a two-fold derepression of phenylalanyl-tRNA synthetase under conditions of phenylalanine deprivation. It is
worth noting that aminoacyl-tRNA synthetases and tRNAs are present in the
E.coli cell at very similar concentrations, of the order of micromolar
concentration[32,33]. Moreover, K_M values for tRNA and aminoacid in reactions of
tRNA aminoacylation, under conditions mimicking in vivo conditions, also are
of the order of μM[34,35] such that in vivo all tRNAs are predominately aminoacylated. Thus under conditions of a reduced Phe-tRNA^{Phe} intracellular concentration due to lack of phenylalanine, a 10-fold increase in phenylalanyl-tRNA
synthetase activity probably could not increase the tRNA^{Phe} aminoacylation
rate better than a 2-fold increase. Moreover, in the case of extreme phenylalanine starvation, stimulation of the expression of phenylalanyl-tRNA
synthetase would have a detrimental effect depriving the cell of necessary
resources which should be directed towards the synthesis of phenylalanine.

The possibility of attenuation in the expression of the pheS-pheT operon
may be only one among several other control mechanisms. The in vivo studies
suggest an effect of the gene products on their own expression. However, it is
not impossible that this autoregulation is related to the presence of an attenuator structure in front of pheS. For example, it is known that the histidine
operon, which is controlled by attenuation, is derepressed by an excess of
histidyl-tRNA synthetase[36,37]. It is not known, however, whether this positive
regulation occurs at the level of the attenuator. Finally, it is worth noting
that, immediately following the attenuator is an open reading frame under

control of a strong translation initiation signal (GGAGG), which is situated within the attenuator RNA structure (Fig. 8). The open reading frame encodes a peptide devoid of phenylalanine residues. Expression of the polypeptide has yet to be shown. One may speculate on the possible role of such a peptide also under control of attenuation. As the coding sequence covers 93 nucleotides out of the 210 nucleotides separating the attenuator from the initiation codon of pheS, one possible role for the translation of such a coding sequence could be that coupled with transcription, it insures maximum RNA elongation in this region, eliminating any premature transcription termination due to polarity. However, deleting the majority of this region would have the same anti-polarity effect ; so may be the hypothetical polypeptide (for which no analogous peptide exists within the amino acid biosynthetic operons) also is involved in control of PheRS expression (or phenylalanine metabolism).

ACKNOWLEDGMENTS

We thank A. Böck, M. Cassadaban and B. Bachmann for the gift of strains. This work was supported by following grants : Centre National de la Recherche Scientifique (G.R. No. 18, L.A. No. 240), Délégation Générale à la Recherche Scientifique et Technique (Conventions 80.E.0872 and 81.E.1207).

REFERENCES

1. Neidhardt, F.C., Parker, J. and Mc Keever, W. (1975) Ann. Rev. Microbiol., 29. 215-250
2. Morgan, S.D. and Söll, D. (1978) Progr. Nucl. Acid Res. and Molec. Biol., 21, 181-207
3. Neidhardt, F.C., Bloch, P.L., Pedersen, S. and Reeh, S. (1977) J. Bacteriol., 129, 378-387
4. Reeh, S., Pedersen, S. and Friesen, J.D. (1976) Molec. Gen. Genet., 149, 279-289
5. Baer, M., Low, K.B. and Söll, D. (1979) J. Bacteriol., 139, 165-175
6. Cheung, A., Morgan, S., Low, K.B. and Söll, D. (1979) J. Bacteriol.. 139. 176-184
7. Putney, S.D. and Schimmel, P. (1981) Nature, 291, 632-635
8. Springer, M., Graffe, M. and Hennecke, H. (1977) Proc. Natl. Acad. Sci. USA, 74, 3970-3974
9. Hennecke, H., Springer, M. and Böck, A. (1977) Molec. Gen. Genet., 152, 205-210
10. Hennecke, H., Böck, A., Thomale, J. and Nass, G. (1977) J. Bacteriol., 131, 943-950
11. Comer, M.M. and Böck, A. (1976) J. Bacteriol., 127, 923-933
12. Springer, M., Graffe, M. and Grunberg-Manago, M. (1979) Molec. Gen. Genet., 169, 337-343
13. Plumbridge, J.A., Springer, M., Graffe, M., Goursot, R. and Grunberg-Manago, M. (1980) Gene, 11, 33-42
14. Sacerdot, C., Fayat, G., Dessen, P., Springer, M., Plumbridge, J.A., Grunberg-Manago, M. and Blanquet, S. (1982) The EMBO Journal 1, in press

15. Springer, M., Plumbridge, J.A., Trudel, M., Graffe, M. and Grunberg-Manago, M. (1982) Molec. Gen. Genet., in press
16. Plumbridge, J.A. and Springer, M. (1980) J. Mol. Biol., 144, 595-600
17. Casadaban, M.J. and Cohen, S.N. (1980 J. Mol. Biol., 138, 179-207
18. Plumbridge, J.A. and Springer, M. (1982) submitted for publication
19. Plumbridge, J.A. and Springer, M. (1982) submitted for publication
20. Nass, G. and Neidhardt, F.C. (1967) Biochim. Biophys. Acta, 134, 347-359
21. Trudel, M. (1982) Thèse de Doctorat de l'Université Paris VI
22. Hilton, J.L., Kearney, P.C. and Ames, B.N. (1965) Arch. Biochem. Biophys., 112, 544-547
23. Yanofsky, C. (1981) Nature, 289, 751-758
24. Miller, J.H. (1972) Experiments in Molecular Genetics, Cold Spring Harbor Laboratory, New-York
25. Neidhardt, F.C., Bloch, P.L. and Smith, D.F. (1974) J. Bacteriol., 119, 736-747
26. Sykes, R.B. and Nordström, K. (1972) Antimicrob. Agents and Chemotherapy, 1, 94-99
27. Shine, J. and Dalgarno, L. (1974) Proc. Natl. Acad. Sci. USA, 71, 1342-1346
28. Fayat, G., Blanquet, S., Dessen, P., Batelier, G. and Waller, J.P. (1974) Biochimie, 56, 33-41
29. Rosenberg, M. and Court, D. (1979) Ann. Rev. Genet., 13, 319-353
30. Pribnow, D. (1975) Proc. Natl. Acad. Sci. USA, 72, 784-789
31. Lee, F. and Yanofsky, C. (1977) Proc. Natl. Acad. Sci. USA, 74, 4365-4369
32. Blanquet, S., Iwatsubo, M. and Waller, J.P. (1973) Eur. J. Biochem., 36, 213-226
33. Neidhardt, F.C., Block, P.L., Pedersen, S. and Reeh, S. (1977) J. Bacteriol., 129, 378-387
34. Lawrence, F., Blanquet, S., Poiret, M., Robert-Géro, M. and Waller, J.P. (1973) Eur. J. Biochem., 36, 234-243
35. Santi, D.V., Danenberg, P.V. and Satterly, P. (1971) Biochemistry, 10, 4804-4812
36. Wyche, J.H., Ely, B., Cebula, T.A., Snead, M.C. and Hartman, P.E. (1974) J. Bacteriol., 117, 708-716
37. Eisenbeis, S.J. and Parker, J. (1981) Molec. Gen. Genet., 183, 115-122

GENES FOR TWO E. COLI AMINOACYL tRNA SYNTHETASES

PAUL SCHIMMEL, TERESA KENG, AND SCOTT PUTNEY
Department of Biology, Massachusetts Institute of Technology, Cambridge, MA
02139, USA

INTRODUCTION

For approximately half of the twenty aminoacyl tRNA synthetases, synthesis
is derepressed under conditions of starvation for the cognate amino acid.[1-5]
Enzyme levels are also subject to metabolic regulation whereby the concentration of each enzyme varies with growth rate.[6,7] Metabolic regulation is also
found for other protein factors used for protein synthesis[8-11] and for stable
RNA species.[8,12] While the general features of both of these regulatory phenomena have been known for some time, only recently have studies been initiated
at the level of the genes for synthetases.

Amino acid dependent regulation is found for at least two other classes of
genes. Most extensively studied are the amino acid biosynthetic operons.[13]
Regulation in these cases is at the level of transcription and involves an
attenuator mechanism. In this case, the region 5' to the coding region of the
first structural gene encodes a short leader peptide which is rich in the cognate amino acid. Following the leader peptide encoding region is an RNA polymerase transcription termination site. Through coupling of translation (of
the leader peptide encoding region) to transcription, an efficient mechanism
operates to regulate the amount of RNA polymerase which escapes premature
transcription termination and, consequently, which transcribes structural genes
for enzymes within the operon.

Genes for amino acid transport proteins are also regulated by amino acid
supply.[14,15] For these systems gene regulatory mechanisms are not well understood, although preliminary data suggest transcriptional control is an important factor.[16]

As for aminoacyl tRNA synthetases, only one system has been characterized in
detail.[17] However, the genes for numerous synthetases are currently under investigation and a more complete picture of regulation of this class of proteins
gradually will emerge. Discussed below are genes for two aminoacyl tRNA syn-
thetases--the alaS and the glyS loci.

GENE FOR Ala-tRNA SYNTHETASE: AUTOGENOUS REGULATION

5'-non-coding region is not analogous to amino acid biosynthetic operons.
The alaS locus is at 56 min of the E. coli chromosome and is closely linked to
recA.[18] The structural gene encodes a polypeptide of 875 amino acids and this
structural gene, and several hundred nucleotides upstream from the start of the
structural gene, have been sequenced.[19, 20] None of the 5'-non-coding region
shows the general nucleotide sequence configuration found in front of structural genes for amino acid biosynthetic operons. That is, there is no leader
peptide encoding segment which is rich in alanine residues and which is followed by an RNA polymerase transcription termination site.

Transcription starts 79 nucleotides upstream from the start of the structural gene. In vitro transcription, with a restriction fragment encompassing
several hundred nucleotides of the 5'-non-coding region and the beginning of
the structural gene, showed transcription starts with pppA 79 nucleotides in
front of the first codon of the structural gene.[19] Shown in Figure 1 is the
nucleotide sequence of the beginning of the structural gene and of 105 nucleotides in front of the structural gene. Immediately preceding the structural
gene segment is a Shine-Dalgarno sequence which is complementary to the 3'-
terminal sequence of 16S rRNA,[21, 22] and just in front of the transcription
initiation point is a sequence (TATCTTA) which fits with the consensus sequence
for bacterial RNA polymerase promoter sites (Pribnow Box).[23, 24]

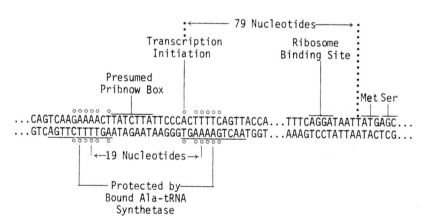

Fig. 1. Characteristics of the 5'-non-coding region of E. coli alaS. The
transcription initiation point, presumed Pribnow Box (shown by overbar, cf 23,
24), ribosome binding site (overbar), and regions protected from DNAase by
bound Ala-tRNA synthetase (underlines) are shown. A palindromic sequence is
indicated by open circles.

Regulation of transcription by Ala-tRNA synthetase and its cognate amino
acid. As stated above, the region in front of the Ala-tRNA synthetase structural gene does not contain nucleotide sequences characteristic of those associated with regulatory regions of amino acid biosynthetic operons. On the
other hand, for Leu-, Ser-, and Val-tRNA synthetases there are mutations which
result in increased enzyme levels, and where the mutations are believed to be
within a cis-acting operator locus.[25-28] To explore whether sequences in the
5'-non-coding region of alaS have a regulatory role, in vitro transcription
studies were carried out in the presence and absence of Ala-tRNA synthetase.
These studies showed that, at in vitro concentrations of 10 μM, the synthetase
represses transcription of its own gene. Other synthetases, and non-specific
proteins such as bovine serum albumin and lysozyme, at a comparable protein
concentration to that required for repression by Ala-tRNA synthetase, have no
effect on in vitro transcription.[17]

These results suggest a role for regulation of transcription by Ala-tRNA
synthetase. On the other hand, it is difficult to rationalize the physiological
significance of this observation. First, concentrations of enzyme required to
achieve repression (10 μM) are substantially above intracellular enzyme concentrations of 0.1 μM to 1 μM.[17] Second, the results do not explain amino acid
dependent repression, because the repression described above was achieved in
the absence of amino acid.

Further experiments showed that addition of alanine, but not other amino
acids, to the transcription mixture greatly intensifies repression caused by
Ala-tRNA synthetase. In particular, using 10-fold lower Ala-tRNA synthetase
concentrations, which is comparable to physiological amounts, addition of alanine gives strong repression of transcription. The amount of alanine required
(approximately 1 mM) is comparable to the synthetase's K_m for that amino acid
as well as to the intracellular concentration of alanine.[29] These results explain amino acid dependent repression of synthetase gene expression (for the
Ala-tRNA synthetase example), and they fit well with physiological considerations.

It is not known whether the cofactor in repression is simply the amino acid
or its aminoacyl adenylate form: transcription of necessity is done in the presence of ATP and the synthetase quickly converts amino acid and ATP to the enzyme bound aminoacyl adenylate. The balance between enzyme-amino acid complex
and enzyme-aminoacyl adenylate complex is determined in part by intracellular
pyrophosphate concentrations; this is because PP_i reacts with enzyme-adenylate

to regenerate ATP and amino acid. In addition to reacting with pyrophosphate, enzyme-bound adenylate reacts with uncharged tRNA. In preliminary experiments, addition of uncharged tRNA^{Ala}, along with amino acid and nucleotide triphosphates, had no obvious effect on the repression efficiency.[17]

Ala-tRNA synthetase binds to a specific DNA sequence which flanks its gene's transcription start site. The results discussed above suggest that the synthetase interacts directly with alaS and that this in turn affects transcription. This was confirmed by DNAase protection experiments[30, 31] which showed that Ala-tRNA synthetase binds to a palindromic sequence which flanks the gene's transcription start site. The location of the DNA sequence protected by Ala-tRNA synthetase, and the palindrome contained therein, are shown in Figure 1. Note that the regions protected by Ala-tRNA synthetase lie on either side of a Pribnow Box. Thus polymerase and synthetase mutually exclude each other from the DNA.

Note also that the centers of the palindromic sequence are 19 base pairs apart. This corresponds almost exactly to two turns of a DNA B helix. Therefore, the spacing of these sequences can accommodate binding of synthetase to one side of the helix, provided that the protein has a two-fold axis of symmetry. It is noteworthy that Ala-tRNA synthetase is an α_4 tetramer and that chemical cross-linking experiments provide evidence that the synthetase is arranged with a two-fold axis of symmetry, in contrast to a four-fold cyclic symmetry arrangement.[32] With a tetramer molecular weight of 380,000, the protein easily can span a distance of 70-100 Å, which is the distance required to explain the DNAase protection experiments.

Physiological implications and in vivo studies. The data described above can explain amino acid dependent repression of enzyme synthesis. Regulation occurs at the level of transcription and occurs at concentrations of enzyme and amino acid which are found in vivo. No additional proteins are needed to explain this aspect of the regulation. But metabolic regulation is not explained by these data, at least not in an obvious way. If the cofactor in repression is the adenylate, and not the amino acid, then repression would be related to ATP levels as well as to amino acid concentrations. However, ATP levels do not appear to offer a satisfactory link to metabolic regulation because these levels are not very sensitive to growth rates.[33, 34]

While a transcriptional regulation mechanism for Ala-tRNA synthetase has been well characterized in the in vitro system, it is one of the enzymes for which little in vivo data exist. It has not been shown that, like some of the

other synthetases, Ala-tRNA synthetase is subject in vivo to amino acid dependent repression. Lambda phage vectors which incorporate the alaS promoter in
front of lacZ[35] are being used to address this question. Host cells which are
lacZ⁻, and which make stable lysogens with the aforementioned recombinant phage
vector, have β-galactosidase synthesis under control of the alaS promoter. In
this way both amino acid dependent repression and metabolic regulation can be
studied through measurements of β-galactosidase levels which are easily assayed.

A second approach involves the use of maxicells.[36] The maxicell system has
not generally been used for studies of cell regulation, and for good reason.
Maxicells are prepared by uv irradiation of cells, and this treatment presumably damages cellular components other than chromosomal DNA. Nevertheless, in
spite of this limitation, preliminary results suggest that the maxicell system
is an effective vehicle for studying expression of Ala-tRNA synthetase under
different conditions.

In addition to studies aimed at confirming and expanding upon regulation of
biosynthesis of this synthetase, the role of the protein structure in binding
to the DNA is also being explored. Previous work showed that the carboxyl-
terminal half of the protein is required to achieve oligomerization of subunits
into the tetramer.[32] The tetrameric structure is presumably required for binding to the DNA. By producing a set of nested deletions in the carboxyl-terminal
coding region of the gene, it has been possible to produce synthetase subunits
of different sizes.[37] These in turn are being investigated for their ability
to oligomerize and to repress transcription.

Gly-tRNA SYNTHETASE: TANDEM CODING REGIONS FOR TWO SUBUNITS AND A SINGLE PROMOTER

Gly-tRNA synthetase was originally characterized as an $\alpha_2\beta_2$ tetrameric enzyme with α- and β-subunit molecular weights, respectively, of 40,000 and
65,000 daltons.[38, 39] The glyS locus is at 79 min on the E. coli chromosome
where it is closely linked to the xylose locus.[40] By complementation of a xyl⁻
host, the glyS locus was originally cloned from a Clarke and Carbon bank of
E. coli chromosomal segments inserted into the ColEl vector.[41] Subsequent subcloning operations in this laboratory localized the gene to a 5.2 kb Hind III
restriction fragment which was cloned into pBR322 to give the recombinant plasmid pTK201. Cells carrying the recombinant plasmid overproduce both subunits
of Gly-tRNA synthetase by several-fold.

Arrangement of genes in the glyS locus. Figure 2 shows a restriction map of the Hind III fragment carrying the glyS locus. In order to locate the subunit coding regions and direction of transcription, Tn5 insertions[42] were made into this region of pTK201. Nine different insertions were mapped with restriction enzymes to a resolution of 0.2 kb within the Hind III fragment. Those that eliminated Gly-tRNA synthetase activity were identified. In addition, maxicells[36] were used to study production of the individual subunits synthesized from plasmids containing various Tn5 insertions.

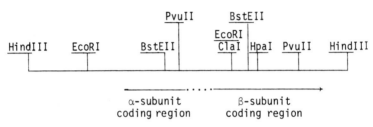

Fig. 2. A 5.2 kb Hind III fragment which contains glyS. The direction of transcriptions and locations of α- and β-subunit coding regions are shown. These features were established by analysis of Tn5 insertions and maxicell protein products (see text). The location of the junction between subunit coding regions is imprecise.

Tn5 insertions into a 1 kb region (of pTK201), centered around the left-most BstEII site, do not allow synthesis of either subunit of glycyl-tRNA synthetase. A second group of insertions, all mapping within a 2 kb region centered around the right-most BstEII site, allow for synthesis of only the α-subunit. There were no insertions which allow for the synthesis of only the β-subunit. From these analyses, it is possible to define the location of the α- and β-subunit coding regions. These are shown in Figure 2. These results also indicate that there is one promoter from which the coding regions of both subunits are transcribed, and that the direction of transcription is from α- to β-subunit coding regions.

Regulation of glyS. Using high resolution two-dimensional protein gels, levels of the β-subunit of gly-tRNA synthetase have been studied under different physiological conditions.[6] These studies show conclusively that the enzyme is subject to metabolic regulation. Studies on amino acid dependent repression have not been reported.

The nucleotide sequence arrangement of the 5'-non-coding region has been established for 500 nucleotides in front of the amino-terminal encoding region of the α-subunit. Like the <u>alaS</u> locus, this locus shows no resemblance to operator regions of amino acid biosynthetic operons. That is, there is no leader peptide encoding region and transcription attenuator arrangement. Preliminary studies of <u>in vitro</u> transcription of <u>glyS</u> are being carried out in this laboratory.

Note that the sum of α-subunit and β-subunit molecular weights (∿100,000) is approximately equal to that for the single subunit of Ala-tRNA synthetase (∿95,000). Other work has indicated, roughly, a two domain structure of the alanine enzyme.[12] These domains may correspond to the two subunits of Gly-tRNA synthetase.

ACKNOWLEDGEMENTS
This work was supported by Grant No. GM15539 from the National Institutes of Health.

REFERENCES
1. Neidhardt, F. C., Parker, J. and McKeever, W. G. (1975) Ann. Rev. Microbiology, 29, 215-250.
2. Brenchley, J. E. and Williams L. S. (1975) Ann. Rev. Microbiology, 29, 251-274.
3. LaRossa, R. and Söll, D. (1978) Transfer RNA, Altman, S. ed., MIT Press, Cambridge, Massachusetts, pp. 136-167.
4. Umbarger, H. E. (1980) Transfer RNA: Biological Aspects, Söll, D., Abelson, J. and Schimmel, P. ed., Cold Spring Harbor Laboratory, New York, pp. 453-467.
5. Morgan, S. D. and Söll, D. (1978) Prog. Nucleic Acid Res. Molec. Biol., 21, 181-207.
6. Pederson, S., Bloch, P. L., Reeh, S. and Niedhardt, F. C. (1978) Cell, 14, 179-190.
7. Neidhardt, F. C., Bloch, P. L., Pedersen, S. and Reeh, S. (1977) J. Bact., 129, 378-387.
8. Kjeldgaard, N. O. and Gausing, K. (1974) Ribosomes, Nomura, M., Tissieres, A. and Lengyel, P. ed., Cold Spring Harbor Laboratory, New York, pp. 369-392.
9. Gordon, J. (1970) Biochemistry, 9, 912-917.
10. Furano, A. V. (1975) Proc. Nat. Acad. Sci. USA, 72, 4780-4784.
11. Herendeen, S., Van Bogelen, R. and Neidhardt, F. (1979) J. Bact., 139, 185-194.
12. Skjold, A. C., Juarez, H. and Hedgcoth, C. (1973) J. Bact., 115, 177-187.
13. Yanofsky, C. (1981) Nature, 289, 751-758.
14. Landick, R., Anderson, J. J., Mayo, M. M., Gunsalus, R. P., Mavromara, P., Daniels, C. J. and Oxender, D. L. (1980) J. Supramol. Struct., 14, 527-537.
15. Higgins, C. F. and Ames, G. F.-L. (1982) Proc. Natl. Acad. Sci. USA, 79, 1082-1087.
16. Landick, R., Mavromara, P., and Oxender, D. L. (1982) Fed. Proc., 41, 2817.
17. Putney, S. D. and Schimmel, P. (1981) Nature, 291, 632-635.
18. Theall, G., Low, K. B. and Söll, D. (1977) Molec. Gen. Genet., 156, 221-227.
19. Putney, S. D., Melendez, D. L. and Schimmel, P. R. (1981) J. Biol. Chem., 256, 205-211.

20. Putney, S. D., Royal, N. J., Neuman de Vegvar, H., Herlihy, W. C., Biemann, K. and Schimmel, P. (1981) Science, 213, 1497-1501.
21. Shine, J. and Dalgarno, L. (1974) Proc. Natl. Acad. Sci. USA, 71, 1342-1346.
22. Steitz, J. A. and Jakes, K. (1975) Proc. Natl. Acad. Sci. USA, 72, 4734-4738.
23. Pribnow, D. (1975) J. Mol. Biol., 99, 419-443.
24. Rosenberg, M. and Court, D. (1979) Annu. Rev. Genet., 13, 319-353.
25. Pizer, L. I., McKitrick, J. and Tosa, T. (1972) Biochem. Biophy. Res. Commun., 49, 1351-1357.
26. Clarke, S. J., Low, B. and Konigsberg, W. (1973) J. Bact., 113, 1096-1103.
27. Baer, M., Low, K. B. and Söll, D. (1979) J. Bact. 139, 165-175.
28. LaRossa, R., Vogeli, G., Low, K. B. and Söll, D. (1977) J. Molec. Biol., 117, 1033-1048.
29. Piperno, J. R. and Oxender, D. L. (1968) J. Biol. Chem., 243, 5914-5920.
30. Galas, D. J. and Schmitz, A. (1978) Nucleic Acids Res., 5, 3157-3170.
31. Johnson, A. D., Meyer, B. J. and Ptashne, M. (1979) Proc. Natl. Acad. Sci. USA, 76, 5061-5065.
32. Putney, S. D., Sauer, R. T. and Schimmel, P. R. (1981) J. Biol. Chem., 256, 198-204.
33. Lowry, O. H., Carter, J., Ward, J. B., and Glaser, L. (1971) J. Biol. Chem., 246, 6511-6521.
34. Franzen, J. S. and Binkley, S. B. (1961) J. Biol. Chem., 236, 515-519.
35. Bassford, P., Beckwith, J., Berman, M., Brickman, E., Casadaban, M., Guarente, L., Saint-Girons, I., Sarthy, A., Schwartz, M., Shuman, H. and Silhavey, T. (1978) in The Operon, Miller, J. and Reznikoff, W. ed., Cold Spring Harbor Laboratory, New York, pp. 245-262.
36. Sancar, A., Mack, A. M. and Rupp, W. D. (1979) J. Bacteriol., 137, 692-693.
37. Jasin, M., Kasper, T., and Schimmel, P., unpublished observations.
38. Ostrem, D. L. and Berg, P. (1970) Proc. Natl. Acad. Sci. USA, 67, 1967-1974.
39. McDonald, T., Breite, L., Pangburn, K. L. W., Horn, S., Manser, J. and Nagel, G. M. (1980) Biochemistry, 19, 1402-1409.
40. Bachmann, B. J. and Low, K. B. (1980) Micro. Reviews, 44, 1-56.
41. Clarke, L. and Carbon, J. (1976) Cell, 9, 91-99.
42. Kleckner, N., Roth, J. and Botstein, D. (1977) J. Mol. Biol., 116, 125-159.

Published 1982 by Elsevier Science Publishing Co., Inc.
Marianne Grunberg-Manago and Brian Safer, editors
INTERACTION OF TRANSLATIONAL AND TRANSCRIPTIONAL
CONTROLS IN THE REGULATION OF GENE EXPRESSION

STRUCTURE AND REGULATION OF E. COLI GLUTAMINYL-tRNA SYNTHETASE

ALICE CHEUNG, PATRICIA HOBEN, K. BROOKS LOW[*], SUSAN MORGAN, NANCY ROYAL[+],
FUMIAKI YAMAO, AND DIETER SöLL
Department of Molecular Biophysics and Biochemistry and [*]Department of
Therapeutic Radiology, Yale University, New Haven, CT 06511 USA and
[+]Department of Chemistry, Massachusetts Institute of Technology,
Cambridge, MA 02139 USA

INTRODUCTION

Aminoacyl-tRNA synthetases are essential enzymes for protein synthesis.

These enzymes catalyse the esterification of an amino acid to its cognate

tRNAs to form aminoacyl-tRNAs.[1] Besides being indispensable enzymes for

protein biosynthesis, aminoacyl-tRNA synthetases are also involved in many

other cellular processes, for example, the regulation of amino acid biosyn-

thesis[2] and amino acid transport.[3] Because of their central role in cellu-

lar metabolism, an understanding of the structure-function relationship of

aminoacyl-tRNA synthetases with their various substrates and the regulatory

mechanisms pertaining to their biosynthesis will undoubtedly further our

knowledge of cellular physiology. In the past two decades, much information

has been collected on the structure and interaction of tRNA species with

their cognate synthetases. However, the structural studies of the enzymes

have yielded less information. Studies on the regulation of the synthetases

centered on the isolation and genetic analysis of mutants[4] and also on

physiological variations of enzyme levels.[5] The latter studies led to the

proposal of metabolic control for some of these enzymes, i.e. their activity

and/or amount varies with physiological environment. Genetic studies led to

the isolation of regulatory mutants with altered level of some synthetases.[4]

Both the physiological and genetic studies yielded no molecular details to

explain the observed effects. The advent of recombinant DNA and nucleic

acid sequencing technology led to the re-examination of these long standing

questions about aminoacyl-tRNA synthetases in a more defined as well as refined level. By now the gene sequence of a number of aminoacyl-tRNA synthetase genes has been determined[6,7,8] and in vitro studies in these genes suggested an autoregulatory mechanism for E. coli alanyl-tRNA synthetase.[9]

Our laboratory has been interested in both the structural and regulatory aspects of aminoacyl-tRNA synthetases for several years. Recently we focused on studying E. coli glutaminyl-tRNA synthetase (GlnRS). E. coli GlnRS is a monomeric polypeptide of 550 amino acids.[10] It offers a polypeptide of convenient size for structure-function analysis. Moreover, glnS, the structural gene for this enzyme has been cloned.[8] Overproduction of GlnRS in E. coli harboring the glnS containing plasmid allows relatively easy access to pure GlnRS in sufficient quantities for structural analysis.[10] Furthermore, structural gene mutations in glnS are available. Analysis of these mutations will help define the structural domains of GlnRS. One of these mutations, glnS*,[14] codes for a mischarging GlnRS which misaminoacylates a noncognate tRNA, namely suppressor-3 $tRNA^{Tyr}$ with glutamine. Analysis of this mutant, and other conceivable ones similar or analogous to glnS*, will shed light on the tRNA recognition region in GlnRS. There is also a temperature sensitive mutation,[11] $glnS^{ts}$, which not only has been useful in the isolation of regulatory mutants for GlnRS and $tRNA^{Gln}$, but will also be useful in defining the structural domains of GlnRS. Analysis of these mutations that affect the level of $GlnRS^{12}$ or $tRNA^{Gln\ 12,13}$ and other possible regulatory mutants to be isolated will enable us to understand better the regulation of GlnRS and $tRNA^{Gln}$ biosynthesis.

ISOLATION OF REGULATORY MUTANTS FOR GlnRS

A temperature sensitive E. coli strain (AB4143) harboring a mutation in the GlnRS gene ($glnS^{ts}$) was isolated and characterized many years ago.[11] In an attempt to obtain regulatory mutants for GlnRS, spontaneous temperature resistant revertants which still contain the thermolabile enzyme were

isolated from strain AB4143.[12,13] Three interesting revertants were characterized. Strain KL361, viable at 42°C causes a four-fold overproduction of
$tRNA_1^{Gln}$, one of the two glutamine isoacceptors.[13] This mutation, glnT, is
unstable and was shown to be a duplication mutation. Strain KL477, viable
at 45°C, causes a two-fold overproduction of both $tRNA^{Gln}$ isoacceptors and a
1.5 to two-fold overproduction of several other tRNA species including
$tRNA^{Met}$, $tRNA^{Ile}$, $tRNA^{Pro}$, $tRNA^{Trp}$, $tRNA^{Cys}$, and $tRNA^{Glu}$. This mutation,
glnU, is mapped[12] and there is no known clustering of tRNA genes around that
locus. A third revertant, KL476, also viable at 45°C, causes a five-fold
overproduction of the thermolabile GlnRS.[12] This mutation, glnR, maps away
from the structural gene glnS. The genetic location of these mutations is
shown in Fig. 1. The molecular characterization of the glnR and glnU mutations awaits their isolation and characterization by cloning and in vitro
studies of their effect on the expression of the glnS gene and the genes for
the tRNAs affected.

E. coli circular linkage map

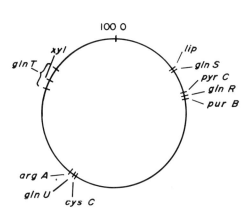

Fig. 1. An E. coli
circular linkage map
showing the position
of glnS, glnT, glnR,
and glnU.

ISOLATION OF THE glnS GENE AND ITS MUTANT DERIVATIVES

In order to study regulation of GlnRS biosynthesis and the structure-function relationship of this enzyme, it was imperative to have its structural gene (glnS) cloned. By using in vivo genetic engineering and complementation of the temperature-sensitive mutation in strain AB4143, a λ-transducing phage carrying glnS (λglnS$^+$) was isolated. From the DNA of this phage the glnS gene was recloned into the pBR322 plasmid (pYY105).[8] In order to map other structural gene mutations, deletion mutants in pYY105 were constructed in vitro. The ability of these deletion mutations to complement the glnS^{ts} mutation to give temperature resistant transformants, or to recombine with the glnS^{ts} gene to give recombinants viable at high temperature was determined. By this way the glnS^{ts} mutation was localized to a 300 bp region in the middle of glnS. Using similar manipulations except with a different selection scheme, glnS*, a spontaneous mutation from glnS was recloned into pBR322. The glnS* mutation is located in the 3'-terminal half of the glnS gene. Comparison of the structure of glnS^{ts}, and glnS* with that of glnS will shed light on the structural changes in GlnRS leading to the observed characteristics of each mutant enzyme.

CHARACTERIZATION OF THE glnS GENE AND ITS PRODUCT

The nucleotide sequence of the entire glnS gene with its 5'- and 3'-flanking regions was determined by the Maxam-Gilbert method.[15] The structural region that codes for a protein consisting of 550 amino acids is shown in Fig. 2. To ensure the accuracy of the DNA sequence, a partial amino acid sequence analysis of GlnRS was performed. Manual Edman degradation was used to obtain an oligopeptide for the amino terminus of GlnRS, carboxypeptidase B digestion determined the dipeptide at the carboxyl-terminus of GlnRS. Some internal tri-, tetra- and pentapeptide sequences, covering about 20% of GlnRS, were obtained by a gas chromatography-mass spectroscopy technique[16]

61

ESCHERICHIA COLI GLUTAMINYL-tRNA SYNTHETASE

```
                        10                        20                        30
AGT GAG GCA GAA GCC CGT TCG ACT AAC TTT ATC CGT CAG ATC ATC GAT GAA GAT CTG GCC AGT GGT AAG CAC ACC ACA GTA CAC ACC CGT
Ser Glu Ala Glu Ala Arg Ser Thr Asn Phe Ile Arg Gln Ile Ile Asp Glu Asp Leu Ala Ser Gly Lys His Thr Thr Val His Thr Arg
                        40                        50                        60
TTC CCG CCG GAG CCG AAT GGC TAT CTG CAT ATT GGC CAT GCG AAA TCT ATC TGC CTG AAC TTC GGG ATC GCC CAG GAC TAT AAA GGC CAG
Phe Pro Pro Glu Pro Asn Gly Tyr Leu His Ile Gly His Ala Lys Ser Ile Cys Leu Asn Phe Gly Ile Ala Gln Asp Tyr Lys Gly Gln
                        70                        80                        90
TGC AAC CTG CGT TTC GAC GAC ACT AAC CCG GTA AAA GAA GAT ATC GAG TAT GTT GAG TCG ATC AAA AAC GAC GTA GAG TGG TTA GGT TTT
Cys Asn Leu Arg Phe Asp Asp Thr Asn Pro Val Lys Glu Asp Ile Glu Tyr Val Glu Ser Ile Lys Asn Asp Val Glu Trp Leu Gly Phe
                        100                       110                       120
CAC TGG TCT GGT AAC GTC CGT TAC TCC TCC GAT TAT TTT GAT CAG CTC CAC GCC TAT GCG ATC GAA CTG ATC AAT AAA GGC CTG GCG TAC
His Trp Ser Gly Asn Val Arg Tyr Ser Ser Asp Tyr Phe Asp Gln Leu His Ala Tyr Ala Ile Glu Leu Ile Asn Lys Gly Leu Ala Tyr
                        130                       140                       150
GTT GAT GAA CTG ACG CCG GAA CAG ATC CGC GAA TAC CGC GGC ACC CTG ACG CAA CCG GGT AAA AAC CCG TAC CGC GAC CGC AGC GTT
Val Asp Glu Leu Thr Pro Glu Gln Ile Arg Glu Tyr Arg Gly Thr Leu Thr Gln Pro Gly Lys Asn Ser Pro Tyr Arg Asp Arg Ser Val
                        160                       170                       180
GAA GAG AAC CTG GCG CTG TTC GAA AAA CGT GCC GGT GGT TTT GAA GAA GGT AAA GCC TGC CTG CGT GCG AAA ATC GAC ATG GCT TCA
Glu Glu Asn Leu Ala Leu Phe Glu Lys Met Arg Ala Gly Gly Phe Glu Glu Gly Lys Ala Cys Leu Arg Ala Lys Ile Asp Met Ala Ser
                        190                       200                       210
CCG TTT ATC GTG ATG CGC GAT CCG GTG CTG TAC CGT ATT AAA TTT GCT GAA CAC CAC CAG ACT GGC AAC AAG TGG TGC ATC TAC CCG ATG
Pro Phe Ile Val Met Arg Asp Pro Val Leu Tyr Arg Ile Lys Phe Ala Glu His His Gln Thr Gly Asn Lys Trp Cys Ile Tyr Pro Met
                        220                       230                       240
TAC GAC TTC ACC CAC TGC ATC AGC GAT GCC CTG GAA GGT ATT ACG CAC TCT CTG TGT ACG CTT GAG TTC CAG GAC AAC CGT CGT CTG TAC
Tyr Asp Phe Thr His Cys Ile Ser Asp Ala Leu Glu Gly Ile Thr His Ser Leu Cys Thr Leu Glu Phe Gln Asp Asn Arg Arg Leu Tyr
                        250                       260                       270
GAC TGG GTA CTG GAC AAC ATC ACG ATT CCT GTT CAC CCG CGC CAG TAT GAG TTC TCG CGT CTG AAT CTG GAA TAC ACC GTG ATG TCC AAG
Asp Trp Val Leu Asp Asn Ile Thr Ile Pro Val His Pro Arg Gln Tyr Glu Phe Ser Arg Leu Asn Leu Glu Tyr Thr Val Met Ser Lys
                        280                       290                       300
CGT AAG TTG AAC CTG CTG GTG ACC GAC AAG CAC GTT GAA GGC TGG GAT GAC CCG ATG CCG ACC ATT TCC GGT CTG CGT CGT CGT GGT
Arg Lys Leu Asn Leu Leu Val Thr Asp Lys His Val Glu Gly Trp Asp Asp Pro Arg Met Pro Thr Ile Ser Gly Leu Arg Arg Arg Gly
                        310                       320                       330
TAC ACT GCG GCT TCT ATT CGT GAG TTC TGC AAA CGC ATC GGC GTG ACC AAG CAG GAC AAC ACC ATT GAG ATG GCG TCG CTG GAA TCC TGC
Tyr Thr Ala Ala Ser Ile Arg Glu Phe Cys Lys Arg Ile Gly Val Thr Lys Gln Asp Asn Thr Ile Glu Met Ala Ser Leu Glu Ser Cys
                        340                       350                       360
ATC CGT GAA GAT CTC AAC GAA AAT GCG CCG CGC GCA ATG GCG GTT ATC GAT CCG GTG AAA CTG GTT ATC GAA AAC TAT CAG GGC GAA GGC
Ile Arg Glu Asp Leu Asn Glu Asn Ala Pro Arg Ala Met Ala Val Ile Asp Pro Val Lys Leu Val Ile Glu Asn Tyr Gln Gly Glu Gly
                        370                       380                       390
GAA ATG GTT ACC ATG CCG AAC CAT CCG AAC AAA CCG GAA ATG GGC AGC CGT CAG GTG CCG TTT AGC GGT GAG ATT TGG ATT GAT CGC GCC
Glu Met Val Thr Met Pro Asn His Pro Asn Lys Pro Glu Met Gly Ser Arg Gln Val Pro Phe Ser Gly Glu Ile Trp Ile Asp Arg Ala
                        400                       410                       420
GAT TTC CGC GAA GAA GCT AAC AAG CAG TAC AAA CGT CTG GTG CTG GGT AAA GAA GTG CGT CTG CGT AAT GCT TAT GTG ATT AAG GCA GAA
Asp Phe Arg Glu Glu Ala Asn Lys Gln Tyr Lys Arg Leu Val Leu Gly Lys Glu Val Arg Leu Arg Asn Ala Tyr Val Ile Lys Ala Glu
                        430                       440                       450
CGC GTG GAA AAA GAT CCG GAA GGT AAT ATC ACC ACC ATC TTC TGT ACT TAT GAC GCC GAT ACC TTA AGC AAA GAT CCG GCA GAT GGT CGT
Arg Val Glu Lys Asp Pro Glu Gly Asn Ile Thr Thr Ile Phe Cys Thr Tyr Asp Ala Asp Thr Leu Ser Lys Asp Pro Ala Asp Gly Arg
                        460                       470                       480
AAA GTC AAA GGT GTT ATT CAC TGG GTG AGC GCG GCA CAT GCG CTG CCG GTT GAA ATC CGT TTG TAT GAC CGT CTG TTC AGC GTG CCT AAC
Lys Val Lys Gly Val Ile His Trp Val Ser Ala Ala His Ala Leu Pro Val Glu Ile Arg Leu Tyr Asp Arg Leu Phe Ser Val Pro Asn
                        490                       500                       510
CCA GGT GCT GCG GAT GAT TTC CTG TCG GTG ATT AAC CCG GAA TCG CTG GTG ATC AAA CAG GGC TTT GCT GAA CCG TCG CTG AAA GAT GCG
Pro Gly Ala Ala Asp Asp Phe Leu Ser Val Ile Asn Pro Glu Ser Leu Val Ile Lys Gln Gly Phe Ala Glu Pro Ser Leu Lys Asp Ala
                        520                       530                       540
GTT GCG GGT AAA GCA TTC CAG TTT GAG CGT GAA GGT TAC TTC TGC CTC GAT AGC CGC CAT TCT ACG GCG GAA AAA CCG GTA TTT AAC CGC
Val Ala Gly Lys Ala Phe Gln Phe Glu Arg Glu Gly Tyr Phe Cys Leu Asp Ser Arg His Ser Thr Ala Glu Lys Pro Val Phe Asn Arg
                        550
ACC GTT GGG CTG CGT GAT ACT GGG CGA AAG-3'
Thr Val Gly Leu Arg Asp Thr Gly Arg Lys-COOH
```

Fig. 2. Complete nucleotide sequence for the structural gene region of
glnS and amino acid sequence of GlnRS. The underlined peptides were determined by gas chromatography-mass spectroscopy analysis of peptide sequences. The NH₂-terminus was determined by manual Edman degradation.
The COOH-terminal dipeptide was obtained by carboxypeptidase B digestion. The numbers above the nucleotide sequence indicate the number of
amino acid from the NH₂-terminus.

developed especially for rapid partial amino acid sequencing to complement

DNA sequencing data. The complete amino acid sequence for GlnRS, obtained

by a combination of the above methods and by deduction from the glnS DNA

sequence is shown below the nucleotide sequence in Fig. 2. The amino acid

sequence of GlnRS does not reveal any internal repeats as had been proposed

for some intermediate to large aminoacyl-tRNA synthetases.[17] Homology com-

parison between GlnRS,[6] AlaRS,[7] TrpRS, and TyrRS (cited in reference 6) E.

coli revealed short, interrupted stretches of homologous regions. The sig-

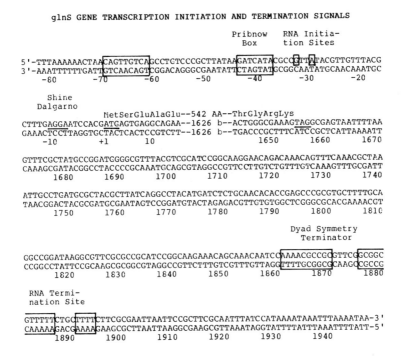

Fig. 3. Partial nucleotide sequence of the glnS gene showing the essential elements for its expression. The first and the last nucleotides for the structural region of glnS are designated as nucleotides +1 and +1653 respectively. Regions for glnS transcription: the Pribnow boxes, the initiation and the termination sites are boxed. Regions for glnS mRNA translation: the Shine-Dalgarno sequence, the initiation and the termination codons are underlined.

nificance of these homologies cannot be evaluated at this point. A more definite assignment of the functional regions in these synthetases by physical and biochemical methods as well as the knowledge of primary structures for more members of the synthetase family is needed for a clear evaluation.

EXPRESSION OF THE glnS GENE

The regulatory regions essential for the expression of glnS are shown in Fig. 3. The structural gene region of glnS is defined by a translational initiation codon ATG starting at nucleotide +1 and by a translational termination codon TAG beginning after nucleotide +1653. Upstream (8-11 basepairs) from the translation start is a Shine-Dalgarno sequence for ribosome binding. Around nucleotide -40 upstream from ATG is a Pribnow box structure serving as the promoter for glnS transcription. About 220 basepairs beyond the translation termination codon and centered at nucleotide +1873 is a dyad symmetry structure followed by a stretch of T residues, typical for transcription termination in procaryotes.

In vitro transcription by E. coli RNA polymerase of DNA fragments containing the glnS promoter gave defined RNA transcripts (Fig. 4A). Nucleotide analysis of the 5'-terminus and 5' end S1 nuclease mapping[18] of these in vitro transcripts showed that glnS transcription initiates predominantly at a G residue (nucleotide -33), 6 bases downstream from the constant T (nucleotide -39) in the Pribnow box. Fusing a DNA fragment containing the 3'-region of glnS with the dyad-symmetry transcription termination signal to the E. coli tryptophan operon promoter[14] provided an easy assay for glnS transcription termination. The length of this fused in vitro transcript and an oligonucleotide fingerprinting of this RNA indicated that glnS transcription indeed terminates at the stretch of T's following the dyad symmetry structure. 5'-end and 3'-end S1 nuclease mapping for glnS mRNA in a

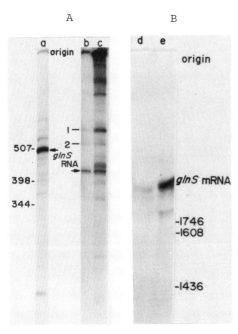

Fig. 4A. Autoradiogram of a 8M urea-polyacrylamide gel analysis of transcription products from DNA fragments containing the glnS promoter. Lane a, template DNA extends from nucleotide -244 to nucleotide +485; lane b, template DNA extends from nucleotide -255 to nucleotide +400 (see Fig. 3 for nucleotide numbering system).

Fig. 4B. Autoradiogram of a Northern blot of total E. coli RNA hybridized by ^{32}P-labeled glnS DNA. The numbers indicate positions of single-stranded DNA molecular weight markers. Lane d and e are 2 different amounts of E. coli RNA.

mixture of in vivo isolated E. coli RNAs yielded results that are in accord with the in vitro studies. These experiments predict that the glnS mRNA should be around 1920 bases long. Northern blot analysis[20] of in vivo RNA probed with glnS DNA suggests a length of 1910 bases for the glnS mRNA (Fig. 4B).

The in vitro transcription from the glnS promoter appears not to require factors other than E. coli RNA polymerase. Addition of GlnRS, glutamine and tRNAs does not affect the in vitro transcription from the glnS promoter. However, the possible presence of regulatory mechanisms for glnS expression involving these molecules in vivo cannot be precluded. Furthermore, data thus far, from both earlier physiological studies and the more recent in vitro biochemical studies (see articles by Schimmel, Blanquet and Springer in this book) lead one to believe that the regulatory mechanisms controlling aminoacyl-tRNA synthetase biosynthesis are very diverse.

IN VIVO REGULATION OF GLUTAMINYL-tRNA SYNTHETASE

In vivo studies on the expression of GlnRS have so far centered on the
metabolic regulation of this enzyme. It has long been known that changes in
the growth environment for E. coli, for example by using different compounds
as the sole carbon source in the growth medium, leads to changes in the
growth rates. Some E. coli aminoacyl-tRNA synthetases have been shown to
vary in their activity and/or amount in accordance to the changes in the
growth rates.[5] This phenomenon is termed metabolic regulation.[5] Metabolic
regulation of GlnRS has been studied in the wild type strain ($glnS^+$ $glnR^+$),
the temperature sensitive strain ($glnS^-$ $glnR^-$) and the temperature resistant
revertant strain ($glnS^-$ $glnR^-$) (Fig. 1). These cells were grown in minimal
salt media with either glucose, glycerol or acetate as the sole carbon
source. The growth rates, the level of GlnRS and the level of glnS mRNA are
measured for all three growth conditions. Data obtained for the wild-type
strain are illustrated in Fig. 5. Changing from acetate to glycerol and to
glucose as the carbon source caused a 3.7 fold increase in the growth rate
for this strain. The specific activity for GlnRS (which is known to reflect
the actual amount of this protein) increased by 2.1 fold. The level of
glnS mRNA was determined by hybridizing total cellular RNA from this strain
grown under the three conditions to an excess of a ^{32}P-labeled glnS DNA
fragment, digesting the hybrid by S1 nuclease, and analyzing the protected
DNA on polyacrylamide gel. The amount of DNA protected was a reflection of
the amount of specific glnS mRNA. Since the ratio of total RNA to total
protein also increased during the range of growth rate (from 0.29 to 0.5),
the amount of DNA protected was normalized to that protected by RNA from a
fixed number of cells (10^9). The result indicated a five-fold increase in
glnS mRNA level. Similar results were also obtained from the other two

66

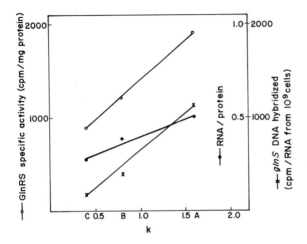

Fig. 5. Metabolic regulation of glnS expression of E. coli strain glnS+ glnR+. A,B,C are media conditions: minimal salt media with glucose (.4%), glycerol (.4%) and acetate (.4%) as carbon source. k is the number of doublings per hour.

strains studied (data not shown). However, it is not appropriate to conclude that the measured increase in the level of mRNA reflected the true increase, since the observed hybridization might be augmented by increase in hybridization efficiency by the presence of more hybridizable species. Nevertheless, the data allowed one to conclude that GlnRS is metabolically controlled. Further experiments are needed to show whether this regulation is solely transcriptional or both transcriptional and post-transcriptional.

FUTURE PERSPECTIVES

In order to gain a more detailed understanding of the substrate binding sites of GlnRS, it is necessary to have comparative data from many mutations. To this end, our laboratory is working on the isolation of more mutations in the structural gene of glnS by both in vivo and in vitro mutagenesis. As an additional advantage, GlnRS is involved in informational suppression through the su+2 amber suppressor tRNA^{Gln}.[21] This provides an easily scorable phenotype to select for mutants specific for tRNA recognition. If we find mutations affecting GlnRS function in different ways (e.g. amino acid specificity, tRNA specificity) it should be possible to delineate the various functional domains of this enzyme and to elucidate the modes of interaction between GlnRS and its substrates.

To facilitate our understanding on the regulation of GlnRS biosynthesis, an in phase fusion between the promoter containing region of glnS and the structural region of lacZ has been constructed. Fusion genes of this nature offer an easy screen for regulatory mutations affecting the activity at the particular promoter of interest[22] which would otherwise be more laborious.

We would like to close with the notion that although aaRS are a family of related important enzymes, study on each aaRS may unveil features unique to each enzyme. Thus, each aaRS merits its own attention to considerable extent as an individual system. Multiple efforts on the studies of aaRS will ultimately prove necessary and rewarding.

ACKNOWLEDGEMENTS

This work was supported by grants from the National Institutes of Health and the National Science Foundation.

REFERENCES

1. Schimmel, P. and Söll, D. (1979) Ann. Rev. Biochem. 48, 601-648.
2. Umbarger, H.E. (1980) in Transfer RNA: Biological Aspects, Söll, D., Abelson, J. and Schimmel, P., eds., Cold Spring Harbor Laboratory, New York, pp. 453-467.
3. Quay, S.C. and Oxender, D.K. (1980) in Transfer RNA: Biological Aspects, Söll, D., Abelson, J. and Schimmel, P., eds., Cold Spring Harbor Laboratory, New York, pp. 481-491.
4. Morgan, S. and Söll, D. (1978) Prog. Nucl. Acid Res. and Mol. Biol. 21, 181-207.
5. Neidhardt, F.C., Parker, J. and McKeever, W.G. (1975) Ann. Rev. Microbiol. 29, 215-250.
6. Putney, S.D., Royal, N.J., Neuman de Vegvar, H., Herlihy, W.C., Biemann, K. and Schimmel, P.R. (1981) Science 213, 1497-1501.
7. Hall, C.V., VanCleemput, M., Muench, K.H. and Yanofsky, C. (1982) J. Biol. Chem. 257, in press.

68

8. Yamao, F., Inokuchi, H., Cheung, A., Ozeki, H. and Söll, D. (1982) J. Biol. Chem. 257, in press.
9. Putney, S. and Schimmel, P. (1981) Nature 291, 632-635.
10. Hoben, P., Royal, N., Cheung, A., Yamao, F., Biemann, K. and Söll, D. (1982) J. Biol. Chem. 257, in press.
11. Körner, A., Magee, B., Liska, B., Low, K.B., Adelberg, E.A. and Söll, D. (1974) J. Bacteriol. 120, 154-158.
12. Cheung, A., Morgan, S., Low, K.B., and Söll, D. (1979) J. Bacteriol. 139, 176-184.
13. Morgan, S., Körner, A., Low, K.B., and Söll, D. (1977) J. Mol. Biol. 117, 1013-1031.
14. Ozeki, H., Inokuchi, H., Yamao, F., Kodaira, M., Sakano, H., Ikemura, T., and Shimura, Y. (1980) in Transfer RNA: Biological Aspects, Söll, D., Abelson, J. and Schimmel, P., eds., Cold Spring Harbor Laboratory, New York, pp. 341-362.
15. Maxam, A.M. and Gilbert, W. (1980) Methods Enzymol. 65, 499-560.
16. Nau, H. and Biemann, K. (1976) Anal. Biochem. 73, 139-153.
17. Kisselev, L.L. and Favorova, O.O. (1974) Adv. Enzymol. Relat. Areas Mol. Biol. 40, 141-238.
18. Berk, A. and Sharp, P. (1977) Cell 12, 721-732.
19. Wu, A.M., Christie, G.E. and Platt, T. (1981) Proc. Natl. Acad. Sci. USA 78, 2913-2917.
20. Alwine, J.C., Kemp, D.J. and Stark, G.R. (1977) Proc. Natl. Acad. Sci. USA 74, 5350-5355.
21. Steege, D. and Söll, D. (1979) in Biological Regulation and Control, Goldberger, R.F., ed., Plenum Publishing Co., New York, pp. 433-485.
22. Casadeban, M.J., Chou, J. and Cohen, S.N. (1980) J. Bacteriol. 143, 971-980.

Published 1982 by Elsevier Science Publishing Co., Inc.
Marianne Grunberg-Manago and Brian Safer, editors
INTERACTION OF TRANSLATIONAL AND TRANSCRIPTIONAL
CONTROLS IN THE REGULATION OF GENE EXPRESSION

69

CONSTRUCTION OF THE BACTERIOPHAGE T4 REPLICATION MACHINE: REGULATION OF SYNTHESIS OF COMPONENT PROTEINS

Kristine Campbell and Larry Gold
Department of Molecular, Cellular and Developmental Biology
University of Colorado
Boulder, Colorado 80309

INTRODUCTION

Bacteriophage T4 employs a complex, inter-related system of controls to ensure a proper sequence of gene expression throughout infection. The phage regA gene, which codes for a protein of approximately 10,000 daltons[1-3], is responsible for the post-transcriptional shut-off of a subset of early genes[1,4-7]. Lack of repression of these genes in a regA⁻ infection leads to overproduction of these gene products relative to a regA⁺ infection. RegA is autogenously controlled[1]; that is, regA itself is included in the list of genes whose products are overproduced in a regA⁻ infection (Table 1). The mechanism of repression of translation by the regA protein is, as of yet, not certain. One reasonable model is that the regA protein binds to the messenger RNAs of the controlled genes and prevents ribosomal access to the initiation region[1,8]. Alternatively, the regA protein could recognize a sequence or structure on messenger RNAs and act as a ribonuclease. Since regA is unique in its ability to repress the translation of many (but not all) phage mRNAs transcribed from different, unlinked genes, the study of regA-modulated repression should prove fascinating, regardless of the specific mechanism of action.

The T4 rIIB gene has been studied as a representative regA-controlled gene because of the extensive classical collection of rIIB mutants[9,10], as well as a more recently acquired collection of mutants in the rIIB translational initiation region[11]. The determination of the sequence of this region[12] and the characterization of the mutants mapping here[13] has contributed to the understanding of the translation of the rIIB message, and has provided an excellent starting point for the study of translational repression of rIIB mRNA by regA[8].

We report here data further defining the site of translational repression of rIIB expression by the regA protein. We then discuss the possible roles played by the regA protein during T4 development. We conclude that the regA protein participates in the economical and orderly construction of the T4 replication machine, and that translational repression of target T4 mRNAs is a secondary function of the protein.

TABLE 1

GENES CONTROLLED BY THE RegA PROTEIN

Gene	Function
1[a]	deoxyribonucleotide kinase
cd[a]	deoxycytidylate deaminase
rIIA[b]	in DNA replication complex?
rIIB[b]	in DNA replication complex?
56[a]	deoxycytidine/deoxyuridine di- and tri-phosphatase
42[a]	deoxycytidylate hydroxymethylase
44[b]	in DNA replication complex
62[b]	in DNA replication complex
45[b]	in DNA replication complex
regA[c]	translational regulator; in DNA replication complex?

[a] From references 4 and 6. Gene e was omitted from this list because the pattern of its apparent regulation by regA is abnormal; there is no enhancement of the regA effect under DNA⁻ conditions. Gene 46 was omitted because, in our hands, it is not regA-controlled.

[b] From references 5 and 7. Gene 63 is not included in this list because, in our hands, it is not regA-controlled.

[c] From reference 1.

Note: Because of technical problems with the identification of low molecular weight proteins on SDS gels, this list may not yet be complete. See low molecular weight region in Figure 1.

MATERIALS AND METHODS

Procedures for radioactive labeling of T4-infected E. coli, SDS polyacrylamide gel electrophoresis, and quantitation of autoradiographic films have been previously described[8]. The sequencing of rIIB mutations will be described elsewhere[14].

RESULTS

An example of the overproduction of early gene products in a regA⁻ infection is shown in Figure 1. Most of the regA-controlled genes have been discovered, since the number of overproduced bands in Figure 1 is close to the number of regA-controlled genes identified in Table 1.

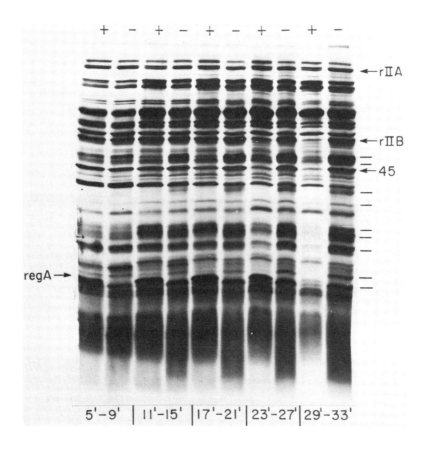

Figure 1. OVERPRODUCTION OF EARLY GENE PRODUCTS IN A RegA‾ INFECTION.
Radioactive amino acid pulse labelings of phage infected E. coli B^E were done
at 30° at the times indicated under DNA‾ conditions. The regA⁺ strain was 42am
N122/44am E4408(+); the regA‾ strain was 42am N122/44am E4408/regAR9(-).

The site on the rIIB messenger RNA recognized by regA has been defined by the screening of rIIB mutants for regA sensitivity[8]. A mutation is defined as being in the regA recognition site if it diminishes the sensitivity of the expression of rIIB to regA. In these constitutive mutants, there is slight or no repression of rIIB expression in a regA[+] infection, relative to a regA[−] infection. Determination of the nucleotide sequence of the regA site has come from sequencing known constitutive mutations. Figure 2 shows the sequences of mutations near the rIIB initiator AUG[14] and the effects of these mutations on the expression of rIIB. All the known constitutive mutations are located in the region defined by deletion 326, in a stretch of nine nucleotides approximately starting with the AUG and extending 3' to it. This, in conjunction with sequence homology with the initiation region of another regA-controlled gene, gene 45[15], leads us to suggest that rIIB's regA recognition site is AUGUACAAU. This sequence does not seem to be involved in any stable structures within the rIIB message[16], making it unlikely that regA recognizes messenger RNA structural determinants. Direct binding experiments between the regA protein and T4 mRNAs are in progress in our lab.

DISCUSSION

Why would an organism control the expression of some genes at the level of translation? There is a rationale for translational control in eukaryotes, where, due to messenger RNA stability and the temporal "distance" between nucleus and cytoplasm, translational regulation could allow rapid adaptation to changing conditions. Prokaryotic messages are generally short-lived, and a quick response to environmental changes can be effected by stopping transcription and allowing messenger decay. Indeed, this is the most common means of negatively modulating gene expression in prokaryotic organisms. Control at the level of translation presents the disadvantage of allowing the unnecessary expenditure of energy during the transcription of mRNAs whose fate is to remain untranslated. However, we will develop a rationale for the utility of prokaryotic translational regulation. We will discuss the ribosomal protein genes of Escherichia coli and gene 32 of phage T4, and extrapolate from these systems to a notion about the T4 regA protein.

In several ribosomal protein operons of E. coli, translation of the polycistronic message is inhibited by one of the proteins of the operon[17-31]. By hypothesis, structural similarities exist between the initiation region of the target mRNAs and the proteins' binding domain on the ribosomal RNA. If the

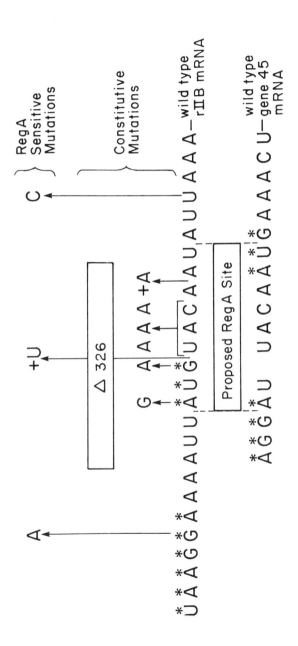

Figure 2. THE PROPOSED RegA RECOGNITION SITE. All but one of the assignments of mutants to the regA sensitive or constitutive class were previously made[8]. We have recently found that the A to G change in the AUG (zAP10) is constitutive[14]. Two of the previously deduced positions of mutations were slightly incorrect and are corrected here[4]; these are the substitution of G for A in the AUG (zAP10) and the insertion of the U after the AUG (FC302). The inclusion of the last U in the proposed site is for two reasons: (1) homology with the gene 45 sequence; (2) the constitutive nature of the A insertion immediately preceeding the U (the only effect of this mutation on the sequence is to move the U one base rightward, which destroys the site). The fact that the U insertion in the middle of the regA site leaves rIIB expression regA sensitive suggests flexibility in what the regA protein may inspect. We imagine that, when all regA recognition sequences are known, they will cluster around a consensus sequence, exhibiting variations from gene to gene. The present catalogue of regA sites includes AUGUACAAU, AUGUUACAAU, and AUUACAAU.

affinity of the protein were greater for the rRNA than mRNA, repression would not occur until the primary ligand, rRNA, was saturated and free repressor protein became available to bind to the secondary ligand, mRNA. (It has been suggested[32] that the ribosomal proteins bind preferentially to rRNA over mRNA partly because of cooperative interactions with other ribosomal proteins in the assembly complex.) This feedback repression is a way of maintaining appropriate free ribosomal protein concentration, thus ensuring control of the synthesis of ribosomes. The inclusion of these proteins in the ribosome is required for proper ribosome function; the repressor proteins are not merely used to "count" particles under construction.

Gene 32 protein is a single-stranded DNA binding protein involved in DNA replication, recombination, and repair during phage T4 infection of E. coli[33].Gene 32 protein binds double-stranded nucleic acids very weakly and will not denature duplex DNA at physiological concentrations of protein[32]. This protein also represses its own translation[34,35]. Gene 32 protein binds first to single-stranded DNA (the primary ligand) and, upon saturation of this lattice, binds to its own messenger RNA (the secondary ligand), thus preventing translation[34,35]. A quantitative analysis of the relative binding affinities of 32 protein for single-stranded DNA and unstructured RNA (the presumptive translational "operator") supports the model[32]. The specificity in this system comes not from the recognition of similar structures but from the recognition of lack of structure. The sequence[36] of gene 32 reveals a long stretch of nucleotides, surrounding the initiation codon, that are unlikely to be involved in intramolecular base-pairing[32]. The model predicts that other mRNAs that are not repressed by 32 protein have some kind of structure near the ribosome binding site that shortens single-stranded domains and prevents recognition and binding of gene 32 protein.

The major function of these autogenous translational repressor proteins is to participate in macromolecular complexes; the ribosome, in the case of the ribosomal proteins, and the T4 "replisome" in the case of the gene 32 protein. Translational repression is a secondary function based on resemblances between the repressed messages and the primary ligands. Expression is turned off only after levels of free protein adequate for complex formation and function have been established. In an evolutionary sense, the ability to bind primary ligands probably preceeded the fine-tuning of translational feedback repression. Message structure could have evolved later to match the structure of the primary ligands. Feedback via transcriptional regulation would have necessitated the evolution of additional binding capacity for duplex DNA, while requiring the retention of binding capacity for the primary ligand. The autogenous aspects of these two control schemes also offer advantages. Autogeny saves the organism the trouble of evolving and maintaining a new gene to function solely as a repressor. In situations where the proteins of interest are components of macromolecular complexes, autogeny allows direct measurement of the number of complexes being assembled and therefore the amount of other protein components needed (see reference 37 for a similar discussion of the advantages of transcriptional autogenous control).

The regA gene of bacteriophage T4 encodes a protein whose only known function is the translational repression of a subset of early T4 genes, including itself. We have been trying to understand the true role of regA during infection from the point of view outlined above; that is, we assume that translational repression is a secondary function of the regA protein. If we are correct, both the non-autogenous target mRNAs and regA mRNA are secondary targets. By analogy to the ribosomal proteins and gene 32 protein, we suspect that the regA protein is a component of a multi-molecular aggregate in which the regA protein binds to something resembling the several messages that it represses.

The first issue, then, is what phage-encoded multi-molecular aggregate might include the regA protein and the proteins encoded by the regA-sensitive genes? An examination of the list of regA-controlled genes (Table 1) reveals that most are concerned with deoxyribonucleotide metabolism and DNA replication. In particular, gene products 45, 44, and 62 are members of the DNA replication complex[38] and gene products 1, cd, 42, and 56 are components of the nucleotide synthesizing machinery[39-52]. These two complexes are physically coupled in the T4 "replisome"[39-52]; dNTPs are directly shunted, after biosynthesis, into DNA. There is evidence that the rIIA and rIIB proteins are also associated with these enzyme complexes[53-55]. Thus genes regulated by the regA protein probably act in concert during T4 DNA synthesis. Since the regA gene is autogenously regulated, we propose that this gene product is also part of the T4 replisome. Co-regulation has been observed previously for enzymes in other bacterial pathways[56-58]. Further evidence for an involvement of regA in T4 DNA replication is the proximity of the regA gene to gene 43 (DNA polymerase) and genes 45, 44, and 62 (DNA replication complex components) on the T4 chromosome[59]. Gene 43 is autogenously controlled[60] and regA controls the other four genes in this cluster.

In Figure 3 we tentatively depict the core components of the T4 replisome, and show as well an interface between substrate biosynthesis and replication. We have noted that HMdCTP biosynthesis is regulated most completely by the regA protein. Although proteins encoded by genes 56 and 1 are regA-controlled and participate in dTTP biosynthesis, td and thymidine kinase (which is not shown) are regA-insensitive. Bacteriophage T4 coordinates HMdCTP biosynthesis with DNA replication for two important reasons: (1) if cytosine replaces the modified base in T4 DNA, T4-encoded cytosine-specific endonucleases will destroy the genome, and (2) excessive biosynthesis of HMdCTP (and dTTP) is energy intensive, and might compete for the capacity to do other energy intensive reactions, such as the protein biosynthetic reactions leading to the phage structural components. The first problem could be solved by synthesis of <u>excess</u> substrate biosynthetic capacity, while the second problem could be solved by <u>no excess</u> substrate biosynthetic capacity. Underexpression of these enzymes would be lethal, since it could lead to incorporation of dCTP rather than the modified substrate. Precise <u>feedback regulation</u> is a solution to such a predicament. If regA is placed, by hypothesis, into the replisome, feedback regulation could be achieved.

What kind of role could the regA protein actually play in DNA replication? If its primary ligand structurally resembles the mRNAs that are repressed, the target for the regA protein must be either single-stranded DNA or RNA. Since most single-stranded DNA at a replication fork is coated by gene 32 protein, we find the possible interactions with RNA more intriguing (although there is no reason that "special" sequences or structures in single-stranded DNA couldn't be recognized by the regA protein). If we assume that the regA protein binds to RNA[1,6,8], the most likely primary ligand for the protein is an RNA primer used to initiate DNA replication. RNA primers may be necessary for the initiation of DNA replication[61]. Strong evidence exists for participation by E. coli RNA polymerase in T4 DNA synthesis[62,63]. Imagine that the T4 "replisome" is constructed at the 3' end of an RNA primer <u>if and only if</u> the phage DNA polymerase binds appropriately to the DNA and the primer RNA prior to the loss of the primer from phage DNA. A cooperative interaction between the proteins encoded by genes 45, 44, 62, and regA <u>at a regA site on the primer</u> RNA could facilitate appropriate binding of DNA polymerase prior to primer loss. Such a simple scenario is depicted in Figure 4. The model has two features which please us: a cluster of T4 genes (45-44-62-regA-43) encode proteins that interact, and "replisome" construction could easily continue upon this

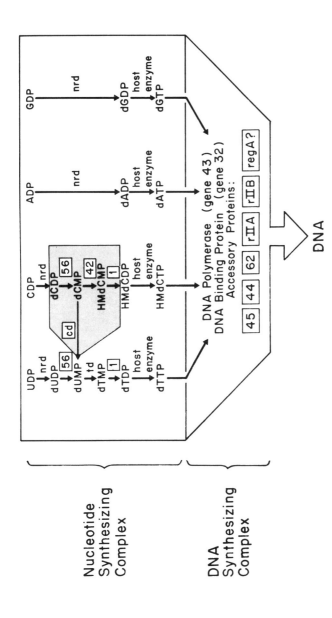

Figure 3. THE INTERFACE OF NUCLEOTIDE SYNTHESIS AND DNA REPLICATION. Boxes around the gene names indicate that these gene products are regA-controlled. Both gene 43 and gene 32 are autogenously controlled[34,35,66]. Note that nrd, td, and the host kinase (at the interface) are not regA-controlled. HMdCTP is the modified nucleotide hydroxymethyl dCTP. The mottled portion of the biosynthetic pathway represents our sense of the most important regA-controlled reactions.

five-component core. Most importantly, inclusion of the regA protein in the replisome assures feedback on the expression of the nucleotide synthesizing enzymes. The initial stages of replisome construction would be accompanied by derepression of genes 56, 1, 42 and cd.

Some other possibilities for the actual nature of the interaction of regA protein with primer RNA are as follows: (1) the regA protein could protect the RNA primer from premature or spurious degradation; (2) the regA protein could be directly involved in primer processing. This possibility would be especially attractive if the action of the regA protein on the sensitive mRNAs turns out to be nucleolytic; (3) an appropriate and necessary primer RNA conformation could be stabilized by the regA protein; (4) the primer, as an RNA transcript, may be available for ribosome binding, and the function of regA is to prevent this. This would be particularly elegant if the origins of DNA replication were situated such that transcripts for regA-sensitive genes were used for priming.

The non-essential nature of the regA gene (in contrast to the essential nature of genes 45, 44, 62 and 43) could be rationalized by the proposed kinetic role for the regA protein. For example, the overproduction of the 45, 44 and 62 proteins in a regA$^-$ infection could allow the DNA polymerase to win the race for the primer, even in the absence of the nucleating protein, regA. Alternatively, the regA protein may only participate in replication that originates from specific primers. Since T4 can use other forms of initiation[62], the regA gene might be non-essential. The regA gene could also be non-essential because it only binds primer RNA in order to "count" replisomes and not to function in the replication process. By analogy to both gene 32 and the ribosomal proteins, which perform actual functions in their respective complexes, this possibility is unlikely. We note that some growth conditions do exist that make the regA protein essential[4,5].

We have made a precise suggestion for the primary ligand inspected by the T4 regA protein. That ligand is a primer RNA containing a regA site. Recognition of that site by the regA protein, coupled with an appropriately placed 3' terminus on the primer RNA, can lead to initiation of DNA synthesis by a T4 replisome, containing stoichiometric quantities of the enzymes needed for HMdCTP biosynthesis. We have been led by our thinking about primary and secondary ligands in autogenous translational repression, and by the phenomenology surrounding the regA gene of bacteriophage T4. In addition to performing its well-characterized role of coordinately controlling elements of the T4 replication complex, the regA protein itself may participate directly in T4 DNA

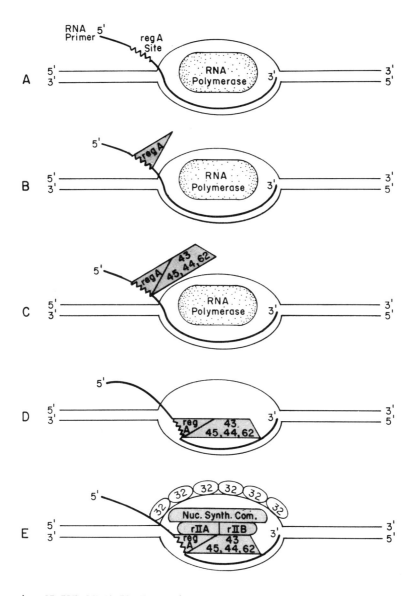

Figure 4. SIMPLE SCENARIO FOR RegA INVOLVEMENT IN REPLISOME CONSTRUCTION. A.
RNA polymerase synthesizes an RNA primer containing a regA recognitionsite. B.
RegA binds to that site. C. RegA protein bound to primer provides a nucleation
site for the assembly of the core replisome. D. The regA-core protein complex
replaces RNA polymerase so that the DNA polymerase (encoded by gene 43) now has
access to the 3' end of the primer. E. The other replisome components join the
complex: rIIA, rIIB, 32 and the nucleotide synthesizing complex. DNA can now be
synthesized from the 3' end of the primer.

replication. We have developed some specific models for this involvement and hope to test them soon.

ACKNOWLEDGEMENTS

Many thanks to Peter Gauss and Nancy Guild for numerous enlightening conversations. David Pribnow graciously allowed us the use of unpublished data, and for this we are grateful. We thank Carol Avery for expert technical assistance. K.C. was a NIH post-doctoral fellow. This work was supported by a grant from the NIH (GM 28685).

REFERENCES

1. Cardillo, T., Landry, E. and Wiberg, J. (1979) J. Virol., 32, 905-916.
2. Karam, J. Personal communication.
3. Avery, C. and Campbell, K. Unpublished data.
4. Wiberg, J., Mendelsohn, S., Warner, V., Hercules, K., Aldrich, C., and Munro, J. (1973) J. Virol., 12, 775-792.
5. Karam, J., and Bowles, M. (1974) J. Virol., 13, 428-438.
6. Trimble, R. and Maley, F. (1976) J. Virol., 17, 538-549.
7. Karam, J., McCulley, C., and Leach, M. (1977) Virology, 76, 685-700.
8. Karam, J., Gold, L., Singer, B. and Dawson, M. (1981) Proc. Nat. Acad. Sci. USA, 78, 4669-4673.
9. Benzer, S. (1961) Proc. Nat. Acad. Sci. USA, 47, 403-415.
10. Barnett, L., Brenner, S., Crick, F., Shulman, R. and Watts-Tobin, R. (1967) Phil. Trans. Royal Soc. London, Series B, 252, 487-560.
11. Nelson, M., Singer, B., Gold, L. and Pribnow, D. (1981) J. Mol. Biol., 149, 377-403.
12. Pribnow, D., Sigurdson, D., Gold, L., Singer, B., Brosius, J., Dull, T. and Noller, H. (1981) J. Mol. Biol., 149, 337-376.
13. Singer, B., Gold, L., Shinedling, S., Hunter, L., Pribnow, D. and Nelson, M. (1981) J. Mol. Biol., 149, 405-432.
14. Pribnow, D. et al., Manuscripts in preparation.
15. Spicer, E. and Konigsberg, W. Personal communication.
16. Gold, L., Pribnow, D., Schneider, T., Shinedling, S., Singer, B. and Stormo, G. (1981) Ann. Rev. Microbiol., 35, 365-403.
17. Dennis, P. and Fiil, N. (1979) J. Biol. Chem., 254, 7540-7547.
18. Goldberg, G., Zarucki-Schultz, T., Caldwell, P., Weissbach, H. and Brot, N. (1979) BBRC, 91, 1453-1461.
19. Brot, N., Caldwell, P., and Weissbach, H. (1980) Proc. Nat. Acad. Sci. USA, 77, 2592-2595.
20. Dean, D. and Nomura, M. (1980) Proc. Nat. Acad. Sci. USA, 77, 3590-3594.
21. Fukada, R. (1980) Molec. Gen. Genet., 178, 483-486.
22. Nomura, M., Yates, J., Dean, D., and Post, L. (1980) Proc. Nat. Acad. Sci. USA, 77, 7084-7088.
23. Wirth, R. and Bock, A. (1980) Molec. Gen. Genet., 178, 479-481.
24. Yates, J., Arfsten, A. and Nomura, M. (1980) Proc. Nat. Acad. Sci. USA, 77, 1837-1841.
25. Zengel, J., Mueckl, D. and Lindahl, L. (1980) Cell, 21, 523-536.
26. Dean, D., Yates, J. and Nomura, M. (1981) Cell, 24, 413-419.
27. Dean, D., Yates, J. and Nomura, M. (1981) Nature, 289, 89-91.
28. Olins, P. and Nomura, M. (1981) Cell, 26, 205-211.
29. Olins, P. and Nomura, M. (1981) Nucleic Acids Research, 9, 1757-1764.

30. Yates, J. and Nomura, M. (1981) Cell, 24, 243-249.31. Yates, J., Dean, D. Strychartz, W. and Nomura, M. (1981) Nature, 294, 190-192.
32. Von Hippel, P., Kowalczykowski, S., Lonberg, N., Newport, J., Paul, L. Stormo, G. and Gold, L. Manuscript submitted.
33. Doherty, D., Gauss, P. and Gold, L. (1982) Multifunctional Proteins: Regulatory and Catalytic/Structural, Kane, J., ed., CRC Press, Cleveland. In press.
34. Russel, M., Gold, L., Morrissett, H. and O'Farrell, P. (1976) J. Biol. Chem., 251, 7263-7270.
35. Lemaire, G., Gold, L., and Yarus, M. (1978) J. Mol Biol., 126, 73-90.
36. Krisch, H., Duvoisin, R., Allet, B. and Epstein, R. (1980) Mechanistic Studies of DNA Replication and Recombination, Alberts, B., ed., Academic Press, N.Y., pp.517-526.
37. Menzel, R. and Roth, J. (1981) J. Mol. Biol., 148, 21-44.
38. Liu, C., Burke, R., Hibner, V., Barry, J. and Alberts, B. (1978) Cold Spring Harbor Symp. Quant. Biol., 43, 469-487.
39. Wovcha, M., Tomich, P., Chiu, C-S., and Greenberg, G.R. (1973) Proc. Nat. Acad. Sci. USA, 70, 2196-2200.
40. Collingsworth, W. and Matthews, C. (1974) J. Virol., 13, 908-915.
41. Tomich, P., Chiu, C-S., Wovcha, M. and Greenberg, G.R. (1974) J. Biol. Chem., 249, 7613-7622.
42. Chiu, C-S., Tomich, P. and Greenberg, G.R. (1976) Proc. Nat. Acad. Sci. USA, 73, 757-761.
43. Wovcha, M., Chiu, C-S., Tomich, P. and Greenberg, G.R. (1976) J. Virol., 20, 142-156.
44. Chiu, C-S., Ruettinger, T., Flanegan, J. and Greenberg, G.R., (1977) J. Biol. Chem., 252, 8603-8608.
45. Flanegan, J. and Greenberg, G.R. (1977) J. Biol. Chem., 252, 3019-3027.
46. Reddy, G.P.V., Singh, A., Stafford, M. and Matthews, C. (1977) Proc. Nat. Acad. Sci. USA, 74, 3152-3156.
47. Reddy, G.P.V., and Matthews, C. (1978) J. Biol. Chem., 253, 3461-3467.
48. Matthews. C., North, T., and Reddy, G.P.V. (1979) Adv. Enz. Regulation, 17, 133-156.
49. Allen, J., Reddy, G.P.V., Lasser, G. and Matthews, C. (1980) J. Biol. Chem., 255, 7583-7588.
50. Chiu, C-S., Cox, S., and Greenberg, G.R. (1980) J. Biol. Chem., 255, 2747-2751.
51. Wirak, D. and Greenberg, G.R. (1980) J. Biol. Chem., 255, 1896-1904.
52. Matthews, C. and Sinha, N. (1982) Biochem., 79, 302-306.
53. Greenberg, G.R., Personal communication.
54. Huang, W., and Buchanan, J. (1974) Proc. Nat. Acad. Sci. USA, 71, 2226-2230.
55. Manoil, C., Sinha, N., and Alberts, B. (1977) J. Biol. Chem., 252, 2734-2741.
56. Dixon, R., Eady, R.R. Espin, G., Hill, S., Iaccarino, M., Kahn, D. and Merrick, M. (1980) Nature, 286, 128-132.
57. Kenyon, C. and Walker, G. (1980) Proc. Nat. Acad. Sci. USA, 77, 2819-2823.
58. Komeda, Y. and Iino, T. (1979) J. Bacteriol., 139, 721-729.
59. Wiberg, J., Mendelsohn, S., Warner, V., Aldrich, C. and Cardillo, T. (1977) J. Virol., 22, 742-749.
60. Krisch, H., VanHowe, G., Belin, D., Gibbs, W. and Epstein, R. (1977) Virology, 78, 87-98.
61. Tomizawa, J. and Selzer, G. (1979) Ann. Rev. Biochem., 48, 999-1034.
62. Luder, A. and Mosig, G. (1982) Proc. Nat. Acad. Sci. USA, 79, 1101-1105.
63. Snyder, L. (1974) Virology, 62, 184-196.

Published 1982 by Elsevier Science Publishing Co., Inc.
Marianne Grunberg-Manago and Brian Safer, editors
INTERACTION OF TRANSLATIONAL AND TRANSCRIPTIONAL
CONTROLS IN THE REGULATION OF GENE EXPRESSION

CONTROL OF TRANSLATION BY THE REG A GENE OF T4 BACTERIOPHAGE

JIM KARAM, MYRA DAWSON, WILLIAM GERALD, MARIA TROJANOWSKA, and
CHARLENE ALFORD
Medical University of South Carolina, Department of Biochemistry, 171 Ashley Avenue,
Charleston, South Carolina 29425, USA

INTRODUCTION

In normal infections of Escherichia coli with bacteriophage T4, transcription of the T4 genome begins immediately after its entry into the host cell. Initially, the host RNA polymerase recognizes a number of promoters on the T4 DNA and transcribes sets of genes which encode a variety of enzymes and other proteins that support viral DNA replication and that activate the transcription of other sets of phage genes (see ref.1 for a recent review). Some of the proteins that are made during the "early" stages of phage growth become associated with the host RNA polymerase[2,3] and endow it with the specificity to initiate transcription of those phage genes that control maturation functions, i.e. "late" gene expression. Expression of the "late" functions also requires DNA replication and the direct participation of at least some of the phage-induced DNA replication enzymes[4,5].

The protein product of the regA gene is an "early" phage function that plays a role in the control of utilization of a subpopulation of T4 "early" transcripts, including the mRNA that encodes the regA protein itself[6-9]. Loss of this phage function results in functional stabilization of these "early" mRNAs and in hyperproduction of their translational products. We have been investigating two possible models for regA protein activity in this type of control of specific translation. Since many regA-controlled transcripts exhibit high functional stability in T4 regA⁻ infections, it is possible that the regA protein functions as a "processing" enzyme that initiates breakdown of the regulated mRNAs. Alternatively, this protein may interact with certain mRNAs and prevent their translation with message decay ensuing as a consequence of translational inhibition. Both models propose the participation of a "target" on mRNA in regA-related translational repression. Such a "target" was identified recently in genetic studies with the T4 rIIB cistron, a phage gene function known to be under regA-mediated translational control[10]. The "target" mapped within the ribosome-binding domain for initiation of rIIB mRNA translation and overlapped the first 3 codons of the mRNA. We report here on our attempts to identify and characterize regA translational control "targets" on other T4 mRNAs. We will also describe results on intracellular localization of the T4 regA gene product and on its possible association with RNA.

84

METHODS AND RESULTS

Methods. Most of the methods used in the studies to be described have been detailed elsewhere[7,8,10]. Additional details are included in the sections to follow.

T4 regA genetics. A large number of T4 regA mutants have been isolated as suppressors of leaky amber and temperature-sensitive mutants of T4 genes 62 (ref. 7 and 11) and 44 (ref. 8). The products of both of these phage genes are subject to regA-mediated control[12]. Most T4 regA mutants are abnormally sensitive to hydroxyurea (HU[S]) because they are defective in T4-induced host DNA breakdown[6]. The relationship between the HU-sensitivity and posttranscriptional effects seen in T4 regA mutant infections is unclear, although the HU[S] phenotype has provided a convenient marker for mapping regA mutations by simple plating methods[7,11]. As depicted in Figure 1, T4 gene regA maps within a cluster of essential phage DNA replication genes. T4 regA mutants, however, appear to replicate normally despite exhibiting alterations in the molar ratios of the enzymes that service replication. The regA gene function is also involved in regulation of other T4-induced enzymes of nucleic acid metabolism, e.g. the gene 63 protein (RNA ligase)[13], the rIIA and rIIB proteins[7,8], and some of the enzymes involved in

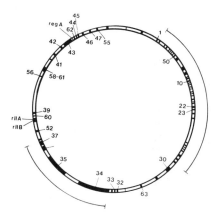

Fig. 1. A schematic diagram of the circular T4 genetic map (after Wood and Revel[22]). The positions of genes known to respond to regA-mediated translational control are indicated on the outer side of the circular map. Those known to be indifferent to the T4 regA function are indicated on the inner side. The bracketed arcs designate clusters of "late" phage gene functions. Genes 30,32,58-61,41,43,62,44, and 45 encode DNA replication enzymes.

the biosynthesis of DNA precursors[6]. No T4 "late" functions are known to be under regA gene control[8]. In summary, translational control by the T4 regA gene is specific and affects the expression of several underlined "early" T4 genes.

The "target" for T4 regA-mediated translational control. In collaboration with L. Gold's laboratory (University of Colorado), we recently identified the sequence 5'(AUG)UACAAU3' in the initiation region of T4 rIIB mRNA as being part of the "target" for regA-mediated translational control[10]. Lesions within this sequence rendered rIIB protein synthesis constitutive and insensitive to a regA[+] genetic background. A similar sequence, 5'AUUACA(AUG)3', was also found in the initiation region of the regA-regulated T4 gene 45, but no genetic evidence is available in this case to implicate the sequence in translational control[14]. It is reasonable to expect that regA-sensitive targets will have similar sequences, but it is almost certain that these targets will exhibit differences as well. An examination of a variety of regA mutants revealed that some differ in the degree to which they affect the synthesis of individual T4-induced proteins. Figure 2 shows some examples. In this experiment (Figure 2) several T4-induced proteins were affected differently by the regA lesions md67 and md73. Such differences probably reflect differential responses of the mRNA sites for regulation by the regA gene function.

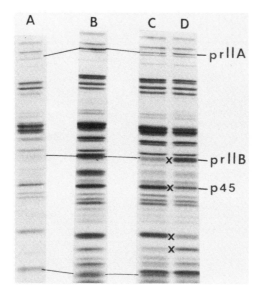

Fig. 2. Patterns of T4-induced protein synthesis in infections of E. coli B[E]str[r] with T4 regA[+] phage (A) and with the regA mutants R9 (B), md67 (C), and md73 (D). Cultures were labeled with ^{14}C-amino acids at a late time (30-35 min) postinfection at 30°C and extracts were analyzed by SDS-gel electrophoresis and autoradiography (see ref. 8). The "X" marks indicate differences between the md67 and md73 infections.

The isolation of "regA target" mutations. Because the T4 rIIB gene is nonessential for phage growth in most E. coli hosts, it was possible to utilize major alterations (additions and deletions) within its "regA target" region[10]. Other target regions, particularly those in essential T4-induced mRNAs, are more difficult to characterize because mutations in sites located 3' to the initiation codon can alter amino acid sequence and biological activity, and those located 5' to the initiator are likely to depress the level of total translation for an essential protein and lead to lethality[15]. Despite these limitations, we may have succeeded in obtaining "regA target" mutations for the T4 gene 44. We isolated a class of cis-acting mutations that mapped near the beginning of gene 44 and that overproduced the gene 44 protein without affecting gene 45 expression[16]. Preliminary studies with one of these cis-dominant mutations (named hp6 in ref. 16) suggested that it affected transcription rather than translation of the gene 44 transcript. Unlike typical regA-lesions, hp6 had little or no effect on the functional stability of gene 44 mRNA; it did, however, cause gene 44 protein synthesis to continue long past the time of its normal shut off in phage-infected cells[16]. More recently, we observed that the effects of hp6 and regA⁻ lesion on synthesis of the gene 44 protein are not additive. Figure 3 shows some results. Infections with phage mutants carrying both hp6 and a regA⁻ lesion yielded approximately the same level of gene 44 protein as did infections with phage mutants carrying only the regA-lesion. These levels were about 15-fold higher than those seen in

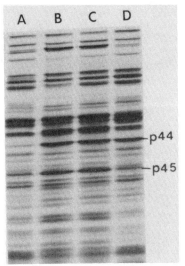

Fig. 3. T4 gene 44 protein synthesis at 42°C in infections of E. coli BᴱstrʳÂ with phage mutants carrying lesions in either or both of the regulatory sites, regA and hp6. Phage-induced proteins were labeled with ¹⁴C-amino acids during the 5-20 min interval after infection. A = control infection (regA⁺ and hp6⁺), B = T4 regAR9 (hp6⁺) infection, C = T4 regAR9-hp6 infection, and D = T4 hp6 (regA⁺) infection. Analyses of protein synthetic rates were carried out as described previously[12].

regA$^+$-hp6$^+$ infections. The hp6 lesions alone enhanced gene 44 protein synthesis by 10-15 fold.

It is possible that control of mRNA stability and translational repression are separable phenotypes of the regA gene function, and that the hp6 lesion alters the level of "repression" at the gene 44 "regA target" without influencing mRNA stability. Message destruction may be signaled by other elements within the "regA target". In such a model, the effects of "derepression" (hp6) would not be expected to add to the effects of total removal of the repressor (regA$^-$).

The T4 regA protein. The T4 regA gene encodes a polypeptide of about 12,000 daltons as determined by gel electrophoretic assays[9]. The "regA target" also seems small[10,14] and we suspect that regA-mediated control of translation may involve a direct interaction between the small protein and its target. The results shown in Figure 4 suggest that regA protein is an RNA-binding protein. In extractions of T4-induced proteins from regA$^+$-infected cultures the regA gene product coisolated with membrane fractions and was easily released from the membrane by pancreatic RNase, but not by DNase treatment (Figure 4). We have also observed that release of the regA protein from its membrane association can be brought about by treatment with salt (results not shown).

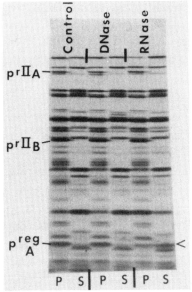

Fig. 4. Extraction of the regA protein from T4 regA$^+$-infected B^E cells. Aliquots of a lysozyme/EDTA-treated[8] suspension of $^{35}SO_4$-labeled infected cells (Control) were further treated for 5 min at 20°C with 2 μg/ml of either pancreatic RNase or DNase. The rechilled "Control" and nuclease-treated suspensions were then fractionated into pellets (P) and supernatants (S) by centrifugation at 12,000 x g for 20 min and the P and S fractions were subsequently analyzed by SDS-gel electrophoresis and autoradiography.

DISCUSSION

It is difficult to know at this stage whether or not the T4 regA protein is unique among the known translational regulatory peptides. In each of several well studied cases involving RNA recognition by proteins, i.e. the RNA phages, the T4 gene 32 protein, and others (reviewed in ref. 17), recognition of the RNA sequence has proven to include recognition of RNA conformation as well. In contrast, genetic evidence seems to imply that recognition of the "regA target" in regA-regulated mRNAs involves only sequence recognition[10]. It should be noted, however, that a sequence as short as the one identified for T4 rIIB mRNA (6-9 nucleotides long) is likely to occur frequently in mRNA molecules, including those that do not respond to regA-mediated control. Possibly, unwanted target-like sequences are normally sequestered by secondary and tertiary structures within the mRNA molecule. It is also possible that other elements, e.g. effector molecules, are involved in directing the regA protein to the translation initiation regions of certain mRNAs. Such factors may also be responsible for differential effects by the regA protein on different targets (e.g. Figure 2).

Another intriguing question relates to whether or not the T4 regA protein inactivates translation of certain mRNAs by cleaving or irreversibly modifying these transcripts at their target sequences. In this regard it should be pointed out that Trimble and Maley[18] observed that one T4-induced mRNA (the gene 1 HMdCMP kinase transcript) was intrinsically stable and yet sensitive to inhibition by regA[+] function. So, for some mRNAs degradation may not be a necessary consequence of inhibition of translation. A detailed nucleotide sequence analysis of the cis-acting T4 hp6 lesion, which may reside in a "regA target" and which may affect translational repression without affecting mRNA degradation (Figure 3), as well as the analysis of several additional "regA target" sequences would be desirable and should yield useful insights about the relationship between mRNA decay and regulation of translation.

What physiological role does the T4 regA protein play in the phage growth cycle? Rapid transcription of the T4 genome may necessitate the use of translational fine-tuning mechanisms for optimal utilization of genetic information. Much of the "early" phase of T4 development in infected E. coli is related to DNA replicative events. The major role of the regA protein may be in regulating the relative concentrations of the various proteins that form the multienzyme complexes needed for DNA replication[19,20] and the biosynthesis and utilization of DNA precursors[20,21]. Some of the events involved are membrane-associated and it may not be surprising that certain T4-induced membrane proteins (e.g. the rII proteins) are regulated by the regA function as well.

ACKNOWLEDGMENTS

This work was supported by Grant No. 5RO1 GM18842 from the National Institute of General Medical Sciences, National Institutes of Health, USA. M.T. was supported in part by a postdoctoral fellowship from the College of Graduate Studies, Medical University of South Carolina. We thank L. Gold, E. Spicer, and W. Konigsberg for discussions and for communicating unpublished results.

REFERENCES

1. Rabussay, D. and Geiduschek, E.P. (1977) in Comprehensive Virology, Fraenkel-Conrat, H. and Wagner, R.R. ed., Plenum, New York, Vol. 8, pp. 1-196.
2. Horvitz, R. (1973) Nature New Biol. 244, 137-140.
3. Stevens, A. (1972) Proc. Nat. Acad. Sci. 69, 603-607.
4. Wu, R. and Geiduschek, E.P. (1975) J. Mol. Biol. 96, 513-538.
5. Coppo, A., Manzi, A., Pulitzer, J.F., and Takahashi, H. (1975) J. Mol. Biol. 96, 579-600.
6. Wiberg, J.S., Mendelsohn, S., Warner, V., Hercules, K., Aldrich, C., and Munro, J.L. (1973) J. Virol. 12, 775-792.
7. Karam, J.D. and Bowles, M.G. (1974) J. Viol. 13, 428-438.
8. Karam, J., McCulley, C., and Leach, M. (1977) Virology 76, 685-700.
9. Cardillo, T.S., Landry, E.F., and Wiberg, J.S. (1979) J. Virol. 32, 905-916.
10. Karam, J., Gold, L., Singer, B.S., and Dawson, M. (1981) Proc. Nat. Acad. Sci. 78, 4669-4673.
11. Wiberg, J.S., Mendelsohn, S.L., Warner, V., Aldrich, C., and Cardillo, T.S. (1977) J. Virol. 22, 742-749.
12. Karam, J., Bowles, M., and Leach, M. (1979) Virology 94, 192-203.
13. Higgins, N.P., Geballe, A.P., Snopek, T.J., Sugino, A., and Cozzarelli, N.R. (1977) Nucleic Acid Res. 9, 3175-3186.
14. Spicer, E.K., Noble, J.A., Nossal, N.G., Konigsberg, W.H., and Williams, K.R. (1982) J. Biol. Chem., in press.
15. Singer, B.S., Gold, L., Shinedling, S.T., Colkitt, M., Hunter, L.R., Pribnow, D., and Nelson, M.A. (1981) J. Mol. Biol. 149, 405-432.
16. Bowles, M. and Karam, J. (1979) Virology 94, 204-207.
17. Gold, L., Pribnow, D., Schneider, T., Shinedling, S., Singer, B.S., and Stormo, G. (1981) Ann. Rev. Microbiol. 35, 365-403.
18. Trimble, R.B. and Maley, F. (1976) J. Virol. 17, 538-549.
19. Liu, C.C., Burke, R.L., Hibner, U., Barry, J., and Alberts, B. (1978) Cold Spring Harbor Symp. Quant. Biol. 43, 469-487.
20. Tomich, P.K., Chiu, C.S., Wovcha, M.G., and Greenberg, G.R. (1974) J. Biol. Chem. 249, 7613-7622.
21. Mathews, C.K., North, T.W., and Reddy, G.P.V. (1979) in Advances in Enzyme Regulation, Weber, G. ed., Pergamon, Oxford and New York, vol. 17, pp. 133-156.
22. Wood, W.B. and Revel, H.R. (1976) Bacteriol. Rev. 40, 847-868.

REGULATION OF RIBOSOME BIOSYNTHESIS IN *ESCHERICHIA COLI*

Masayasu Nomura, Sue Jinks-Robertson and Akiko Miura
Institute for Enzyme Research and the Departments of Genetics and Biochemistry,
University of Wisconsin, Madison, WI 53706 USA

INTRODUCTION

The rate of ribosome biosynthesis in *Escherichia coli* is regulated and is directly proportional to cellular growth rate.[1] Rapidly growing cells synthesize more ribosomes per unit amount of total protein than do slowly growing cells, and the rate of ribosome biosynthesis can be adjusted rapidly in response to environmental changes. The molecular mechanisms involved in regulating the rate of ribosome accumulation have been the subject of intensive study for a number of years, but remain unknown. Nevertheless, these studies have led to an understanding of the genetic organization of ribosomal components, and have revealed some novel mechanisms of gene regulation.

The *E. coli* ribosome is composed of three species of RNA (rRNA) and approximately 52 proteins (r-proteins). The synthesis rates of all the ribosomal components are balanced and, like ribosomes, respond coordinately to changes in environmental conditions.[2] One approach to studying the control of ribosome biosynthesis has been to examine the regulation of the individual components. Studies in our laboratory during the last few years have been devoted to understanding the mechanisms responsible for coordinating the synthetic rates of the 52 r-proteins. Our studies and those of others have led to the proposal of a translational feedback regulation model for coordinating the synthesis of r-proteins. In this article, we discuss the essential features of the model and the experimental data that support it. In addition, we discuss the relevance of this model to the overall regulation of ribosome biosynthesis, and suggest another level of feedback regulation which might be involved in the regulation of ribosome accumulation.

THE TRANSLATIONAL FEEDBACK REGULATION MODEL

The genes for the 52 r-proteins are subdivided into approximately 16 transcription units (operons) that contain from one to eleven genes (for reviews, see Refs. 3 and 4). We have studied the operons contained within the two major clusters of r-protein genes on the *E. coli* chromosome: the cluster in the *str-spc* region at 72 min and the cluster in the *rif* region at 89 min. The *str-spc* region contains four operons which encode 27 r-proteins, elongation factors Tu and G, and RNA polymerase subunit α. The *rif* region contains two operons which encode four r-proteins and RNA polymerase subunits β and β'. The organization of these genes within their respective operons is shown in Figure 1.

The translational feedback regulation model of r-protein synthesis was originally based on experiments which analyzed the effect of gene dosage on the synthesis rate of r-protein mRNA and the synthesis rates of r-proteins. We found that in strains merodiploid for r-protein genes in the *str-spc* region, the rate of r-protein mRNA synthesis increases in proportion to the increase in gene dosage, yet the rates of synthesis of the corresponding r-proteins do not increase in proportion to gene copy.[18] Similar gene dosage experiments were also carried out by other investigators.[19,20,21] To explain these results, we proposed a model of post-transcriptional regulation of r-protein synthesis. We suggested that r-protein synthesis and ribosome assembly might be coupled so that when the r-protein synthesis rates exceed those needed for ribosome assembly, "free" r-proteins would interact with their respective mRNA to inhibit further translation.[18]

At the time the model was proposed, there was no information to indicate whether all or only some of the r-proteins might function as repressors of mRNA translation. Subsequent experiments, however, have identified specific r-proteins (L1, L4, L10, S4, S7, and S8) as translational repressor proteins which regulate their own synthesis and the synthesis of some or all of the r-proteins that are co-transcribed with the repressor. Both *in vitro* experiments utilizing a DNA dependent protein synthesis system and *in vivo* experiments utilizing gene fusion plasmids were used to identify the translational repressors and their regulatory units, and the results of these studies are summarized in Figure 1.

The following points are what we believe to be the essential features of translational feedback regulation of r-protein gene expression: (1) only certain r-proteins can act as inhibitors to block the translation of their respective mRNA; (2) there are units of translational regulation and each unit contains the gene for its own unique translational repressor. There may be one or more regulatory units **within** each operon, but these regulatory units do not overlap; (3) multicistronic units are regulated by action of the repressor molecule at a single site near the translation initiation site for the first cistron in the unit. Translation of the distal cistrons in a unit is dependent on translation of the previous cistrons ("translational coupling" or "sequential translation", see below); (4) the coordinate regulation of all r-proteins is accomplished by competition between rRNA (or rRNA in assembly intermediates) and mRNA for repressor r-proteins; (5) repressor r-proteins act in *trans* as well as

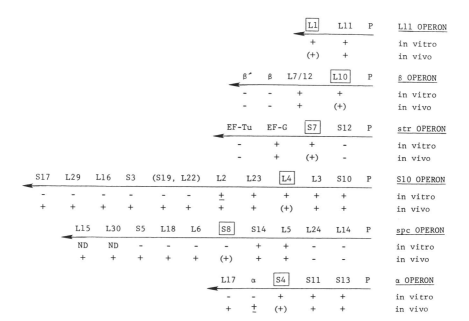

Figure 1. Organization and regulation of genes contained within the *str-spc* and *rif* regions of the bacterial chromosome. Genes are represented by the protein product. For each operon, the direction of transcription from the promoter (P) is indicated by an arrow. It should be noted that it has recently been shown that the L11 and β operons are probably a single operon. That is, the L11 promoter functions as the major promoter for all genes contained within the L11 and β operons in exponentially growing cells (Ref. 5; Catherine Squires, personal communication). It appears that the β operon promoter previously identified is used only when transcription from the major upstream promoter ceases (see the discussion in Ref. 6). Regulatory r-proteins are indicated by the boxes. Effects of the boxed proteins on the *in vitro* or *in vivo* synthesis of proteins from the same operon are shown. Results obtained *in vivo* are those from experiments using hybrid plasmids to achieve overproduction of the regulatory protein. *In vitro* experiments have identified r-proteins L1,[7,8] L10,[9,10,11] L4,[12] S7,[13] S8,[14] and S4.[7] r-Proteins demonstrated to be repressors in *in vivo* experiments are: L1,[15] L10,[11] S7,[13] L4,[16,17] S8,[14] and S4 (Ref. 15; D. Bedwell, S. Jinks-Robertson, and M. Nomura, unpublished experiments). +, specific inhibition of synthesis; -, no significant effect on synthesis; ±, weak inhibition of synthesis; (+), inhibition presumed to occur *in vivo*; ND, not determined. It has not been established how the regulation of the synthesis of r-proteins S12, L14, or L24 is achieved. There is some preliminary evidence which suggests that each of these proteins feedback regulates the translation of its own message (Dean and Nomura, unpublished experiments).

in *cis*; and (6) translational feedback regulation plays a significant role in controlling r-protein synthesis rates.

The experimental evidence that supports these conclusions is described in the original papers (see legend to Figure 1 for references) and has been reviewed previously (see Refs. 22 & 23). Here, we make only some specific comments to clarify the points made above.

The specific inhibition of the expression of a group of r-protein genes takes place at the translation step. This conclusion was demonstrated in the *in vitro* experiments by uncoupling the transcription and translation steps.[7,9,12,30] Even so, there have been suggestions that the mechanism of feedback regulation by repressor r-proteins *in vivo* might involve inhibition of transcription initiation at the promoter site.[16,24] We think that any observed effects on transcription are probably indirect. It should be noted that in both *in vitro* and *in vivo* experiments, r-proteins S7 and S8 do not inhibit the synthesis of promoter proximal proteins; the inhibitory action of S7 starts at the second cistron in the *str* operon,[13] and that of S8 starts at the third cistron in the *spc* operon[14] (see Figure 1). Clearly these two r-proteins do not act at the level of transcription initiation. In addition, we have observed that overproduction of repressor r-proteins *in vivo* leads to stimulation rather than inhibition of the transcription initiated at most r-protein promoters, including the promoter for the operon encoding genes regulated by the overproduced repressor (see below for discussion).

Repressor r-proteins act at a single target site. Most of the repressor r-proteins so far identified inhibit the translation of more than one cistron, and there are two possible mechanisms which could achieve this type of polycistronic translational regulation. One possibility is that each of the sensitive cistrons has a target site for the repressor r-protein. An alternative possibility is that repressor molecules interact with mRNA at a single target site and affect the translation of all the sensitive cistrons as a consequence of this single interaction. In the case of the L11 operon, *in vitro* experiments with DNA templates carrying the intact operon have clearly demonstrated that L1 inhibits translation of itself. In contrast, when the synthesis of L1 is directed by a DNA template lacking the N-terminal part of the L11 cistron, the synthesis of L1 is not inhibited by L1. Conversely, a hybrid plasmid carrying the N-terminal part of the L11 cistron directs synthesis of a small polypeptide which contains an N-terminal portion of L11, and this synthesis is inhibited by L1. In this way, it has been shown that L1 acts at a single site within the first 160 bases of the bicistronic L11-L1 mRNA, near or at the translation initiation site of the L11 cistron.[25] There is similar *in vitro* evidence for the β operon that L10 inhibits the synthesis of both L10 and L7/L12 by acting at a single site near or at the translation initiation site of the L10 cistron.[11] As shown in Figure 1, S8 and L4 inhibit the synthesis of some proteins coming from the distal parts of the respective regulatory units *in vivo*, but not *in vitro*. These seemingly contradictory results are consistent with the conclusion that these distal cistrons themselves do not have the target sites for repressors, and that the inhibition observed *in vivo* is a consequence of the direct inhibition of the first cistron in the regulatory unit. We think that the apparent escape of the distal cistrons from the inhibition *in vitro* is due to physical separation of distal mRNA from the target site, as could be caused by nonspecific nucleolytic cleavage of mRNA in the *in vitro* system (see the discussion in Ref. 12).

The only indication of exceptions to the single target site mechanism is the regulation of α operon protein synthesis (see Figure 1). Our recent experiments have shown that the synthesis of L17 is regulated by S4 as is that of S13, S11, and S4. The regulation of the synthesis of α is in many respects different from that of the four co-transcribed r-proteins and, therefore, α does not appear to be regulated by S4 (D. Bedwell, C. S. Jinks-Robertson and M. Nomura, unpublished experiments). One possible explanation of α operon regulation is that the repressor r-protein S4 has two target sites: one at or near the beginning of the S13 cistron and the other at or near the beginning of the L17 cistron. This possibility is currently under investigation.

Regulation of translational units is probably achieved by translational coupling. As discussed above, repressor r-proteins apparently inhibit the translation of several cistrons by acting at a single upstream target site. How can the inhibition of the translation of one cistron prevent the translation of downstream cistrons? We suggest that independent translation of downstream cistrons does not occur in the absence of the translation of the first cistron in a translationally regulated unit. Experiments which localized the target site for repressor action on the bicistronic L11-L1 mRNA[25] (see above) support this suggestion, and there is evidence for a similar situation in the translation of other polycistronic mRNAs. Oppenheim and Yanofsky,[26] for example, have shown that translation of the *trpD* cistron requires the translation of the preceding *trpE* cistron and have called this phenomenon "translational coupling." In the case of most of the r-protein regulatory units (except the L10-L7/L12 unit), we suggest that only the translation initiation site for the first cistron of the regulatory unit can serve as the entry site for free ribosomes to begin translation. Upon termination of the translation of the first cistron, the same ribosomes (or ribosome components) would then initiate and translate the second cistron, and the process would continue to the end of the regulatory unit of mRNA. This hypothetical process (called "sequential translation" as a specific case of translational coupling) is attractive because it ensures equimolar synthesis of all the proteins in a regulatory unit, but it has not yet been experimentally verified.

Alternative mechanisms to achieve regulation of translational units. Two other mechanisms have been suggested to explain how inhibition of translation of the first cistron of a regulatory unit leads to the inhibition of the expression of distal cistrons in the unit.[12,17] First, mRNA complexed with a repressor r-protein might be recognized by a specific nuclease and be selectively degraded, leading to the inhibition of distal gene expression. In the gene dosage experiments using strains diploid for r-protein genes, it was found that the excess r-protein mRNA synthesized in the diploid strains is preferentially degraded.[18,21] Specific inhibition of both L1 and L11 synthesis from the L11 operon *in vitro*, however, can take place without mRNA degradation.[9,25] It appears that the selective mRNA degradation observed *in vivo* is a consequence of translational repression, and that mRNA degradation is probably not the primary mechanism responsible for the regulation of several cistrons in a regulatory unit.

The second alternative mechanism is that inhibition of translation of the first cistron in a regulatory unit by the repressor r-protein causes termination of transcription, thus preventing the expression of all the downstream cistrons in the operon. In support of this mechanism, it has been reported that overproduction of L4 *in vivo* causes a marked decrease in the rate of mRNA synthesis from the S10 operon.[17] We do not believe this to be the primary mechanism for regulating distal gene expression for the following reasons. First, the gene dosage experiments mentioned above clearly demonstrate that translational regulation can take place without transcription being inhibited. This conclusion can be made with respect to the *spc* and α operons,[18,21] and the β operon.[19] Second, overproduction of L10 *in vivo* does not cause any significant decrease in the synthesis of β and β' from the distal part of the operon,[11] suggesting that translational repression is not necessarily accompanied by transcription termination. As was found for the α,[7] L11,[7] *spc*,[7] and *str*[13] operons, it should be noted that in experiments analyzing *in vitro* regulation of the S10 operon, specific translational repression of the first four to five cistrons by r-protein L4 was observed in the absence of transcription.[12] Clearly, such *in vitro* effects cannot be explained by a transcription termination mechanism. We think that the inhibition of translation of a promoter-proximal cistron might cause some transcription termination *in vivo*, but we do not believe that this is the primary mechanism responsible for the regulation.

Competition between rRNA and r-protein mRNA target sites. Most of the repressor r-proteins identified thus far are known to have strong and specific
binding to rRNA under the conditions of *in vitro* ribosome assembly.[27,28] Based
on this, we have suggested that translational feedback regulation can be regarded as competition between rRNA and mRNA for repressor r-proteins.[29] The postulated competition has been demonstrated in *in vitro* experiments (Refs. 25, 30;
M. Johnsen, T. Christensen, and N. Fiil, personal communication), and further
support comes from the presence of homologies in nucleotide sequence and in potential secondary structure between r-protein binding regions on rRNA and the
presumptive target sites on mRNA. Homologies have been identified for r-proteins S4,[29] S7,[29] S8,[31] L4,[32] and L1.[33,34] As an example, the homologous
structures for the S8 binding site on rRNA and mRNA are shown in Figure 2. A
feature that should be noted is that the homologous regions include the ribosome
binding site (inferred from the Shine-Dalgarno sequence and the translation initiation codon) on mRNA, suggesting that the repressor function involves inhibition of ribosome binding and/or translation initiation. Recently, Weissbach and
his co-workers have shown that this is indeed the case for the repressor action
of L10 *in vitro*.[35]

It should be noted that the association of r-proteins with rRNA is a highly
cooperative process,[27,28] and that this cooperativity should help ensure that
the repressor r-proteins take part in ribosome assembly preferentially to mRNA
binding. It is conceivable that r-proteins that have a rather low affinity for
rRNA and participate in ribosome assembly mainly through protein-protein interactions could still have a sufficiently strong interaction with their mRNA for
feedback inhibition of translation to occur. Consequently, the cooperative
interactions of r-proteins during the assembly process might be very important
in sequestering the regulatory r-proteins, thereby preventing the unnecessary
inhibition of r-protein mRNA translation.

Repressor r-proteins act in trans. There are two possibilities as to how the
translational regulation by repressor r-proteins is coupled with ribosome assembly. One possibility is that under steady-state growth conditions, "free" repressor r-proteins are released from mRNA and then interact with either rRNA (or
assembly intermediates) or an mRNA target site, depending on the availability of
rRNA (or assembly intermediates). This implies that repressor r-proteins should
function not only in cis (on their own specific mRNA) but also in trans on any
homologous mRNA with the proper target site. Another possibility is that repressor r-proteins interact only with the target site on their own specific mRNA
and that prevention of such an interaction requires competition for repressor r-

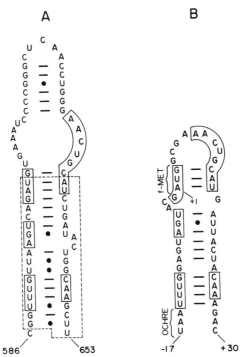

Figure 2. Model of the secondary structure of S8 binding sites on 16S rRNA (A) and on mRNA (B). Homologies which are considered to be significant are boxed. The structure shown in (A) is taken from Woese *et al.*[36] The binding site for S8[37, 38] is indicated by a broken line. In (B) the L24 coding region ends at -18, and the L5 coding region begins at +4. Since S8 inhibits translation of the L5 and distal cistrons, but not translation of L14 and L24 cistrons,the mRNA. around the L24-L5 intercistronic region shown here is the likely target site for S8.[31]

proteins by ribosome assembly. This second possibility implies that repressor r-proteins act only in cis and not in trans. We have recently examined this question using derivatives of an S4 mutant strain (loss of S4 repressor func-tion) which overproduces α operon r-proteins. It has been found that the wild type S4 synthesized from a transducing phage genome functions as a translational repressor in trans to regulate the synthesis of the chromosomal mutant S4.[39] It is therefore likely that release of repressor r-proteins from the mRNA templates is not coupled with ribosome assembly, and that repressor r-proteins in general can act in trans.

The significance of r-protein translational feedback regulation *in vivo*. There are several experimental observations which support our belief that the translational feedback regulation discussed above plays a significant role in controlling r-protein synthesis *in vivo*. First, it is known that strains which have mutational alterations in repressor r-proteins specifically overproduce r-proteins in the corresponding regulatory unit.[39,40,41,42] Such observations indicate that the synthesis of r-protein mRNA is in excess of that required by the cell to meet the need for ribosome biosynthesis, and that the overproduction of r-proteins from the excess mRNA is normally prevented by repressor r-proteins. Second, our recent analysis of the expression of the *galK* gene (or *lacZ* gene) fused to and under control of one of the r-protein promoters has indicated that the characteristic increase in the relative synthesis rates of r-proteins with increasing growth rate is not determined by the activity of r-protein promoters. The implication is that a post-transcriptional mechanism(s) must play an important role in the growth rate dependent regulation of r-protein synthesis.[43] Since the relationship between rRNA synthesis rate and growth rate is similar to that between r-protein synthesis rates and growth rate in the normal growth rate range (μ=0.5 to 3.0; μ is the growth rate in doublings per hour), the growth rate dependent regulation of r-protein synthesis can be formally explained as being a consequence of the regulation of rRNA synthesis combined with the feedback inhibition of r-protein mRNA translation.

TRANSCRIPTIONAL REGULATION OF RIBOSOMAL PROTEIN GENE EXPRESSION

It should be emphasized that the presence of a transcriptional feedback regulatory mechanism does not imply that there is no transcriptional regulation of r-protein gene expression. In fact, it would be surprising if the regulation of such a large and important group of genes does not involve any other mechanism in addition to the translational regulation discussed above. It is probable that under some conditions, transcriptional regulatory mechanisms may play the dominant role in regulating r-protein synthesis rates.

The first example of proposed transcriptional regulation is the stringent control of r-protein synthesis. In wild type *E. coli*, amino acid starvation elicits a stringent response which is characterized by a dramatic and specific decrease in the synthesis rates of rRNA, tRNA, and r-proteins. It was initially established that the amount of r-protein mRNA is regulated by the stringent control system,[44] and subsequent pulse-labeling experiments have indicated that stringent regulation probably occurs at the level of transcription of r-protein genes.[45] The discovery of translational feedback regulation of r-protein synthesis, however, suggested another possible explanation for these observations. It seemed possible that stringent control might act primarily on the transcription of rRNA operons, and that the resultant change in the rRNA synthesis combined with the feedback regulation of r-protein synthesis might indirectly cause the observed stringent control of r-protein synthesis. The decrease in the amount of r-protein mRNA (and perhaps the decrease in apparent synthesis rate measured by pulse-labeling mRNA) could be explained by rapid degradation of translationally repressed mRNA. We have studied this possibility using a *trp-lac* fusion operon which is under control of an r-protein promoter. Since the target site for the repressor r-protein is almost certainly missing in the fusion operon used, synthesis of *trp-lac* mRNA from the r-protein promoter should not be stringently controlled if the above alternative explanation is correct. Our preliminary data, however, have indicated that the explanation given above is not correct, and that the stringent control system probably acts directly on r-protein promoters and affects the transcription of r-protein mRNA (J. R. Cole and M. Nomura, unpublished experiments).

A second example of transcriptional regulation was recently discovered while examining the *in vivo* effects of inhibiting ribosome assembly. We have found that when ribosome assembly is inhibited by the specific overproduction of repressor r-proteins, transcriptional activities of r-protein promoters are stimulated, and the degree of the stimulation increases with time (A. Miura and M. Nomura, manuscript in preparation). It should be emphasized that this observation is in contrast to the earlier suggestion by some workers that feedback regulation by repressor r-proteins is achieved by specifically inhibiting transcription initiation of operons encoding the overproduced repressor r-protein. First of all, we observe stimulation rather than inhibition of transcription after overproduction of repressor r-proteins. Second, not only is transcriptional activity of the operon encoding the overproduced r-protein stimulated, but the transcriptional activities of all other r-protein promoters are stimulated. We suggest that *E. coli* cells detect a deficiency of ribosomes which is caused by the inhibition of ribosome assembly, and try to compensate for this deficiency by increasing the transcriptional activities of all r-protein genes (and rRNA genes, see below). Studies are under way to define this regulatory phenomenon more precisely and to elucidate the molecular mechanisms involved.

REGULATION OF RIBOSOME SYNTHESIS

The translational feedback regulation model can account for the coordinated and balanced synthesis of all the ribosomal components, but it does not answer the question of how global ribosome accumulation is regulated. The rate of ribosome accumulation is presumably determined by the transcriptional activity of

a rate-limiting promoter(s) among the many promoters for the ribosomal components. With the r-protein translational feedback mechanism operating, the regulation of this rate-limiting promoter would, in principle, be sufficient to
regulate the accumulation of ribosomes. We suspect that, under normal growth
conditions, the rate-limiting promoters are those for rRNA operons. [Under the
conditions of very slow growth where extensive degradation of newly synthesized
rRNA has been observed,[46] the rate-limiting promoter(s) could be one of the r-
protein promoters. Alternatively, some unknown mechanism might be operating to
cause the specific degradation of rRNA under such conditions.] It is known that
under normal growth conditions, the number of ribosomes per unit amount of total
cellular protein is directly proportional to the growth rate of the cells and
that almost all of the ribosomes are engaged in protein synthesis. Bacterial
cells can adjust the rate of ribosome accumulation depending on the supply of
carbon and energy sources in the growth media, so that all the ribosomes are
utilized and the maximum growth rate is achieved. One can ask what kind of
mechanisms might be involved in determining the "correct" amount of ribosomes to
be synthesized under a given set of growth conditions. An attractive model is
that the rate of ribosome accumulation is subject to a feedback regulatory mechanism. Since ribosomes are not present in excess under normal growth conditions,
one can imagine that whenever ribosomes are overproduced, the nontranslating ribosomes prevent further biosynthesis and accumulation of ribosome particles by
inhibiting transcription from the proposed rate-limiting promoter(s). This
could occur by the direct interaction of ribosome particles with the rate-limiting promoter, or the effect could be indirect. Ribosome particles could, for
example, exert an indirect regulatory effect by binding to a positive regulator
which is necessary for transcription of the rate-limiting promoter. Such a hypothetical positive regulator would be expected to bind strongly to free ribosome
particles but not to translating particles.

We have been testing this model using several experimental approaches, and
there is some evidence to support this idea. One approach is to specifically inhibit ribosome assembly. If the model is correct, then the synthesis of ribosomal components in cells which are growing at intermediate growth rates should
be limited by the proposed feedback mechanism. Such cells should have a higher
capacity for making ribosomal components, and one would predict that inhibiting
ribosome assembly without inhibiting macromolecular synthesis should lead to
overproduction of ribosomal components. Presumably, blocking ribosome assembly
should deplete the cell of free ribosomes, which should in turn stimulate transcription from the rate-limiting promoter. In agreement with this prediction,

we have found that when ribosome assembly is inhibited by a specific overproduction of repressor r-proteins, the transcription of rRNA operons is specifically stimulated together with that of r-protein operons (A. Miura and M. Nomura,
manuscript in preparation). In addition, we are re-examining the effect of gene
dosage on the synthesis rate of rRNA and our prelminary data suggest that the
rate of rRNA synthesis is regulated and is not strongly influenced by changes in
the number of rRNA operons per genome (C. S. Jinks-Robertson and M. Nomura, unpublished experiments). This observation is also consistent with the notion
that rRNA synthesis is feedback regulated by excess ribosomes. Several known
features of the regulation of ribosome synthesis, such as selective stimulation
(or inhibition) of ribosome synthesis after nutritional shift-up (or shift-down),
could also be explained by the suggested model. It should be noted that our
efforts to show any possible direct regulatory effects of ribosomes on the transcription of ribosomal promoters *in vitro* have been negative so far (R. Sharrock,
J. L. Yates, and M. Nomura, unpublished experiments). Although our model of
global ribosome regulation is very attractive and can explain *in vivo* observations, it needs further experimental verification.

In summary, we believe that the presence of translational feedback regulation
of r-protein synthesis has now been firmly established, although details of the
molecular mechanisms involved need to be further studied for each of the many
regulatatory units identified. In the past, it has been repeatedly emphasized
that the fraction of energy and matter used in making ribosomes is high and that
the regulation of ribosome biosynthesis has to be considered in conjunction with
other biosynthetic activities of cells.[47,48] One clear feature which has emerged
from the recent studies on the translational feedback regulation of r-proteins is that all of the operons encoding components for the protein synthesizing machinery (rRNA, r-proteins and other protein factors) are interconnected
by feedback regulatory loops and are coordinately regulated. As discussed above,
it is tempting to think that the biosynthesis of ribosomes and accessory proteins is connected to the cell's other biosynthetic activities through another
loop of a feedback regulation.

ACKNOWLEDGMENTS

The work in this paper was supported in part by the College of Agriculture
and Life Sciences, University of Wisconsin-Madison, by grant GM-20427 from the
National Institutes of Health, and by grant PCM79-10616 from the National Science Foundation. This is paper number 2588 from the Laboratory of Genetics,
University of Wisconsin-Madison.

REFERENCES

1. Maaløe, O. and Kjeldgaard, N.O. (1966) "Control of Macromolecular Synthesis," Benjamin Press, New York, 284 pp.
2. Gausing, K. (1980) In "Ribosomes: Structure, Function and Genetics" (G. Chambliss, G.R. Craven, J. Davies, K. Davis, L. Kahan, and M. Nomura, eds.), University Park Press, Baltimire, pp. 693-718.
3. Nomura, M., Morgan, E.A., and Jaskunas, S.R. (1977) *Ann. Rev. Genet. 11*, 297-347.
4. Lindahl, L. and Zengel, J. (1982) "Advances in Genetics, Vol. 21" (E.W. Caspari, ed.) Academic Press, New York, pp. 53-111.
5. Brückner, R. and Matzura, H. (1981) *Mol. Gen. Genet. 183*, 277-282.
6. Post, L.E., Strycharz, G.D., Nomura, M., Lewis, H., and Dennis, P.P. (1979) *Proc. Natl. Acad. Sci. USA 76*, 1697-1701.
7. Yates, J.L., Arfsten, A.E., and Nomura, M. (1980) *Proc. Natl. Acad. Sci. USA 77*, 1837-1841.
8. Brot, N., Caldwell, P., and Weissbach, H. (1981) *Arch Biochem. Biophys. 206*, 51-53.
9. Brot, N., Caldwell, P., and Weissbach, H. (1980) *Proc. Natl. Acad. Sci. USA 77*, 2592-2595.
10. Fukuda, R. (1980) *Mol. Gen. Genet. 172*, 483-486.
11. Yates, J.L., Dean, D., Strycharz, W.A., and Nomura, M. (1981) *Nature 294*, 190-192.
12. Yates, J.L. and Nomura, M. (1980) *Cell 21*, 517-522.
13. Dean, D., Yates, J.L., and Nomura, M. (1981) *Cell 24*, 413-419.
14. Dean, D., Yates, J.L., and Nomura, M. (1981) *Nature 289*, 89-91.
15. Dean, D. and Nomura, M. (1980) *Proc. Natl. Acad. Sci. USA 77*, 3590-3594.
16. Lindahl, L. and Zengel, J. (1979) *Proc. Natl. Acad. Sci. USA 76*, 6542-6546.
17. Zengel, J.M., Mueckl, D. and Lindahl, L. (1980) *Cell 21*, 523-535.
18. Fallon, A.M., Jinks, C.S., Strycharz, G.D., and Nomura, M. (1979) *Proc. Natl. Acad. Sci. USA 76*, 3411-3415.
19. Dennis, P.P. and Fiil, N.P. (1979) *J. Biol. Chem. 254*, 7540-7547.
20. Geyl, D. and Böck, A. (1977) *Mol. Gen. Genet. 154*, 327-334.
21. Olsson, M.O. and Gausing, K. (1980) *Nature 283*, 599-600.
22. Nomura, M., Dean, D., and Yates, J.L. (1982) *Trends in Biochemical Sciences 7*, 92-95.
23. Nomura, M. and Dean, D. In "AMBO Symposium - The Future of Nucleic Acid Research," (I. Watanabe, ed.), Academic Press, Tokyo, in press.
24. Holowachuck, E.W., Friesen, J.D., and Fiil, N.P. (1980) *Proc. Natl. Acad. Sci. USA 77*, 2124-2128.
25. Yates, J.L. and Nomura, M. (1981) *Cell 24*, 243-249.
26. Oppenheim, D.S. and Yanofsky, C. (1980) *Genetics 95*, 785-795.
27. Nomura, M. and Held, W.A. (1974) In "Ribosomes" (M. Nomura, A. Tissières and P. Lengyel, eds.), Cold Spring Harbor Laboratory, Cold Spring Harbor, New York, pp. 193-223.
28. Nierhaus, K.H. (1980) In "Ribosomes: Structure, Function and Genetics" (G. Chambliss, G.R. Craven, J. Davies, K. Davis, L. Kahan, and M. Nomura, eds.), University Park Press, Baltimore, pp. 267-294.
29. Nomura, M., Yates, J.L., Dean, D., and Post, L. (1980) *Proc. Natl. Acad. Sci. USA 77*, 7084-7088.
30. Wirth, R., Kohles, V., and Böck, A. (1981) *Eur. J. Biochem. 114*, 429-437.
31. Olins, P.O. and Nomura, M. (1981) *Nucleic Acids Res. 9*, 1757-1764.
32. Olins, P.O. and Nomura, M. (1981) *Cell 26*, 205-211.
33. Gourse, R.L., Thurlow, D.L., Gerbi, S.A. and Zimmerman, R.A. (1981) *Proc. Natl. Acad. Sci. USA 78*, 2722-2726.
34. Branlant, L., Krol, A., Machatt, A., and Ebel, J.-P. (1981) *Nucleic Acids Res. 9*, 293-307.
35. Robakis, N., Meza-Basso, L., Brot, N., and Weissbach, H. (1981) *Proc. Natl.*

Acad. Sci. USA 78, 4261-4264.

36. Woese, C.R., Magrum, L.J., Gupta, R., Siegel, R.B., Stahl, D.A., Kop, J. Crawford, N., Brosius, J., Gutell, R., Hogen, J.J., and Noller, H.F. (1980) *Nucleic Acids Res. 8*, 2275-2293.

37. Ungewickell, E., Garrett, R., Ehresmann, C., Stiegler, P. and Fellner, P. (1975) *Eur. J. Biochem. 51*, 165-180.

38. Zimmerman, R.A., Mackie, R.A., Muto, A., Garrett, R.A., Ungewickell, E., Ehresmann, C., Stiegler, P., Ebel, J.-P., and Fellner, P. (1975) *Nucleic Acids Res. 2*, 279-302.

39. Jinks-Robertson, S. and Nomura, M. (1982) *J. Bacteriol.*, in press.

40. Jinks-Robertson, S. and Nomura, M. (1981) *J. Bacteriol. 145*, 1445-1447.

41. Olsson, M.O. and Isaksson, L.A. (1979) *Mol. Gen. Genet. 169*, 271-278.

42. Stöffler, G., Hasenbank, R., and Dabbs, E.R. (1981) *Mol. Gen. Genet. 181*, 164-168.

43. Miura, A., Krueger, J.H., Itoh, S., de Boer, H. and Nomura, M. (1981) *Cell, 25*, 773-782.

44. Dennis, P.P. and Nomura, M. (1975) *Nature 255*, 460-465.

45. Maher, D.L. and Dennis, P.P. (1977) *Mol. Gen. Genet. 155*, 203-211.

46. Gausing, K. (1977) *J. Mol. Biol. 115*, 335-354.

47. Maaløe, O. (1969) *Dev. Biol. Suppl. 3*, 33-58.

48. Maaløe, O. (1979) In "Biological Regulation and Development, Vol. 1" (R.F. Goldberger, ed.), Plenum Publishing Corporation, New York, pp. 487-542.

REGULATION OF AN ELEVEN GENE RIBOSOMAL PROTEIN OPERON

LASSE LINDAHL, RICHARD H. ARCHER AND JANICE M. ZENGEL
Department of Biology, University of Rochester, Rochester, New York, 14627 USA

INTRODUCTION

In Escherichia coli, the rate of ribosome formation is correlated with the rate of cell mass synthesis: with increasing growth rates, the cell devotes an increasing fraction of its total synthetic capacity to the synthesis of ribosomes and their ancillary components. This "growth rate dependent regulation" suggests that the adjustment of ribosome accumulation is an integral part of the regulation of cell growth rate. However, the molecular basis for this regulation remains obscure.

A central question in the growth rate dependent regulation is how the bacterial cell coordinates the expression of the various ribosomal protein (r-protein) genes. The 52 different r-proteins are synthesized in equimolar amounts (except L7/L12) even though their genes are organized into at least 16 different transcription units (here called operons). Our approach to elucidate the molecular basis for coordinating r-protein synthesis has been to develop a system in which we can specifically and conditionally perturb this coordination by inducing the oversynthesis of one or several r-proteins. Using this approach, we found that the S10 r-protein operon is regulated autogenously by an r-protein, L4, encoded by the operon. That is, in addition to its role as a structural component of the ribosome, the protein L4 also performs a regulatory function by inhibiting the expression of its own operon when it accumulates in excess. Experiments in other laboratories have led to similar models for regulation of a number of other r-protein operons (for a review, see Ref 1). Thus, operon-specific autogenous regulation appears to be a general mechanism by which the bacterial cell coordinates the expression of the various r-protein genes.

Recent studies in our laboratory have focused on the molecular basis for autogenous regulation of the S10 operon. As reported here, we have found that in vivo the autogenous response coincides with a reduced synthesis of mRNA from the S10 operon. Thus, in addition to the translational regulation of the operon observed in in vitro studies of the autogenous response,[2] regulation of transcription may also play a role in the autogenous regulation of the S10 operon.

MATERIALS AND METHODS

Bacteria and Phages. All strains used for the experiments reported in the text were derivatives of LL308 (E. coli K12 F' laci^q/(Δlac-pro) recA)[3] harboring the indicated plasmids. λspc1 is a transducing λ bacteriophage carrying the α and spc r-protein operons from the E. coli chromosome[4]. λCh3F17i$^\lambda$ (here called λS10) is a transducing bacteriophage carrying the 1.2kb EcoRI fragment shown in Fig. 3[2].

Plasmids. pLL133 is a derivative of pBR322 containing a hybrid operon consisting of the lac operator-promoter, a partial lacz gene and the gene for r-protein L4. The expression of the plasmid borne L4 gene is controlled by the lac operator-promoter[3]. pJZ101 is a plasmid derived from pSC101[5] carrying the same lac promoter-operator and lacz region as pLL133, plus the genes for r-proteins L4, L23 and L2. pLL36 was obtained by inserting the 1.2kb EcoRI fragment shown in Fig. 3 into the EcoRI site of pSC101[3]. pLL36ΔSac2 (pLL39) was constructed from pLL36 as follows: Aliquots of pLL36 DNA (1.7μg each) were opened at the unique SacI site in the 1.2kb EcoRI fragment and then treated with various concentrations of S1 nuclease (in 100μl 25mM Na-acetate, pH 4.4/ 150mM NaCl/0.1mM ZnSO$_4$/12mM β-mercaptoethanol) for 90 minutes at 37°C. The remainder of the procedure was performed using the DNA treated with the lowest S1 nuclease concentration which prevented formation of detectable amounts of closed DNA circles (as analyzed by gel electrophoresis) upon treatment of a 0.5μg aliquot with 75 units (cohesive end assay, New England Biolabs) of T4 DNA ligase at 15°C for four hours. A 0.2μg aliquot of this DNA was then incubated with 250 units (cohesive end assay) of T4 DNA ligase at 15°C for 40 hours. The ligation mixture was used to transform E. coli K12 to tetracycline resistance. Plasmids were prepared by the "quick technique" (described below) from 20 tetracycline resistant transformants. Eight of the tested transformants contained plasmids which were resistant to digestion with SacI. One such mutant plasmid, called pLL36ΔSac2, was analyzed in more detail (see text).

Quick Technique for Small-Scale Plasmid Purification. Plasmid DNA from overnight cultures (1.4 to 10ml) was prepared by a lysozyme/EDTA/sodium dodecylsulfate/NaCl technique[6]. The supernatant, enriched for plasmid DNA, was incubated at 68°C for 30 minutes and centrifuged at 10,000 rpm for 15 minutes. The supernatant was mixed with one-sixth volume 8M potassium acetate, incubated on ice for 30 minutes and again centrifuged at 10,000 rpm for 15 minutes. The DNA was precipitated from the supernatant with two to three volumes ethanol and

collected by centrifugation. The pellet was dried and redissolved in 200μl
10mM Tris-HCl, pH 7.4/0.1mM EDTA, mixed with one-tenth volume 1%
Na-diethylpyrocarbonate in ethanol and incubated at 37°C for ten minutes. The
solution was then treated with RNase A (20μg/ml) for one to two hours, and the
diethylpyrocarbonate treatment was repeated. Finally, the DNA was again
precipitated with ethanol, and the pellet was dissolved in 20 to 100μl of the
appropriate buffer. Typical yields were 1μg DNA per ml of culture for
ColE1 derivatives and about 0.2μg/per ml of culture for pSC101 derivatives.

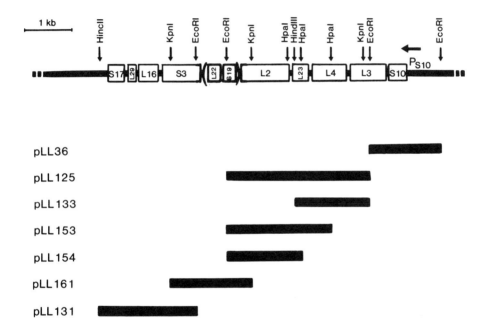

Figure 1. Map of the S10 operon, showing regions of the operon cloned on
plasmids. The genes in the S10 operon are identified by the names of their
gene products and indicated by boxes whose lengths are approximately
proportional to the expected gene lengths. For further details of the mapping
of the S10 operon, see Refs. 3 and 7. We have shown all sites in or near the
S10 operon for restriction endonucleases EcoRI, KpnI, HpaI and HindIII. Only
one of several HindII sites has been indicated. P_{S10} designates the position
of the promoter, with the direction of transcription indicated by the small
horizontal arrow. The solid bars below the map indicate the DNA fragments
carried by each of the plasmids named on the left. In all plasmids except
pLL36, the cloned r-protein genes are now under control of the lac promoter.
In the case of pLL36, r-protein S10 is synthesized constitutively from the
natural S10 promoter also carried on the plasmid. This figure is modified from
Zengel et al.[3] (copyright by the MIT press).

RESULTS

Effect of oversynthesis of L4 on r-protein synthesis

The S10 operon codes for eleven ribosomal proteins: S10, L3, L4, L23, L2, (S19, L22), S3, L16, L29, and S17 (Fig. 1). To analyze the regulation of the S10 operon, we cloned various genes from the operon onto a plasmid vector which carries lacOP. The r-protein genes, removed from their natural promoter, were inserted in the correct orientation so that they could be transcribed from the lac promoter. Since the cells carrying the constructed plasmids contain the normal complement of r-protein genes on the chromosome, r-protein synthesis follows the normal, coordinated pattern when no inducer is present. However, addition of the lac inducer isopropylthiogalactoside (IPTG) activates the plasmid-borne r-protein genes, resulting in rapid and specific oversynthesis of the corresponding protein products. By measuring the synthesis of individual r-proteins after induction of the oversynthesis of one or several specific r-proteins, we could follow the regulatory response of the cell.

Using this in vivo technique, we found that the oversynthesis of a single protein from the S10 operon, L4, strongly inhibits the synthesis of all proteins encoded by the operon[3],[8] (Table 1). The synthesis of proteins

TABLE 1.

RELATIVE SYNTHESIS RATES OF INDIVIDUAL R-PROTEINS AFTER ENHANCED SYNTHESIS OF SPECIFIC PROTEINS FROM THE OPERON.

Protein	Plasmids					
	pLL125 (L4,L23,L2)	pLL133 (L4)	pLL153 (L23,L2)	pLL154 (L2)	pLL161 (S19,L22)	pLL131 (L16,L29,L17)
S10 operon						
L3	0.09	0.10	0.37	1.16	0.86	0.92
L4	2.97	3.81	1.00	0.95		
L23	8.74	0.23	10.33	1.19	1.11	0.98
L2	10.00	0.09	10.00	11.90	0.95	0.78
S19	0.10	0.24	0.67	1.08	3.50	0.92
L22	0.22	0.23	0.67	1.17	3.47	0.94
S3	0.14	0.16	0.77	1.24	0.82	0.76
L16	0.14	0.16	0.59	1.06	0.81	8.28
L29	0.23	0.25	0.74	1.03	0.77	7.84
S17						3.63
Other r-proteins						
average	0.98	0.99	0.88	1.07	1.03	0.87

Cells containing the indicated plasmids were pulse-labeled with ^{35}S-methionine before or 10 min after addition of IPTG. The relative rates of synthesis of the indicated proteins were calculated as described[3]. The table gives the ratio of the relative synthesis rate of a given protein after induction to the relative synthesis rate of the same protein before induction. This table is modified from Zengel et al.[3] (copyright by the MIT press).

from other r-protein operons was largely unaffected by excess L4[3]. Moreover, oversynthesis of none of the other eight r-proteins tested (S10, L23, L2, S19, L22, L16, L29 and S17) had any effect on the expression of the S10 operon[3] (Table 1). These results indicated that the S10 operon is autogenously regulated, and that a single r-protein, L4, functions as the autogenous regulator as well as a structural component of the ribosome.

Effect of L4 oversynthesis on mRNA synthesis

To characterize the mechanism by which L4 regulates the S10 operon, we investigated if the rate of transcription of the operon is changed in response to L4 oversynthesis. Relative rates of messenger RNA synthesis were determined by hybridizing RNA extracted from cells pulse-labeled with ^3H uridine to DNA from the proximal or the distal regions of the operon. Proximal DNA was obtained from λCh3F17i$^\lambda$ (hereafter called λS10), whose genome includes a 1.2kb EcoRI fragment carrying the promoter for the S10 operon, the S10 gene and a portion of the L3 gene[2] (Fig. 3). Distal DNA was obtained from a plasmid carrying the EcoRI-HincII fragment from the distal end of the operon[3] (Fig. 1). As a control we also measured the synthesis of mRNA from two other r-protein operons (the α and spc operons) by hybridizing to DNA from λspc1[4]. Since L4 specifically affects expression of the S10 operon (Table 1), we have normalized all measurements of incorporation into mRNA from the S10 operon to the incorporation into α + spc mRNA.

The results from these experiments show that excessive accumulation of L4 leads to a two- to three-fold inhibition of incorporation of radioactivity into both the proximal mRNA (Table 2) and the distal mRNA from the S10 operon[3] during a one minute pulse. These results suggest that excess L4 leads to decreased transcription of the entire S10 operon.

We also analyzed the effect of excess L4 on synthesis of S10 messenger RNA in cells carrying multiple copies of the proximal part of the S10 operon. These cells contain a multicopy plasmid, pLL36, carrying the S10 promoter and the first one and a half genes of the S10 operon[3] (Fig. 3). Cells harboring pLL36 constitutively oversynthesize S10-L3' mRNA by 6- to 10-fold (Table 2). Induction of excess L4 synthesis (from a P_{lac}-L4 hybrid operon on pLL133 which is also carried by this strain) inhibits the synthesis of this mRNA to approximately the same degree as we observed in cells containing only the chromosomal S10 gene (Table 2). Therefore, it appears that the expression of the partial S10 operon on the plasmid is regulated like the chromosomal S10 operon in response to L4 oversynthesis. Since the cells carrying pLL36 synthesize more S10-L3' message than the haploid strain, we have used this strain

TABLE 2.

THE EFFECT OF EXCESS L4 ON THE SYNTHESIS OF S10 mRNA

		Input RNA	RNA Hybridized (% input)			+IPTG/-IPTG	
Plasmids	IPTG	(cpm x 10^{-5})	λspc	λS10	$\dfrac{\lambda S10}{\lambda spc}$	S10 mRNA	S10 protein
(1)	(2)	(3)	(4)	(5)	(6)	(7)	(8)
pJZ101	-	2.6	1.9	0.14	0.073	0.44	0.41
	+	2.3	1.7	0.055	0.032		
pLL36,	-	3.2	1.2	0.91	0.76	0.33	0.25
pLL133	+	3.9	1.1	0.28	0.25		

Cultures of strain LL308 containing the indicated plasmids were labeled for 1 min with ^{3}H-uridine or with ^{35}S-methionine. Where indicated (column 2), IPTG (1mM) was added to the culture 10 min prior to the labeling. RNA was purified from the ^{3}H-labeled cells and hybridized to immobilized DNA from λS10 (carrying the S10 gene and part of the L3 gene; see Fig. 3) or from λspc1 (carrying the α and spc operons). The amount of radioactive RNA binding to each type of DNA was calculated as the % of the input radioactivity (columns 4 and 5). The amount of radioactive S10 mRNA was then normalized to the amount of radioactive mRNA from the α and spc operons (column 6). The normalized amount of S10 mRNA synthesized in cells after induction with IPTG was then divided by the normalized S10 mRNA synthesized before induction (column 7) (for further technical details, see Ref. 3). Proteins were extracted from the ^{35}S-labeled cells and fractionated by two-dimensional gel electrophoresis. The rate of S10 protein synthesis after induction with IPTG relative to the rate of S10 synthesis before induction (column 8) was determined as described previously[8].

in all subsequent experiments to increase the sensitivity of the hybridization measurements.

The hybridization results reported above indicate that overproduction of L4 leads to inhibition of mRNA synthesis from the entire S10 operon in vivo. However, experiments by Yates and Nomura[2] have shown that the in vitro effect of L4 is to inhibit the translation of mRNA from the S10, L3, L4 and L23 genes, with no apparent effect on transcription. This disparity suggested that our estimates of mRNA synthesis, based on one minute labeling times, might be incorrect. One obvious possibility is that the L4-mediated inhibition of mRNA translation might change the stability of the mRNA, so that significant turn-over of radioactive mRNA takes place even during a pulse as short as one minute. We therefore varied the length of the pulse, labeling for times as short as 0.1 minute. Fig. 2a shows the relative incorporation into mRNA from the proximal part of the S10 operon as a function of pulse length. Extrapolation of these graphs to zero pulse time should give an estimate of the relative S10-L3' mRNA synthesis irrespective of differences in mRNA stability. That is, if message stability, but not synthesis, is affected by excess L4, then the values for the

IPTG induced cells should approach the values for the uninduced cells as the pulse time approaches zero. The results shown in Fig. 2a indicate that this is not the case: the amount of radioactive S10-L3' mRNA is decreased about two-fold after induction of L4 oversynthesis (Fig. 2a, squares) even after a 0.1 minute pulse. Unless the S10-L3' mRNA consists of two components, one of which turns over extremely rapidly in response to L4 accummulation (a hypothesis which is impossible to test experimentally), it appears that the change in mRNA synthesis in response to excess L4 is genuine.

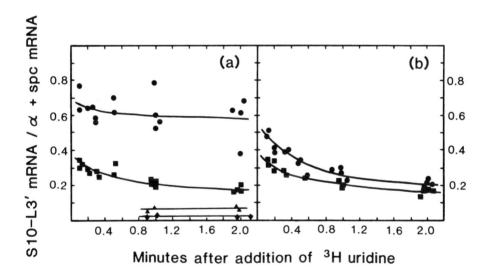

Minutes after addition of ^{3}H uridine

Figure 2. Synthesis of S10-L3' mRNA. Cultures of LL308 carrying the indicated plasmids were mixed with ^{3}H-uridine before or 10 min after addition of IPTG (added to induce L4 oversynthesis). At various times after addition of ^{3}H-uridine, samples were withdrawn and immediately lysed. RNA was extracted and hybridized as explained in the legend to Table 2. The graph shows the ratio between ^{3}H cpm hybridizing to λS10 and ^{3}H cpm hybridizing to λspc1, plotted as a function of labeling time. (a) upper curves: Strain LL308 carrying pLL36 (see Fig. 3) and pLL133 (a ColE1 derivative carrying a P$_{lac}$-L4 hybrid operon, see Table 1) in the absence (circles) or presence (squares) of IPTG. Lower curves: Strain LL308 carrying only pLL133 in the absence (triangles) or presence (diamonds) of IPTG. (b) Strain LL308 carrying pLL36ΔSac2 (see text) and pLL133 in the absence (circles) or presence (squares) of IPTG.

It should be noted that the ratio of radioactivity in S10-L3' mRNA after induction of L4 oversynthesis (Fig. 2a, squares) to the radioactivity in S10-L3' mRNA before L4 induction (Fig. 2a, circles) may decline slowly with increasing pulse times suggesting that the stability of the mRNA from the S10 operon decreases a little during L4 accumulation. However, this change in stability cannot explain the difference in incorporation during the very short pulses.

Another possible complication with these hybridization experiments is that our hybridization probe contains sequences outside the S10 operon (see Fig. 3). No gene has been mapped upstream from the S10 promoter. Nevertheless we were concerned that transcription of this upstream sequence could influence our hybridization measurements. To determine if significant amounts of mRNA hybridize to the region upstream from P_{S10} we "mapped" the mRNA:DNA hybrids using the protocol of Berk and Sharp[10]. Nonradioactive RNA from the haploid strain or the pLL36-containing strain was hybridized to an excess of ^{32}P-labeled 1.2kb EcoRI fragment. The hybrids were treated with the single strand specific nuclease S1, and the nuclease resistant hybrids were analyzed by gel electrophoresis. Only one hybrid was found in significant amounts (data not shown). The mRNA forming this hybrid was mapped to the S10 operon, since hybridization to a DNA fragment shortened by cleavage in the L3' gene with BalI (see Fig. 3) produced a slightly shorter hybrid (data not shown). Moreover, the amount of the mRNA:DNA hybrid was strongly reduced after induction of L4 oversynthesis indicating that L4 reduces the concentration of S10-L3' mRNA as expected from the measurements of the synthesis rates (Fig. 2a).

Transcription regulation versus translation regulation

As already mentioned, mRNA synthesis was not inhibited by addition of L4 to an in vitro DNA-dependent protein synthesizing system. In this case, L4 inhibited the translation but not the transcription of the S10, L3, L4, L23 and possibly L2 genes[2]. One way to account for the differences in the in vitro and in vivo results is to assume that transcription is coupled to translation. That is, the primary effect of L4 may be an inhibition of mRNA translation (as shown by the in vitro system); this translational inhibition, in turn, may lead to an inhibition of transcription. Implied in this hypothesis is the assumption that such a coupling of transcription and translation fails to function in vitro. This is a reasonable assumption, since polarity of nonsense mutations, presumably due to a similar coupling of transcription and translation[11], is usually not observed in vitro.

One prediction of our coupling hypothesis is that transcription should also be reduced if the translation of the S10 gene is inhibited by means other than L4 overproduction. To test this hypothesis we examined how the transcription of the S10 operon is affected by deletion of the ribosome binding site in front of the S10 cistron. Because the products of the chromosomal S10 operon are indispensable to the cell, we used the plasmid pLL36 carrying the proximal portion of the S10 operon (described above) for the targeted mutagenesis of the ribosome binding site. The sequence of the 1.2kb EcoRI fragment cloned on this plasmid has been determined by Olins and Nomura[9]. Our analysis of this sequence showed that the predicted Shine-Dalgarno sequence [12] for the S10 structural gene overlaps a unique site on the plasmid for the restriction endonuclease SacI (Fig. 3). Therefore, to delete this region of the S10 ribosome binding site (called SD$_{S10}$), we digested pLL36 with SacI, treated the linearized DNA with S1 nuclease to remove the single-stranded ends, and ligated the now blunt ends to recircularize the plasmid DNA. After transformation, clones were screened for plasmids lacking the SacI site. One such plasmid, called pLL36ΔSac2, was analyzed by the Maxam-Gilbert sequencing technique[16] and found to have a fifteen base pair deletion which removes SD$_{S10}$ but leaves intact the S10 structural gene and the promoter for the S10 operon (Fig. 3).

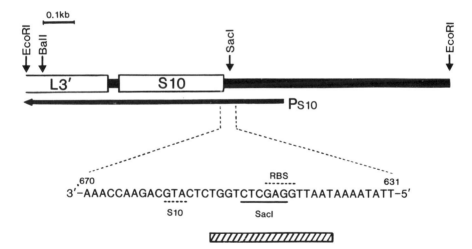

Figure 3. Map of the 1.2 kb EcoRI fragment carrying the promoter-proximal portion of the S10 operon. This EcoRI fragment was inserted into pSC101 to construct pLL36, and is also carried on the λS10 phage. P$_{S10}$ indicates the position of the promoter for the S10 operon. The direction of transcription is indicated by the horizontal arrow. Sites for restriction endonucleases EcoRI, BalI and SacI are shown. The relevant portion of the DNA sequence[9] is given. The ATG initiation codon for the S10 structural gene is shown by dashed underlining. Transcription of S10 mRNA begins approximately 175 bases upstream from this start codon. The SacI site is indicated by solid underlining and the partially overlapping Shine-Dalgarno sequence (RBS) is shown by dashed overlining. pLL36ΔSac2 has a 15 base pair deletion in the 16 base pair interval indicated by the hatched bar.

To verify that the deletion in pLL36ΔSac2 inhibits the translation of the S10 gene on the plasmid, we measured the rate of synthesis of S10 in strains carrying the deletion plasmid or the wild type pLL36. S10 was overproduced in cells carrying pLL36 (see Ref. 3), but no oversynthesis of S10 was observed in cells carrying pLL36ΔSac2 (data not shown). We then used the deletion plasmid to determine what effect translational inhibition resulting from an altered S10 ribosome binding site had on transcription of the S10 operon. The results are shown in Fig. 2b. In the absence of excess L4, the incorporation of radioactive uridine into S10-L3' messenger in a pulse of 0.5 min or longer is reduced 2- to 3-fold in the pLL36ΔSac2 containing cells (Fig. 2b, circles) compared to the pLL36-containing cells (Fig. 2a, circles). However, as the pulse time is decreased, it appears that the incorporation of radioactivity into S10-L3' mRNA synthesized from pLL36ΔSac2 approaches, and at the ordinate may equal, the values obtained for mRNA synthesized from pLL36. These kinetics are different from the incorporation kinetics in pLL36-containing cells oversynthesizing L4 (Fig. 2a, squares). In the latter cells, the values even at very short labeling times are significantly reduced. Thus, the effect on transcription of an altered ribosome binding site is apparently not equivalent to the effect on transcription of excess L4. Oversynthesis of L4 in cells carrying pLL36 is correlated with a decreased rate of S10-L3' mRNA synthesis, while decreased translation of S10 in cells carrying pLL36ΔSac2 appears to make the message less stable. The synthesis of S10-L3' mRNA from pLL36ΔSac2 is, however, decreased by excess L4 (Fig. 2b, squares). It appears then that the mechanism responsible for L4-mediated decreased transcription of the S10 operon is not as simple as suggested in our coupling hypothesis above.

DISCUSSION

We have shown that the S10 operon encoding eleven r-proteins is regulated by an autogenous mechanism: excessive accumulation of one of the proteins encoded by the operon (L4) specifically inhibits the expression of the entire operon [3,8] (Table 1). Experiments by Yates and Nomura[2] have demonstrated that in vitro L4 also regulates the synthesis of proteins from the S10 operon. Since L4 present in ribosomes has no effect on expression of the S10 operon, the critical parameter appears to be the concentration of free L4 in the cell. Because the newly synthesized L4 is normally incorporated into ribosomes in equimolar amounts with other r-proteins and rRNA, the pool size of free L4 may provide the signal for balancing the synthesis of proteins from the S10 operon with the synthesis of other r-proteins. Ribosomal r-proteins encoded by other r-protein operons apparently serve the same function for their respective operons (see review by Lindahl and Zengel[1]). Thus ribosome assembly, in conjunction with autogenous regulation, is probably responsible for coordinating the expression of the individual r-protein operons.

Our experiments show that oversynthesis of L4 leads to a decreased concentration of mRNA from both the proximal (Table 2, Fig. 2a) and the distal[3] parts of the S10 operon. Detailed kinetic measurements (Fig. 2a) suggest that this change in message concentration is due to a decrease in the synthesis (rather than the stability) of mRNA from the S10 operon. Our in vivo data indicate therefore that L4 regulates the transcription of the S10 operon. On the other hand, the in vitro experiments of Yates and Nomura[2] clearly demonstrated that, in the cell-free system, L4 regulates translation (but not transcription) of mRNA from the proximal genes of the operon. The apparent discrepancy between the in vivo and the in vitro results suggests two possible mechanisms for L4 regulation of the S10 operon: first, L4 may inhibit transcription and translation by two independent mechanisms, and second, L4 may regulate both transcription and translation simultaneously by a single regulatory event.

One simple model for the simultaneous regulation of both transcription and translation is that inhibition of translation of the first gene of the operon leads to a reduced rate of transcription of the entire operon. For example, the translational inhibition might provoke a premature termination of transcription in a manner similar to what has been found with polar nonsense mutations[11]. To test this "coupling" model we constructed a deletion mutant lacking the Shine-Dalgarno sequence in front of the S10 cistron. Our simple model would predict that the resulting reduction in the translation of the S10 gene should mimic the effect of excess L4: a reduction in the synthesis of S10 mRNA. However, a detailed kinetic analysis of mRNA synthesis in the deletion mutant (Fig. 2b) indicates that synthesis is affected little, and perhaps not at all, by the deletion of the Shine-Dalgarno sequence. Rather, the primary effect appears to be an increased turnover rate for the S10 message. Therefore, our simple coupling model is inadequate to explain the effect of L4 on both transcription and translation. Further experiments will hopefully clarify the mechanism by which L4 inhibits transcription of the S10 operon.

It is somewhat surprising that both transcription and translation of the S10 operon seem to be regulated by L4. One might argue, teleologically, that translational regulation of protein synthesis is more advantageous to the cell. Inhibition of translation gives an immediate response, while inhibition of transcription still allows several minutes of residual protein synthesis from pre-existing messenger RNA molecules. On the other hand, it is possible that translational regulation is not sufficient to ensure effective regulation of long multicistronic operons. Experiments with the L11[13] and β[14] operons have led to the suggestion that multicistronic r-protein operons are sequentially translated, i.e. translation of a given cistron requires the prior translation of the upstream cistron[13]. The message from the S10 operon probably contains a single target for L4 regulation, upstream of the S10 cistron[9]; presumably L4 binding to this target directly inhibits translation of the S10 cistron and indirectly, by virtue of the sequential translation mechanism, prevents translation of the 10 downstream cistrons. The details of such a mechanism remain to be investigated, but it is conceivable that translation of a given cistron is "leaky"; that is, it may not be absolutely dependent on prior translation of upstream message. For example, if the translation of each cistron downstream of the S10 gene were only 95% inhibited when the upstream sequence was not translated, total inhibition of S10 synthesis would lead to only 50% inhibition of the synthesis of the most distal (S17) gene. Results from in vitro protein synthesis experiments with the S10 operon[2] are consistent with this possibility: translation of only the proximal 4 or 5 genes was regulated in the cell-free system. Although the lack of regulation of downstream cistrons could be due to other causes (for example, artificial fragmentation in vitro of the mRNA transcript[2]), the data could also reflect an inefficient sequential translation mechanism. In vivo studies of a strain containing a nonsense mutation in the L3 gene[15] also suggest that inhibition of the translation of an upstream cistron does not efficiently inhibit the translation of all downstream cistrons. This nonsense mutation apparently affects the translation of only the next three cistrons; i.e., the inhibition of translation was not propagated all the way through the operon.

In conclusion, we would argue that while regulation of translation is a rapid means of regulating protein synthesis, it might not be efficient for regulating expression of the most distal genes in a long multicistronic operon. Therefore, tight control of the expression of an operon like the S10 operon might require the addition of a second mechanism, operating at the level of transcription. Experiments are in progress to determine whether transcriptional control is an essential aspect of L4-mediated regulation of the S10 operon.

ACKNOWLEDGEMENTS

We are grateful to P. Olins and M. Nomura for generously making the sequence of the 1.2kb EcoRI fragment available to us prior to publication. We thank C. Squires and G. Barry for advice on the "Berk-Sharp" hybridizations. This work was supported by a grant from the National Institute for Allergy and Infectious Diseases.

REFERENCES

1. Lindahl, L. and Zengel, J. M (1982) Adv. in Genetics 21, 53-121.
2. Yates, J. L. and Nomura, M. (1980) Cell 21, 517-522.
3. Zengel, J. M., Mueckl, D. and Lindahl, L. (1980) Cell 21, 523-535.
4. Jaskunas, S. R. and Nomura, M. (1977) J. Biol. Chem. 252, 7337-7343
5. Cohen, S. N. and Chang, A. C. Y. (1973) Proc. Natl. Acad. Sci. 70, 1293-1297.
6. Guerry, P., LeBlanc, D. J. and Falkow, S. (1973) J. Bacteriol. 116, 1064-1066.
7. Nomura, M. and Post, L. E. (1980) in Ribosomes: Structure, Function and Genetics, Chambliss, G., Craven, G. R., Davies, J., Davis, K., Kahan, L., and Nomura, M., eds., Univ. Park Press, Baltimore, Md., pp. 671-692.
8. Lindahl, L. and Zengel, J. M. (1979) Proc. Natl. Acad. Sci. 76, 6542-6546.
9. Olins, P. and Nomura, M. (1981) Cell 26, 205-211.
10. Berk, A. J. and Sharp, P. A. (1978) Proc. Natl. Acad. Sci. 77, 3331-3335.
11. Rosenberg, M. and Court, D. (1979) Ann. Rev. Genet. 13, 319-353.
12. Gold, L., Pribnow, D., Schneider, T., Shinedling, S., Singer, B. S. and Stormo, G. (1981) Ann. Rev. Microbiol. 35, 365-403.
13. Yates, J. L. and Nomura, M. (1981) Cell 24, 243-249.
14. Yates, J. L., Dean, D., Strycharz, W. A. and Nomura, M. (1981) Nature 294, 190-192.
15. Cabezon, T., Delcuve, G., Faelen, M., Desmet, L. and Bollen, A. (1980) J. Bacteriol. 141, 41-51.
16. Maxam, A. M. and Gilbert, W. (1980) Methods in Enzymology 65, 499-560.

Marianne Grunberg-Manago and Brian Safer, editors
INTERACTION OF TRANSLATIONAL AND TRANSCRIPTIONAL
CONTROLS IN THE REGULATION OF GENE EXPRESSION

RIBOSOMAL PROTEIN L10 AND S1 CONTROL THEIR OWN SYNTHESIS

STEEN PEDERSEN, JAN SKOUV, TOVE CHRISTENSEN, MORTEN JOHNSEN AND
NIELS FIIL
University Institute of Microbiology, Øster Farimagsgade 2A,
1353 Copenhagen K., Denmark

INTRODUCTION

One model for coordination of the biosynthesis of r-proteins
and of rRNA[1; 2] suggests that each ribosomal protein operon
contains a gene for one of the r-proteins that can bind directly
to rRNA[3]. The mRNA has a site homologous to the binding site on
rRNA. Binding of this r-protein to the site on the mRNA blocks
translation of mRNA, and competition between the binding site on
mRNA and on rRNA adjusts the translation of mRNA to match the
amount of synthesized rRNA. Furthermore, regulation can take
place at the transcriptional level[4; 5].

We are investigating two operons, namely (1) the rplJ operon,
carrying the L10, L12, β and β' genes, and (2) the rpsA operon.
The rplJ operon has an mRNA processing site and an attenuator
site[6], and the proteins are synthesized in relative amounts of
1:4:0.2:0.2. In certain experiments the rpsA operon responds
differently from typical ribosomal operons, suggesting a
different molecular mechanism for S1 regulation.

THE L10 OPERON

The model requires establishment of binding between one of
the operon's proteins and the leader part of the mRNA. Previous
work by Tove Christensen and by Brot et al.[7] and Fukuda[8] has
shown the complex between L10 and L12 to inhibit L10 as well as
L12 synthesis in vitro in a coupled protein-synthesizing system.
We have now positioned the binding site to be on the rplJ leader
mRNA between bases no. 1533 and no. 1602 (numeration according
to Post et al.[9]).

We synthesized ^{32}P-labelled RNA in vitro using the Pst-Pst
fragment of 922 bases, spanning base pairs 869 to 1791, and
including α-^{32}P UTP in the synthesis reaction. This RNA was
bound to the L10-L12 complex and digested with ribonuclease T1.
The complex between L10-L12 and RNA was isolated by membrane

filtration, phenol-extracted and once again digested with T1.
The resulting mixture was analyzed on a gel as shown in Figure 1.

Fig. 1. Electrophoresis of the
ribonuclease T1 resistent
fragments arising from L10-L12
complex-protected rplJ mRNA on
a 20% DNA sequencing gel.

Fragments of 5, 6 and 15 bases can be clearly identified, using
the contaminant bands as molecular weight markers. The bands were
then cut out of the gel and subjected to alkaline hydrolysis
followed by thin layer chromatography of the ^{32}P-labelled
nucleotides. This treatment transfers ^{32}P from UMP residues to
the 5' neigboring base.

In Table 1 these results are compared with the expected result
if the protected piece originates from base no. 1533 to base no.
1579. No other region from the 922 bp template would give this
result. We observed a weaker protection of the GMP residue no.
1523. Protection extending to residue no. 1600 would not be
discovered because the region 1579 to 1602 does not give rise to
characteristic fragments.

TABLE 1

NEAREST NEIGHBOR ANALYSIS OF THE L10-L12 PROTECTED RNA

Number of bases in fragment	Radioactivity (counts/100 min) in		
	GMP	UMP	AMP+CMP
5	<10 (0)	1170 (2)	<10 (0)
6	465 (1)	<10 (0)	523 (1)
15	<10 (0)	625 (1)	1852 (3)

Note: The numbers in parentheses are the expected molar amount if the protected piece is from residue no. 1533 to no. 1579.

The binding site on L10 leader mRNA shows a high degree of similarity to the part of 16S RNA which binds the L10-L12 complex[3; 10], and secondary structures could be imagined which in the presence of L10 mask the ribosome binding site of L10. In this region point mutations affecting L10 translation have been isolated[11].

The molar ratio of L10 to L12 proteins in the cell is 1:4, and the question arises whether rplJ, rplL mRNA is synthesized normally and has a four-times more efficient ribosome-binding site in front of rplL; or, whether the rplJ, rplL mRNA synthesis is four times elevated compared to other ribosomal mRNAs with L10 translation specifically reduced. Previously the amount of ribosomal protein mRNA from the rif region was measured to be the same as the average r-protein mRNA[5]; but in the light of our findings, we measured this again with probes specific for several regions in the L10-L12 operon. Our results are shown in Table 2, where the synthesis rates of L10 and L12 mRNA are about four times that of average ribosomal protein mRNA.

Conclusion

We have established the existence of a binding site for the L10-L12 complex on the leader part of rplJ mRNA. rplJ mRNA is synthesized about four times as frequent as the average ribosomal protein mRNA. Thus, by binding to the leader part of the mRNA, the L10 or L10-L12 complex adjusts the syntheses of L10 and L12 to each other, and to the amount of rRNA in the cell.

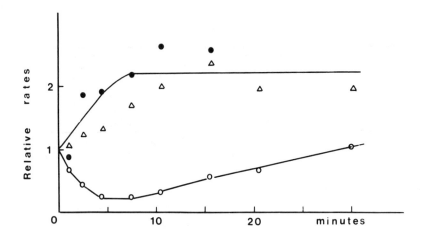

Fig. 2. Rate of S1,O , L12,△,and S6,●, syntheses relative to the rate
of total protein synthesis in E. coli AS19 valS[ts] relA1 after
amino acid starvation. The absolute rate of total protein synthesis
was lowered to about 25% after the temperature shift from 30°C to
39°C. The synthesis rates were determined using a one min ^{35}S
methionine pulse, followed by a 3 min chase at 30°C to complete
all labelled peptides. Proteins were analyzed by 2D gel
electrophoresis[12].

After the establishment of the genetic location of rpsA[17], we
constructed a λ phage carrying rpsA by selecting for the neighbor
marker aspC[13]. A similar phage has been constructed selecting for
serC[18]. We then cloned rpsA on either pACYC184 or pBR322.

The gene orientation was known[19], and wanting to define the
promoter, we created deletions starting in the Sal site, and
extending further and further towards the structural gene. This
was done by Bal31 digestion, and several of the resulting
deletions are shown on Figure 3. The plasmids were then
transferred to 'maxicells' and the expression of the cat and rpsA
genes analyzed (see Fig. 4). The largest deletion still
expressing rpsA normally was no. 59; no. 48 produced a faint S1
band, and longer deletions no bands at all. These results locate
an essential part of the rpsA promoter region between 113 bp and
240 bp from the start of the structural gene.

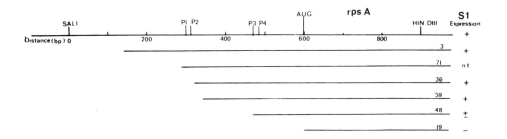

Fig. 3. Map of the rpsA promoter region. As described [13], the Sal-Bam fragment from λaspC has the rpsA promoter. The figure shows the 600 bp in front of the AUG of rpsA and the first part of the structural gene. The distance between the SalI site and the HindIII is about 900 bp. The position of the putative rpsA promoters, P1, P2, P3, and P4 (see text) is indicated. Also shown are the 6 deletions generated by Bal-31 digestion from the Sal site, followed by religation in the presence of Xho linkers. Deletion 48 has no Xho linker, and the size of this deletion was determined by DNA sequencing. Also shown is a summary of the S1 expression data from Figure 4.

Fig. 4. The deletion plasmids numbered at the foot of the picture were transferred to 'maxicells' and the labelled proteins were analyzed on a 10% SDS gel[13]. The upper bands are S1 as shown by 2D gel electrophoresis[13]; the lower strong band might represent the CAM resistance gene. We believe the weak bands to be breakdown products of S1, because they have acid isoelectric points similar to that of S1.

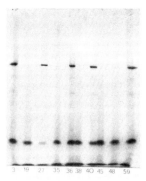

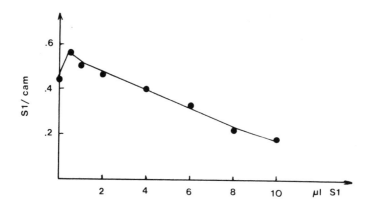

Fig. 5. The effect of purified S1 on the chloramphenicol-
resistance protein and S1 synthesis in a coupled protein-
synthesizing system. The wild-type plasmid having the whole
promoter region was used as template. Synthesis was performed
in a volume of 0.1 ml. The magnesium concentration was 7 mM.
Purified S1 was a gift from Dr. A. Subramanian, Berlin, and
was used at a concentration of 0.025 mg/ml.

To verify the location of the rpsA promoter and to define the
transcription start more precisely, we 5'-labelled an AvaI-AvaI
fragment, spanning the entire promoter region. After digestion
with Sal we isolated the Sal-AvaI fragment, labelled only on the
anti-sense strand 57 bases inside the structural gene. mRNA:DNA
hybrids were constructed with mRNA from a strain having rpsA on
pACYC184. These hybrids were then treated with nuclease S1 and
analyzed on a sequencing gel together with the A+G DNA sequencing
reaction of the same fragment[20]. The strongest start sites were
found 276 and 283 bases from the AUG; only weaker bands were seen
102 and 113 bases from the start of the structural gene. At all
four positions, Pribnow boxes and corresponding -35 regions are
found. The DNA sequence was determined by Schnier and Isono
(pers. comm.) and later by ourselves. The position of the four
promoters are indicated on Figure 3.
    The short RNA pieces observed after nuclease S1 digestion
of mRNA:DNA hybrids might be artifacts, or, possibly, the results

TABLE 2

SYNTHESIS RATES OF mRNA FROM rplK, rplJ, rpsA and spc OPERONS

Condition of culture	Species of mRNA hybridized				
	L11-L1	L10	L12	S1	S5-L15
Exponential growth	1.3	3.0	3.3	0.9	1
Amino acid starvation	2.0	4.2	3.8	1.5	1

E. coli AS19 valS[ts] relA1 was pulse-labelled for one min with 10 µCi/ml carrier free ³H uridine at 28°C, and at 7 min after a temperature shift to 39°C which starts amino acid starvation[12]. The RNA preparation and hybridization technique has been described[13]. As hybridization probes we: used: L11-L1, the 617 bp EcoRI-BglII fragment; L10, the 363 bp PstI-HindIII fragment; L12, the 290 bp HindIII-EcoRI fragment; S1, the 720 bp XhoI-HindIII fragment from deletion 3 in Figure 3; S5-L15, the 1500 bp Pst-Sma fragment. These fragments were cloned on M13mp8 or M13mp9 phage DNA, and one µg single-stranded DNA from the phage carrying the anti-sense strand was loaded onto the filters. This amount should be in ten-fold molar excess to the RNA. Hybridization backgrounds to M13mp9 were subtracted before the synthesis rates were calculated.

The hybridized fraction of total labelled RNA was determined from a plot of hybridized cpm versus cpm in the hybridization vial. Three different RNA concentrations were used. mRNA relative to the spec region is shown. Hybridization per base pair probe is similar in the two experiments for each mRNA.

THE S1 OPERON

S1 is regulated as a typical ribosomal protein during exponential growth and is also a unit protein; i.e. one molecule per ribosome[14]. The synthesis rate of several proteins was followed after the onset of amino acid starvation in the relA strain. S1 synthesis was found to be transiently repressed[12], whereas other ribosomal proteins were derepressed[12; 15], also the amount of ribosomal protein mRNA in the cell was increased[16]. The synthesis of rRNA increases under these conditions and according to the model for ribosomal protein regulation, one would expect the derepression observed for the majority of ribosomal proteins[15], but not the repression of S1. These observations are shown on Figure 2 and led us to study the regulation of S1, because it might have unique molecular mechanisms.

of mRNA degradation and/or processing. Our deletion analysis
shows that the region, containing P3 and P4, can give as
efficient expression of S1 synthesis as the whole region both in
'maxicells' (Fig. 4) and in a coupled protein synthesizing
system. P1+P2 and P3+P4 are found on separate HpaII fragments.
Both these fragments show strong promoter activity in the
expected orientation, when inserted into a promoter-cloning
vehicle, constructed by Dr. T. Atlung in this laboratory. In our
experiment the promoter activity of each separate fragment was
similar to the promoter activity of the complete region.

These results open the possibility of a complicated regulation
of rpsA at the transcriptional level. We are now examining the
effect of different in vivo conditions on the expression of S1
from the promoters we have identified.

Addition of purified S1 to the coupled protein synthesizing
system repressed S1 synthesis relative to the synthesis of the
chloramphenicol resistance protein (Fig. 5), but we have not yet
established which of the two genes is affected by S1.

By cloning rpsA on plasmids and on phage M13 we have measured
rpsA transcription directly and investigated the effect of
increased rpsA dosage. A cell having rpsA on a multicopy plasmid
increases rpsA transcription 40-fold, but S1 translation only
2.4 times[13].

Figure 2 shows rpsA as transiently repressed in the relA
strain during amino acid starvation. At the time of maximal
repression, measurement of the rate of synthesis of rpsA mRNA
in this strain gave no indication of a reduced rpsA transcription
(see Table 2). Thus, S1 is regulated at the translational level
also in this case.

After addition of streptolydigin or rifampicin, the functional
half-life of individual mRNAs was measured[21]. Except in the case
of S1, the same half-lives were observed using one or the other
of the antibiotics. For rpsA mRNA it was 40 sec (stl) and 2.5 min
(rif), and this points to a specific interference with S1
synthesis, most likely at the post-transcriptional level.

Conclusion

Measurements of rates of S1 protein and mRNA synthesis have in
two cases shown the regulation to take place at the translational

level. The gene dosage and the in vitro experiments both indicate that S1 is involved in its own control.

In contrast to our findings for rplJ operon and to the results with other ribosomal operons[2], rpsA regulation does not seem to involve competition between homologous structures on rRNA and rpsA leader mRNA. Thus derepression of rRNA synthesis fails to derepress S1 synthesis. Direct binding of S1 to 16S rRNA has been reported[22], but several studies, most recently[23], seem to show that S1 binds to ribosomes mainly through protein-protein interactions.

The regulation of S1 synthesis may involve binding of S1 to a specific secondary structure on rpsA leader mRNA. The DNA sequence is compatible with several very elaborate structures, but the possible biological relevance of these structures remains to be analyzed. We consider the present work a first step in this direction.

ACKNOWLEDGMENT

This work was supported by Carlsberg Foundation, the NOVO Foundation and the Danish Natural Science Research Council (grants nos. 11-1772 and 11-3114).

REFERENCES

1. Fiil, N.P., Friesen, J.D. and Dennis, P.P. (1980) in Genetics and Evolution of RNA Polymerase, tRNA and Ribosomes, Osawa, S. Ozeki, H., Uchida, H. and Yura, T., eds. University of Tokyo Press. Tokyo. pp. 33-50.
2. Nomura, M., Yates, J.L., Dean, D. and Post, L.E. (1980) Proc. Nat. Acad. Sci. 77, 7084-7088.
3. Zimmermann, R.A. (1980) in RIBOSOMES Structure, Function, and Genetics, Chambliss, G., Craven, G.R., Davies, J., Davis, K., Kahan, L. and Nomura, M., eds., University Park Press, Baltimore, pp. 135-169.
4. Zengel, J.M., Mueckl, D. and Lindahl, L. (1980) Cell 21, 523-535.
5. Dennis, P.P. and Fiil, N.P. (1979) J. Biol. Chem. 254, 7540-7547.
6. Barry, G., Squires, C. and Squires, C.L. (1980) Proc. Nat. Acad. Sci. 77, 3331-3335.
7. Brot, N., Caldwell, P. and Weissbach, H. (1980) Proc. Nat. Acad. Sci. 77, 2592-2595.
8. Fukuda, R. (1980) Mol. Gen. Genet. 178, 483-486.
9. Post, L.E., Strycharz, G.D., Nomura, M., Lewis, H. and Dennis, P.P. (1979) Proc. Nat. Acad. Sci. 76, 1697-1701.

128

10. Noller, H.F. (1980) in RIBOSOMES Structure, Function, and Genetics, Chambliss, G., Craven, G.R., Davies, J., Davis, K., Kahan, L. and Nomura, M., eds., University Park Press, Baltimore, pp. 3-22.
11. Fiil, N.P., Friesen, J.D., Downing, W.L. and Dennis, P.P. (1980) Cell 19, 837-844.
12. Reeh, S., Pedersen, S. and Friesen, J.D. (1976) Mol. Gen. Genet. 149, 279-289.
13. Christiansen, L. and Pedersen, S. (1981) Mol. Gen. Genet. 181, 548-551.
14. Pedersen, S., Bloch, P.L., Reeh, S. and Neidhardt, F.C. (1978) Cell 14, 179-190.
15. Dennis, P.P. and Nomura, M. (1974) Proc. Nat. Acad. Sci. 71, 3819-3823.
16. Dennis, P.P. and Nomura, M. (1975) Nature 255, 460-465.
17. Ono, M., Kuwano, M. and Mizushima, S. (1979) Mol. Gen. Genet. 174, 11-15.
18. Kitakawa, M., Blumenthal, L. and Isono, K. (1980) Mol. Gen. Genet. 180, 343-349.
19. Schnier, J., Kimura, M., Foulaki, K., Subramanian, A.-R., Isono, K. and Wittmann-Liebold, B. (1982) Proc. Nat. Acad. Sci. 79, 1008-1011.
20. Maxam, A.M. and Gilbert, W. (1980). Methods in Enzymology 65, 499-560.
21. Pedersen, S., Reeh, S. and Friesen, J.D. (1978) Mol. Gen. Genet. 166, 329-336.
22. Dahlberg, A.E. and Dahlberg, J.E. (1975) Proc. Nat. Acad. Sci. 72, 2940-2944.
23. Boni, I.V., Zlatkin, I.V. and Budowsky, E.I. (1982) Europ. J. Biochem. 121, 371-376.

USE OF DIPEPTIDE SYNTHESIS TO STUDY THE *IN VITRO* EXPRESSION OF THE Ll0 (β)
OPERON

NIKOLAOS ROBAKIS, YVES CENATIEMPO, SUSAN PEACOCK, NATHAN BROT AND HERBERT
WEISSBACH[+]
[+]Roche Institute of Molecular Biology, Nutley, New Jersey, 07110, USA

INTRODUCTION

Previous studies from this laboratory have described the use of a highly
defined DNA-directed *in vitro* protein synthesis system to study gene expression
in *E. coli*.[1,2] A goal of these studies is to better understand the mechanisms
for regulating the expression of genes involved in transcription and translation. The Ll0 operon (β operon) which contains the genes for ribosomal proteins Ll0 and Ll2, as well as the genes for the β and β' subunits of RNA polymerase[3] (Fig. 1) has been of special interest. *In vitro* and *in vivo*
results[2,4,5] have shown that the ratio of synthesis of Ll0:Ll2:β:β' is
1:4:∿0.1:∿0.1. How the second gene product (Ll2) is overproduced relative to
Ll0 and the mechanism of shift-down for the synthesis of the β and β' subunits
are still not fully understood. However, as summarized in Fig. 1, several

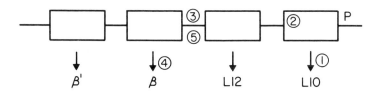

① LlO INHIBITS SYNTHESIS OF LlO.
② PROMOTER FOR Ll2 AND ββ' IN THE LlO GENE.
③ TRANSCRIPTION ATTENUATOR SITE FOR
   ββ' SYNTHESIS.
④ RNA POLYMERASE INHIBITS ββ' SYNTHESIS.
⑤ L FACTOR STIMULATES ββ' SYNTHESIS.

Fig. 1. Transcriptional and translational regulation of the β operon (Ll0
operon). Genes are represented by the protein product. Direction of transcription is from right to left. P, primary promoter.

types of regulation have been described. It is now established that ribosomal protein L10 inhibits its own synthesis (autoregulation) at the level of translation.[6,7] A secondary promoter located within the L10 gene may account for part of the overproduction of L12,[8] and a strong attenuator site in the intracistronic region between the L12 and the β subunit genes is largely responsible for the shift-down in the synthesis of the β and β' subunits.[9] RNA polymerase holoenzyme inhibits the synthesis of the β and β' subunits[2,10,11] and L factor,[12] which is known to modulate transcription pause sites[13], stimulates β,β' synthesis. Thus, both transcriptional and translational mechanisms are involved in the regulation of expression of this operon. Since most of the regulation of gene expression occurs either at the initiation of transcription or translation, it would be advantageous to specifically examine the initial steps in protein synthesis. To accomplish this, we have recently developed a simplified *in vitro* system to study gene expression based on the formation of the first di- or tripeptide of the gene product.[14,15,16] A description of this system and its use to study the regulation of the expression of the genes in the β operon are reported here.

MATERIALS AND METHODS

DNA and RNA templates

    For most of the studies presented here, plasmid pNF1337, which was kindly supplied by Dr. J. Friesen, University of Toronto, was used as template.[17] The bacterial insert in this plasmid is shown in Fig. 2. It contains part of the

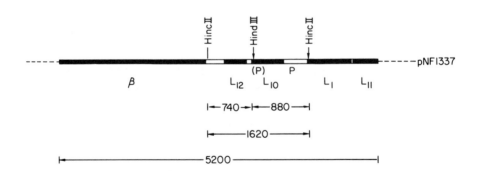

Fig. 2. Partial restriction map of the bacterial DNA inserted into plasmid pBR322. Direction of transcription is from right to left. P, primary promoter; (P), secondary promoter.

L11 gene, the genes for L1, L10, L12 and part of the β gene. Since the promoter required for L11 and L1 expression is lacking, only L10, L12 and a part of the β subunit are synthesized when this segment is expressed. For some of the experiments, pNF1337 was digested with Hinc II endonuclease to yield a 1.6 kbp fragment containing the L10 and L12 genes (Fig. 2). This fragment could be further digested with Hind III endonuclease to yield two fragments.[16] The larger one (880 bp) contains the promoter for the L10 gene and most of the structural gene whereas the smaller piece (740 bp) contains the 3'-end of the L10 gene plus all of the L12 gene. However, since there is no promoter for L12 in this latter fragment, it should not be expressed. mRNA was prepared using the Hinc II and the Hind III restriction fragments as described elsewhere.[16]

Since only a limited number of genes are expressed from pNF1337, it was reasoned that the synthesis of a di- or tripeptide could be used to distinguish the different gene products and also to provide a system to study regulation at the level of initiation of either transcription or translation. Table 1 shows the gene products, the N-terminal amino acid sequences and the codons for the initial tripeptides of the proteins of interest present on pNF1337. It can be seen that the use of di- or tripeptide formation to assay for the products would present a problem only in the case of L12 and β-lactamase synthesis since both proteins begin with fMet-Ser-Ile. As shown below, these gene products can be distinguished by using different tRNA^{Ser} isoacceptor species.

Assays

DNA and mRNA-directed dipeptide synthesis. The incubation conditions used for the *in vitro* synthesis of dipeptides have been described previously.[14] The reaction mixture (35 μl) contained IF-1, IF-2, IF-3, EF-Tu, RNA polymerase, 70S ribosomes, fMet-tRNA$_f^{Met}$ (labeled with [^{35}S] or unlabeled), and either

TABLE 1

THE AMINO TERMINAL TRIPEPTIDES AND NUCLEOTIDE CODING SEQUENCES OF pNF1337 GENE PRODUCTS

Gene Product	NH$_2$-Terminal Sequence	Nucleotide Coding Sequence
Ribosomal protein L10	fMet-Ala-Leu	AUG GCU UUA
Ribosomal protein L12	fMet-Ser-Ile	AUG UCU AUC
β Subunit RNA polymerase	fMet-Val-Tyr	AUG GUU UAC
β-Lactamase	fMet-Ser-Ile	AUG AGU AUU

unfractionated *E. coli* tRNA or an isoacceptor tRNA species (kindly supplied by
Dr. B. Reid, Univ. of Washington, Seattle) charged with the appropriate labeled
amino acid, e.g. Ala-tRNA (protein L10). The reaction was initiated by the
addition of DNA or mRNA and the mixture was incubated at 37°C for 60 min. The
reaction was stopped by the addition of 2.5 µl of 1 N NaOH. After an additional 10 min at 37°C, 20 µg of carrier peptide were added and the mixture was
acidified with 4 µl of 2 N HCl. The precipitated material was removed by
centrifugation and the dipeptide formed was isolated after thin layer chromatography.[14] Under these conditions, using a suitable solvent system and appropriate aminoacyl-tRNA species, the synthesis of two dipeptides could be assayed
in the same incubation. As an example, fMet-Ala and fMet-Ser are well resolved
from each other when the thin layer plate is developed with ethyl acetate,
hexanes, glacial acetic acid (8:3:1, v/v).

Although the above assay was useful since it provided chromatographic
identification of the product it was time consuming and somewhat tedious.
Therefore, a more simple assay was developed based on the extraction of the
dipeptide into ethyl acetate.[16] For this assay the incubations were performed
as described above except that non-labeled fMet-tRNA$_f^{Met}$ was employed. After
the NaOH addition and subsequent incubation, 0.5 ml of 0.5 N HCl was added,
followed by 3 ml of ethyl acetate. The mixture was shaken, centrifuged and an
aliquot of the organic phase which contained the dipeptide was removed and
assayed for radioactivity. For quantitation of the amount of dipeptide synthesized the extraction coefficient for each dipeptide was taken into account.

Synthesis of the tripeptide fMet-Ala-Leu. For the synthesis of this tripeptide, the following modifications of the incubation for dipeptide synthesis
were made: 0.1 µg of EF-G and 10 pmol of [^{3}H]Leu-tRNA$_4^{Leu}$ were added and nonradioactive Ala-tRNA$_3^{Ala}$ was used. The synthesized radioactive tripeptides were
detected and isolated by thin layer chromatography or the ethyl acetate extraction as described above for the assay of dipeptides.

Assay of fMet-tRNA binding to ribosomes. Reaction mixtures (35 µl) contained 50 mM Tris-HCl, pH 7.4, 100 mM NH$_4$Cl, 5 mM Mg-acetate, 2 mM β-mercaptoethanol, 0.4 mM GTP, 10 pmol of f[^{35}S]Met-tRNA$_f^{Met}$, 1.0 µg IF-2, 0.5 µg IF-3, 1
A$_{260}$ unit of 70S ribosomes or 0.5 A$_{260}$ unit of 30S ribosomal subunits and 0.5
µg of Hind III mRNA. The mixture was incubated for 10 min at 37°C, and then
diluted to 0.3 ml with cold buffer containing 50 mM Tris-HCl, pH 7.4, 100 mM
NH$_4$Cl, 10 mM Mg-acetate and 2 mM β-mercaptoethanol. The mixture was rapidly

filtered through a nitrocellulose membrane and the filters were washed three
times with 2 ml of the dilution buffer, dissolved in Bray's scintillation fluid
and counted in a liquid scintillation spectrometer.

RESULTS

Expression of the L10 gene as measured by dipeptide formation

   In order to characterize the system, the initial experiments were designed
to study the *in vitro* expression of *E. coli* ribosomal protein L10 gene by
measuring the synthesis of the first dipeptide fMet-Ala (see Table 1). Using
f[^{35}S]Met-tRNA$_f^{Met}$, [^{3}H]Ala-tRNA and pNF1337 as template the rate of synthesis
of fMet-Ala was determined under the incubation conditions described above and
reported elsewhere.[14]   The radioactive product comigrated on thin layer chromatography with non-labeled chemically synthesized carrier fMet-Ala, and the
product contained stoichiometric amounts of [^{35}S] and [^{3}H] (Fig. 3). Dipeptide

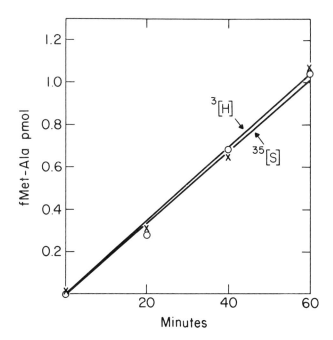

Fig. 3.   Kinetics of fMet-Ala synthesis in the dipeptide system.   Incubation
conditions are described in the text.   From Robakis *et al.*[14]

synthesis was linear for up to 1 hr of incubation and relatively large amounts
of dipeptide were synthesized. These observations suggested that the dipeptide
was being released from the mRNA permitting the mRNA to recycle (see below).
Table 2 shows that there is an absolute requirement for all of the protein
components used in the dipeptide synthesis system with the exception of IF-1.
There was no fMet-Ala synthesized when pBR322 DNA was used as template or when
DNA from plasmid pNF1341, which lacks the coding region for the first 26 amino
acids of L10, was employed. These data provide strong evidence that the
synthesis of fMet-Ala results from the proper transcription and initiation of
translation of the L10 gene.

Use of tRNA isoacceptors to discriminate between the expression of plasmid
genes

In most cases, the dipeptide synthesizing system cannot be used if two gene
products encoded by the same plasmid have an identical initial dipeptide (e.g.
fMet-Ser for L12 and β-lactamase). However, in some instances, dipeptide
synthesis can distinguish the gene products if different tRNA isoacceptor
species are required for the second amino acid. Another solution to the problem is to modify the system and synthesize a tripeptide if the third amino acid
of the gene products is different.

An example of the use of different isoacceptor species is seen in the case
of L12 and β-lactamase.[18] As shown in Table 1, L12 and β-lactamase have the
same initial di- and tripeptides, although the serine codons specify different
tRNA isoacceptor species. The synthesis of the β-lactamase dipeptide should
require the $tRNA^{Ser}_3$ isoacceptor species whereas the L12 dipeptide should need

TABLE 2

DEPENDENCIES FOR THE SYNTHESIS OF fMet-Ala

Omission	fMet-Ala
	pmol
None	1.7
RNA polymerase	0
Ribosomes	0
IF-1	0.7
IF-2	0
IF-3	0
EF-Tu	0
pNF1337 + pNF1341 or pBR322	0

Incubation conditions are described in the text. From Robakis et al.[14]

tRNA$_1^{Ser}$. This difference has been used to distinguish between the synthesis of
the two gene products as shown in Table 3. With pBR322 DNA as template which
contains the β-lactamase gene, but not the Ll2 gene, only tRNA$_3^{Ser}$ was active in
fMet-Ser synthesis. In contrast tRNA$_1^{Ser}$ and not tRNA$_3^{Ser}$ was active when a 1.6
kbp fragment, generated by restriction of pNF1337 DNA with Hinc II endonuclease
was used (see Fig. 2). The latter DNA fragment contains the Ll2 gene. With
pNF1337, each isoacceptor species yielded product since both the β-lactamase
and Ll2 genes are present. These results demonstrate that tRNA$_1^{Ser}$ is the
isoacceptor species used for fMet-Ser synthesis from the Ll2 gene whereas
tRNA$_3^{Ser}$ is used for the formation of fMet-Ser from the β-lactamase gene. A
similar result has been found using the plasmid pJEA4 (derived from pBR322)
which contains the genes for the large subunit (LS) of ribulose bisphosphate
carboxylase and β-lactamase. It has been shown *in vitro* that the synthesis of
fMet-Ser from the LS gene (serine codon UCA) can be distinguished from β-
lactamase expression since the LS fMet-Ser synthesis requires the tRNA$_1^{Ser}$
isoacceptor species.[18]

DNA-directed tripeptide synthesis *in vitro*

As indicated above, an assay was developed for tripeptide synthesis in order
to discriminate between gene products with the same initial dipeptide sequence.
The dipeptide system used in previous experiments was modified to allow the
synthesis of the initial tripeptide of L10, fmet-Ala-Leu, using pNF1337 DNA as
template. As shown schematically in Fig. 4, formation of fMet-Ala-Leu should
require all of the components for dipeptide synthesis in addition to Leu-tRNA
and EF-G. A comparison of di- and tripeptide synthesis *in vitro* is shown in
Table 4. When Leu-tRNA$_4^{Leu}$ and EF-G are omitted, 2 pmol of dipeptide are formed
with no detectable tripeptide. In the presence of Leu-tRNA$_4^{Leu}$ and EF-G, about

TABLE 3

UTILIZATION OF TWO tRNA^{Ser} ISOACCEPTOR SPECIES IN THE DIPEPTIDE ASSAY

| Template | Protein products containing fMet-Ser | Synthesis of fMet-Ser | |
		tRNA$_1^{Ser}$	tRNA$_3^{Ser}$
		pmol	
pBR322	β-lactamase	0	4.0
Hinc II fragment	Ll2	1.2	0
pNF1337	Ll2, β-lactamase	2.4	1.6

The reactions were carried out as described in the text, using 2 μg of plasmid
DNA (pBR322, pNF1337) or 0.2 μg of the 1.6 kbp Hinc II DNA fragment. Adapted
from Cenatiempo et al.[18]

136

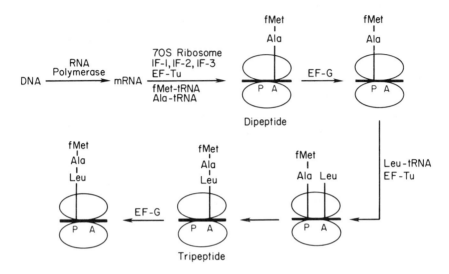

Fig. 4. Scheme of di- and tripeptide formation. From Cenatiempo et al.[15]

0.6 pmol of tripeptide is formed with no significant amount of dipeptide.
Thus, the lower level of tripeptide formed is not due to accumulation of dipeptide in the incubations, but could be attributed to the differences in the
stability of the di- and tripeptidyl-tRNA·ribosome·mRNA complexes. The larger
amount of dipeptide formed may result from the instability of the dipeptidyl-
tRNA·ribosome·mRNA complex permitting the mRNA to recycle and function catalytically.

  It was also noticed that low levels of EF-Tu and EF-G were required to
saturate the system. As little as 0.07 µg of EF-Tu and 0.09 µg of EF-G gave a

TABLE 4

COMPARISON OF DI- AND TRIPEPTIDE SYNTHESIS

	fMet-Ala	fMet-Ala-Leu
	pmol	
Dipeptide system	2.0	0
Tripeptide system	0	0.6

Di- and tripeptides were detected and separated by chromatography as described elsewhere.[16] Other conditions are described in the text. The tripeptide system contained the components required for dipeptide synthesis
plus EF-G and Leu-tRNA$_4^{Leu}$. Taken from Cenatiempo et al.[15]

maximum effect (data not shown). This level of EF-Tu is about 100-fold less than that required in a coupled defined *in vitro* system in which the entire gene product was synthesized.[1,2] The difference in amounts of EF-Tu needed for the two systems appears to be related to the different amount of acylated-tRNA present in the system.

Tripeptide formation as a measure of functional mRNA

If a di- or tripeptide formed a stable complex on the ribosome, the amount of peptide made could be used as a measure of active mRNA present. Some of the data presented above indicated that in the case of dipeptide formation a stable complex was not present, and there was recycling of the mRNA. In contrast, under the conditions used, evidence has been obtained that the tripeptide forms a stable complex on the ribosome and allows, as shown below, quantitation of the amount of active mRNA present. Using an uncoupled *in vitro* system (in which transcription was separated from translation) a marked difference was observed in the kinetics of di- and tripeptide formation. Dipeptide synthesis was linear for at least 60 min after transcription was stopped, whereas tripeptide formation was linear for only 20 min (data not shown). These results can be interpreted as being due to differences in the stability of the di- or tripeptide on the mRNA·ribosome complex. More direct evidence for this has come from filter binding experiments. Since a stable peptidyl-tRNA·ribosome·mRNA complex should be retained by a nitrocellulose filter, the di- and tripeptide incubation mixtures were filtered through a nitrocellulose membrane. The results in Table 5 show that no dipeptide synthesis is detected by a nitrocellulose filter binding assay although significant amounts are measured by thin layer chromatography. In contrast, almost the same amount of tripeptide was found by either the filter binding assay or thin layer chromatography. These results indicate that the dipeptidyl-tRNA·ribosome·mRNA complex is extremely labile compared to the tripeptidyl-tRNA·ribosome·mRNA complex. In

TABLE 5

DIPEPTIDE AND TRIPEPTIDE SYNTHESIS BY DIFFERENT ASSAYS

	fMet-Ala	fMet-Ala-Leu
	pmol	
Thin layer	2.5	0.9
Nitrocellulose filter	<0.1	0.8

The assay systems and incubation conditions are described in the text. Adapted from Cenatiempo *et al.*[15]

order to prove that the amount of tripeptide formed could be used to measure the amount of functional mRNA present, it was necessary to study a single mRNA species. A Hind III restriction fragment (see Fig. 2) was used to direct the synthesis of mRNA that should consist of only L10 mRNA. The level of functional mRNA was quantitated by measuring the amount of fMet-tRNA$_f^{Met}$ bound to the ribosome·mRNA complex since this mRNA should contain only one initiation site. Fig. 5 shows the effect of mRNA on fMet-tRNA$_f^{Met}$ binding and tripeptide synthesis. It is seen that the amount of tripeptide synthesized from this mRNA template is equal to the amount of fMet-tRNA$_f^{Met}$ bound. These results indicate that tripeptide formation can be used as a valid and rapid measure of functional mRNA. It should be emphasized that the tripeptide assay provides a new way to distinguish a specific active mRNA in a mixture of mRNA species.

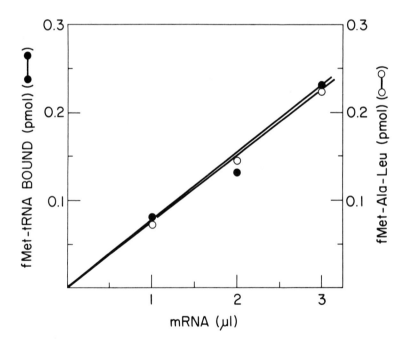

Fig. 5. fMet-tRNA binding of 70S ribosomes and tripeptide formation in the presence of various amounts of L10 mRNA. The mRNA (1.2 pmol/μl) was prepared from the Hind III restriction digest of the 1.6 kbp DNA fragment (see text and Fig. 2). Details of the incubation mixtures and binding experiments are described elsewhere.[15,16] From Cenatiempo et al.[15]

Use of dipeptide formation to study the regulation of L10 synthesis

Effect of guanosine-5'-diphosphate-3'-diphosphate (ppGpp) on *in vitro* dipeptide synthesis. The nucleotide ppGpp which accumulates when cells are starved for an amino acid is known to inhibit ribosomal protein synthesis both *in vivo* and *in vitro*.[19] The effect of ppGpp on the expression of the L10 gene, as measured by dipeptide formation, was determined. When 100 µM ppGpp is present in the incubation mixtures with pNF1337 DNA as template, a 70% inhibition of fMet-Ala synthesis is observed (Fig. 6) whereas GDP has no effect (data not shown). In other experiments ppGpp was found to inhibit the expression of the L12 and β-lactamase genes, i.e. fMet-Ser formation, directed by pNF1337.[18]

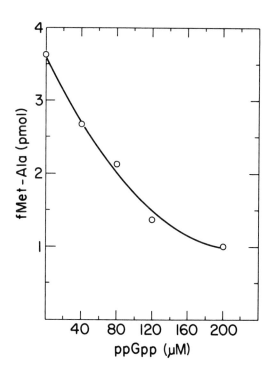

Fig. 6. Effect of ppGpp on the synthesis of the L10 N-terminal dipeptide, fMet-Ala. pNF1337 DNA was used as template. Taken from Robakis et al.[14]

140

Autoregulation of L10 synthesis by L10. Previous studies have demonstrated
that the *in vitro* synthesis of ribosomal protein L10 is autogenously inhibited
by exogenous L10 at the level of translation,[6] although the site of inhibition
has not been determined. In an attempt to pinpoint the site of L10 regulation,
we have studied the effect of L10 on the DNA- and mRNA-dependent synthesis of
fMet-Ala. As shown in Figure 7, using pNF1337 DNA, 150-200 pmoles of L10
inhibits the synthesis of fMet-Ala by about 60-70%. This effect was shown to
be specific since ribosomal protein L12 had no effect on the synthesis of fMet-
Ala and L10 did not inhibit fMet-Ser formation (first dipeptide of L12 and β-
lactamase).[14] These results show that the L10 synthesis is regulated by a step
prior to the formation of the first peptide bond. Because the inhibition by
L10 is very specific, it is unlikely that L10 interferes with either the EF-Tu-
dependent binding of Ala-tRNA to the ribosome or the subsequent peptidyl trans-
ferase reaction. The most likely explanation is that L10 exerts its effect at
the level of initiation of translation. We have recently investigated the
effect of L10 on the formation of the fMet-tRNA$_f^{Met}$·70S and 30S initiation

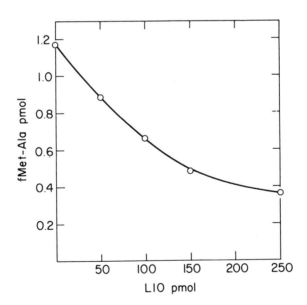

Fig. 7. Effect of L10 on fMet-Ala synthesis. For details, see text. From
Robakis *et al.*[14]

complexes using mRNA prepared by the transcription of a Hind III restriction
digest of the 1.6 kbp Hinc II DNA fragment (see Fig. 2). This mRNA preparation
directs the synthesis of fMet-Ala but not fMet-Ser indicating that only the
mRNA coding for L10 is functional. With this mRNA, it was found that L10
inhibits the formation of both the 70S (data not shown) and 30S initiation
complexes (Fig. 8). Also, L10 has no effect on initiation complex formation
when MS2 RNA is used as template (Fig. 8). These results indicate that L10
inhibits the formation of the 30S initiation complex on the L10 mRNA, and are
consistent with recent reports by Nomura and coworkers[20-22] and Fukuda.[7] They
suggested that the leader region of the mRNA is involved in the autogenous

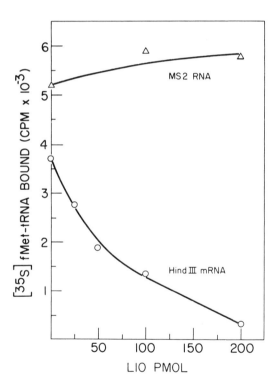

Fig. 8. Effect of L10 on the formation of the 30S initiation complex. The
mRNA was prepared from the Hind III restriction digest of the 1.6 kbp DNA fragment. Details of the incubation mixture, mRNA preparation and 30S complex
formation are described in the text and elsewhere.[16]

control of the synthesis of ribosomal proteins S4, S7, L1, and L10. In support of this suggestion is the observation that the binding sites for S4 and S7 on 16S RNA and L1 on 23S RNA have a striking homology to nucleotide sequences at, or close to, the initiation codon of their respective mRNAs.[22] Studies are in progress to further define the site on the mRNA where L10 interacts.

Studies on β subunit synthesis

Dipeptide synthesis has also been used to investigate the expression of the β subunit gene. Previous studies have shown that the low expression of the β and β' genes, relative to the expression of the L10 and L12 genes, was primarily due to a strong attenuator site located in the intracistronic region between the L12 and β genes.[9] In addition, the synthesis of the β subunit is stimulated by L factor, a 68,000 dalton protein that has been identified as the nus A gene product.[23] Since L factor is known to modulate RNA polymerase activity at pause sites,[13] it seemed reasonable that L factor could function by altering the extent of attenuation in the L12-β intracistronic region. RNA polymerase holoenzyme has also been shown to inhibit β subunit synthesis.[2,10,11] There is some evidence that this effect is at the level of translation[10] although the previous results were not conclusive. As seen in Table 1, the initial dipeptide in the β subunit is fMet-Val. Figure 9 shows the effect of pNF1337 on the synthesis of fMet-Val in the dipeptide system. About 2 μg of

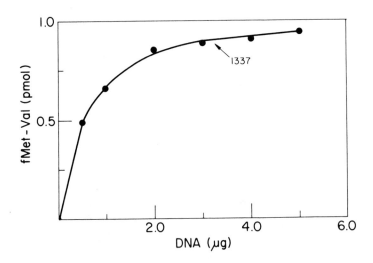

Fig. 9. Effect of pNF1337 DNA concentration on the synthesis of fMet-Val. Incubation conditions are described in the text. From Peacock et al.[24]

DNA saturated the reaction and the *in vitro* synthesis was linear for at least 40 min.

Effect of RNA polymerase and L factor on fMet-Val synthesis. Dipeptide synthesis was used to determine whether RNA polymerase and L factor exerted their effects before the formation of the first dipeptide. Figure 10 shows the effect of RNA polymerase on the synthesis of fMet-Val using pNF1337 as template. At levels of RNA polymerase above 10 µg per incubation significant inhibition (∿50%) of fMet-Val synthesis is observed. λrif^d18 DNA, which also contains the L10 (β) operon, could also be used as template for fMet-Val synthesis in these experiments and a similar effect of RNA polymerase holoenzyme was obtained. As also seen in the figure, there is no effect of RNA polymerase on fMet-Ser synthesis (L12) directed by pNF1337 indicating that the inhibitory effect of higher levels of RNA polymerase holoenzyme is specific.

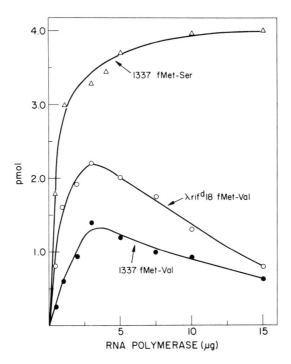

Fig. 10. Effect of RNA polymerase concentration on fMet-Val and fMet-Ser synthesis using λrif^d18 DNA or pNF1337 as template. Conditions are described in the text. From Peacock *et al.*[24]

Figure 11 shows the effect of L factor on fMet-Val and fMet-Ser synthesis
using pNF1337. L factor (0.1-0.3 µg) stimulates β subunit synthesis by 2-3
fold but has no effect on fMet-Ser formation from the same DNA template.

By using an uncoupled system (to separate transcription from translation) an
attempt was made to determine whether RNA polymerase and L factor exerted their
effects at the transcription or translation level of protein synthesis. The
results are summarized in Fig. 12. When RNA polymerase was added either before
transcription or at the beginning of translation, the same extent of inhibition
of fMet-Val synthesis was observed. These results indicate that RNA polymerase
is inhibiting translation before the synthesis of the first dipeptide. In
contrast, L factor stimulates fMet-Val synthesis only if it is added before
transcription indicating that L factor stimulates the transcription of the β
subunit gene.

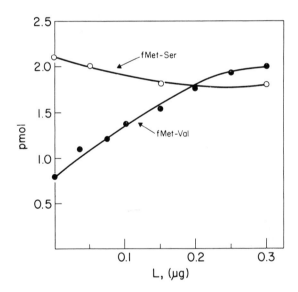

Fig. 11. Effect of L factor on fMet-Val and fMet-Ser synthesis using pNF1337
as template. Incubation conditions are described in the text. From Peacock
et al.[24]

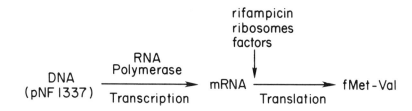

Fig. 12. Effect of RNA polymerase and L factor on fMet-Val synthesis using
an uncoupled system. For details, see text.

SUMMARY

   The expression of the genes on the L10 (β) operon has been investigated in a
defined *in vitro* system by measuring the formation of the initial dipeptide of
the gene products. Dipeptide synthesis provides a rapid, gene specific and
quantitative assay to measure gene expression and is ideally suited to studies
on regulation. Using fMet-Ala as a measure of the expression of the L10 gene,
it has been shown that L10 inhibits its own synthesis (autogenous regulation)
before the formation of the 30S initiation complex. The regulation of β
subunit synthesis has also been examined by dipeptide formation (fMet-Val).
The synthesis of the β subunit is autogenously regulated by RNA polymerase
holoenzyme and is stimulated by L factor. RNA polymerase holoenzyme inhibits
fMet-Val synthesis at the level of translation, whereas L factor stimulates
transcription of the β subunit gene. Gene products with similar N-terminal
dipeptides have been distinguished by using different tRNA isoacceptor species.
The system has also been modified to synthesize tripeptides. Although under
the conditions used the dipeptidyl-tRNA·mRNA·ribosome complex readily dissociates, the corresponding tripeptide complex is stable. Tripeptide formation can
therefore be used in this system as a measure of the amount of active mRNA
present.

146

REFERENCES

1. Kung, H.F., Redfield, B., Treadwell, B.V., Eskin, B., Spears, C. and Weisssbach, H. (1977) J. Biol. Chem. 252, 6889-6894.
2. Zarucki-Schulz, T., Jerez, C., Goldberg, G., Kung, H.F., Huang, K.H., Brot, N. and Weissbach, H. (1979) Proc. Natl. Acad. Sci. USA 76, 6115-6119.
3. Post, L.E., Strycharz, G.D., Nomura, M., Lewis, H. and Dennis, P.P. (1979) Proc. Natl. Acad. Sci. USA 76, 1697-1701.
4. Dennis, P.P. and Fiil, N.P. (1979) J. Biol. Chem. 254, 7540-7547.
5. Dennis, P.P. (1977) J. Mol. Biol. 115, 603-625.
6. Brot, N., Caldwell, P. and Weissbach, H. (1980) Proc. Natl. Acad. Sci. USA 77, 2592-2595.
7. Fukuda, R. (1980) Mol. Gen. Genet. 178, 483-486.
8. Goldberg, G., Zarucki-Schulz, T., Caldwell, P., Weissbach, H. and Brot, N. (1979) Biochem. Biophys. Res. Commun. 91, 1453-1461.
9. Barry, G., Squires, C. and Squires, C.L. (1980) Proc. Natl. Acad. Sci. USA 77, 3331-3335.
10. Kajitani, M., Fukuda, R. and Ishihama, A. (1980) Mol. Gen. Genet. 179, 489-496.
11. Lang-Yang, H. and Zubay, G. (1981) Mol. Gen. Genet. 183, 514-517.
12. Kung, H.-F., Spears, C. and Weissbach, H. (1975) J. Biol. Chem. 250, 1556-1562.
13. Kassavetis, G.A. and Chamberlin, M.J. (1981) J. Biol. Chem. 256, 2777-2786.
14. Robakis, N., Meza-Basso, L., Brot, N. and Weissbach, H. (1981) Proc. Natl. Acad. Sci. USA 78, 4261-4264.
15. Cenatiempo, Y., Robakis, N., Reid, B.R., Weissbach, H. and Brot, N., submitted for publication.
16. Robakis, N., Cenatiempo, Y., Meza-Basso, L., Brot, N. and Weissbach, H., submitted for publication.
17. Fiil, N.P., Bendick, D., Collin, J. and Friesen, J.D. (1979) Mol. Gen. Genet. 173, 39-50.
18. Cenatiempo, Y., Robakis, N., Meza-Basso, L., Brot, N., Weissbach, H. and Reid, B.R. (1982) Proc. Natl. Acad. Sci. USA 79, 1466-1468.
19. Gallant, J.A. (1979) Ann. Rev. Genet. 13, 393-415.
20. Yates, J.L., Arfsten, A.E. and Nomura, M. (1980) Proc. Natl. Acad. Sci. USA 77, 1837-1841.
21. Dean, D., Yates, J.L. and Nomura, M. (1981) Cell 24, 413-420.
22. Nomura, M., Yates, J.L., Dean, D. and Post, L.E. (1980) Proc. Natl. Acad. Sci. USA 77, 7084-7088.
23. Greenblatt, J., Li, J., Adhya, S., Friedman, D.I., Baron, L.S., Redfield, B., Kung, H.F. and Weissbach, H. (1980) Proc. Natl. Acad. Sci. USA 77, 1991-1994.
24. Peacock, S., Cenatiempo, Y., Robakis, N., Brot, N. and Weissbach, H., unpublished results.

THE STRUCTURE AND EXPRESSION OF INITIATION FACTOR GENES

John W. B. Hershey*, J. Greg Howe*, Jacqueline A. Plumbridge†, Mathias Springer†
and Marianne Grunberg-Manago†
*Department of Biological Chemistry, School of Medicine, University of Cali-
fornia, Davis, CA 95616, USA; †Institut de Biologie Physico-Chimique 13, rue
Pierre et Marie Curie, 75005 Paris, France

INTRODUCTION

The synthesis and levels of ribosomal components and elongation factors are coordinately regulated in exponentially growing bacteria.[1,2] The molar amounts of these 55 proteins and 3 RNAs are stoichiometric except for L7/L12, which occur in 4 copies per ribosome, and EF-Tu, which is present in a 5 to 7-fold molar excess. In addition, the number of ribosomes per cell genome varies in proportion to the growth rate (μ), except in very slowly growing cells. Since degradation appears to play no role in establishing the levels of these proteins and RNAs, synthesis rates must be regulated to provide for stoichio-metric expression and sensitivity to growth rate. The absolute synthesis rates for ribosomal RNAs and proteins are proportional to the square of the growth rate (μ^2). For ribosomal RNA operons, this means that transcription becomes relatively more efficient compared to other operons as the growth rate increases, a phenomenon called metabolic control. For many ribosomal protein operons, relative transcription rates are not more efficient in fast growing cells; the absolute rates of mRNA synthesis are proportional to μ, not μ^2, and thus resemble the bulk of cellular gene expression.[3] Instead, these ribosomal protein operons are regulated at the translational level by a feedback mechanism, called autogenous regulation.[4] When a protein product of the operon is present in molar excess of ribosomal RNA, the protein binds to its own mRNA and inhibits its translation. Autogenous regulation provides an explanation for how ribosomal RNA and protein levels are coordinated stoichiometrically, but we do not yet understand the molecular mechanism of metabolic control of ribosomal RNA operons.

In addition to the translational components above, there are three proteins, called initiation factors IF1, IF2, and IF3, which are uniquely involved in the initiation phase of protein synthesis.[5] IF1 (8,016) and IF3 (20,668) occur as single molecular forms in crude lysates of E. coli strain MRE600,[6] although in some strains a second, shorter form of IF3 has been observed. IF2 occurs in cells in at least two molecular weight forms[6], IF2α (115,000) and IF2β (90,000), which are closely related in primary sequence. We are interested in determining how initiation factor levels relate to one another and to ribosomal levels, and how their genes are organized and expressed. Since initiation factor genes have not been found in the ribosomal protein operons so far studied, any coordinate expression must presumably involve different operons. Regulation of initiation factor levels is also interesting because their levels may be of importance in controlling protein synthesis. For example, mRNAs with extensive Shine/Dalgarno complementarity to 16S rRNA do not exhibit so strong a requirement for IF3 as do mRNAs with little complementarity[7,8]. Therefore, if IF3 levels were to change, there could be a shift in translation to a different class of mRNAs. We describe here the determination of initiation factor levels in cells growing in different physiological states, the cloning of the genes for IF2 and IF3, and beginning studies of their expression. Our preliminary results suggest that initiation factor genes may be regulated by metabolic control mechanisms, not by autogenous regulation.

DETERMINATION OF INITIATION FACTOR LEVELS

We sought methods for quantitating amounts of initiation factors which would accurately reflect their levels in intact cells. Procedures which require fractionation of cell extracts prior to analysis introduce ambiguities in the determinations, since recoveries of factors are difficult to estimate. Methods which rely on initiation factor activities as determined by in vitro assays also are unsatisfactory since other components in the cell extracts may influence the assays. Over the past few years we have developed two immunochemical methods and a two-dimensional gel procedure which are suitable for measuring protein levels in crude cell lysates.

Radioimmunoassay (RIA). The procedure utilizes monospecific rabbit antisera against IF1, IF2 and IF3, highly purified and radio-labeled initiation factors, and the classical methodologies of the RIA.[9] Standard displacement curves are sensitive in the range of 30-300 ng factor and analyses of crude cell lysates gave linear results for all three factors.

Immunoblotting. Using the same antisera above, we developed a quantitative method which involves fractionation of cell lysate proteins in SDS/polyacryl-amide gels, electrotransfer to nitrocellulose sheets, treatment with specific antisera, and quantitation of the antibodies attached to the nitrocellulose by using ^{125}I-labeled second antibody.[10] Immunoblotting enjoys three major advan-tages over the RIA: it is more rapid and simpler to execute; it is more sensitive by an order of magnitude, measuring as little as 1 to 3 ng protein; and it is capable of distinguishing multiple forms of the initiation factors.

NEPHGE/SDS PAGE. A totally independent method also was developed which utilizes the two-dimensional gel procedure of O'Farrell et al.[11] to separate initiation factors from other proteins in cell lysates. A similar procedure has been used by Neidhardt and coworkers[12] to analyze a large number of cell proteins, especially those involved in translation. The method involves growing the bacteria in the presence of [35S]sulfate, separating proteins by NEPHGE/SDS PAGE, excising identified spots, and determining their radio-activity. The advantage of the method is that absolute levels of many different proteins can be determined from a single gel if their location on the gel and their sulfur content are known. This allows us to study the initiation factors and compare their levels with those of other proteins involved in protein synthesis. A disadvantage is that IF1 and IF2β are not resolved from other proteins and therefore cannot be measured by this method.

The three methods above have given comparable results, which are summarized as follows. The three initiation factors are found in approximately equimolar amounts in cells growing exponentially in enriched medium, although IF1 levels may be somewhat lower than the levels of the other two factors. The molar ratio of factors to ribosomes is in the range 0.15 to 0.2. This indicates that the absolute levels of initiation factors differ from those of ribosomal proteins and elongation factors. Since there are about 30,000 ribosomes per genome in such cells[13], there must be about 5,000 molecules of each of the initiation factors per genome. It appears that the number of initiation factors exceeds the number of native ribosomal subunits by about 2-fold.

INITIATION FACTOR LEVELS VS. GROWTH RATE

We asked whether initiation factor levels remain constant as a function of growth rate, like most other proteins, or whether they increase along with ribosomes as the growth rate increases. Different rates of balanced growth were achieved by using a defined minimal medium supplemented with [35S]sulfate and acetate (k = 0.15); glycerol (k = 0.60); glucose (k = 0.90); or glucose plus amino acids (k = 1.28). Cells were grown in exponential phase for at least 10 generations, harvested in exponential phase, and analyzed by the RIA or by NEPHGE/SDS PAGE. Initiation factor and ribosome levels are plotted as a function of growth rate, as shown in Figure 1. The data show that initiation factor levels are coordinately controlled and rise as a function of increasing growth rate. The factor curves are similar to that for ribosomal RNA but are somewhat less steep.   The results suggest that initiation factors may be under metabolic control.

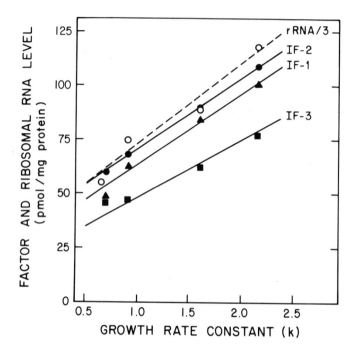

Fig. 1.   Initiation factor levels as a function of cell growth rate.

INITIATION FACTOR GENE STRUCTURE

A mutant strain of E. coli with a thermosensitive IF3 has been identified. The strain was screened by a laborious in vitro assay procedure from a collection of nitrosoguanidine mutants enriched for thermosensitivity by the tritiated amino acid suicide method[14].   Using this mutant we were able to isolate a lambda transducing phage (λp2) which complements the thermosensitivity.[15]   The λp2 phage expresses five E. coli genes after infection of U.V.-irradiated lysogenic cells: the infC gene for IF3; pheS and pheT for the small and large subunits of phenylalanyl-tRNA synthetase; thrS for threonyl-tRNA synthetase; and an unidentified gene which codes for a 12,000 dalton protein, p12.   These genes map at 38 minutes on the E. coli genome in the order: thrS infC p12 pheS pheT.[16]   They were cloned into a plasmid, pB1, for further characterization (see Figure 2).   Analysis of the direction of transcription of this cluster of genes showed that all 5 genes are expressed

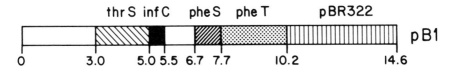

Fig. 2. Genetic map of the infC-carrying plasmid, pB1.

thrS to pheT.[17] Expression of the synthetase genes is described elsewhere in this volume. By studying cloned restriction enzyme fragments of λp2 or pB1, we showed that infC expression is independent of both thrS and the pheS-pheT operon.[18] Thus the IF3 gene appears to possess its own promoter and may exist alone or possibly with p12 in the operon. The DNA sequence of infC and surrounding regions has been determined,[19] and compares well with the amino acid sequence for IF3. Most noteworthy is the fact that the initiation codon is AUU. A strong Shine/Dalgarno sequence, GGAGG, exists upsteam from the AUU.

No mutations in the genes for IF1 (infA) or IF2 (infB) have ever been identified. In order to clone these genes we devised a novel method which does not rely on genetic complementation of a mutation. We argued that strains carrying multiple copies of such genes might be identified by their overproduction of the initiation factor. The quantitative immunoblotting procedure was used to screen about 300 cloned cell lines from a cosmid library of total E. coli DNA for overproduction of the three factors. We found two clones which overproduced IF2 and one clone which overproduced IF3.[20] The DNA in the cosmid of the IF3-overproducing clone also expresses the threonine and phenylalanine synthetase genes and gives the same restriction map as the λp2 or pB1 DNAs, thus providing independent confirmation that the region around the infC gene as analyzed above reflects that in the genome. A higher variability in the levels of IF1 prevented identification of possible overproducers of this factor. Proof that the IF2 clones contained the infB gene in the cosmids was obtained by identification of IF2 synthesized in the "maxicell" assay of Sancar et al.[21] A 4.8 kb HindIII-BamH1 fragment carrying infB (and nusA) was subcloned and inserted into an integration deficient recombinant lambda phage. The point of lambda lysogenization was mapped on the E. coli chromosome at 68 min, very close to argG, nusA, rpsO and pnp. A restriction map of one of the IF2 cosmids is shown in Figure 3, along with an indication of which fragments code for the genes analyzed. From the map positions of IF2 and IF3, it is obvious that the infB and infC genes are widely separated on the genome. Neither the location of the infA gene for IF1 nor the direction of transcription and operon structure for infB have been determined yet.

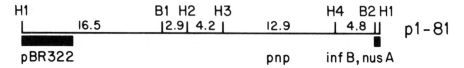

Fig. 3.   Restriction map of the   cosmid p1-81 carrying the infB gene.   The
figure shows the sites of cleavage of the restriction enzymes HindIII (H) and
BamH1 (B); the numbers indicate the lengths of the fragments in kilobase pairs.
Genes are denoted below the fragments in which they occur.   Plasmid pBR322 DNA
is shown by the solid horizontal bars.

INITIATION FACTOR GENE EXPRESSION

From the evidence cited above, we observe that both IF2 and IF3 are
overproduced in strains carrying their respective genes in plasmids.   This
suggests that initiation factor levels are not determined by autogenous
regulatory mechanisms but rather are sensitive to gene dosage.   The quantitative effects of infC gene dosage on IF3 levels was measured directly.
Strains haploid, diploid, triploid and polyploid for infC were grown and
analyzed by immunoblotting with anti-IF3.[22]   Figure 4 shows that the level of
IF3 is roughly proportional to the number of infC genes.   Similarly, in strains

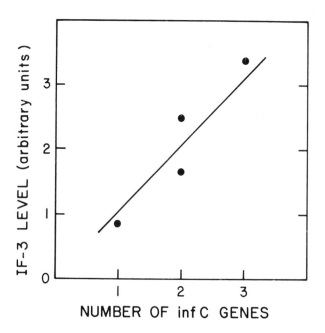

Fig. 4.   IF3 levels as a function of infC gene dosage.   The data is taken from
experiments described by Lestienne et al.[22].

carrying about 5 infB genes, IF2 is overproduced to a similar extent, but more precise studies have not yet been carried out. The synthesis of IF3 in a DNA-coupled cell-free transcription-translation system was not inhibited by the addition of a large excess of exogenous IF3.[22] The results argue strongly that IF3 does not specifically inhibit its own synthesis.

In many strains of E. coli, more than one molecular weight form of IF2 and IF3 have been observed.[6] For IF3, two forms have been purified and sequenced;[23] the shorter form, IF3β (Mr = 19,997) lacks a hexapeptide from the N-terminus of the larger form, IF3α (Mr = 20,668). We can ask whether these closely related proteins are coded by different genes, are produced by differential transcription and translation from a single gene, or are converted post-translationally to two forms by limited proteolysis. Analysis of lysates prepared from freshly grown cells in SDS buffer showed only the IF3α form, either from haploid or polyploid strains.[22] In contrast, if the lysates were made in non-denaturing buffer and then analyzed, both IF3α and IF3β forms were detected. In the cell-free synthesizing system for IF3, IF3α levels increase more rapidly than IF3β levels, but then decrease following cessation of protein synthesis while IF3β levels continue to increase. These results are consistent with the view that IF3β is produced by proteolytic cleavage of IF3α. The cleavage does not appear to occur in exponentially growing cells but rather is an artifact of analysis following cell lysis.

Multiple forms of IF2 also have been reported. IF2α (Mr = 115,000) and IF2β (Mr = 90,000) have been purified and are closely related in primary structure and immunochemical crossreactivity.[5] However, in the case of IF2, both forms are detected in fresh lysates made with SDS buffers,[6] suggesting that IF2α and IF2β are present in about equal amounts in intact cells. Lysates made with non-denaturing buffers show even more forms of IF2, the additional forms likely being artifacts of sample preparation. The limited cloning results described above indicate that both IF2 forms are coded by the same gene. In the strain which carries the 4.8 kb HindIII-BamH1 fragment, both IF2α and IF2β are overproduced. Since the gene for nusA (a 78,000 dalton protein) also is present on the fragment, there is not enough DNA to code for all three proteins. It appears likely that IF2α is cleaved post-translationally to produce IF2β in vivo, but differential transcription or translation of the infB gene cannot be ruled out at this time.

Initiation factor genes are coordinately expressed and mimic ribosomal RNA genes, and yet are located in widely separated operons. We are interested in elucidating what kinds of mechanisms control their expression. Are the genes under the same metabolic controls as the rrn genes and if so, what are the common features which explain their similar behavior. Are the initiation factors under stringent control? Preliminary experiments[22] show that IF3 synthesis in vitro is inhibited by ppGpp whereas Reeh et al.[24] report that IF2β synthesis is not under stringent control. Experiments are in progess to measure the rates of initiation factor synthesis and degradation in cells growing in different physiological states. We plan to investigate further the control of the infB and infC promoters by recombinant DNA techniques and to clone and map the IF1 gene. We hope to explain how initiation factor gene expression is coordinated with that of other translational components and to elucidate the fundamental mechanisms involved.

ACKNOWLEDGEMENTS

We thank Beverley Haskins for expert typing of the manuscript. This work was supported in part by grants from the American Cancer Society (to J.H.) and from the C.N.R.S. (to M.G.M.).

REFERENCES

1. Maaloe, O. (1979) in Biological Regulation and Development, Goldberger, R. E., ed., Vol. 1, Plenum Press, New York, pp. 487-542.
2. Gausing, K. (1980) in Ribosomes: Structure, Function and Genetics Chambliss, G., Craven, G. R., Davies, J., Davis, K., Kahan, L. and Nomura, M. eds., University Park Press, Baltimore, pp. 693-718.
3. Miura, A., Krueger, J. H., Itoh, S., deBoer, H. A., and Nomura, M. (1981) Cell 25, 773-782.
4. Nomura, M. and Post, L. E. (1980) in Ribosomes: Structure, Function and Genetics, Chambliss, G., Craven, G. R., Davies, J., Davis, K., Kahan, L. and Nomura, M. eds., University Park Press, Baltimore, pp. 671-691.
5. Grunberg-Manago, M. (1980) in Ribosomes: Structure, Function and Genetics Chambliss, G., Craven, R., Davies, J., Davis, K., Kahan, L. and Nomura, M. eds., University Park Press, Baltimore, pp. 445-477.
6. Howe, J. G. and Hershey, J. W. B. (1982) Arch. Biochem. Biophys. 214, 446-451.
7. Benne, R., and Pouwels, P. H. (1975) Molec. gen. Genet. 139, 311-319.
8. McLaughlin, J. R., Murray, C. L., and Rabinowitz, J. C. (1981) Proc. Natl. Acad. Sci. USA 78, 4912-4916.
9. Howe, J. G., Yanov, J., Meyer, L., Johnston, K., and Hershey, J. W. B. (1978) Arch. Biochem. Biophys. 191, 813-820.
10. Howe, J. G., and Hershey, J. W. B. (1981) J. Biol. Chem. 257, 12836-12839.
11. O'Farrell, P. Z. O., Goodman, H. M., and O'Farrell, P. H. (1977) Cell 12, 1133-1142.
12. Pedersen, S., Block, P. L. Reeh, S., and Neidhardt, F. C. (1978) Cell 14, 179-190.
13. Kjeldgaard, N. O. and Gausing, K., (1974) in Ribosomes, Nomura, M., Tissieres, A. and Lengyel, P. eds., Cold Spring Harbor Laboratory, New York, pp. 369-392.

14. Springer, M., Graffe, M. and Grunberg-Manago, M. (1977) Molec. gen. Genet. 151, 17-26.
15. Springer, M., Graffe, M. and Hennecke, H. (1977) Proc. Natl. Acad. Sci. USA 74, 3970-3974.
16. Springer, M., Graffe, M. and Grunberg-Manago, M. (1979) Molec. gen. Genet. 169, 337-343.
17. Plumbridge, J. A. and Springer, M. (1980) J. Mol. Biol. 144, 595-600.
18. Springer, M., Plumbridge, J. A., Trudel, M., Graffe, M. and Grunberg-Manago, M. (1982) Molec. gen Genet., in press.
19. Sacerdot, C., Fayat, G., Springer, M., Plumbridge, J. A., Grunberg-Manago, M. and Blanquet. S. (1982) EMBO J., in press.
20. Plumbridge, J. A., Howe, J. G., Springer, M., Touati-Schwartz, D., Hershey, J. W. B. and Grunberg-Manago, M. (1982) Proc. Natl. Acad Sci. USA, in press.
21. Sancar, A., Hach, A. M. and Rupp, W. D. (1979) J. Bacteriol. 137, 692-693.
22. Lestienne, P., Dondon, J. Plumbridge, J. A., Howe, J. G., Mayaux, J.-F., Springer, M., Blanquet, S., Hershey, J. W. B. and Grunberg-Manago, M. (1982) Eur. J. Biochem. 123, 483-488.
23. Brauer, D. and Wittmann-Liebold, B. (1977) FEBS Lett. 79, 269-275.
24. Reeh, S., Pedersen, S. and Friesen, J. D. (1976) Molec. gen. Genet. 116, 192-198.

REGULATION OF THE EXPRESSION OF *tufA* AND *tufB*, THE TWO GENES ENCODING THE
ELONGATION FACTOR EF-Tu IN *E. COLI*

LEENDERT BOSCH AND PETER H. VAN DER MEIDE
Department of Biochemistry, State University of Leiden, Wassenaarseweg 64,
2333 AL Leiden, The Netherlands

INTRODUCTION

The elongation factor EF-Tu fulfils an essential function in the bacterial
cell: it mediates the binding of aminoacyl-tRNA to the ribosomes during protein
biosynthesis.[1,2] In this context it appears appropriate that the intracellular
concentrations of EF-Tu and aminoacyl-tRNA are about the same.[3,4] It implicates that EF-Tu is one of the most abundant proteins in the bacterial cell
(ref. 4-7 and this paper).

In *E. coli* two genes, designated *tufA* and *tufB*, code for EF-Tu. *TufA* is
located at 73 min on the *E. coli* linkage map[8] and is the terminal gene of the
str ribosomal protein operon.[9,10] The four genes coding for S12, S7, EF-G and
EF-Tu are transcribed from a single promoter.[9] *TufB* lies near 88 min[8] and is
cotranscribed with four upstream tRNA genes.[11-13] The nucleotide sequences
of *tufA* and *tufB* differ at 13 positions only[14,15] and the corresponding gene
products EF-TuA and EF-TuB are identical except for the C-terminal amino acid
residue.[16,17] No functional differences between the two proteins have been
reported.[36] The question therefore arises how the expression of *tufA* and *tufB*
is regulated. In particular the arrangement of *tufB* in an operon which is
transcribed into structural (tRNA) and informational RNA[11-13] poses an interesting problem.[13]

We have studied the regulation of the expression of the two *tuf* genes by
determining the intracellular concentrations of EF-TuA and EF-TuB.[18] Due to the
great similarity in structure of these proteins we have taken advantage of
structural differences caused by mutations. A prerequisite of this approach is
that these mutations do not alter the regulation of the expression of the
genes. Mutations of this type have been described[18-23] and analyses of the
intracellular *tuf* products revealed a constant ratio between the synthesis
rates of EF-TuA and EF-TuB[23] and between the intracellular concentrations of
these proteins (ref. 18 and this paper) at all growth rates studied. Here we
address the question what mechanism underlies this coordinate expression of
tufA and *tufB*. Our results indicate that the EF-Tu protein itself is involved
in the expression of *tufB*, presumably at a posttranscriptional level.

MATERIALS AND METHODS

EF-Tu·GDP and EF-Ts were isolated by affinity chromatography.[24] Antibodies against EF-Tu were prepared according to Van de Klundert[25] and antibodies against EF-Ts according to Carroll *et al.*[26] For details and minor modifications see ref. 18 and 27. These references should also be consulted for the determination of intracellular EF-Tu and EF-Ts concentrations by rocket immuno-electrophoresis and for the separation of EF-TuA and EF-TuB$_O$ by isoelectric focusing.

Bacterial strains, plasmids and media

The genotypes and origins of the bacterial strains are listed in table 1. Complete medium (LC) and minimal medium (VB) supplemented with different carbon sources have been described previously.[18] Strains C 600 and LBE 2012 harbouring respectively pTuA$_1$ and pTuB$_1$ were obtained from Dr. Y. Kaziro.[35] Plasmid pPLa 2311 and pcI 857 were obtained from Dr. W. Fiers.[40] For the isolation of pTuA$_1$ and pTuB$_1$, 40 ml cultures were grown in LC medium and the plasmid was amplified by the addition of 170 µg/ml of chloramphenicol.[37] Plasmid DNA was isolated by the cleared lysate technique described by Birnboim and Doly.[38]

Construction of plasmids pGp 81 and pGp 82

Plasmid DNA was digested in 100 mM Tris pH 7.6, 10 mM MgCl$_2$ and 50 mM NaCl. The ligase buffer used was 50 mM Tris pH 7.8, 10 mM MgCl$_2$, 20 mM DTT and 1 mM ATP. The plasmid DNA's (pTuA$_1$ and pPLa 2311) were cleaved separately at 37°C with *Eco*R1 endonuclease and the restriction enzymes inactivated by heating for 10 min at 65°C. Digested pPLa 2311 DNA was ligated for 16 h at 14°C with digested DNA from pTuA$_1$ using T4 DNA ligase in a total volume of 50 µl. Strain M 5219[40] was transformed with the ligated DNA using the method of Lederberg and Cohen.[39] The transformed cells were incubated for 1 h at 28°C before plating these cells on LC agar supplemented with 50 µg/ml kanamycin. Ten transformants were analyzed. The plasmid DNA of 4 transformants appeared to contain the 4 kb *Eco*R1 fragment from pTuA$_1$. The orientation of the inserted fragment was studied by its cleavage pattern with *Sma*I and *Eco*R1, using the two enzymes simultaneously in the reaction mixture.

Transformation and selection

Bacteria were transformed according to the procedure of Lederberg and Cohen.[39] Strains were grown in LC medium[18] to an A560 of 0.2. Cells (10 ml) were centrifuged, washed in 7 ml of ice-cold 100 mM MgCl$_2$, resuspended in 5 ml of 100 mM CaCl$_2$ (0°C) and kept at 0°C for 20 min. They were sedimented and resuspended in 1 ml 100 mM $CaCl_2$ (0^OC). After 15 min 2-5 µg of plasmid DNA was added. After 30 min at 0^OC the temperature was raised to 42^OC for 2 min where-after the mixture was kept at 0^OC for 15 min. The cells were then inoculated into 10 ml of LC medium and permitted to grow for 1 h at 37^OC. They were con-centrated by centrifugation, resuspended in 1 ml LC medium and 100 µl of this suspension was spread on LC plates supplemented with ampicillin and/or kana-mycin (50 µg/ml).

RESULTS

Mutants of *E. coli* altered in *tufA* and *tufB*

In the past we have isolated a series of *E. coli* mutants that are resistant to the antibiotic kirromycin.[18-22,25] They are derived from the kirromycin resistant mutant LBE 2012 which harbours a *tufA* coding for a kirromycin resis-tant EF-TuA and a mutated *tufB* coding for an EF-TuB, sensitive to the antibio-tic (cf. table 1). The *tufB* mutation of LBE 2012 is recessive with respect to that in *tufA* and the corresponding EF-TuB (designated EF-TuB$_O$) does not block the ribosome on the messenger after binding of kirromycin.[20-22] The kirromycin

TABLE 1

MUTANTS OF *E. COLI* USED IN THIS STUDY

Strain	EF-Tu[a] symbols	Genotype	Phenotype[b]
LBE 1001	A_SB_S	wild type	
LBE 2020	A_SB_O	*tufB, rpoB*	Rif^R
LBE 12020	A_SB_O	*tufB, rpoB, recA56*	Rif^R, UV^S
LBE 2021	A_RB_O	*tufA, tufB, rpoB*	Rif^R, Kir^R
LBE 12021	A_RB_O	*tufA, tufB, rpoB, recA56*	Rif^R, Kir^R, UV^S
PM 505	A_S	*tufB::(Mu), rpoB*	Rif^R
PM 1505	A_S	*tufB::(Mu), rpoB, recA56*	Rif^R, UV^S
PM 455	A_R	*tufA, tufB::(Mu), rpoB*	Rif^R, Kir^R
PM 1455	A_R	*tufA, tufB::(Mu), rpoB, recA56*	Rif^R, Kir^R, UV^S
PM 816	A_RB_S	*tufA, fus*	Fus^R
LBE 2012	A_RB_O	*xyl, tufA, tufB*	Kir^R

For the isolation of these mutants see ref. 18. For the introduction of the $recA^-$ allele, bacteria were treated with trimethoprim to select Thy^- cells as described by Miller.[43] Thy^- cells were subsequently crossed with Hfr strain KA 273 (Hfr, *recA*56, thr, ile). Selection was for Thy^+ and screening for UV sensitivity.
[a]The designations A_S, A_R, B_S and B_O refer respectively to a wild type *tufA* pro-duct, a kirromycin resistant *tufA* product, a wild type *tufB* product and an altered *tufB* product, which properties have been described previously.[20-22]
[b]Kir^R is kirromycin resistance; Rif^R is rifampicin resistance; Fus^R is fusidic acid resistance and UV^S is UV sensitivity.

resistant EF-TuA, designated EF-TuA$_R$, displays a strongly reduced affinity for kirromycin.[21] Here we refer to each strain by their serial number supplemented with the corresponding EF-Tu symbols as indicated in table 1.

Some of the strains harbour an insertion of bacteriophage Mu into *tufB* (cf. PM 505,A$_S$ and PM 455,A$_R$; table 1). They produce EF-TuA as a single gene product derived from *tufA* and no EF-TuB.[18-20] An important aspect of these mutant strains is that they are virtually isogenic except for the *tuf* genes under study.[18]

Intracellular concentrations of EF-TuA and EF-TuB under varying growth conditions

The intracellular contents of EF-Tu were determined using rocket immuno-electrophoresis of total crude bacterial extracts.[18,27] The relative amounts of EF-TuA and EF-TuB were assayed by submitting ribosome free supernatants from the strains LBE 2020,A$_S$B$_O$ and LBE 2021,A$_R$B$_O$ to isoelectric focusing on poly-acrylamide gels. The *tuf* products of these strains differ 0.1 pH unit in iso-electric point. After separation, the two EF-Tu species were analyzed by rocket immuno-electrophoresis as described.[18] The ratios thus obtained for ribosome free supernatants and the total EF-Tu contents of crude bacterial extracts were used to calculate the intracellular amounts of EF-TuA and EF-TuB (control experiments[27] show that this calculation is justified). Variation in growth rate was achieved by growing cells in different media according to table 2.

TABLE 2

GROWTH RATES OF THE VARIOUS MUTANT STRAINS OF *E. COLI* K12 ALTERED IN *tufA* AND/OR *tufB*

Strain	EF-Tu[a] symbols	Generation times (doublings per h) in different media				
		LC	cas.a.a.	glucose	rhamnose	acetate
LBE 1001	A$_S$B$_S$	N.D.[b]	2.30	0.99	0.68	0.24
LBE 2020	A$_S$B$_O$	2.25	1.50	0.89	0.52	0.30
LBE 2021	A$_R$B$_O$	1.95	1.16	0.63	0.43	N.G.[c]
PM 505	A$_S$	N.D.[b]	1.85	0.92	0.65	0.30
PM 455	A$_R$	2.11	1.37	0.88	0.48	N.G.[c]
PM 816	A$_R$B$_S$	1.98	1.39	0.89	0.63	0.36

For details concerning culturing of cells see ref. 18.
[a]See table 1.
[b]Not done.
[c]No growth possible.

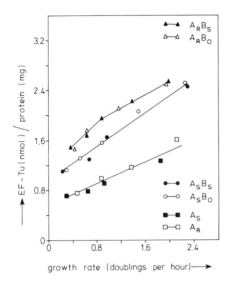

Fig. 1. Intracellular concentrations of EF-Tu in various strains of *E. coli* altered in *tufA* and/or *tufB* cultured at various growth rates. For experimental details see ref. 18.

Figure 1 shows that the total cellular EF-Tu content, expressed as nmol EF-Tu per mg of crude extract protein, varies linearly with the growth rate under steady state growth conditions as has been demonstrated by other investigators.[4,6,7,28,29] Slight deviations from linearity are seen with strains LBE 2021,$A_R B_O$ and PM 816,$A_R B_S$. The results of figure 1 further illustrate that inactivation of *tufB* by Mu insertion causes a reduction of 43% in EF-Tu content (compare the EF-Tu contents of LBE 1001,$A_S B_S$ and PM 505,A_S). This reduction is constant at all growth rates studied.

A highly interesting observation (cf. figure 2) is that the intracellular content of EF-TuA_S from strain PM 505,A_S is identical to that of cells from LBE 2020,$A_S B_O$ at comparable growth rates. This means that cells from PM 505,A_S do not compensate for the functional loss of the *tufB* gene by an enhanced expression of *tufA*. The same conclusion can be drawn from a comparison of the EF-TuA_R content of PM 455,A_R cells and that of LBE 2021,$A_R B_O$ cells (cf. figure 3). Apparently inactivation of *tufB* has no influence on the expression of *tufA* at all growth rates studied. Recently Young and Furano[30] also investigated the

effect of Mu insertion into *tufB*, using a strain (KB 31) originally constructed
in our laboratory and comparable to PM 505,A_S. Contrary to our results they
concluded that these cells compensate for the loss of a functional *tufB* by enhanced expression of *tufA*. This conclusion is unwarranted, however, since it
was based on comparisons of non-isogenic strains. The data of figure 1 also
show that the total amounts of EF-Tu in cells from LBE 1001,$A_S B_S$ are identical
to those of cells from LBE 2020,$A_S B_O$. This indicates that the single site mutation in *tufB* has no effect on the expression of *tufA*

Disturbance of the coordinate expression of *tufA* and *tufB* by a

specific single site mutation in *tufA*

In cells from the strain LBE 2020,$A_S B_O$ the two EF-Tu species EF-TuA$_S$ and
EF-TuB$_O$ occur in a constant molar ratio of 1.33 at all growth rates studied
(figure 2). Apparently the expression of *tufA* and *tufB* is coordinately regulated. Most likely this is also the case for wild type cells from LBE 1001,$A_S B_S$
since the total EF-Tu levels of the latter strain and those of LBE 2020,$A_S B_O$
are the same.

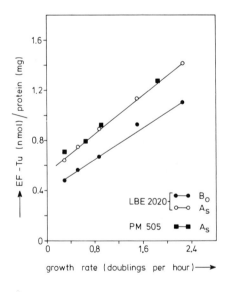

Fig. 2. Intracellular concentrations of EF-TuA$_S$ and
EF-TuB$_O$ of strains LBE 2020,
$A_S B_O$ and PM 505,A_S cultured
at various growth rates. For
experimental details see
ref. 18.

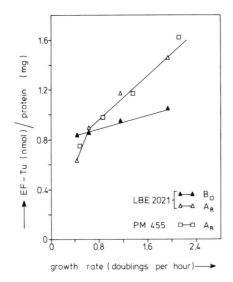

Fig. 3. Intracellular concentrations of EF-TuA$_R$ and EF-TuB$_O$ of strains LBE 2021, A$_R$B$_O$ and PM 455,A$_R$ cultured at various growth rates. For experimental details see ref. 18.

Coordination is completely lost in cells from the mutant strain LBE 2021,A$_R$B$_O$ (figure 3). It may be concluded, therefore, that this loss is due to the specific single site mutation of *tufA*, rendering the EF-TuA product resistant to the antibiotic kirromycin. The mutant cells show an enhanced expression of *tufB* as compared to the wild type cells (figures 2 and 3). This enhancement becomes more pronounced at lower growth rates and is also apparent in figure 1 which shows that the total EF-Tu levels of LBE 2021,A$_R$B$_O$ and PM 816,A$_R$B$_S$ are higher than wild type levels.

That we are dealing with a highly specific effect of the mutation of *tufA* is illustrated by the finding that the expression of the *tsf* gene coding for EF-Ts and that of the ribosomal genes is not affected under all environmental conditions (not shown). The possibility of an indirect effect cannot be excluded rigidly but the data are more readily explained by a direct role of the EF-Tu protein itself in the regulation of *tufB* expression. The nature of *tufA* mutation has been characterized in our laboratory by Duisterwinkel *et al.*[49] who showed that alanine in position 375 of the protein chain is replaced by threonine.

EF-TuA$_R$·GTP has a lowered affinity for aminoacyl-tRNA

Further substantiation for a direct regulatory role of EF-Tu in the expression of *tufB* has been sought by elevating the intracellular EF-Tu level via plasmids harbouring either *tufA* or *tufB*. The results to be described below, are in agreement with EF-Tu affecting the expression of *tufB*, not that of *tufA*. Such a differential effect of EF-Tu on the expression of the two *tuf* genes may be correlated with the different organization of *tufA* and *tufB* in the *str* and the *rif* region, respectively. The fact that *tufB* is cotranscribed with four upstream tRNA genes[13] may be relevant in this respect. The possibility can be envisaged that the tRNA elements on the primary transcript act as potential binding sites for EF-Tu. If so the experiments of figure 3 would indicate a lowered affinity of EF-TuA$_R$ towards these binding sites as compared to wild type EF-TuA$_S$ and EF-TuB$_O$. As a preliminary approach we have studied the relative affinities of the EF-Tu species for aminoacyl-tRNA. To this end we analyzed the ternary complexes formed between EF-Tu·GTP and a mixture of aminoacyl-tRNA's. The reaction mixtures were filtered through nitrocellulose filters and the filtrates were supplemented with alcohol. Precipitates were collected and after resuspending submitted to isoelectric focusing in a pH gradient (pH 5-7). This analysis takes advantage of the fact that EF-Tu·GTP bound to aminoacyl-tRNA passes through nitrocellulose filters but unbound EF-Tu·GTP does not. Furthermore the difference in isoelectric point between EF-TuB$_O$ and the other EF-Tu species enabled the use of EF-TuB$_O$ as an internal standard. Figure 4A shows the results of an experiment in which the aminoacyl-tRNA's were incubated with a mixture of EF-TuA$_S$·GTP and EF-TuB$_O$·GTP. In figure 4B a similar experiment was performed with a mixture of EF-TuA$_R$·GTP and EF-TuB$_O$·GTP. The ratio between aminoacyl-tRNA and EF-Tu was varied from 0.5 to 5.0 in lanes 3 to 7 of figure 4A and from 0.2 to 5.0 in lanes 3 to 9 of figure 4B. When aminoacyl-tRNA was omitted from the reaction mixtures (lanes 2 of figures 4A and 4B) no EF-Tu appeared in the filtrate. Isoelectric focusing of the mixtures EF-TuA$_S$ and EF-TuB$_O$, which had not been passed through millipore, showed that the two EF-Tu species were present in a ratio of 4 : 3, respectively (lane 1 of figure 4A). A similar ratio was found for the mixture EF-TuA$_R$ and EF-TuB$_O$ (lane 1 of figure 4B). After reacting with increasing amounts of aminoacyl-tRNA and passing the reaction mixtures through the filters the relative amounts of EF-TuA$_S$ and EF-TuB$_O$ remained the same. This can be concluded from figure 4A and from table 3 in which the ratios based on the results of scanning profiles of isoelectric focusing gels are presented. By contrast the relative amounts of EF-TuA$_R$ and EF-TuB$_O$ appearing in the ternary complexes, varied considerably with the amino-

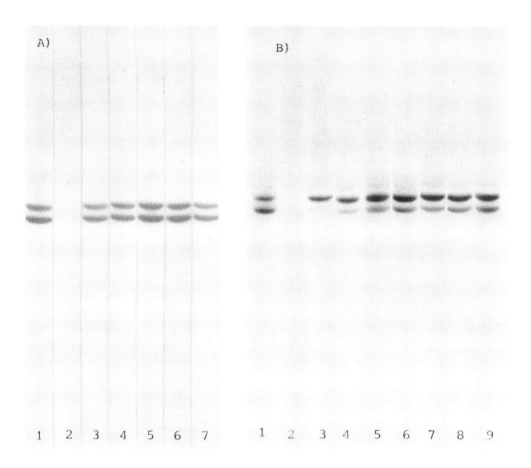

Fig. 4. Analyses of ternary complexes formed from aminoacyl-tRNA and various mutant species of EF-Tu·GTP. (A) Ternary complexes with EF-TuA$_S$ and EF-TuB$_O$. (B) Ternary complexes with EF-TuA$_R$ and EF-TuB$_O$. The conditions were as described elsewhere[22] with the exception that soaking of the nitrocellulose filters in 0.5 M KOH was omitted. For further details see the text.

acyl-tRNA/EF-Tu·GTP ratios in the reaction mixtures (figure 4B, table 3). At a low aminoacyl-tRNA/EF-Tu·GTP ratio (lane 3) almost no EF-TuA$_R$ appeared in the ternary complex and the preponderance of EF-TuB$_O$ persisted at aminoacyl-tRNA/ EF-Tu·GTP ratios up to 5.0. The conclusion can therefore be drawn that the affinity of EF-TuA$_R$·GTP towards aminoacyl-tRNA is greatly reduced compared to that of EF-TuA$_S$·GTP (wild type) and EF-TuB$_O$·GTP. (We do not know whether this reduction concerns all species of aminoacyl-tRNA (cf. also ref. 22)).

TABLE 3

ANALYSES OF THE TERNARY COMPLEXES FORMED FROM AMINOACYL-tRNA AND
VARIOUS MUTANT SPECIES OF EF-Tu·GTP

EF-Tu species in reaction mixture	Ratio aminoacyl-tRNA/ EF-Tu·GTP (mol/mol) in reaction mixture	Ratio EF-TuA/EF-TuB in ternary complex
EF-TuA$_S$ + EF-TuB$_O$	0.5	1.4
	1.0	1.3
	2.0	1.3
	3.0	1.4
	5.0	1.3
EF-TuA$_R$ + EF-TuB$_O$	0.2	0.1
	0.5	0.3
	0.8	0.3
	1.0	0.5
	2.0	0.4
	3.0	0.6
	5.0	0.7

The experiments were carried out as described in the text, ref. 22 and in the
legend to figure 4.

A lowered affinity of EF-TuA$_R$ for aminoacyl-tRNA does not necessarily reflect an impaired binding of this mutant EF-Tu to putative target sites on the
primary transcript of the tRNA-tufB operon. TPCK labeling studies by Jonak et
al.[31] are in agreement with the assumption, however, that aminoacylated and
non-aminoacylated-tRNA interact with the same site on the EF-Tu protein. The
experiments of figure 4 and table 3, although not proving the occurrence of
target sites for EF-Tu binding, lend suggestive support for their existence.

Plasmids harbouring tufA and tufB

In wild type cells aminoacyl-tRNA and EF-Tu are present in approximately
equimolar amounts[3,4] and thus presumably are taken up completely in ternary
complexes. It may be of interest, therefore, to raise the intracellular EF-Tu
concentration above the aminoacyl-tRNA level and study the effect on tufB expression. In principle this can be achieved by introducing into the cell a
multicopy plasmid harbouring tufA or tufB. Recombinant plasmids of this type
have been constructed.[32-34] Shibuya et al.[33] reported the cloning of the tufA
gene from λfus3 DNA into a ColE1 derivative plasmid. The hybrid plasmid designated pTuA$_1$ contains the C-terminal portion of the fus gene (EF-G), the intercistronic region between fus and tufA, and the complete structural gene for
tufA.[41] This cloned tufA gene is only weakly expressed in a cell-free system
due to a reported lack[33] of the natural promoter. In whole cells of LBE 2012,

it is also poorly expressed (35, this paper). By contrast, plasmid pTuB$_1$ constructed by Miyajima *et al.*[32] and containing part of *rrnB*, four tRNA genes, *tufB*, the gene coding for an unidentified protein U and part of *rplK*, permits a good expression of *tufB* both in a cell-free system[32] and in whole cells of the kirromycin resistant mutant LBE 2012.[35]

Although the copy number of pTuB$_1$ in the transformants was about 20, the rate of EF-Tu synthesis was not appreciably increased.[35] The authors suggested the presence of a regulatory mechanism that maintains the normal cellular level of EF-Tu in the cell. In the light of our hypothesis this may not be surprising since it would mean that transformation with pTuB$_1$ introduces *tufB*, together with the regulatory elements of the primary transcript.

We have made use of both plasmids (kindly provided by Dr. Y. Kaziro) for studying expression of the plasmid borne *tuf* genes. In order to compensate for the reported lack of the natural promoter of the cloned *tufA* we have introduced the 4 kb *Eco*R1 fragment of pTuA$_1$ into the unique *Eco*R1 site of the vector pPLa 2311.[40] This vector was originally constructed by Remaut *et al.*[40] and contains the powerful major leftward promoter (P$_L$) of phage λ. The *Eco*R1 site is located 150 bp downstream of the P$_L$ promoter. Two plasmids designated pGp81 and pGp82 and harbouring *tufA* in both orientations were obtained. In order to control transcription of the plasmid borne *tufA*, a second plasmid compatible with pGp81 and pGp82 was introduced. This plasmid (designated pcI857) bearing a cI$_{ts}$ gene of λ, codes for a temperature sensitive repressor. Transcription from P$_L$ can be switched off at low temperature in the presence of this cI$_{ts}$ gene. In order to prevent recombination between cloned and chromosomal *tuf* genes the *recA*$^-$ allele of strain KA 273 (see table 1) was crossed into the mutants.

Expression of plasmid borne *tufA in vivo*

Restriction analysis (cf. Materials and Methods) showed that pGp81 contains *tufA* in the sense orientation with regard to the P$_L$ promoter whereas pGp82 contains this gene in the opposite orientation. Transformants harbouring pGp81 are unable to survive on selective plates at 37°C in the absence of the plasmid pcI in contrast to transformants harbouring pGp82 which do survive under these conditions. Apparently transcription of *tufA* from the P$_L$ promoter in the absence of the λ repressor is incompatible with growth. In the presence of the repressor the steady state growth rate of LBE 12020 transformed with pGp81 is only slightly reduced as compared to that of the parental cells (figure 5 and table 4A). The same holds true for cells transformed with pGp82 but lacking

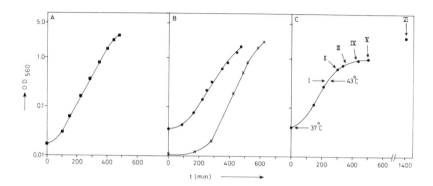

Fig. 5. Growth curves of cells from LBE 12020,A_SB_0 transformed with different plasmids. (A) With pcI. (B) With pcI and pGp81 (upper curve); with pGp82 (lower curve). (C) With pcI, pGp81. For further details see tables 4A and 4B.

pcI. It should be noted, however, that cells from LBE 12020, pcI, pGp81 show a considerable lag phase (figure 5) which is further prolonged in LBE 12020, pGp82. In CAS medium both transformants grow very slowly essentially due to a considerably extended lag phase (not shown).

Expression of the plasmid borne *tufA* was studied in two kirromycin resistant strains: LBE 12021 and PM 1455 (table 5). Since sensitivity to the antibiotic dominates resistance[19,20] expression in the transformants is revealed by a change in phenotype. The data presented in table 5 illustrate that pGp81 and pGp82 are expressed in both strains. This indicates that transcription of the plasmid borne *tufA* can proceed independent of the P_L promoter, presumably from a secondary promoter for the *tufA* gene. Yokota *et al.*[41] reported that *tufA* probably has a weak secondary promoter, since *tufA* of $pTuA_1$ is weakly expressed both *in vivo* and *in vitro*. Table 5 shows that expression of $pTuA_1$ in LBE 12021 and PM 1455 is hardly detectable. By contrast expression of pGp81 and pGp82 in these strains is much more efficient. Since transcription of *tufA* on all three plasmids is assumed to occur from the same secondary promoter, the expression of *tufA* in the genetic environment of $pTuA_1$ is reduced for unknown reasons. We conclude from the efficient expression of pGp81 and pGp82 that a secondary promoter must exist, in addition to the major promoter of the *str* operon. Apparently this secondary promoter has significant activity to initiate transcription of the *tufA* gene. Very recently An *et al.*[44] reached a similar conclusion based on studies of plasmids carrying *tufA* but lacking the major promoter for the *str* operon. They indicated that the secondary promoter is located within the coding region of *fus*, approximately 120 basepairs from the start codon of *tufA*. A possible binding site for RNA polymerase (TATGATG) was suggested in this position by Yokota *et al.*[41]

After having established the expression of the plasmid borne *tufA*, the intracellular EF-Tu concentrations were determined in transformant and paren-tal strains grown in rich (LC) medium at 37°C. It can be concluded from table 4A that cells from LBE 12020 transformed either with pGp81 and pcI or with pGp82 displayed an almost two-fold elevated EF-Tu level as compared with the parental strain (LBE 12020,pcI). This increase appeared to be quite specific as the levels of EF-Ts (table 4A) and ribosomes (not illustrated) re-mained unaltered. Transformants harbouring $pTuA_1$ showed no demonstrable in-crease in EF-Tu content in accordance with the low expression of the plasmid borne *tufA*.

The EF-Tu level in transformants harbouring pGp81 and pcI could be further increased by raising the temperature from 37°C to 43°C (compare figure 5). Due to inactivation of the λ repressor the EF-Tu content reached values up to 30% of the cellular protein (table 4B). That this increase is rather specific and restricted to EF-Tu can also be seen in the table showing that the EF-Ts level remains virtually constant under these conditions. Soon after the temperature shift the growth rate starts to decline (figure 5C). After about 24 h at 43°C (growth phase VI, figure 5C and table 4B) cells had lost substantial amounts of EF-Tu and EF-Ts but not their viability.

We are particularly interested in the effect of the elevated intracellular EF-Tu levels on the expression of the chromosomal *tufB*. Unfortunately the assay of $EF-TuB_O$ in transformants of LBE 12021,A_SB_O has to occur in the pre-sence of the strongly accumulated EF-TuA which makes high demands upon the separation of the two EF-Tu species by isoelectric focusing. Up till now this assay has to await improvement of the technique.

Nonetheless the investigations with pGp81 and pGp82, although not yet com-plete, have led to the conclusion that transcription of the plasmid borne *tufA* gene from the secondary promoter results in intracellular EF-Tu levels which amount to as much as 17% of the bacterial protein. This contrasts, as we will see below, with the transcription of the *tufB* gene on the plasmid $pTuB_1$ carry-ing the major promoter for the tRNA-*tufB* operon. We come back to this differ-ence in the Discussion.

TABLE 4A

INTRACELLULAR AMOUNTS OF EF-Tu AND EF-Ts IN TRANSFORMANTS OF
LBE 12020,A_SB_O HARBOURING DIFFERENT PLASMIDS

Strain	Medium	Growth rate	EF-Tu content nmol/mg protein	EF-Tu content %	EF-Ts content nmol/mg protein
LBE 12020,A_SB_O pcI	LC	1.1	2.34	10.1	0.35
LBE 12020,A_SB_O pcI, pGp81	LC	0.9	3.96	17.0	0.34
LBE 12020,A_SB_O pGp82	LC	1.0	4.10	17.6	0.34
LBE 12020,A_SB_O pTuA$_1$	LC	1.2	2.21	9.5	0.36

Cells were grown at 37^OC in LC medium containing 50 μg/ml kanamycin (LBE 12020, pcI) or 50 μg/ml ampicillin (LBE 12020, pTuA$_1$) or in the presence of both anti-biotics (LBE 12020, pcI, pGp81; LBE 12020, pGp82). The analyses were performed as described previously.[18]

TABLE 4B

INTRACELLULAR AMOUNTS OF EF-Tu AND EF-Ts IN TRANSFORMANT LBE 12020,A_SB_O, pcI pGp81 DURING DIFFERENT STAGES OF GROWTH AFTER A TEMPERATURE SHIFT UP FROM 37^OC TO 43^OC

Growth phase[a]	EF-Tu content nmol/mg protein	EF-Tu content %	EF-Ts content nmol/mg protein
I	3.85	16.5	0.33
II	4.36	18.7	0.29
III	6.88	29.6	0.31
IV	6.92	29.8	0.29
V	6.59	28.3	0.30
VI	1.90	8.2	0.23

The culture was shifted to 43^OC as indicated in figure 5C. At different times 100 ml samples were taken from the culture (I-VI) and analysed for EF-Tu and EF-Ts content (for experimental details see ref. 18).
[a]See figure 5C.

Expression of plasmid borne *tufB in vivo*

Expression of *tufB* on pTuB$_1$ in transformants of LBE 12021,A_RB_O and PM 1455, A_R is illustrated in table 5, showing a change in phenotype from kirromycin resistance to sensitivity. These data confirm earlier results reported by Miyajima and Kaziro[35] for the strain LBE 2012,A_RB_O. The interesting observa-tion of the latter authors that the rate of EF-Tu synthesis was not notably in-

TABLE 5

MINIMAL KIRROMYCIN CONCENTRATIONS CAUSING GROWTH INHIBITION IN VARIOUS
TRANSFORMANT AND PARENTAL STRAINS OF *E. COLI* K12

Transformant and parental strains	Kirromycin concentration µg/ml
LBE 1001, $A_S B_S$	< 20
LBE 12021, $A_R B_O$	> 4.000
PM 1455, A_R	> 4.000
LBE 12021, $A_R B_O pA_S$, pTuA$_1$	> 4.000[a]
PM 1455, $A_R pA_S$, pTuA$_1$	> 4.000[a]
LBE 12021, $A_R B_O pA_S$, pTuB$_1$	< 20
PM 1455, $A_R pB_S$, pTuB$_1$	< 40
LBE 12021, $A_R B_O pA_S$, pcI, pGp81	< 20
PM 1455, $A_R pA_S$, pcI, pGp81	< 40
LBE 12021, $A_R B_O pA_S$, pGp82	< 20
PM 1455, $A_R pA_S$, pGp82	< 30

Transformed cells were grown at 37°C in LC medium containing 50 µg/ml kanamycin
and/or 50 µg/ml ampicillin. Parental strains were grown under identical conditions in the absence of the antibiotica. Cells (2 x 10^6) were plated on VB agar
which contained 10^{-3} M EDTA (to make cells permeable for kirromycin), glucose
1%, casamino acids 0.5%, varying concentrations of kirromycin and 50 µg/ml kanamycin and/or ampicillin (in the case of transformants).
[a]Some growth retardation.

creased in transformants of LBE 2012, $A_R B_O$, harbouring about 20 copies of pTuB$_1$
prompted a more detailed analysis of the intracellular amounts of EF-Tu derived
from both the chromosome and plasmid borne genes in transformants of LBE 12020,
$A_S B_O$ and PM 1505, A_S. As illustrated in table 1, LBE 12020, $A_S B_O$ harbours two
active chromosomal *tuf* genes whereas PM 1505, A_S lacks an active chromosomal
tufB gene due to an insertion of bacteriophage Mu. Transformant and parental
strains were grown under steady state growth conditions at varying rates. The
EF-Tu levels are presented in table 6. As was to be expected, the EF-Tu contents of the parental cells increased with the growth rate. After transformation of LBE 12020, $A_S B_O$, the EF-Tu content already reached a level of approximately 12.5% of the bacterial protein at the lowest growth rate (0.5 doublings
per h) and this content did not increase further upon raising the growth rate.
By contrast transformed cells from the strain PM 1505, A_S respond to a nutritional shift up with an increase in their EF-Tu contents. These contents do not
reach the maximum of 12.5% even at the highest growth rate.

TABLE 6

INTRACELLULAR AMOUNTS OF TOTAL EF-Tu, $EF\text{-}TuB_O$ AND EF-Ts BEFORE AND AFTER TRANSFORMATION WITH $pTuB_1$

Strain	Medium	Growth rate		EF-Tu content nmol/mg protein		EF-Tu content %		$EF\text{-}TuB_O$ content nmol/mg protein		EF-Ts content nmol/mg protein	
		$+p^a$	$-p^a$	+p	-p	+p	-p	+p	-p	+p	-p
LBE 12020,A_SB_O	LC	1.2	1.7	2.94	2.22	12.6	9.5	0.63	0.95	0.34	0.34
LBE 12020,A_SB_O	CAS	0.9	1.0	3.01	2.04	12.9	8.8	0.75	0.88	0.32	0.33
LBE 12020,A_SB_O	Gluc	0.5	0.6	2.84	1.60	12.2	6.9	0.64	0.69	0.23	0.22
PM 1505,A_S	LC	1.7	1.8	2.71	1.49	11.7	6.4	-	-	0.37	0.36
PM 1505,A_S	CAS	1.7	1.6	2.26	1.24	9.7	5.3	-	-	0.33	0.32
PM 1505,A_S	Gluc	1.1	1.0	1.68	1.10	7.2	4.7	-	-	0.25	0.26

Cells were grown in different media to vary generation time.[18] The total amounts of EF-Tu and EF-Ts were determined in crude bacterial extracts by means of rocket immuno-electrophoresis.[18] The values for $EF\text{-}TuB_O$ content shown in this table were assayed by a combined procedure of isoelectric focusing and rocket immuno-electrophoresis (see text).
[a]With (+) and without (-) $pTuB_1$.

Apparently it is the lack or the presence of the chromosomal *tufB* product ($EF\text{-}TuB_O$ in this case) which determines whether these transformant cells will respond to an enrichment of the medium with an increase in their EF-Tu content (see also the Discussion).

The analytical techniques developed for this investigation also enabled the separate assay of the intracellular concentrations of the products of the chromosome and plasmid borne *tuf* genes in transformant and parental strains. In figure 6 these concentrations are plotted graphically. The intracellular amounts of $EF\text{-}TuB_O$ were determined after separating this mutant species of EF-TuB from the other EF-Tu species in ribosome-free S100 supernatants by isoelectric focusing. From the ratio between $EF\text{-}TuB_O$ and the other EF-Tu species thus determined by rocket immuno-electrophoresis of the separated bands and the total EF-Tu contents of crude bacterial extracts, the absolute amounts of $EF\text{-}TuB_O$ in the cell could be calculated (table 6). Virtually the same ratios between the $EF\text{-}TuB_O$ and the other EF-Tu species were found when EF-Tu, isolated and purified to apparent homogeneity by affinity chromatography[24], was submitted to isoelectric focusing (figure 7). This illustrates the reliability of the assay.

The amount of EF-TuA in transformed cells was assumed to be identical to that in the parental cells grown under the same nutritional conditions. This

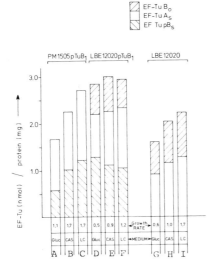

Fig. 6. Intracellular concentrations of the EF-Tu products of chromosome and plasmid borne tuf genes in transformant (pTuB$_1$) and parental strains of $E.$ $coli.$ The amounts (expressed as nmol EF-Tu per milligram of protein) of chromosomal EF-TuA and plasmid derived EF-TuB in PM 1505 were assayed by comparing the total amounts of EF-Tu in parental and transformant strains under varying growth conditions. The amount of chromosomal EF-TuB$_O$ in strain LBE 12020 and its transformant was determined as described [18]. For further details see the text.

assumption is based on the finding that inactivation of chromosomal $tufB$ did not affect the expression of $tufA$ (figure 2) and on preliminary results with a plasmid derivative of pTuB$_1$ showing that an increase in EF-Tu level did not affect $tufA$ expression either (see Discussion).

Finally the concentration of the EF-Tu species derived from the plasmid borne $tufB$ gene can be calculated from the total EF-Tu concentration and the sum of the chromosomal tuf products. The contributions of each EF-Tu species are plotted as parts of the bars of figure 6.

It can be seen in this figure that transformants of LBE 12020,A$_S$B$_O$ respond to a nutritional shift up with a decrease in the expression of the plasmid borne $tufB$ (bars D, E and F), whereas transformants of PM 1505,A$_S$ respond with an increase (bars A, B and C). As will be discussed below (Discussion) this means that the intracellular concentration of EF-Tu in the transformants determines the response of the plasmid borne $tufB$ gene to nutritional changes. (It may be noted that the expression of the plasmid $tufB$ on a per copy basis is much lower than that of the chromosomal $tufB$, a finding which we cannot explain).

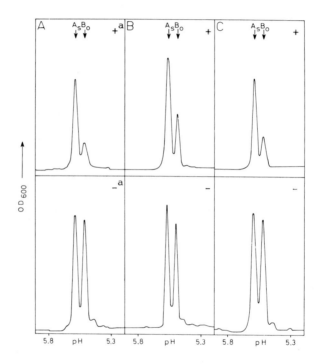

Fig. 7. Scanning profiles of IEF gels loaded with pure EF-Tu isolated by affinity chromatography from transformant (LBE 12020,A_SB_O, pTuB$_1$) and parental
(LBE 12020,A_SB_O) strain in three different media. The analyses were performed
as described previously.[18] The letters above the bands signify the two EF-Tu
species present in the EF-Tu preparations.
(A). Pure EF-Tu isolated from cells cultured in rich medium LC. (B). Pure EF-Tu
isolated from cells cultured in CAS medium. (C). Pure EF-Tu isolated from cells
cultured in minimal medium (Gluc).
[a]With (+) pTuB$_1$ and without (-) plasmid.

A comparison of the amounts of EF-TuB$_O$ in the transformant cells from strain

LBE 12020,A_SB_O with those of the parental cells shows (cf. table 6 and figure 6)

that introduction of pTuB$_1$ causes a drop in the expression of the chromosomal
tufB$_0$. This drop becomes more pronounced at higher growth rates. It is, at
least in part, due to the introduction of a great number of plasmid borne tufB
genes together with their regulatory elements. The chromosomal tufB gene now
has to compete with the plasmid borne tufB genes for the transcriptional and
translational apparatuses. The reduction in chromosomal tufB expression can
therefore not readily be interpreted in terms of autogenous repression.

Previously, we[18,27] and others[4,6,7,23,28,30,46-48] have shown that at growth
rates exceeding 0.5 doublings per h the intracellular concentrations of EF-TuA,
EF-TuB$_0$, EF-Ts and ribosomes increase with the growth rate in a constant ratio.
The question must therefore be considered here, whether the variations in tufB
expression discussed above are the consequence of this growth retardation
rather than that of the variations in the level of total intracellular EF-Tu.
It may be noted from table 6 that transformation with pTuB$_1$ is not accompanied
by a decrease in EF-Ts content. The same holds true for the cellular contents
of ribosomes (not illustrated). This demonstrates the specificity of the dosage
effect of intracellular EF-Tu (compare also the Discussion).

DISCUSSION

The expression of the two EF-Tu encoding genes, which are distantly located
on the E. coli chromosome and are positioned in two different transcription
units, is regulated coordinately (cf. figure 1 and ref. 18). This coordination
is not affected by a specific point mutation in tufB (tufB$_0$).

Remarkably inactivation of tufB by the insertion of bacteriophage Mu into
the coding region of tufB, leaves the expression of tufA unaltered. This inactivation reduces the cellular EF-Tu content with 43%. Also an increase of
the EF-Tu content does not affect the expression of tufA as was shown by preliminary experiments (not described above) with the plasmid pTuB$_0$. This plasmid
is identical to pTuB$_1$ except that the tufB gene is replaced by tufB$_0$. After
transformation with pTuB$_0$ of cells, harbouring a chromosomal tufB$_0$, the intracellular amount of the tufA product could be assayed directly with the procedures described in this paper (see also ref. 18). No change in the EF-TuA
level was detectable. It is concluded that the expression of tufA is independent of the intracellular EF-Tu level.

A specific single site mutation of tufA, rendering EF-TuA resistant to the
antibiotic kirromycin, disturbs the coordinate expression of tufA and tufB and
enhances the expression of tufB, particularly at lower growth rates. These data

suggest that the EF-Tu protein itself is directly involved in the control of
tufB expression.

Support for this suggestion has been derived from experiments with the plasmid pTuB$_1$, which contains a 8.9 kb *Eco*R1 fragment of the *rif* region harbouring
the entire tRNA-*tufB* operon. Two strains, differing only in the activity of the
chromosomal *tufB*, were used for transformation with pTuB$_1$: PM 1505,A$_S$ with an
interruption of the coding region of the chromosomal *tufB* due to Mu insertion
and LBE 12020,A$_S$B$_O$ with *tufB* uninterrupted. Restriction analysis of the chromosomal *tufB*[30] indicates that the regions controlling expression of the tRNA-*tufB*
operon in PM 1505,A$_S$ are still intact. The two hosts for pTuB$_1$ differ therefore
only in their intracellular EF-Tu concentration.

After transformation these strains responded quite differently to a nutritional shift up of their medium. Transformants of PM 1505,A$_S$ displayed an increase
in EF-Tu content whereas transformants of LBE 12020,A$_S$B$_O$ maintained a level of
about 12.5% which could not be augmented by an improvement of the nutritional
conditions. The latter cells apparently had reached a maximum EF-Tu content
which could not be exceeded whereas transformants of PM 1505,A$_S$ did not reach
this maximum even in the richest medium. Apparently it depended on the amount
of EF-Tu which was already present in the cell, whether the EF-Tu level was permitted to increase further or not.

The intracellular concentration of EF-Tu determines in particular how the
expression of *tufB* on the plasmid pTuB$_1$ will respond to changes in the nutritional environment. This was demonstrated by assaying the products of the chromosome and plasmid borne *tuf* genes separately (figure 6). When the EF-Tu was at
its maximum (in LBE 12020,A$_S$B$_O$, pTuB$_1$) the expression of plasmid *tufB* declined;
when it was submaximal (in PM 1505,A$_S$, pTuB$_1$), plasmid *tufB* expression increased upon enrichment of the medium. Under the latter conditions the expression of
chromosomal *tufA* increased in both transformant strains. In PM 1505,A$_S$, pTuB$_1$,
which had not reached its maximum EF-Tu level, there was still room for the
plasmid derived EF-TuB to increase. By contrast in LBE 12020,A$_S$B$_O$, pTuB$_1$ the
increase in chromosomal *tufA* expression and the nearly constant level of chromosomal EF-TuB$_O$ forced the expression of the plasmid *tufB* to decline.

The question may be raised at which level EF-Tu may control the expression
of *tufB*. A regulatory function at the level of transcription may not be excluded (see also Travers[45]) but it is attractive to consider a control function
at a posttranscriptional level.

Above we have already hypothesized that EF-Tu can bind to the tRNA elements
on the primary transcript of the *tufB* operon thus interfering with the processing of this transcript. The lowered affinity of EF-TuA$_R$ for aminoacyl-tRNA as

compared to wild type EF-Tu is in accordance with this concept. In this respect it is of interest that transformation with pTuB$_1$ has introduced the entire tRNA-*tufB* operon into the cell. This allows for the posttranscriptional control of plasmid borne *tufB* as postulated above and for the regulation of the transcription of the entire transcription unit.

A different situation exists in cells transformed with the plasmids pGp81 and pGp82 which harbour *tufA*. Here the *tuf* gene has been cloned together with the C-terminal region of the *fus* gene and the intergenic region. These plasmids lack the major promoter of the *str* operon. Our data show, however, that they harbour a secondary promoter which, according to Yokota *et al.*[41] and An *et al.*[44] is located in the *fus* gene, approximately 120 basepairs from the start codon of *tufA*. Preliminary experiments (not shown above) indicate that the expression of *tufA* on pGp82 is not influenced by the intracellular amount of the EF-Tu protein under conditions that transcription proceeds from this secondary promoter. Apparently the plasmid borne *tufA* differs in this respect from the plasmid borne *tufB* (on pTuB$_1$) which is in accordance with the absence on pGp82 of similar targets for the putative posttranscriptional regulation by EF-Tu.

An *et al.*[44] reported that the secondary promoter for *tufA* transcription is of approximately equal activity as the major promoter of the tRNA-*tufB* operon. Table 4A shows that transcription from the secondary *tufA* promoter on both pGp81 and pGp82 in transformants of LBE 12020,A$_S$B$_O$ results in an intracellular EF-Tu concentration of approximately 17.5%, which is far above that of pTuB$_1$ transformants (12.5%). The increment in EF-Tu level due to transformation with pGp81 and pGp82 is 75%, that due to transformation with pTuB$_1$ is 25%. This difference may be ascribed to a difference in translational efficiency of *tufA* and *tufB*. It is equally well possible, if not more likely, that pGp81 and pGp82 lack the putative target sites for the controlling action of EF-Tu.

The hypothesis that EF-Tu exerts this regulatory function implicates that binding of the factor protein to the tRNA targets on the primary transcript of the tRNA-*tufB* operon will also affect the processing of the tRNA precursor. Since the four tRNA genes involved have not been detected elsewhere on the chromosome, such an interference with the processing may also affect the growth of the cell. Further experiments are necessary to answer the question whether elevation of the intracellular EF-Tu level depresses specifically the concentration of certain tRNA species while leaving that of others unaltered. In this context the long lag in the growth of pGp81 and pGp82 transformants of LBE 12020,A$_S$B$_O$ and its prolongation in CAS media is of interest.

The retardation of growth observed during various experiments described above also raises the question of the specificity of the variations in *tufB* expression. In order to study this specificity the cellular contents of EF-Ts and of the ribosomes were studied. Unvariably the expression of *tsf* and of the ribosomal genes were found unaltered. Even when transcription of the pGp81 borne *tufA* occurred from the P_L promoter (after inactivating the temperature sensitive repressor) and the cellular EF-Tu content rose to values of 30%, the EF-Ts content remained essentially the same (table 4B). The cells lost their ability to grow but not their viability, neither when kept under these conditions for 10 h.

Finally the hypothesis proposed above is reminiscent of that suggested for the regulation of the expression of ribosomal protein genes. Various authors have reported (for a review see ref. 50) that certain key ribosomal proteins act as negative feedback regulators, inhibiting the translation of mRNA coding for themselves and certain other proteins in the same transcription unit. An analogous mechanism for the regulation of *tufB* expression would implicate that the main part of the EF-Tu population is present in ternary complexes with aminoacyl-tRNA and GTP while only free EF-Tu in excess of the ternary complexes would exert its regulatory action.

ACKNOWLEDGMENTS

We are grateful to Dr. R.A. Kastelein for his assistance in the construction of plasmids pGp81 and pGp82, to Erik Vijgenboom and Marcel Dicke for their technical assistance. Collaboration with Dr. J. Sedláček in the preparation of antibodies is gratefully acknowledged.

REFERENCES

1. Miller, D.L. and Weissbach, H. (1977) in Molecular Mechanisms in Protein Biosynthesis, Weissbach, H. and Pestka, S. ed., Academic Press, New York, pp. 323-373.
2. Kaziro, Y. (1978) Biochim. Biophys. Acta 505, 95-127.
3. Rosset, R., Julien, J. and Monier, R. (1966) J. Mol. Biol. 18, 308-320.
4. Furano, A.V. (1975) Proc. Nat. Acad. Sci. 72, 4780-4784.
5. Jacobson, G.R. and Rosenbusch, J.P. (1976) Nature 261, 23-26.
6. Miyajima, A. and Kaziro, Y. (1978) J. Biochem. 83, 453-462.
7. Pedersen, S., Bloch, P.L., Reeh, S. and Neidhardt, F.C. (1978) Cell 14, 179-190.
8. Bachmann, B.J. and Low, K.B. (1980) Microbiol. Rev. 44, 1-56.
9. Jaskunas, S.R., Lindahl, L., Nomura, M. and Burgess, R.R. (1975) Nature 257, 458-462.
10. Jaskunas, S.R., Fallon, A.M. and Nomura, M. (1977) J. Biol. Chem. 252, 7323-7336.
11. Lee, J.S., An, G., Friesen, J.D. and Fiil, N.P. (1981) Cell 25, 251-258.

12. Miyajima, A., Shibuya, M., Kuchino, Y. and Kaziro, Y. (1981) Mol.Gen. Genet. 183, 13-19.
13. Hudson, L., Rossi, J. and Landy, A. (1981) Nature 294, 422-427.
14. Yokota, T., Sugisaki, H., Takanami, M. and Kaziro, Y. (1980) Gene 12, 25-31.
15. An, G. and Friesen, J.D. (1980) Gene 12, 33-39.
16. Arai, K.-I., Clark, B.F.C., Duffy, L., Jones, M.D., Kaziro, Y., Laursen, R.A., L'Italien, J., Miller, D.L., Nagarkatti, S., Nakamura, S., Nielsen, K.M., Petersen, T.E., Takanashi, K. and Wade, M. (1980) Proc. Nat. Acad. Sci. 77, 1326-1330.
17. Jones, M.D., Petersen, T.E., Nielsen, K.M., Magnusson, S., Sottrup-Jensen, L., Gausing, K. and Clark, B.F.C. (1980) Eur. J. Biochem. 108, 507-526.
18. Van der Meide, P.H., Vijgenboom, E., Dicke, M. and Bosch, L. (1982) FEBS Letters, in press.
19. Van de Klundert, J.A.M., Van der Meide, P.H., Van de Putte, P. and Bosch, L. (1978) Proc. Nat. Acad. Sci. 75, 4470-4473.
20. Van der Meide, P.H., Borman, T.H., Van Kimmenade, A.M.A., Van de Putte, P. and Bosch, L. (1980) Proc. Nat. Acad. Sci. 77, 3922-3926.
21. Van der Meide, P.H., Duisterwinkel, F.J., De Graaf, J.M., Kraal, B., Bosch, L., Douglass, J. and Blumenthal, T. (1981) Eur. J. Biochem. 117, 1-6.
22. Duisterwinkel, F.J., De Graaf, J.M., Schretlen, P.J.M., Kraal, B. and Bosch, L. (1981) Eur. J. Biochem. 117, 7-12.
23. Reeh, S. and Pedersen, S. (1977) in Gene Expression, Clark, B.F.C. *et al.* ed., 11th FEBS Meeting, Copenhagen, pp. 89-98.
24. Jacobson, G.R. and Rosenbusch, J.P. (1977) FEBS Letters 79, 8-10.
25. Van de Klundert, J.A.M. (1978) Ph.D. Thesis, State University of Leiden, The Netherlands.
26. Carroll, R.B., Goldfine, S.M. and Melero, J.A. (1978) Virology 87, 194-198.
27. Van der Meide, P.H., Vijgenboom, E. and Bosch, L., to be published.
28. Neidhardt, F.C., Bloch, P.L., Pedersen, S. and Reeh, S. (1977) J. Bacteriol. 129, 378-387.
29. Maaløe, O. (1979) in Biological Regulation and Development, Goldberger, R.F., ed., Plenum Press, New York, pp. 487-542.
30. Young, F.S. and Furano, A.V. (1981) Cell 24, 695-706.
31. Jonák, J., Smrt, J., Holý, A. and Rychlík, A. (1980) Eur. J. Biochem. 105, 315-320.
32. Miyajima, A., Shibuya, M. and Kaziro, Y. (1979) FEBS Letters 102, 207-210.
33. Shibuya, M., Naskimoto, H. and Kaziro, Y. (1979) Mol. Gen. Genet. 170, 231-234.
34. An, G. and Friesen, J.D. (1979) J. Bacteriol. 140, 400-410.
35. Miyajima, A. and Kaziro, Y. (1980) FEBS Letters 119, 215-218.
36. Miller, D.L., Nagarkatti, S., Laursen, R.A., Parker, J. and Friesen, J.D. (1978) Mol. Gen. Genet. 159, 57-62.
37. Clewell, D.B. (1972) J. Bacteriol. 110, 667-676.
38. Birnboim, H.C. and Doly, J. (1979) Nucleic Acids Res. 7, 1513-1523.
39. Lederberg, E.M. and Cohen, S.N. (1974) J. Bacteriol. 119, 1072-1074.
40. Remaut. E., Stanssens, P. and Fiers, W. (1981) Gene 15, 81-93.
41. Yokota, T., Sugisaki, H., Takanami, M. and Kaziro, Y. (1980) Gene 12, 25-31.
42. Zengel, J.M. and Lindahl, L. (1982) J. Bacteriol. 149, 793-797.
43. Miller, J.H. (1972) in Experiments in Molecular Genetics, Cold Spring Harbor Laboratory, pp. 218-220.
44. An, G., Lee, J.S. and Friesen, J.D. (1982) J. Bacteriol. 149, 548-553.
45. Travers, A. (1976) Nature 263, 641-646.
46. Gordon, J. (1970) Biochemistry 9, 912-917.
47. Gordon, J. and Weissbach, H. (1970) Biochemistry 9, 4233-4236.
48. Pedersen, S., Reeh, S.V., Parker, J., Watson, R.J., Driesen, J.D. and Fiil, N.P. (1976) Mol. Gen. Genet. 144, 339-343.

49. Duisterwinkel, F.J., De Graaf, J.M., Kraal, B. and Bosch, L. (1981) FEBS Letters 131, 89-93.
50. Lindahl, L. and Zengel, J.M. (1982) Adv. in Genetics 21, 53-121.

DETERMINANTS OF TRANSLATIONAL SPECIFICITY IN Bacillus subtilis.

JESSE C. RABINOWITZ
Dept. of Biochemistry, University of California, Berkeley, CA.
94720, U.S.A.

PROTEIN SYNTHESIS BY AN IN VITRO CLOSTRIDIAL SYSTEM

About 10 years ago, I became interested in working on the biosynthesis of a Clostridial protein and attempted to obtain an in vitro protein synthesizing system from Clostridial cells. The ribosomal system prepared from E. coli was active in incorporating amino acids into acid insoluble material in response to f2 RNA or T4 early mRNA, but systems containing Clostridial ribosomes were inactive[1]. After much difficulty, it was found that the Clostridial ribosomes could form protein when a crude mRNA preparation derived from Clostridial cells was used[2,3].

TRANSLATIONAL SPECIFICITY IN PROKARYOTES

Ribosomes and mRNA preparations were prepared from a variety of bacterial species, and homologous and heterologous combinations were tested for protein synthesis activity. Ribosomes derived from E. coli and Pseudomonas responded to all mRNA preparations tested. Ribosomal systems derived from Clostridia, Bacillus and several other Gram-positive organisms only responded to the mRNA from the Gram-positive species, and not to the mRNA preparations derived from Gram-negative species or phages associated with them[4,5]. These observations provided very clear evidence of the occurrence of translational specificity in prokaryotes and suggested to us that its molecular basis might differ from that involved in the determination of the efficiency of expression of proteins encoded by a prokaryotic gene. We therefore attempted to determine the molecular basis for this "Translational Specificity."

RIBOSOME BINDING SITE OF THE β-LACTAMASE GENE

We were able to demonstrate that the determinants of this specificity were associated with the 30S ribosomal subunit and the mRNA[6]. One of the most prominent interactions involving these components occurs at the "Shine-Dalgarno" site[7]. When we started this work, 123 such ribosome binding sites

in E. coli had been sequenced[8], but not a single one was known for a mRNA
derived from a Gram-positive organism. We therefore decided to determine such
a ribosome binding site sequence, and chose to do it for the β-lactamase, or
penicillinase, from the Staphylococcus aureus plasmid PC1[9]. This work was
done by Jane McLaughlin as a graduate student in my laboratory[10,11]. Our
ability to recognize the Shine-Dalgarno sequence in the DNA sequence that she
determined was possible because the amino acid sequence of the penicillinase
was known[12]. We could therefore locate the DNA sequence coding for the
excreted protein[11]. From the size of the protein formed in vitro, we deduced
that it probably includes a leader sequence of 20-25 amino acids. We therefore looked for an initiation codon in that region, and for a Shine-Dalgarno
sequence 3-11 bases from it. A perfect Shine-Dalgarno sequence was found,
GGAGG, that base paired completely with the full pyrimidine rich sequence,
CCUCC, of the B. subtilis or E. coli 16S ribosomal RNA. This sequence is conserved in all prokaryotic 16S rRNAs examined (C.R. Woese, personal communication), although base changes do occur in the positions adjacent to this
sequence. UUG, normally a codon for Leu, was shown to be the initiation
codon, and to code for Met in this case[10]. These two features distinguished
this binding site and initiation site from that which normally occurs in E.
coli, where only 3 or 4 bases are usually involved in the Shine-Dalgarno
base-pairing, and AUG is normally the initiation codon. In vitro experiments
with [^{35}S]methionine, [^{3}H]leucine and [^{3}H]lysine established the start of the
leader sequence with the amino acid methionine, and UUG as its codon, as well
as the leader amino acid sequence. The sequences proposed for the "-35
sequence" and the Pribnow box are not distinguished from those recognized by
the E. coli polymerase.

ø29 TRANSCRIPTION/TRANSLATION

In order to determine whether these sequences were characteristic of ribosome binding sites associated with Gram-positive organisms, we chose to determine the ribosome binding sites of several proteins synthesized in response to
the Bacillus phage ø29, one of the simplest of the Bacillus phages. However,
since ø29, like all other phages associated with Gram-positive organisms, is a
DNA phage, it was necessary to develop a preparation of DNA-dependent RNA
polymerase from B. subtilis[13] and to develop a coupled
transcriptional/translational system from this organism[10]. It was shown that

the translational specificity was maintained by the coupled system. ∅29 DNA
directs the synthesis of three major proteins in the B. subtilis system. They
have molecular weights of 22.4, 13.9 and 10.5 kd. Their synthesis is relatively independent of the presence of initiation factors[14]. The 13.9 kd protein is unique among those that we have encountered in our studies of mRNA
translation. The mRNA for this protein is the only one translated by the system derived from B. subtilis and not by the system derived from E. coli.

We have succeeded in determining the base sequences of the ribosome binding
sites of the 22.4 and the 13.9 kd proteins[15]. This was not as straightforward
as was the case for the β-lactamase because the amino acid sequences of these

proteins were not known. From considerations of the size of the 13.9 kd protein and the size of the mRNA formed from a particular restriction fragment
coding for that protein, it could be calculated that the initiation of the
13.9 kd protein might be between nucleotides 300 and 400 of this fragment.
The first initiation codon to occur in this region preceded by a plausible
Shine-Dalgarno sequence was at the AUG sequence starting at residue 345. To
determine whether this was the initiation codon for the 13.9 kd protein, an in
vitro synthesis of the protein was carried out in the presence of [^{35}S]Met.
The isolated 13.9 kd protein was subjected to Edman degradation and the
labeled amino acid Met was found in the expected positions, 1, 4, and 5[15].
The sequence of the 22.4 kd protein that was formed from mRNA derived from
another restriction fragment was confirmed in a similar manner[15].

SUMMARY OF GRAM-POSITIVE RIBOSOME BINDING SITES

A summary of the base sequences of the ribosome binding sites and initiation sites of proteins derived from sources related to Gram-positive organisms
is shown in Fig. 1.

(a) S. aureus β-lactamase   5′ ...UACAACUGUAAUAUCGGAGGGUUUAUUUUGAAAAGUUAAUAUUUUUAAUUGUAAUUGCUU... 3′

(b) B. licheniformis β-lactamase   ...AUAUUCAAACGGAGGGAGACGAUUUUGAUGAAAUUAUGGUUCAGUACUUUAAAACUGAAAA...

(c) Phage φ29 22.4K protein   ...GACAACCAAUCAUAGGAGGAAUUACACAUGAAUAACUAUCAAUUAACUAUCAAUGAGGUAA...

(d) Phage φ29 13.9K protein   ...AAAUAUAAAUAGAAAGUGGGACGAAGAAAUGGCAAAAAUGAUGCAGAGAGAAAUCACAAA...

Fig. 1.   Initiation sites recognized by Bacillus subtilis ribosomes.

Three of the four sites shown here contain the sequence GGAGG and extend
complementarity on either side of this Shine-Dalgarno pairing region. This is
unusual relative to the sites present in E. coli which usually contain only 3
or 4 bases complementary to the Shine-Dalgarno sequence[8]. Furthermore, 4 of
these 5 base pairs are GC pairs. GC stacking leads to unusually high free
energies of binding relative to other base pairs[16]. The free energies of the
Shine-Dalgarno regions for the 123 E. coli examples[8], are shown in Fig. 2
(open circles). They form a distribution with an average ΔG of about -11.5
kcal/mol; whereas, the values for the 8 ribosome binding sites related

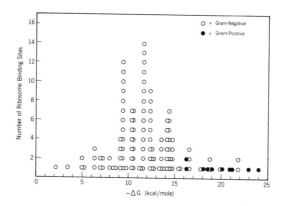

Fig. 2 Distribution of free energies of binding at Gram-negative and Gram
positive ribosome binding sites.

to Gram-positive organisms, that were known when this figure was prepared,
shown by the closed circles, fall in the region of -16 to -24 kcal/mol.

From these results, we have proposed that the molecular basis of the
"Translational Specificity" shown by mRNAs derived from Gram-positive sources
is related to the relatively high free energy of base pairing at the ribosome
binding site compared to that of mRNAs derived from Gram-negative sources.
This generally results from the involvment of not only more of the bases in
the mRNA complementary to the Shine-Dalgarno sequence CCUCC but also those
bases of the 16S rRNA adjacent to either end of this sequence. The limited
number of examples available at the present time does not indicate any

features of the "window" between the end of the ribosome binding site and the initiation codon, or the initiation codon itself that are unique among the mRNAs derived from Gram-positive sources relative to those derived from Gram-negative sources. This theory also provides an explanation for the relative initiation factor independence of translation of mRNAs derived from Gram-positive organisms by protein synthesizing systems derived from either Gram-positive or -negative sources[14].

ACKNOWLEDGMENTS

This work was supported by Grant AM2109 from the National Institutes of Arthritis, Metabolic, and Digestive Diseases.

References

1. Himes, R.H., Stallcup, M.R., and Rabinowitz, J.C. J. Bacteriol. 112, 1057-1069 (1972).

2. Stallcup, M.R. and Rabinowitz, J.C. J. Biol. Chem. 248, 3209-3215 (1973).

3. Stallcup, M.R. and Rabinowitz, J.C. J. Biol. Chem. 248, 3216-3219 (1973).

4. Stallcup, M.R., Sharrock, W.J., and Rabinowitz, J.C. Biochem. Biophys. Res. Commun. 58, 92-98 (1974).

5. Stallcup, M.R., Sharrock, W.J., and Rabinowitz, J.C. J. Biol. Chem. 251, 2499-2510 (1976).

6. Sharrock, W.J., Gold, B.M., and Rabinowitz, J.C. J. Mol. Biol. 135, 627-638 (1979).

7. Shine, J. and Dalgarno, L. Proc. Natl. Acad. Sci. U.S.A. 71, 1342-1346 (1974).

8. Gold, L., Pribnow, D., Schneider, T., Shinedling, S., Singer, B.S., and Stormo, G. Annu. Rev. Microbiol. 35, 365-403 (1981).

9. Murphy, E. and Novick, R.P J. Bacteriol. 141, 316-326 (1980).

10. McLaughlin, J.R., Murray, C.L., and Rabinowitz, J.C. J. Biol. Chem. 256, 11273-11282 (1981).

11. McLaughlin, J.R., Murray, C.L., and Rabinowitz, J.C. J. Biol. Chem. 256, 11283-11291 (1981).

12. Ambler, R.P. Biochem. J. 151, 197-218 (1975).

13. Davison, B.L., Leighton, T., and Rabinowitz, J.C. J. Biol. Chem. 254, 9220-9226 (1979).

14. McLaughlin, J.R., Murray, C.L., and Rabinowitz, J.C. Proc. Natl. Acad. Sci. U.S.A. 78, 4912-4916 (1981).

15. Murray, C.L. and Rabinowitz, J.C. J. Biol. Chem. 257, 1053-1062 (1982).

16. Tinoco, I., Borer, P.N., Dengler, B., Levine, M.D., Uhlenbeck, O.C., Crothers, D.M., and Gralla, J. Nature New Biol. 246, 40-41 (1973).

SECTION II: CODON CONTEXT AND TRANSLATIONAL FIDELITY

SITE DIRECTED MUTAGENESIS OF THE ANTICODON REGION:

THE "UNIVERSAL U" IS NOT ESSENTIAL TO tRNA SYNTHESIS AND FUNCTION

ROBERT C. THOMPSON†, STEVEN W. CLINE†† AND MICHAEL YARUS††
†Department of Chemistry, Temple University, Philadelphia, Pennsylvania
19122; ††Department of Molecular, Cellular and Developmental Biology,
University of Colorado, Campus Box 347, Boulder, Colorado 80309, Telephone
(303) 492-8376.

INTRODUCTION

Transfer RNA's (tRNA's) are of central importance to the speed and
accuracy of translation, and also serve as co-regulators of gene
expression. In order to discriminate among the myriad of molecular
interactions undergone by tRNA's, we propose to create a homologous series
of tRNA molecules, which differ in sequence by single nucleotides. These
tRNA's may then be employed to dissect the structural requirements for tRNA
gene function, and for translation and regulation by tRNA itself.

We have chosen to do this by altering termination-suppressor forms of
the tRNA^{Trp} (E. coli) gene (=Su$^+$7, supU, or trpT 175). By working at the
level of the gene, and recloning altered genes, we can observe tRNA gene
function and tRNA function in a steady-state environment in vivo.
Similarly, the use of a termination suppressor makes possible quantitative
evaluation of tRNA function at the naturally cycling ribosomal A site in
vivo. We expect these varied values for the relative effectiveness of
tRNA's to be very informative in themselves, and very useful when the tRNA
activities are later dissected into individual biochemical steps in vitro.

In Yarus et al[1] we have described the construction of a cloning system
which allows our synthetic tRNA genes to be incorporated in a small,
defined transcription unit under lac operator-promoter control. We also
described the creation of a novel suppressor tRNA gene by intercalation of
a new anticodon region into the tRNA^{Trp} gene.

In this current communication, we wish to introduce a new site-directed
mutagenesis protocol which offers many experimental advantages over our
previous method. Foremost among these is its generality: it allows the
site-directed creation of a series of tRNA's which differ in the anticodon
loop and stem, a region vital to translational activity.

To illustrate this method, we describe the construction and activity of
a tRNA gene which is altered in a way not previously seen. The "universal

U," U[33], which immediately preceeds the anticodon of every known elongator
tRNA, is converted to a C.

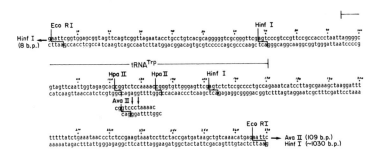

Fig. 1. Sequence of the 264 bp tRNA gene fragment. The tRNA[Trp] gene is
overlined, and the central Hpa II fragment which contains the anticodon
region is indicated. The positions of several other restriction sites of
significance to the analysis (see text) are also indicated. Below the main
sequence, the changed sequence of the mutant tRNA gene created in this work
is shown.

Preparation of a new tRNA gene by site-directed mutagenesis

Oligonucleotide synthesis. The deoxyoligonucleotide used in this work,
5' CGGTCCCTAAAAC 3' (C-oligonucleotide), was synthesized starting with a
nucleotide linked to macroporous silica through a base-labile linker on its
3' hydroxyl.[2] The chain was extended in a 5' direction using 5'
dimethoxytrityl nucleoside 3' phosphoramidites.[3] After each cycle the new
phosphite ester bond was oxidized with iodine, failure sequences were
capped with acetic anhydride and the 5' dimethoxyltrityl group removed with
ZnBr$_2$. The overall yield of the tridecanucleotide
CpGpGpTpCpCpCpTpApApApApC-silica, based on yields of dimethoxyltrityl
cation at each step, was 8.5%. We thank Bill Efcavitch and Marvin
Caruthers for schooling us in their DNA synthetic methods.

After removal from the silica the oligonucleotide was purified on a G-50
gel filtration column and the early fractions containing about one third of
the total A$_{260}$ material were phosphorylated using T4 polynucleotide kinase
and [gamma-^{32}P] ATP. The labeled material was fractionated by gel
electrophoresis. An autoradiogram enabled us to find the region of the gel

corresponding to the phosphorylated tridecanucleotide. This region was
extracted with buffer, desalted, and used in the ligation reaction.

Construction and cloning of the new gene. We have cloned the Su⁻7
tRNA^Trp gene of E. coli into the Eco RI site of pOP203 UV5 as part of a
264 bp insert which also includes a tRNA^Asp gene and the rrnC
terminator.[1] The sequence of the 264 bp insert from the resulting plasmid
(pMY231) is shown in Figure 1. A derivative of pMY231, pMY242, lacking the
13 bp Hpa II fragment coding for the anticodon loop-and-stem of tRNA^Trp
was prepared by Hpa II cleavage of the Eco RI insert, religation of the two
large fragments (2-121 and 135-265), and recloning into pOP203 UV5. This
"anticodonless" clone (pMY242) provides a convenient supply of tRNA^Trp
gene fragments free from any indwelling anticodon loop-and-stem information
(Figures 1 and 2).

The mutagenesis scheme takes advantage of the fact that partial duplexes
and the Eco RI insert of pMY231 Su⁻7 and the Hpa II fragments of the Eco RI
insert of pMY242 are single stranded in the 13-base region coding for the
anticodon loop and stem of tRNA^Trp (see Figure 2). To prepare such
partial duplexes, 0.8 pmol of Hpa II-cleaved Eco RI insert from pMY242 and
0.6 pmol of Eco RI insert from pMY231 Su⁻7 were denatured together and
annealed at 37°C in 4x SSC/50% formamide. The DNA was then precipitated.

The mutation is achieved by ligating a synthetic oligonucleotide, which
is only partially complementary to the single stranded region, into the gap
at the anticodon region of the partial duplex. To do this, 6 to 12 pmol of
phosphorylated synthetic tridecanucleotide was introduced into the partial
duplex with T4 DNA ligase (15°C for 14 hours). Then Eco RI was added to
cut the ligated oligomeric insert into monomers (Figure 2).

To clone the altered tRNA gene, this heteroduplex was ligated to 1.3
pmol of Eco RI-cleaved, dephosphorylated pOP203 lac OP UV5 (this plasmid
was a gift from Forrest Fuller) and used to transform a Tet^S leu_am Su⁻
strain of E. coli containing 3 mutations which enhance termination
suppressor activity (MY118).[4] The overall efficiency of transformation to
tet^R was 10³ to 10⁴ per ug vehicle, of which 0.5% (87 colonies) were leu⁺
trp⁺ and therefore Su⁺. Plasmids carrying replicas of the original
"instructive strand" (Figure 2) (template for the new gene) are easily
detected or selected against because they lack suppressor activity (they
are Su⁻7).

192

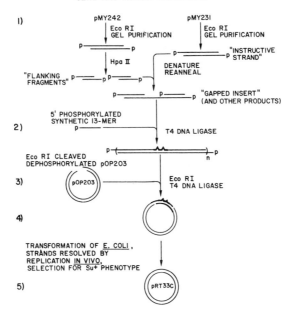

Fig. 2. A schematic for the construction of the cloned tRNA gene. The plasmid pRT33C Su⁻ was constructed by ligation (step 2) of a synthetic deoxyoligonucleotide (heavy line, step 2) into a "gapped insert" destined for cloning adjacent to the lacOP region of pOP203 (step 4). The specific gap is formed when denatured "flanking fragments" hybridize with a continuous "instructive strand" which contains the intact Su⁻7 tRNA^{Trp} gene sequence (step 1). The heteroduplex (deviations in the heavy line, steps 2 and 4) is resolved by replication into the original Su⁻7 gene and a new potential termination suppressor (step 5).

To purify the plasmid containing the mutant tRNA gene, plasmids were isolated from several Su⁺ colonies and used to transform a tet^S his$_{am}$

trp$_{am}$ recA$^-$ strain of E. coli containing F'laci^Q (MY75). This strain contains no suppressor enhancer mutations; the new suppressor was nevertheless quite functional (see below). Plasmids were isolated from Su$^+$ transformants by sarkosyl and alkaline SDS procedures, and were characterized as to size, restriction pattern, and DNA sequence.

The plasmid carrying the C^{33} tRNA gene will be called pRT33C (Figure 2); it is otherwise the same in structure as pMY228 Su$^+$7.[1]

TABLE 1

BACTERIAL STRAINS USED IN THIS WORK

LAB NUMBER	PREVIOUS NOMENCLATURE	SOURCE	GENOTYPE
MY50	--	Lab collection	W3110 trpR$^-$ trpA9605$_{am}$ his29$_{am}$ ilvl tdr Su$^+$1
MY51	--	Lab collection	as MY50, but Su$^+$2
MY52	--	Lab collection	as MY50, but Su$^+$3
MY75	--	by mating LS286, from Larry Soll	W3110 trpR$^-$ trpA9605$_{am}$ his29$_{am}$ ilvl recA56/ F'laci^Q
MY118	--	Su$^-$ transductant of MX653 from Max Oeschger	F$^-$ MetBl leu$_{am}$ trp$_{am}$ lacZ$_{am}$ galK$_{am}$ galE sueA sueB sueC tsx relA sm^R
MY347	SPOC	Jeffrey Miller	ara$^-$ (lac pro) strA/ F'laci^Q-lacZ fusion pro$^+$
MY349	SPOC A24	Jeffrey Miller	ara$^-$ (lac pro)strA/ F'laci^Q A24$_{am}$$^-$ lacZ fusion pro$^+$

Characterization of pRT33C

Plasmids derived from 3 independent Su$^+$ isolates from the C-oligonucleotide mutagenesis were shown by agarose gel electrophoresis to have a size not appreciably different from that of their ancestor pMY231.

We conclude that they contain a single copy of the pOP203 cloning vehicle, and probably a single copy of the Eco RI insert. This latter fact was confirmed by a Hinf I digestion of these plasmids (Figure 3), which also showed that the insert was oriented productively with respect to the lac UV5 promoter of the vehicle. One of the two 75 bp fragments seen in these digests results from cleavage of the Hinf I site at base 67 of the insert and Hinf I site at base 8 upstream from the Eco RI site (cf. Figure 1).

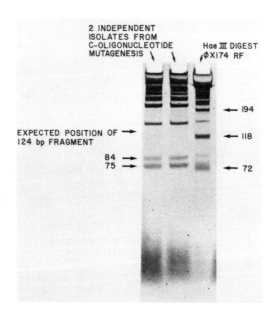

Fig. 3. The Hinf I digest of pRT33C determines that the tRNA genes are correctly oriented singlets.

Fragments of 124 bp, characteristic of a backwards orientation of the insert, and 180 bp, characteristic of multiple inserts, were conspicuous by their absence. The plasmid pRT33C therefore contains a single set of tRNA genes, oriented for correct transcription from lac P.

The presence of one of the desired mutations, $T^{127} \longrightarrow C^{127}$ in the Eco RI insert of pRT33C can be demonstrated by an Ava II digest of the plasmid (Figure 4). A 256 bp fragment was noted when pRT33C was digested with this enzyme, but was not seen when a homologous plasmid carrying only the Su^+7 amber anticodon mutation (pMY228) was treated similarly. Instead a fragment larger than 1500 bp was observed (1678 bp is predicted). The mutation described converts the GGTCT sequence to a new <u>Ava</u> II site, GGTCC, at base 123 of the insert (see Figure 1). Since no <u>Ava</u> II site occurs within 600 bp upstream of this site the new 256 bp fragment must result from another <u>Ava</u> II site downstream. The existence of an <u>Ava</u> II site 112 bp downstream from the vehicle-insert junction, in fact, is predicted by comparision with the sequence of pAO2. This latter plasmid, whose sequence is homologous to pOP203 in this region, was constructed by Oka et al.[5]

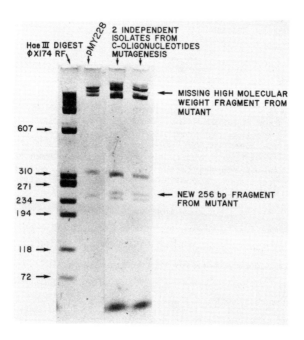

Fig. 4. The digest of pRT33C shows that a predicted <u>Ava</u> II site has been created in the anticodon region.

Final confirmation that the desired mutations had been produced was obtained by sequencing of the Eco RI insert from pRT33C. For this purpose the <u>Eco</u> RI fragment of the plasmid was first cloned into M13mp7 by the methods of Messing et al.[6] and sequenced by the dideoxy method of Sanger.[7] The part of a sequence gel corresponding to the suppressor tRNA gene is shown in Figure 5. Under the gel lanes the sequence of the critical region is lettered in. The wild type sequence is connected by arrows to the mutated sequence, as in Figure 6. Note that the sequence of the tRNA complementary strand is shown in Figure 5 (lower strand in Figure 1).

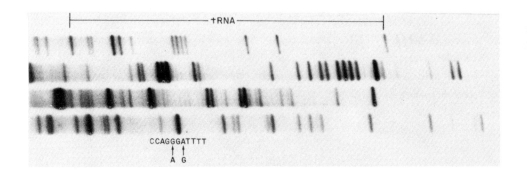

Fig. 5. A section of a dideoxy sequencing gel[7] which shows the complement of the new tRNA gene sequence. The tRNA region is overlined, and the site-directed alterations are indicated below the figure. Continuous lettering shows the observed, mutant sequence, and the arrows show the changes from the wild type. GAT is the backwards complement of the new anticodon; that is, it represents the new amber codon, UAG (cf. Figure 1).

From the top, the lanes show oligonucleotides specifically terminated at T, G, C, and A (bottommost).

The structure of the tRNA gene inferred is shown in the standard cloverleaf conformation in Figure 6. The information for tRNA nucleotides 28 through 40 has been replaced in the tRNA^{Trp} gene, with the result that the anticodon is now complementary to the amber codon ($C^{35} \longrightarrow U$), and the "universal U" has become a C ($U^{33} \longrightarrow C$).

Function of the mutant tRNA gene and gene product

Three systems have been used to determine whether the mutant gene is expressed and whether the gene product is a functional tRNA. Firstly, the mutant gene product was tested for its ability to suppress amber codons in phage lambda by spotting the phages on MY75 (pRT33C). The set of lambda ambers is known to respond characteristically to different suppressors. The results of this test are shown in Table 2 along with those of analogous experiments with E. coli containing other well-characterized amber suppressors. By comparison of the suppression pattern to characterized suppressors which are strong or weak, and which insert various amino acids, a rough idea of the strength and specificity of the new suppressor can be gained.

The results of these tests suggest that Su$^+$7 (C33), like Su$^+$7, is a glutamine inserter. When fully induced (+IPTG) the pattern of suppression of the new tRNA parallels that of Su$^+$7 with the single exception that it shows weak suppression of Q_t57. This single exception may indicate a

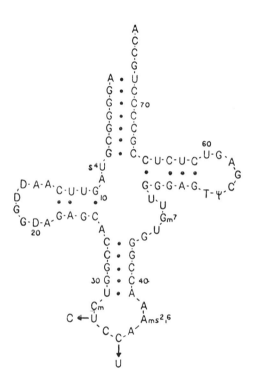

Fig. 6. The Su$^-$7 tRNA^{Trp} sequence,[8] folded into a planar cloverleaf. The wild type tRNA gene product[8] is shown as a continuous sequence. The sequence alterations produced by our site-specific mutagenesis protocol (Figure 5) are shown at the termini of the arrows.

context effect or other unknown circumstance. In contrast, the suppression pattern of Su$^+$7 (C33) differs from that of Su$^+$3 (a tyrosine inserter) in

four cases. Su⁺3 is unable to suppress phage ambers E13, N7, P3 and R216.
Su⁺7 (C33) differs from Su⁺1 (inserts serine) for seven of the mutants
tested, and is therefore clearly distinguished.

A comparison of the abilities of Su⁺7 (C33) and Su⁺7, when not fully
induced, to suppress A32, B10 and P3 indicates that the mutant gene tRNA is
a weaker suppressor than Su⁺7 itself. Similarly, tRNA Su⁺7 (C33), when
fully induced, suppresses lambda ambers J27 and C42, suggesting that it is
a stronger suppressor than Su⁺2, which is a weak glutamine inserting
suppressor than Su⁺7. We therefore expect Su⁺7 (C33) to be a
glutamine-tRNA of intermediate suppressor efficiency. This interpretation
is sustained by the more quantitative tests of suppressor activity
described below.

TABLE 2

SUPPRESSION OF LAMBDA AMBER MUTANT PHAGE BY VARIOUS SUPPRESSORS

Phage Amber	Su^+7		$Su^+7(C33)$		Su^-7	Su^+1	Su^+2	Su^+3
	+IPTG	−IPTG	+IPTG	−IPTG				
A32	+	±	+	0	0	±	+	+
B10	+	±	+	0	0	+	+	+
E13	+	0	+	0	0	±	0	0
J27	+	0	+	0	0	0	0	+
N7	+	0	±	0	0	0	+	0
029	+	0	+	0	0	0	+	+
P3	+	+	+	0	0	0	+	0
Qt57	0	0	±	0	0	0	0	+
R60	+	+	+	+	0	±	+	+
R216	+	+	+	+	0	0	+	0
S7	0	0	0	0	0	0	0	0
C42	+	0	+	0	0	0	0	±
CI⁺	+	+	+	+	+	+	+	+

Legend: About 10⁵ PFU of the phage in 10 ul are spotted on lawns of MY75
carrying Su⁺7, Su⁺7 (C33), orSu⁻7 tRNA genes in plasmids pMY228, pRT33C, or
pMY231 respectively. Su⁺1, Su⁺2, and Su⁺3 are chromosomal, in MY50, MY51,
or MY52 respectively. Plates[1] are incubated for 16 hr. at 42°C. + = total
clearing; ± = partial clearing; 0 = no lysis or only occasional phage
revertant plaques.

The efficiency of Su$^+$7 (C33) may be evaluated quantitatively in vivo by measuring its ability to promote TrpA protein synthesis in MY75, which contains an amber mutation (trpA9605$_{am}$). The details of this assay have been published[1] and indicate that, when fully induced with IPTG, Su$^+$7 has a transmission efficiency of 80-94% in this cell. That is, the amber codon is read most of the time by Su$^+$7 and only in the minority of the cases by release factors. When fully repressed in the presence of laci^Q, Su$^+$7 is 7-10% efficient.[1] For Su$^+$7 (C33) we find an efficiency of 30% when fully induced (+IPTG) and 2% when not induced. Therefore, the replacement of the universal U seems to have reduced suppressor efficiency by 2.5 (induced) to 4-fold (uninduced). In principle, this difference might be due to differences in aminoacylation, or maturation, or ribosomal function of these tRNAs (Yarus et al., in preparation). Thus, the U33 --> C33 modification reduces overall tRNA function by about 3-fold. Alternatively, the "universal U" must be dispensable for all of these steps.

The third comparison we can make between Su$^+$7 (C33) and Su$^+$7 comes from the levels of beta-galactosidase present in cells carrying F' lacI$^Q_{am}$lacZ fusions (SPOCA24). To synthesize an enzyme capable to hydrolyzing O-nitrophenylgalactoside, these strains must initiate translation at the lacI^Q initiation codon. They must then suppress the intervening amber mutation (Miller, personal communication) using the tRNA supplied by the pRT33C suppressor tRNA gene, and proceed into the fused Z gene. Laurel Raftery in our lab has measured the levels of beta-galactosidase activity in these cells carrying both Su$^+$7 and also carrying Su$^+$7 (C33) coding plasmids. When fully induced these levels are 18% and 8% of that found in a comparable lacI^{Q}lacZ fusion strain (SPOC). This result is not unlike the one mentioned above for Trp A protein. The U^{33} --> C^{33} modification has reduced the ability of the tRNA to read an amber codon by a factor of 2. Thus the tRNA created in these experiments is moderately efficient in translation, despite the disruption of an otherwise strictly conserved sequence feature, the "universal U."

DISCUSSION

In vivo, structure-function relations in tRNA have been pursued in the past largely by selection of genetic mutants, usually suppressor tRNA's which have lost their function.[9,10] The sequences of these mutants have adumbrated many essential features of the tRNA structure; indeed, they are responsible for much of what we know about tRNA function in vivo.

Nevertheless, it is likely that site-specific mutagenesis will reveal
additional information. For example, mutants showing partial loss of
function are likely to have been missed in most previous selections. We
have now shown that alteration of the "universal U" may only moderately
reduce function, and this may explain why mutational alteration of this
ubiquitous structure has not previously been observed. Thus, strong
conservation of struture does not imply that the conserved structure is
essential for function. We have previously shown,[11], see also [12] that
the highly conserved TψCG sequence is not necessarily essential. However,
both the "universal U" and TψCG must be present for optimal translational
function.

We may rationalize these findings in the notion that evolution probably
bears very effectively on the translational efficiency of tRNA's. Even
relatively small decrements of function can therefore be selected against.
This statement is much less obvious, however, when put in an alternative
form: the existence of a rigorously conserved natural sequence feature
does not necessarily imply critical effect on function. While suppressor
efficiency assays easily measure the effect of alteration of the "universal
U," it is nevertheless inessential. We imagine that there might be many
puzzling future cases of site-specific mutagenesis of persistent sequence
features which fail to have dramatic phenotypes, because of the ruthless
delicacy of the selection which maintains them. To put this another way,
careful quantitation of the function of altered molecules is indispensible
to effective study using these systematic sequence alteration techniques.

The case of the universal U^{33} is particularly interesting, because
examination of the crystal structure of yeast tRNA^{Phe} had suggested[13]
that U^{33} might have a role in the particular structure of the anticodon
loop. An apparent H-bond from the hydrogen on the U-ring nitrogen to a
phosphate oxygen of residue 36 seemed to help pin the sides of the
anticodon loop together. This pin occurs across a sharp bend in the
backbone which is necessary to e.g., stack the anticodon on the 3-prime
side of the loop. Because C in its predominant tantomeric form cannot form
this H-bond, we can now conclude that the bond is nevertheless not
essential in vivo for any tRNA function; not for coding in the ribosomal
A-site, nor for effective tRNA maturation, or tRNA aminoacylation.
However, it is required for maximal function.

The mutagenesis protocol we have used here offers several advantages
over our previous method.[1] We previously intercalated a small double

stranded anticodon-containing fragment from another tRNA gene between flanking tRNA^{Trp} gene fragments (cf. Figure 2). The current method requires synthesis of only one strand for the region to be altered, making deoxynucleotide synthesis simpler. In addition, the use of a template strand insures that the anticodon region is flanked by two different flanking regions (instead of an inverted repeat), and that the anticodon fragment is in correct orientation to the two flanking regions. Both these features potentially increase the yield of correctly formed mutant genes by two-fold, or four-fold overall. The heteroduplex inevitably formed by this method (Figure 2, section 2) may be repaired after transformation, but before segregation of the desired new tRNA gene. This might be a disadvantage. It may be turned to advantage, however, if the products of partial repair (which is rare in our experiments) are also genes of interest. By way of contrast, intercalation of a double-stranded fragment, if it creates a tRNA gene at all, almost surely creates the desired one. This may be a useful quality if many changes are to be made in one step, or if there is likely to be biological selection against the newly created gene. Finally, the synthetic production of the new sequence allows us to systematically vary the anticodon region. We have used the method of single-strand insertion to make other such alterations, on which we will soon report.

To summarize: The uses of systematically altered tRNA's are only beginning to be apparent, but there are already theoretical reasons and experimental results which suggest that structure-function studies can be carried out at a level of resolution not previously accessible. This should contribute to our understanding of tRNA in particular, and, more generally, to our understanding of the uses of RNA primary, secondary, and tertiary structure.

ACKNOWLEDGEMENTS

This work was supported by NIH research grant GM25627 to M.Y. and NIH Senior Fellowship GM08575 to R.C.T.

REFERENCES

1. Yarus, M., McMillan, D., Cline, S., Bradley, D. and M. Snyder (1980) Proc. Natl. Acad. Sci. 77, 5092-5096.
2. Mateucci, M.D. and M.H. Caruthers (1980) Tet. Lett. 21, 719-722.
3. Beaucage, S.L. and M.H. Caruthers (1981) Tet. Lett. 22, 1859-1862.
4. Oeschger, M.P. and G.T. Wiprud (1980) Mol. Gen. Genet. 178, 293-300.

5. Oka, A., Nomura, N., Morita, M., Sugisaki, H. and K. Sugimoto (1979) Mol. Gen. Genet. 172, 151-159.
6. Messing, J., Crea, R. and P.H. Seeburg (1981) Nucleic Acids Res. 9, 309-323.
7. Sanger, F., Nicklen, S. and A.R. Coulson (1977) Proc. Natl. Acad. Sci. USA 74, 5463-5467.
8. Hirsh, D. (1971) J. Mol. Biol. 58, 439-458.
9. Smith, J.D. (1979) in: Nonsense Mutations and tRNA Suppressors, (J.E. Celis and J.D. Smith, eds.), Academic Press, New York, pp. 109-125.
10. Kurjan, J., Hall, B.D., Gillam, S. and M. Smith (1980) Cell 20, 701-709.
11. Yarus, M. and L. Breeden (1981) Cell 25, 815-823.
12. Kudo, J., Leineweber M. and U.L. Rajbhandary (1981) Proc. Natl. Acad. Sci. USA 78, 4753-4757.
13. Quigley, G. and A. Rich (1976) Science 194, 796-806.

PROPERTIES OF A YEAST tRNA^{Phe} AMBER SUPPRESSOR

A. GREGORY BRUCE AND RAYMOND F. GESTELAND
Department of Biology, Howard Hughes Medical Institute, University of Utah,
Salt Lake City, Utah 84112.

INTRODUCTION

The tRNA population in a cell is by no means an equimolar set of all the
species. There is only one molecule of the most abundant tRNA species for
every two ribosomes in E. coli and minor species are often ten-fold less
abundant[1]. This means that a rate limiting step in translation could be ribosomes searching for needed, rare tRNA species and the choice between a codon
requiring a rare tRNA and a codon specifying the same amino acid but requiring
a common tRNA could result in slow versus rapid synthesis of the protein[2,3].
That tRNA levels might indeed be rate limiting under certain conditions is best
demonstrated by the codon usage of bacteriophage T4. When E. coli is infected
by the bacteriophage, eight phage encoded tRNA genes are expressed and the
products supplement the normal host **complement** of tRNAs. While these T4 genes
can be deleted without affecting growth on most lab strains of E. coli, various
natural bacterial isolates will not support growth of T4 unless the tRNA^{Ile}
gene is functional[4]. This implies that T4 has found it necessary to supplement
certain tRNAs that, due to T4's codon choices, are needed in greater abundance
than the host cell can supply. This evidence, admittedly circumstantial,
suggests that the rate of mRNA translation can be preset by the choice of
codons.

A second and more subtle way in which codon sequence and codon choice may
affect the rate of mRNA translation is through "context effects". It has been
known for some time that the efficiency of suppression of nonsense mutations by
the same suppressor tRNA is very different depending on the surrounding mRNA
information[5]. This has been most dramatically demonstrated by the experiments
of Bossi and Roth[6] which show that changing the first base downstream from an
amber codon has an enormous effect on suppression efficiency. Thus it seems
likely that context information around a codon influences reading efficiency,
but whether this is due to direct interactions with tRNA, to mRNA structure or
to interactions between tRNAs reading adjacent codons is not yet clear. In any
case, context can influence reading rate and hence also serve to set the level
of translation of a message.

Another oddity of translation that has recently been brought to light is

204

that occasional frameshifting at particular sites is probably a normal event.
Experiments using in vitro protein synthesis first showed that altering the
balance of normal tRNAs would greatly increase the rate of frameshifting at
specific codons in MS2 RNA and that only certain normal tRNAs would do this[7].
Moreover a hybrid coat-lysis protein arising from a frameshift event was identified both in vitro and in vivo[7,8]. This led to the idea that some tRNAs
could occasionally slip in at incorrect codons and effectively read 2,4 or 5
nucleotides. Subsequently, it has been shown that a frameshift event during the
translation of MS2 coat protein is needed in order to express the overlapping
lysis gene[9] and that several of the T7 proteins are made in two forms with the
minor product resulting from a frameshift event within the gene[10]. At this
point, it is unclear what allows these frameshifts to occur and what rules
govern these events.

With these questions in mind and with the realization that neither the
rules of the codon-anticodon interaction nor the specific roles of modified
bases in tRNA function are clearly worked out, we decided to study the codon-
anticodon interaction using the synthetic approach of Bruce and Uhlenbeck[11]
that permits construction of tRNAs with anticodon loops of choice. In particular we have replaced the anticodon and the hypermodified Y-base of yeast
tRNA[Phe] with oligonucleotides CpUpApX, where X is any nucleotide. These tRNAs
have an anticodon which should recognize the UAG terminator so that the constructed tRNA functions as an amber suppressor. The activity of these suppressors can be readily measured in an in vitro protein synthesizing system.

CONSTRUCTION OF ALTERED tRNAS
A summary of the method for the construction of yeast tRNA[Phe] molecules
with altered anticodon loops, described in detail by Bruce and Uhlenbeck[11], is
shown in Figure 1. First the hypermodified Y-base is specifically removed by
incubation at pH 2.9 leaving only the ribose in position 37. The chain can
then be cleaved at this position by treatment with aniline to yield intermediate 2. The anticodon is then removed by a partial digestion with ribonuclease A (intermediate 3). A tRNA with a new anticodon can then be constructed by inserting a new oligoribonucleotide. The insertion of CpUpApA is
illustrated in Figure 1. The 3' end of the oligonucleotide is joined to the
5' end of the 3' half molecule using T4 RNA ligase (intermediate 4). Incubation
at pH 6.9 with T4 polynucleotide kinase/phosphatase causes the removal of the
3' phosphates from both half molecules and the transfer of a phosphate from ATP
to the 5' end of the oligonucleotide added in the previous step. The anticodon

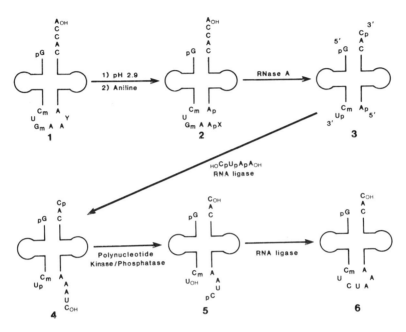

Fig. 1. Summary of the steps required to replace the GmpApApY sequence of yeast tRNAPhe (nucleotides 34-37) with CpUpApA.

loop is then sealed by the addition of T4 RNA ligase and the product is purified on a 20% polyacrylamide-urea gel. Although there are a number of steps, this procedure is relatively simple and yields of greater than 40% are possible. This altered tRNA can be aminoacylated by yeast phenlalanyl-tRNA synthetase but it is a poorer substrate for the enzyme than the native tRNAPhe [12].

IN VITRO PROTEIN SYNTHESIS WITH A CONSTRUCTED tRNA

The suppression of nonsense mutations in an in vitro protein synthesizing system provides a convenient assay for the activity of the constructed tRNAs[13]. In this system, the suppressor tRNA competes with the release factors for the termination codon rather than with another tRNA. This allows the suppressor efficiencies of different tRNAs to be compared directly from a quantitation of the ratio of read-through to termination at a nonsense codon.

The tRNA constructed in Figure 1, tRNA$^{Phe}_{CUAA}$, has the anticodon CpUpA which is complementary to the amber stop codon (UAG) and an adenosine in place of the hypermodified nucleotide. The ability of this tRNA to function as an amber suppressor in the in vitro system using Qβ RNA with an amber mutation in the

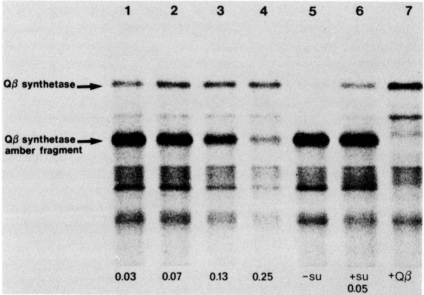

Fig. 2. Autoradiograph of a 10% polyacrylamide-SDS gel analysis of the products of an in vitro protein synthesizing system containing 2 ug (in 12.5 ul reaction) of either wild type Qβ RNA (lane 7) or Qβ RNA with an amber mutation in the synthetase gene (lanes 1-6). Reactions contained: 1-4)tRNA$^{Phe}_{CUAA}$, 5) no added suppressor tRNA and 6) yeast tRNA^{Ser} amber suppressor. The amount of tRNA added (ug) is shown at the bottom of each lane.

synthetase gene as the messenger RNA is demonstrated in Figure 2. Very little read through of the amber codon can be detected when no suppressor tRNA is added (lane 5). When increasing amounts of tRNA$^{Phe}_{CUAA}$ are added, lanes 1-4, increasing amounts of the larger protein which migrates with the wild type Qβ synthetase (lane 7) can be seen. Although the anticodon loop of this tRNA has been drastically altered, it recognizes the amber codon with a high efficiency and, despite the reduced efficiency of aminoacylation by the yeast phenylalanyl-tRNA synthetase, the amino acid inserted by this tRNA is phenylalanine[14]. In a comparison with a yeast tRNA^{Ser} amber suppressor, the constructed tRNA was found to be about 2-fold more efficient at suppressing amber termination codons. The finding that a constructed tRNA can be a better suppressor than one which was isolated should not be surprising since a suppressor which is too efficient in vivo will be lethal. This tRNA has also been shown to suppress the normal amber termination signals for the MS2 synthetase, brome mosaic virus coat protein, R17 synthetase and an amber mutation in the Qβ coat protein (GB-11)[15].

The addition of tRNA^{Phe}, tRNA$^{Phe}_{AAAC}$, tRNA$^{Phe}_{CCCG}$, tRNA$^{Phe}_{GCUA}$, tRNA$^{Phe}_{UCAA}$, tRNA$^{Phe}_{UUAA}$, or the tRNA^{Phe} half molecules sealed with no oligonucleotide inserted in the anti-codon loop did not cause read-through of the amber codons (data not shown). Thus, in this system, a yeast tRNA^{Phe} with an altered anticodon loop will insert its amino acid only in response to its complementary codon.

EFFECT OF ALTERING THE NUCLEOTIDE ADJACENT TO THE ANTICODON

The nucleotide on the 3' side of the anticodon can have a dramatic effect on the codon-anticodon interaction. Suppressor tRNAs isolated from yeast[16] and E. coli[17] which have an amber anticodon, CpUpA, but lack the normal modification on the adenosine adjacent to the anticodon have been shown to be unable to suppress amber mutations in vitro. These results suggest that by inserting different nucleotides in this position the suppressor efficiency of the con-structed tRNAs could be dramatically altered. The ability of tRNAs with C, U, G, or A in this position to suppress the Qβ synthetase amber mutation is shown in Figure 3. In lanes 1-4, the tRNAs containing a pyrimidine show only small amounts of suppression while the tRNAs which contain a purine in this position,

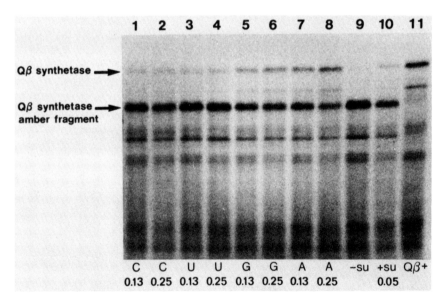

Fig. 3. Autoradiograph of a 10% polyacrylamide-SDS gel analysis of an in vitro protein synthesizing system containing 2 ug (in 12.5 ul reaction) of either wild type Qβ RNA (lane 11) or Qβ RNA with an amber mutation in the synthetase gene (lanes 1-10). Reactions contained: 1-2) tRNA$^{Phe}_{CUAC}$, 3-4) tRNA$^{Phe}_{CUAU}$, 5-6) tRNA$^{Phe}_{CUAG}$, 7-8) tRNA$^{Phe}_{CUAA}$, 9) no suppressor tRNA and 10) yeast tRNA^{Ser} amber suppressor. The amount of tRNA added (ug) is shown at the bottom of each lane.

lanes 5-8, show a high level of suppression. The greater ability of these tRNAs to suppress the amber mutation is not due to a modification of the adjacent purine during the incubation in the extract[14]. In a similiar experiment, the ability of these tRNAs to cause read-through of the normal termination signal of the BMV coat protein was determined (Figure 4). The pyrimidine substituted tRNAs are even less efficient with the BMV RNA than with the Qβ synthetase amber RNA. In the previous case, the $tRNA^{Phe}_{CUAC}$ and the $tRNA^{Phe}_{CUAU}$ were as efficient as the yeast $tRNA^{Ser}$ suppressor. With the BMV RNA, the control tRNA is considerably more efficient. The purine containing tRNAs are still quite efficient with the BMV RNA. In Figure 5, the suppression of an amber mutation at position 17 in the Qβ coat protein gene is shown. In this case, the $tRNA^{Phe}_{CUAU}$ suppresses poorly but the $tRNA^{Phe}_{CUAC}$ shows a significant level of suppression. The purine containing tRNAs are still stronger suppressors with the $tRNA^{Phe}_{CUAA}$ being the most efficient.

Altering the nucleotide on the 3' side of the anticodon of the constructed amber suppressor tRNAs has a large effect on the codon-anticodon interaction, however, the explaination for this effect is not clear. In experiments of this type, one hopes to be able to make significant changes which do not alter the

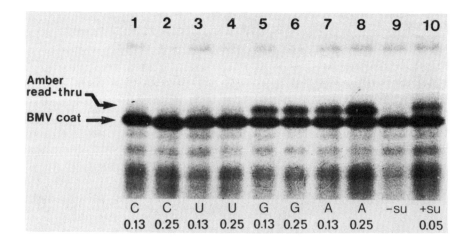

Fig. 4. Autoradiograph of a 15% polyacrylamide-SDS gel analysis of an in vitro protein synthesizing system containing 1 ug (in 12.5 ul reaction) of brome mosaic virus RNA. The BMV coat protein has a molecular weight of 19,000 and an amber terminator codon. The addition of an amber suppressor tRNA produces a slower migrating band on the gel corresponding to a protein with 8 extra amino acids[18]. Reactions contained: 1-2) $tRNA^{Phe}_{CUAC}$, 3-4) $tRNA^{Phe}_{CUAU}$, 5-6) $tRNA^{Phe}_{CUAG}$, 7-8) $tRNA^{Phe}_{CUAA}$, 9) no added suppressor tRNA and 10) yeast $tRNA^{Ser}$ amber suppressor. The amount of tRNA added (ug) is shown at the bottom of each lane.

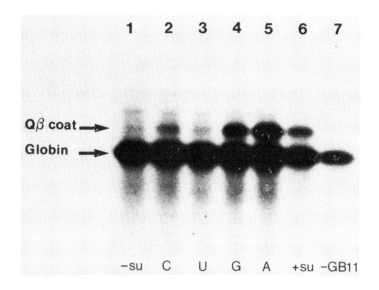

Fig. 5. Autoradiograph of a sodium salicylate treated 15% polyacrylamide-SDS gel analysis of an in vitro protein synthesizing system containing 1 ug (in 12.5 ul reaction) of Qβ RNA with an amber mutation at position 17 in the coat protein gene. The reactions contained: 1) no added suppressor tRNA, 2) tRNA$_{CUAC}^{Phe}$ (0.25 ug), 3) tRNA$_{CUAU}^{Phe}$ (0.25 ug), 4) tRNA$_{CUAG}^{Phe}$ (0.25 ug), 5) tRNA$_{CUAA}^{Phe}$ (0.25 ug), 6) yeast tRNA^{Ser} amber suppressor (0.05 ug) and 7) no added mRNA.

properties of the tRNA or a very subtle change which produces a large effect. In these experiments, the observation that the tRNAs with a purine adjacent to the anticodon are efficient suppressors while those containing a pyrimidine in this position are not may reflect a specific interaction which requires a purine in this position or it could be due to the different stacking properties of these nucleotides causing the anticodon loop to adopt an unfavorable confor- mation for codon recognition. The construction of tRNAs with modified purine nucleotides in this position will help to eliminate this complication.

CONSTRUCTION OF tRNAS CONTAINING MODIFIED NUCLEOTIDES

McCloskey and Nishimura[19] observed that there is a strong correlation between the identity of the nucleotide in the first position of the codon and the modification found on the adjacent nucleotide. When the codon starts with a U, the tRNA contains a hydrophobic modification such as i^{6}A, ms^{2}i^{6}A or m^{2}A. When the codon starts with an A, the tRNA contains a hydrophilic modification such as t^{6}A or mt^{6}A. This suggests that the degree of hydrophobicity of the nucleotide adjacent to the anticodon will influence the stability of the codon-

210

anticodon interaction.

　　To construct tRNAs to test this hypothesis, a variety of modified purine
nucleosides were purchased from either Sigma or PL Biochemicals. The nucleosides were converted to nucleoside 2'(3'),5'-bisphosphates (pNps) by reaction
with pyrophosphoryl chloride[20]. The pNps were then added to the 3' end of the
chemically synthesized[21] trinucleotide CpUpA (a generous gift from T. Neilson)
by incubation with T4 RNA ligase[22]. The tetranucleotides were purified by HPLC
on a Partisil SAX-10 column with a 20 to 300 mM ammonium acetate (pH 5)
gradient. The oligonucleotides were desalted by lyophylization and inserted
into the tRNA as described earlier.

　　The ability of the tRNAs with different modified nucleotides adjacent to
the amber anticodon to cause read-through of the BMV coat termination signal
(UAG) is shown in Figure 6 and the structures of the modified nucleotides are
shown in Figure 7. The tRNA containing purine riboside (Pu) shows a level of
suppression which is almost as low as with the tRNAs containing a pyrimidine in
this position. The purine and the adenosine differ only by the amino group at
the 6 position yet differ dramatically in their effect on the codon-anticodon
interaction suggesting that this functional group plays an important part in
stabilizing the tRNA on the ribosome. This appears to contradicted, however, by
the strong suppression observed with the tRNA containing a guanosine in this
position (see discussion).

　　The nucleotides m^6A, m^6_2A, ϵA, i^6A, zeatin riboside (Z) and aminohexyl-

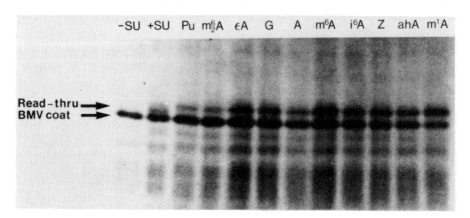

Fig. 6. Autoradiograph of a 15% polyacrylamide-SDS gel analysis of an in vitro
protein synthesizing system containing 1 ug (in 12.5 ul reaction) of brome
mosaic virus RNA. Reactions contained: no added tRNA (-su), 0.05 ug of crude
yeast tRNA^{Ser} amber suppressor (+su) or 0.25 ug of tRNA$^{Phe}_{CUAX}$ where X is the
nucleotide shown at the top of the lane.

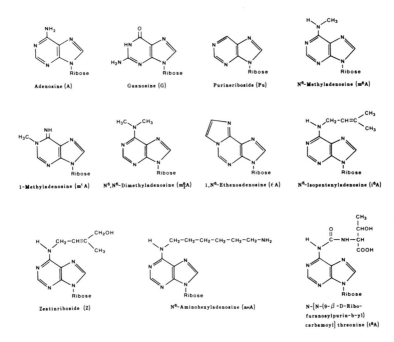

Fig. 7. Structures of the nucleotides adjacent to the amber anticodon in the constructed tRNAs and the hydrophilic nucleotide t^6A.

adenosine (ah-A) were inserted to determine if a hydrophobic side chain would stabilize the codon-anticodon interaction. The tRNAs containing m^6A, i^6A, Z, and ah-A all show higher levels of suppression than the tRNA with A in this position, however, the m_2^6A and εA containing tRNAs show considerably lower levels of suppression. In both of these cases, there is no hydrogen available for hydrogen bonding on the nitrogen in the 6 position, again suggesting the importance of this group. When the compound which was isolated from the reaction of m^1A with pyrophosphoryl chloride was converted to a 5'-monophosphate by incubation with P-1 nuclease and analyzed by two dimensional chromatography on a cellulose TLC[23], it did not migrate in the position predicted for pm^1A. The identity of this compound is not yet known but when inserted into the tRNA it produces an efficient suppressor. The effect of inserting the hydrophilic nucleotide t^6A, shown in Figure 7, has not been determined.

DISCUSSION

The replacement of the anticodon of yeast tRNA^{Phe} with the anticodon CUA produces a tRNA which will suppress amber stop codons in an in vitro protein

synthesizing system. The efficiency of the suppressor is dependent upon the
identity of the nucleotide on the 3' side of the anticodon. The presence of a
hydrophobic group on the amino group in the 6 position of the adenosine adjacent to the anticodon does appear to strengthen the interaction of the tRNA
with its codon as predicted[18] but other features of this nucleotide are also
important. The preliminary data presented here indicates that there is a
specific interaction involving the functional groups of the adenosine or guanosine adjacent to the anticodon which is important for stabilizing the codon-
anticodon interaction in these tRNAs. One possibility is that there is an
interaction with the invarient uridine residue of the tRNA bound in the P-site
of the ribosome. Alternatively, the 16S ribosomal RNA is also positioned near
the anticodon[24] and could also be involved in an interaction with the tRNA. We
feel that further experiments with constructed tRNAs will provide us with a
more detailed understanding of the interactions involved in the decoding of
messenger RNAs and how the choice of codons can be used to regulate the rate
of translation.

REFERENCES

1. Ikemura, T. (1981) J. Mol. Biol. 146, 1-21.
2. Ames, B.N. and Hartman, P.E. (1963) Cold Spring Harb. Symps. Quant. Biol.
   28, 349-356.
3. Stent, G.S. (1964) Science 144, 816-820.
4. Guthrie, C. and McClain, W.H. (1979) Biochemistry 18, 3706-3795.
5. Salser, W. (1969) Molec. Gen. Genet. 105, 125-130.
6. Bossi, L. and Roth, J.R. (1980) Nature 286, 123-127.
7. Atkins, J.F., Gesteland, R.F., Reid, B.R. and Anderson, C.W. (1979) Cell 18,
   1119-1131.
8. Beremand, M.N. and Blumenthal, T. (1979) Cell 18, 257-266.
9. Kastelein, R.A., Remaut, E., Fiers, W. and van Duin, J. (1982) Nature 295,
   35-41.
10. Dunn and Studier, personal communication.
11. Bruce, A.G. and Uhlenbeck, O.C. (1982) Biochemistry 21, 855-861.
12. Bruce, A.G. and Uhlenbeck, O.C. (1982) Biochemistry, in press.
13. Anderson, C.W., Lewis, J.B., Atkins, J.F. and Gesteland, R.F. (1974) Proc.
    Nat. Acad. Sci. USA 71, 2756-2760.
14. Bruce, A.G. Atkins, J.F., Wills, N., Uhlenbeck, O.C. and Gesteland, R.F.
    (in preparation).
15. Capecchi, M.R., Hughes, J.H. and Whal, G.M. (1975) Cell 6, 269-277.
16. Laten, H., Gorman, J. and Bock, R.M. (1980) in Transfer RNA: Biological
    Aspects, eds. Schimmel, P., Soll, D. and Abelson, J.N. (Cold Spring Harbor
    Laboratory, New York), pp. 395-406.
17. Gefter, M.L. and Russel, R.L. (1969) J. Mol. Biol. 39, 145-157.
18. Dasgupta, R., Ahlquist, P. and Kaesberg, P. (1980) Virology 104, 339-346.
19. McCloskey, J.A. and Nishimura, S. (1977) Acc. Chem. Res. 10, 403-410.
20. Bario, J.R.,del Carmen, M., Bario, G., Leonard, N.J., England, T.E. and
    Uhlenbeck, O.C. (1978) Biochemistry 17, 698-706.
21. Werstiuk, E.S. and Neilson, T. (1976) Can. J. Chem. 54, 2689.
22. England, T.E. and Uhlenbeck, O.C. (1978) Biochemistry 17, 2069-2076.

23. Nishimura, S. (1979) in Transfer RNA: Structure, Properties, and Recognition, eds. Schimmel, P., Soll, D. and Abelson, J.N. (Cold Spring Harbor Laboratory, New York) pp. 551-552.
24. Ofengand, J., Liou, R., Kohut, J., Schwartz, I. and Zimmerman, R.A. (1979) Biochemistry 18, 4322-4332.

THE EFFECT OF CODON CONTEXT ON MISTRANSLATION OF MESSENGER RNA

RICHARD H. BUCKINGHAM[+] AND MARTIN J. CARRIER[+]
Institut de Biologie Physico-Chimique, 13, rue Pierre et Marie Curie,
75005 Paris, France

Selection of aminoacyl-tRNA (AA-tRNA) by the messenger-programmed ribosome depends first of all on the nature of the codon bound in the ribosomal A-site and the anticodons present in the population of ternary complexes AA-tRNA: EF-Tu.GTP[1]. However, other factors enter into the selection process. Study of a nonsense suppressor tRNA, the UGA suppressor mutant of tRNA^{Trp} described by Hirsh[2], has, in particular, emphasised that other parts of the tRNA molecule can exercise a profound influence on the overall process of AA-tRNA selection. Bases in the anticodon loop, and their state of post-transcriptional modification, significantly affect the stability of AA-tRNA binding[3,4]. A further factor, which presents some analogy to the influence of bases around the anticodon, is the possible effect exerted by bases around the codon, commonly described as the effect of "codon context".

The first phenomenon which was shown to be sensitive to codon context was that of nonsense codon suppression. Variations of up to tenfold were observed in the efficiency of suppression, depending on the site investigated[5]. Interpretation of this variation was complicated, however, by the difficulty in distinguishing between an effect on recognition of the nonsense codon by the suppressor tRNA, or on the competing process of recognition by termination factors. More recently, a clear indication that context affects suppressor tRNA selection has been provided by the observations of Bossi and Roth[6], who showed that a C to A change in the nucleotide following an amber codon in the *hidD* gene of *Salmonella typhimurium* increased the efficiency of some tRNA suppressors while decreasing the efficiency of another. Such an observation cannot easily be accounted for by codon context acting solely on termination factor recognition.

Another approach to the problem is to study the process of tRNA selection directly, by measuring errors of translation and any effect that differing context might have on the accuracy of translation. We will describe such effects on the translation of the cysteine UGU codon. We have previously demonstrated that UGU codons may be mistranslated as tryptophan in the random U-rich polymer poly(U$_5$,G), both by wild type tRNA^{Trp} and at a frequency several-fold higher by the UGA suppressor tRNA^{Trp} [7]. In this synthetic messenger, UGU codons are flanked by U-rich codons which reflect the statistical codon composition of the polymer.

A different codon context is present in the strictly repeating U,G containing polymer, poly(U-G), where the cysteine codons are always flanked by valine GUG codons. We have therefore measured and compared tryptophan misincorporation in response to the UGU codons present in these two polymers. Such a comparison is complicated by the fact that the polymers contain different numbers of UGU codons. This requires that problems related to reutilisation of the cognate tRNA be taken into account.

The amount of cognate tRNA present in the form of a ternary complex Cys-tRNA.EF-Tu.GTP and thus a substrate for the ribosome is less than the total amount present and depends upon the rate of turnover. Information about the selection error on the ribosome can be obtained only by studying the error of incorporation as a function of the level of amino acid incorporation, and hence of cognate AA-tRNA turnover. This has been done by varying messenger and ribosome concentrations. Extrapolation can then be made to zero level of incorporation, under which conditions almost all cognate AA-tRNA will be in the form of the ternary complex. In order to perform this extrapolation correctly, a simple model for cognate AA-rRNA turnover was developed, which led to the following relationship :

$$\frac{C}{W} \cdot \frac{T_W}{T_c} = \frac{1}{E} - \frac{rm}{T_c}$$

In this equation E, the translational error, is defined as $\frac{W}{C} \cdot \frac{T'_c}{T'_W}$, where W and C are the incorporations of tryptophan and cysteine respectively in response to UGU codons, T'_W and T'_c are the concentrations of corresponding ternary complexes, and T_W and T_c the total concentrations of non-cognate and cognate tRNA (i.e. in all forms : charged, free, bound to ribosome, etc.). T_W is considered equal to T'_W under our experimental conditions. The rate of incorporation of cysteine into polypeptide is r, and m is a constant with the dimensions of time.

In our experiments the rate of incorporation was varied by altering the concentration of messenger added to the *in vitro* system, and in some cases the ribosome concentration was varied also. Under each set of conditions, a range of concentration of tRNA^{Trp}(su$^+$) was added, and the ratio of incorporation of (^{14}C)tryptophan and (^{3}H)cysteine was measured. This ratio ($^W/_C$) was a linear function of suppressor tRNA concentration, passing through the origin in the case of the strictly repeating polymer but with a positive intercept on the ordinate in the case of poly(U_5,G), corresponding to the ratio of UGG codons to UGU codons in the messenger. These results and further details of the experiments will be published shortly[8].

Analysis of these results by linear regression thus leads to values of $\frac{C}{W} \cdot \frac{T_W}{T_C}$ for each polymer. These are presented in Fig. 1 as a function of the cysteine concentration (r), and extrapolation to zero r enables E to be deduced. Values of E for the mistranslation of UGU codons in poly(U-G) and poly(U_5,G) are 4×10^{-3} and 1.75×10^{-2} respectively. This four-fold difference between the two polymers we attribute to effects of the differing sequence surrounding UGU codons in the two messengers.

These observations point clearly to an effect of codon context on tRNA selection. We cannot *a priori* say whether the effect is on cognate or non-cognate tRNA selection, or both. Furthermore, we can say little at present concerning the mechanism of the effect of context. One of the important factors that stabilise the interaction between complementary anticodons in tRNA molecules is that of stacking of adjacent bases on the base paired segment[9,10]. Grosjean and Chantrenne[11] have suggested that the interaction between complementary anticodons is a useful model for tRNA:mRNA interaction on the ribosome. Bases flanking the A-site-bound codon may therefore influence the codon:anticodon interaction to an important degree.

Another hypothesis suggests that A-site binding may be influenced by the nature of the tRNA species which is present in the ribosomal P-site and thus specific to the codon preceding that in the A-site. Other parameters that potentially affect accuracy, such as rate constants associated with EF-Tu-associated GTP hydrolysis, or peptidyl transfer, might also be sensitive to the P-site-bound tRNA species. Although the hypothesis needs much further investigation, the results of one experiment we have performed support this possibility : if tRNA[Val] translating the GUG codons present in poly(U-G) is replaced by photochemically crosslinked tRNA[Val] [12], an increase of about five-fold is observed in the rate of misincorporation of tryptophan at the following cysteine codon[8].

If effects of codon context on misreading prove to be of general occurence in messenger translation, it is readily seen that the need for accurate translation at critical regions of polypeptide chains may impose evolutionary pressure on messenger sequence.

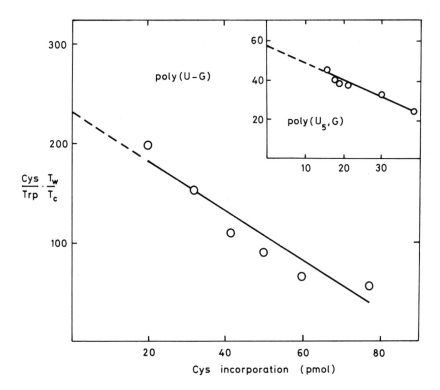

Fig. 1. Accuracy of translation of cysteine UGU codons in poly(U$_5$,G) and poly (U-G) as a function of cysteine incorporation into polypeptide. *In vitro* translation of poly (U$_5$,G) or poly(U-G) was performed under conditions described previously[7] with the following exceptions : overall polymerisation was varied by means of messenger or ribosome concentration in the ranges 0.3 - 6 x 10^{-4} A$_{260}$ units polymer and 0.2 to 2 A$_{260}$ units ribosomes in incubations of 50 µl ; 0.8 A$_{260}$ units total tRNA plus 0 - 42 pmol UGA-su[+] tRNA^{Trp} ; in some experiments with poly(U-G) pure tRNA species only were added - 3.4 pmol tRNA^{Cys}, 30 pmol tRNA^{Val} and 35 pmol tRNA^{fMet}, 0 - 42 pmol tRNA^{Trp} ; (^{3}H)-Cys (1 Ci/mmol, 20 µM) and (^{14}C)-Trp (58 Ci/mol, 16 µM) were present. Incorporation of Cys and Trp in response to UGU codons was measured into hot-trichloroacetic acid - insoluble material (see text), and the ratio Cys/Trp normalised with respect to the total concentrations of tRNA^{Cys} (T$_c$)and tRNA^{Trp} (T$_w$) is shown as a function of Cys incorporation for each polynucleotide messenger.

ACKNOWLEDGMENTS

This work was supported by grants to Professor M. Grunberg-Manago (Groupe de Recherche No. 18) from the Centre National de la Recherche Scientifique, and the Délégation Générale à la Recherche Scientifique et Technique (Convention No. 78.7.1087). Martin J. Carrier thanks the Science Research Council (Great-Britain) and N.A.T.O. for a Fellowship.

REFERENCES

1. Grunberg-Manago, M., Buckingham, R.H., Cooperman, B.S. and Hershey, J.W.B. (1978) Symposia of the Society for General Microbiology, No. XXVIII, Cambridge University Press, U.K., pp. 27-110
2. Hirsh, D. (1971) J. Mol. Biol., 58, 439-458
3. Laten, H., Gorman, J. and Bock, R. (1978) Nucleic Acids Res, 5, 4329-4342
4. Vacher, J., Buckingham, R.H., Houssier, C. and Grosjean, H. (1981) Arch. Internat. Physiol. Biochem., 89, 204-205
5. Feinstein, S.I. and Altman, S. (1978) Genetics, 88, 201-219
6. Bossi, L. and Roth, J.R. (1980) Nature, 286, 123-127
7. Buckingham, R.H. and Kurland, C.G. (1977) Proc. Natl. Acad. Sci. USA., 74, 5496-5498
8. Carrier, M.J. and Buckingham, R.H. (1982) manuscript in preparation
9. Grosjean, H., Söll, D. and Crothers, D.M. (1976) J. Mol. Biol., 103, 499-519
10. Pörschke, D. (1977) Molec. Biol. Biochem. Biophys., 24, 191-210
11. Grosjean, H. and Chantrenne, H. (1980) Molec. Biol. Biochem. Biophysics, 32, (Chapeville, F. and Hanenni, A.L., eds.), Springer-Verlag, pp. 347-367
12. Favre, A., Michelson, A.M. and Yaniv, M. (1971) J. Mol. Biol., 58, 367-379

RIBOSOMAL FRAMESHIFT ERRORS CONTROL THE EXPRESSION OF AN OVERLAPPING GENE
IN RNA PHAGE

Rob Kastelein and Jan van Duin
Biochemisch Laboratorium, Rijksuniversiteit Leiden
Wassenaarseweg 64
2333AL LEIDEN, The Netherlands

INTRODUCTION

The group I single-stranded RNA bacteriophages (including MS2, f2, R17 and
M12) contain the information for at least four proteins.[1-5] The genetic map
of the phage MS2 is presented in Figure 1. It shows that the recently discovered L gene overlaps the coat and synthetase genes in a reading frame
that is +1 with regard to those genes.

The coat gene specifies the major viral protein. The synthetase gene codes
for a protein that together with four host polypeptides, forms the RNA
replicase responsible for multiplying the phage RNA. The A protein is a
minor structural component required for proper phage assembly and infectivity.
The lysis, or L protein, is involved in cell lysis late in infection. The
complete sequence of MS2RNA has been determined by Fiers et al.[6]

The timing and efficiency of phage gene expression in the infected cell is
under phage control, and has been extensively reviewed as far as the three
longer known genes are concerned.[1,7] Some features pertinent to the topic
discussed in this paper will be mentioned here.

Late in infection, the coat protein acts as a translational repressor of
the synthetase gene by binding to its start region. Infection with late coat
amber mutants leaves the synthetase gene unrepressed. Only a few suppressor
strains donate a compatible amino acid at the site of the nonsense mutation,
restoring the repressor function of the coat protein.[7] Similarly, infection

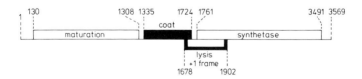

Fig. 1. Genetic map of the RNA phage MS2. Nucleotide numbers are taken
from ref. 6. The reading frame of the lysis gene is +1 with respect to the
coat and synthetase frame.

with coat amber mutants does not lead to cell lysis, and only suppressor
strains that donate a compatible amino acid will sustain productive infection
and lysis.[8] Thus, an intricate relationship exists between cell lysis and the
type of coat protein formed during infection.

The L protein is synthetized late in infection, and its appearance coincides with that of the coat protein.[5] From an anthropomorphic viewpoint,
the position of the L gene within the MS2RNA chain seems an unfortunate
choice. First its ribosome binding site is expected to be never vacant since
translation of the coat gene is very frequent. Secondly, the protein is
needed late in infection, when the synthetase start is blocked by the coat
protein. Thus, ribosomes reading the L cistron must be able to dislodge the
coat protein.

To study the regulation of the L gene we have made use of cloned MS2 DNA.[9]
The recombinant DNA approach has many advantages over studies using real phage
infections. It allows the site specific introduction of almost any kind of
mutation, and the expression of a single phage gene can be examined. In
addition, the synthesis of phage RNA is no longer dependent on successful
infection. It is placed under host control.

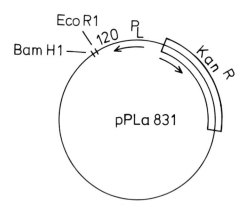

Fig. 2. Schematic representation of the plasmid used to insert and trans-
scribe segments of the MS2 genome. The detailed structure of the inserted
λ DNA region is given in ref. 10.

TABLE 1. HOST CELL LYSIS BY pPLa831-DERIVED PLASMIDS AFTER INDUCTION AT 42°C

Plasmid number	MS2DNA inserted in pPLa831	Cell lysis after induction at 42°C	Schematic representation of MS2 genes present
pMS1	1628-2057	–	
pMS1.9	1628-2057.1221-1736	–	
pMS2	103-2057	+	
pMS9	1221-1736	–	
pMS12	1221-2057	+	
pMS16	869-2057	+	
pMS17	1305-2057	+	
pMS18	869-3569	+	
pMS20	869-2057 (∇1764-1822)	–	
pMS23	103-3569	+	
pMS25	869-3569 (∇1013-1365)	–	
pMS25.9	869-2057 (∇1013-1365).1221-1736	–	
	MS2DNA inserted in pPLa932		
pMS32.00	1221-2057	+	
pMS32.04	1221-2057 1628 1629 (G.AAT.TAA.TTC.)	–	
pMS32.259	1221-2057 (∇1369-1628) GTC.GAA.ATT.C 1369 1629	+	

The plasmid (pPLa831) used in the present study was constructed by Remaut et al.[10] and is depicted in Figure 2. In addition to a kanamycine resistance marker it carries a 240 nucleotide long segment from phage λ, containing the strong inducible promotor pL. At 28°C, this promotor is closed by a thermosensitive repressor encoded in the host chromosome. At 42°C, the repressor is unstable and the promotor is turned on. The plasmid further contains two unique restriction sites some 120 nucleotides downstream from pL, into which various regions of MS2DNA may be inserted. Plasmids carrying MS2DNA sequences are termed pMSm, where m is an arbitrary number referring to a particular MS2 segment (see Table 1).

RESULTS

Cell lysis in MS2DNA transformed cells is due to the product of the L gene

The stringency of temperature control over the pL promotor in pPLa831 derived plasmids is demonstrated in Figure 3. Here Escherichia coli strain M5219[10] was transformed with pMS2 and labeled with radioactive amino acids. At 28°C, no MS2 coat protein is detectable (Fig. 3 lanes 1 and 2), but at 42°C the coat protein of the phage becomes the major product synthesized (lanes 3-7). Similarly at 28°C the MS2DNA information present in pMS23 does not constitute a heavy burden to the cell, since the growth rate in the presence and absence of the MS2DNA insert is the same. However, when the MS2DNA information is transcribed by turning on pL at 42°C, the cells lyse completely. To ascertain that this is due to expression of the L gene, we deleted various parts of the MS2 genome (Table 1). Clearly neither intact

---

LEGEND TO TABLE 1: All nucleotide numbers refer to the primary structure of MS2RNA (Fig. 1; ref. 6). MS2DNA was inserted in pPLa831 according to standard procedures using either the EcoRI site, the BamHI site or both. When the MS2DNA region of interest did not contain a suitable site, EcoRI or BamHI restriction site linker sequences were attached. MS2DNA sites used were: EcoRI (103)(nucleotide number), PvuII (869), SalI (1,013), HaeIII (1,221), XbaI (1,305), SalI (1,366), EcoRI (1,628), HaeIII (1,736), BamHI (2,057). pPLa932 is identical to pPLa831 except that the EcoRI site was removed by cutting with EcoRI, filling in the single-stranded regions with DNA polymerase I (large fragment) and re-sealing with T4 DNA ligase. Thus, pMS32.00 is equivalent to pMS12 except for the EcoRI site in the vector. pMS32.04 was constructed by opening up the EcoRI site of pMS32.00 at position 1,628. The resulting sticky ends were filled with DNA polymerase I (large fragment(and the plasmid re-sealed with T4 DNA ligase. The new sequence around 1,628 is shown. The coat reading frame is indicated by dots. pMS32.259 is constructed by cutting pMS32.00 with EcoRI (1,628) and Sal I (1,366), filling of the sticky ends with Klenow enzyme and re-sealing. The sequence analysis of this clone is shown in Fig. 5. The distance between the Sal I and EcoRI positions is 263 nucleotides. As 4 nucleotides are recovered by the action of the polymerase, the final deletion is 259 nucleotides.

synthetase nor the A protein have to be present to obtain cell lysis (pMS16), but a small Taq deletion in the L gene (pMS20) abolished it. This shows that it is indeed the product of this gene that dissolves the bacteria. Note that the coat protein by itself does not possess any lytic capacity (pMS9). A more detailed account of these experiments is found in reference 11.

The lysis gene has a closed ribosome binding site

At the same time it is clear however that in some way the coat gene is crucial to the expression of L. pMS17 represents the minimum (non mutated) MS2DNA segment that is still able to lyse its host. The introduction of deletions that eliminate the coat initiation region will prevent cell lysis (pMS1 and pMS25). To resolve whether the coat gene is required in _cis_ or in _trans_, we constructed two plasmids bearing the complete coat gene downstream from the L gene (pMS1.9 and pMS25.9). These two clones do not lyse, although the synthesis of the coat protein could be shown in gels of the type presented

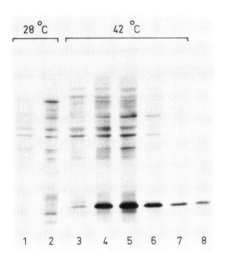

Fig. 3. Expression of MS2 DNA in cells transformed with pMS2. Cultures were grown to 0.2 A_{650}, then labeled with ^{14}C-amino acids for the time intervals indicated below. Proteins were analyzed on polyacrylamide gels containing 0.1% SDS. Lanes 1,2: pattern after labelling at 28°C for 5-25 and 95-115 min, respectively. Lanes 3-7: labeling at 42°C for 5-10, 5-25, 35-55, 65-85 and 95-115 min, respectively. Lane 8: in _vitro_ labeled MS2 coat protein.

in Figure 3. Apparently the coat cistron is required at least in _cis_ to express L. This control mechanism is not new to the phage. The synthetase has long since been shown to be under polar translational control of the coat cistron.[7] Our finding then seems to add one more phage gene whose expression is coupled to the movement of ribosomes over the coat message. This

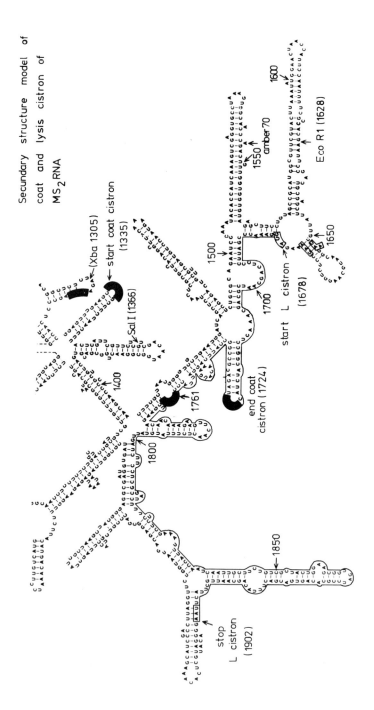

Figure 4. Secondary structure model of the coat and lysis cistron of MS2RNA.[6]

translational coupling observed here also potentially explains why coat amber mutants of RNA phages do not lyse their host.[8]

How far must ribosomes travel over the coat cistron to allow expression of the L gene?

Polar control over the synthetase gene is relieved when a part of the region coding for the 6th through the 50th amino acid of the coat cistron becomes decoded. The molecular basis of this phenomenon is thought to be the observed basepairing of a coat protein coding sequence to the start of the synthetase gene[12] (consult Figure 4). The genetic evidence is that MS2RNA mutants having a nonsense mutation in the position of the 6th amino acid in the coat gene do not produce the synthetase, whereas synthesis is restored when the mutation occurs further downstream at the 50th amino acid.[7]

Starting on the assumption that a similar structure would be responsible for the coupled expression of the coat and the L gene, we began introducing nonsense codons to determine when the polarity would be relieved. Considering the fact that initiation of protein synthesis is slow compared to elongation, we expected that nonsense codons very early in the coat would open up the L cistron (the assumption is that ribosomes passing along the L start will interfere with initiation). Filling in the opened Sal I site at position 1,366 with DNA polymerase I (large fragment) diverts ribosomes starting at the coat gene to the -1 reading frame, causing termination at position 1,415. This clone does not lyse. We therefore turned to the other end of the gene and introduced a stop codon at position 1,628 (pMS32.04) (see Table 1). This position is on a translational time scale only one second away from the L start. To our surprise, this nonsense codon so close to the L start did not yet permit cell lysis.[11] At this point it became difficult to imagine that a ribosome reading further down from position 1,628 would allow enough time for a second ribosome to initiate at the L gene. Thus, we anticipated an unconventional control mechanism.

The introduction of new in phase stop codons still closer to the L start is technically not simple. We therefore exploited an existing +1 frame ochre codon, 6 nucleotides upstream from the L gene. To enable ribosomes, reading the coat message to "see" this codon a deletion in the coat gene was made. This resulted in a reading shift to +1 at the new splice (pMS32.259, Table 1). This new clone lyses rapidly upon induction at 42°C.

This result allows two important conclusions. First, if ribosomes on the coat message shift their reading frame to +1 so as to terminate at position

1,672, L message translaton will follow. Second, the presence or synthesis of the MS2 coat protein is not required per se for cell lysis.

Reading frameshifts in the coat cistron promote synthesis of the lysis protein

To examine the possible general relationship between expression of the L cistron and ribosomal reading frame errors, we produced a large number of deletions in the coat gene around position 1,628. This was done by opening the plasmid pMS32.00 at its unique EcoRI site, treating with exonuclease Bal 31, and recircularizing with T4 DNA ligase. This procedure creates deletions of varying length, which eventually shift the reading frame. A number of

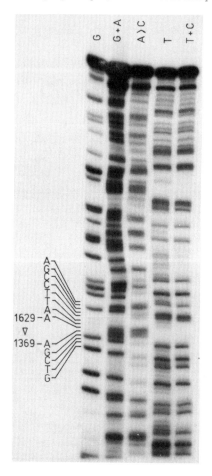

Fig. 5. Autoradiography of a polyacrylamide gel, showing the sequence of pMS32.259 at the junction of the Sal I site (1,366) and the EcoRI site (1,628). Sequencing was carried out as described in ref. 42.

TABLE 2. EFFECTS OF FRAMESHIFTS IN THE COAT PROTEIN GENE ON CELL LYSIS

Number of nucleotides deleted	Mutated Sequence (1620 ... 1650 ... 1680)	Cell lysis at 42°C
0	CAUUCCAAUUUCGCCACGAAUCCGACUGCGAGCUUAUUGUUAAGGCAAUGCAAGGUCUCCUAAAAAGAUGGAAACCC	+
4	CAUUCCAAUUUCGCCACG CCGACUGCGAGCUUAUUGUUAAGGCAAUGCAAGGUCUCCUAAAAAGAUGGAAACCC	+
7	CAUUCCAAUUUCGCCACG ACUGCGAGCUUAUUGUUAAGGCAAUGCAAGGUCUCCUAAAAAGAUGGAAACCC	+
16	CAUUCCAAUUUCGCC AGCUUAUUGUUAAGGCAAUGCAAGGUCUCCUAAAAAGAUGGAAACCC	+
19	CAUU CCGACUGCGAGCUUAUUGUUAAGGCAAUGCAAGGUCUCCUAAAAAGAUGGAAACCC	+
28	CAUUCCAAUUUCGCC AGGCAAUGCAAGGUCUCCUAAAAAGAUGGAAACCC	+
49	CA GCAAGGUCUCCUAAAAAGAUGGAAACCC	+
55	CAUUCC CUAAAAAGAUGGAAACCC	+
85	← deleted till 1580 GUCUCCUAAAAAGAUGGAAACCC	+
88	CAUUCCAAUUUCGCC deleted till 1740 →	−
5	CAUUCCAAUUUCGCCA UCCGACUGCGAGCUUAUUGUUAAGGCAAUGCAAGGUCUCCUAAAAAGAUGGAAACCC	+
17	CAUUCCAAUUUCGC AGCUUAUUGUUAAGGCAAUGCAAGGUCUCCUAAAAAGAUGGAAACCC	+
50	CAUUC GUCUCCUAAAAAGAUGGAAACCC	−
98	← deleted till 1559 AUGCAAACCC	−

The deletions shown were introduced by treating the plasmid pMS32.00 with Bal 31 after cutting at position 1,628 in the MS2DNA insert with EcoRI. The sequences were determined as described in ref. 42. The L cistron is underlined. Stopcodons in frame with the coat protein gene start are overlined. Nucleotide numbers refer to the MS2RNA sequence (fig. 4).

selected clones were sequenced and the nucleotide order around the deletions
are shown in Tables 2 and 3. Table 2 presents those clones where the deletion
results in a frameshift. In the left column, the number of deleted nucleotides is shown. This number defines the new frame downward from the new
splice. The deletion mutants have the general nomenclature pMS32.N, where
32 indicates their derivation from pMS32.00, and N gives the number of
nucleotides deleted. When nucleotides are inserted a 0 is placed before the
number of inserted bases (e.g., pMS32.04).

As shown in Table 2, all +1 frame clones express the L gene as long as the
L gene is still present and termination at position 1,672 can take place.
As expected, the length or position of the deletion (within the above limits)
is irrelevant (see Fig. 6 for a +1 frame lysis curve).

It should be noted that in these clones, the synthesis of L protein still
depends on the flow of ribosomes to the stop codon at nucleotide 1,672,
rather than on the presence of a deletion per se. When this flow is cut off,
the lysis protein is no longer made. This conclusion derives from manipulat-
ing the Sal I site (1,366) in pMS32.55 and pMS32.85 which diverts the flow of
ribosomes to the -1 frame stop codon at position 1,415. This modification

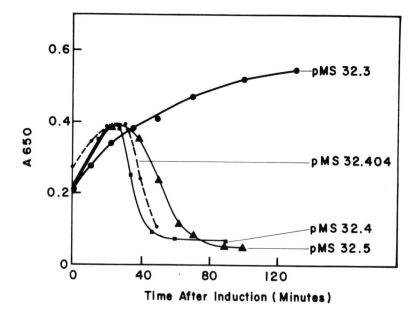

Fig. 6. Growth curves of various clones after induction of the pL promotor
at 42°C. Cultures were grown at 28°C till the A_{650} was 0.2.

TABLE 3. EFFECTS OF 0 FRAME COAT PROTEIN GENE DELETIONS ON CELL LYSIS

Number of nucleotides deleted	Mutated Sequence	Cell lysis at 42°C
	1620 1650 1680	
0	CAUUCCAAUUUUCGCCACGAAUUCCGACUGCGGAGCUUAUUGUUAGGCAAUGCAAGGUCUCCUAAAAAGAUGGAAACCC	+
3	CAUUCCAAUUUUCGCC AAUUCCGACUGCGCGAGCUUAUUGUUAGGCAAUGCAAGGUCUCCUAAAAAGAUGGAAACCC	−
3(a)	CAUUCCAAUUUUCGCCACGAAUU ACUGCGGAGCUUAUUGUUAGGCAAUGCAAGGUCUCCUAAAAAGAUGGAAACCC	+
6(a)	CAUUCCAAUUUUCGCCA CCGACUGCGGAGCUUAUUGUUAGGCAAUGCAAGGUCUCCUAAAAAGAUGGAAACCC	±
6(b)	CAUUCCAAUUUUCGCCACG GACUGCGGAGCUUAUUGUUAGGCAAUGCAAGGUCUCCUAAAAAGAUGGAAACCC	±
15	CAUUCCAA CCGACUGCGGAGCUUAUUGUUAGGCAAUGCAAGGUCUCCUAAAAAGAUGGAAACCC	−
27	CAUUCCAAUUUUGGCC AAGGCAAUGCAAGGUCUCCUAAAAAGAUGGAAACCC	−
51	CAUUCCAAUUUUCGCCAC UGGAAACCC	−

Number of nucleotides inserted		Cell lysis
03	..........ACGAAUUCUUCCGAC..........	±
06	..........ACGAAUUGGAAUUCCGAC..........	−

The deletions shown are introduced as described in the legend to table 2, except pMS32.3 which was obtained by treating pMS32.00 with EcoRI and subsequently with T4 DNA polymerase and T4 DNA ligase. Insertions were obtained using the EcoRI linker GGAATTCC 3'. Details of the procedure will be published elsewhere.

aborts cell lysis.

Clones pMS32.55 and pMS32.85 lack part or all of the Shine and Dalgarno sequence before the L gene. Thus, this sequence does not seem strictly required for reinitiation, although the kinetics of cell lysis are delayed in these cultures.[11]

The lower part of Table 2 shows deletions that lead to a -1 reading mode at the new junction. Such a shift will in general result in termination of protein synthesis at the ochre codon 1,652. As demonstrated by clones pMS32.5 and pMS32.17, termination at this point also elicits an initiation event at the L gene start; these two clones lyse at 42°C although not as rapidly as their +1 frame counterpart (Fig. 6). No lysis is observed for the two -1 phaseshift mutants, pMS32.98 and pMS32.50. They have an intact L gene, but the stop codon at 1,652 has been erased. We conclude that the frameshift itself is not crucial to the readout of the L message, but rather its inevitable consequence, chain termination.

Lysis properties of 0 frame coat gene deletions

A somewhat complicated picture arises when we examine the phenotype of clones that carry deletions and insertions in the coat gene that do not change the reading frame (Table 3). In general, 0 frame mutants display various phenotypes. Some do not lyse at all (pMS32.3, pMS 32.15, pMS32.27, pMS32.06), some very delayed (pMS32.6a, 32.6b, pMS32.03) and others behave as the wild type (pMS32.3a and pMS32.18)(Figures 6 and 7). For one thing, the nonlysing mutants show that the mere passing of ribosomes over the start of the L gene does not result in the reading of that gene.

At first glance, it is difficult to understand why deletions that do not affect the reading frame sometimes cancel out cell lysis. The wild type sequence is after all read in the 0 frame. We suggest two factors that contribute to the observed results.

First, the deletion removes all or part of the region where the frame-shift is supposed to take place. This positions the frameshift region around nucleotide 1,628. Apparently the deletion in pMS32.3a is outside such a frameshift zone. This clone behaves as the wild type (Fig. 7). Two other deletions have been obtained around the Sal I site at position 1,366 by digestion with Bal 31. One, pMS32.18 removes the nucleotides 1,366-1,383 whereas the other, pMS32.111, has lost 37 codons (∇1,339-1,499). Both clones still lyse although a bit delayed (Fig. 7). The deleted regions here are definitely outside the frameshift zone, and the reason for their altered lysis behavior must be sought elsewhere (see below).

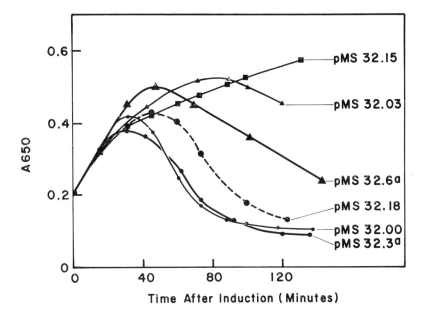

Fig. 7. Growth curves of various clones after induction of the pL promotor at 42°C. Cultures were grown at 28°C till the A_{650} was 0.2.

The notion that lysis expression can disappear in 0 frame mutants because part or all of the frameshift region has been excised also predicts that the insertion of codons in that area will reduce L gene expression. We have constructed clones with 1 or 2 triplet insertions in the region of interest (pMS32.03 and pMS32.06, Table 3). The newly generated sequence greatly retards lysis in pMS32.03 (Fig. 7), and prevents it in pMS32.06. The deletions and insertions seem to define an area in the RNA that can upset the reading frame of the ribosomes.

The second factor that possibly modulates lysis gene expression in 0 frame deletions is the presence of mutant coat protein. There is evidence from infection studies that wild type coat protein can stimulate its own synthesis.[7,13] If this mechanism, whose nature is not understood, also operates in our assay system, this would mean that the frequency of translation of the coat gene and thus of the L gene could be lower in the 0 frame mutants. This effect may explain the altered lysis properties of pMS32.18 and pMS32.111.

Ribosomal access to the L cistron; reinitiation or coupled initiation?

In this section we would like to discuss in some detail how we envisage initiation of protein synthesis at the L gene. As pointed out above, ribosomes have no independent admission to the start of the L gene, but must gain access via the coat gene entry site. We have noticed that termination at nucleotide 1,652 as well as at 1,672 permits translation of the L gene. The protein synthesis starts that are facilitated by termination at 1,652 can be of two kinds. A terminating ribosome at 1,652 may buy the time for a second ribosome to attach to the now temporarily exposed L cistron entry site. The distance between the respective start and stop sites is just large enough to accommodate two ribosomes. Such an event be better termed coupled or conditional initiation. Alternatively, the same ribosome that terminates at 1,652 may reinitiate at 1,678. This type of real reinitiation evidently occurs for those ribosomes that terminate at 1,672. Stop and start codons are too close here to place two ribosomes. From the published data, it seems that both types of translational restarts occur. Sometimes the new start point is far downstream from the nonsense mutation, but in other instances it is very close or even somewhat upstream from the stop codon.[14,15] A true reinitiation event probably takes place at the E-D and B-A gene overlap in the tryptophan operon.[16] These stop and start codons have one nucleotide overlap.

We see no necessary conflict between a model in which termination and reinitiation are done by the same ribosome, and the data recently published by Kaji's lab.[17,18] Ryoji et al.[17] report that in a system depleted of the ribosome releasing (RR) factor, termination at a nonsense codon is followed by resumption of protein synthesis at the first triplet downstream from the stop codon. This restart does not use fMet-tRNA but initiates with the amino acid that is called for by that first triplet. When the RR factor is introduced in their assay system, ribosomes are released from the message at the nonsense codon.[17] This of course does not necessarily mean that release would always occur at every stop codon. The nonsense mutation used by Kaji (amB_2R_{17}RNA) does not have a restart possibility, thus ribosomes must exit at this amber codon when the RR factor is present.

If a potential restart site exists close to the termination signal, we consider the following scenario for true reinitiation. After the peptide chain has been removed by the termination factors, the RR factor first dissociates the 50S subunit (see ref. 19). The 30S subunit then has a choice of being released or restarting at a close by AUG or GUG triplet, the final outcome depending on the particular mRNA sequence covered. The 30S subunit has the capacity to bind fMet-tRNA in response to attached message, but irrespective of the presence of an AUG or GUG codon in its decoding site.[21,22] This initiator tRNA, in turn, will help the subunit to "line up" with any close by start codon and start translation in the usual way. If no potential start codon is available, the complex likely decays. "Lining up" may also occur during normal initiation. It is only an assumption that in the primary complex between a 30S subunit and the messenger, the start codon is already in its proper place.

In our view the important difference between termination in the presence or absence of the RR factor is that in the latter case the 50S subunit is not released.[20] The resulting 70S couple can neither select initiator regions,[23] nor bind fMet-tRNA. Thus its only possibility appears to be to resume translation at the next available codon.

It is interesting that the restart accomplished by 70S ribosomes in the RR free system is in phase.[17]

Other ways to approach the L cistron

One wonders whether the L cistron start has peculiar structural features that make its translation unconditionally dependent on the termination of ribosomes at a close by stop codon. It is very well possible that its inaccessibility may just be due to a particular RNA folding, that masks its ribosomal binding site. To distinguish between the two possibilities we have examined whether deletions upstream from the L gene could create an RNA structure that will permit direct ribosomal access to the L gene. Experimentally, pMS1 was opened at its unique EcoRI site (1,628), treated with Bal 31, and subsequently with T4DNA ligase. Two clones that lysed rapidly

```
         MS2 1258...CGAUGGUCCA|UAAAAGAUG...pMS32.404
           pL49...GCCCUG|AAGGUCUCCUAAAAGAUG...pMS1.34
     pL108...GCAUUGG|AAGGCAAUGCAAGGUCUCCUAAAAGAUG...pMS1.24
    MS2 wild type UGUUAAGGCAAUGCAAGGUCUCCUAAAAGAUG...pMS32.00
```

Fig. 8. Relevant sequence part of clones selected for lysis in the absence of the coat protein gene start. pL nucleotide numbers are taken from reference 10. Number 1 in the pL sequence denotes the first transcribed nucleotide. pMS1.34 and pMS1.24 derive from pMS1. The action of Bal 31 has removed 24 nucleotides from the MS2 sequence and no nucleotides from the pL sequence in pMS1.24. In pMS1.34 34 nucleotides from the MS2 sequence are deleted and 60 from pL. In pMS32.404 an internal MS2 sequence of 404 nucleotides is removed from pMS32.00 (∇1268-1671). The new splice points are indicated with a vertical dotted line. The AUG start codon of the lysis gene is underlined.

were analyzed further; their sequences around position 1,678 are shown in Figure 8 (pMS1.24 and pMS1.34).

Another procedure, designed to obtain clones that lyse independently of a termination event was to delete the large central portion of the coat gene in pMS32.00 (∇1,366-1,628) and digest the new linear plasmid further with Bal 31. After ligation, the DNA was used for transformation. This procedure also yielded several clones that lysed, one of which was sequenced (pMS32.404). Its growth curve is shown in Figure 6 and its sequence in Figure 8. In this new sequence the lysis gene is hooked up to a 3' terminal part of the A protein gene. In these three recombinants (pMS1.24, pMS1.34, pMS32.404), the start of the coat protein cistron is absent. Neither do any known translational start signals exist in the pL sequence present in pMS1.24 and pMS1.34 or in the 3' terminal region of the A protein (pMS32.404). Thus in these clones the L gene is directly open to translation. A reasonable molecular explanation for this observation is that in none of these clones the stem structure of the hairpin 1,651-1,674 can be formed. It is thus possible that this hairpin proposed by Fiers et al.[6] (compare Figure 4) exists in vivo and blocks ribosome access to the L gene. It may be interesting to note that a ribosome, terminating at 1,629 in pMS32.04 is predicted not to affect the structure of the stem; the distance is too large. pMS32.04 does not lyse indeed.

Clone pMS32.404 is instructive also from another viewpoint. Here the Shine and Dalgarno sequence of the L gene is erased by Bal 31, but is replaced by what apparently is the next best thing; a sequence derived from the coding region of the A protein gene showing distinct similarities with the original Shine and Dalgarno sequence. As discussed above, independent initiation seems to rely more on a Shine and Dalgarno sequence than does reinitiation (pMS32.55 and pMS32.85).

We conclude that the L gene is not intrinsically different from nonoverlapping genes, but as its expression obviously needs careful control it has evolved in a gene context that permits translation only under strict circumstances.

Other overlapping genes

The list of overlapping genes is growing and the question arises whether their expression is controlled as described here for the MS2 phage. Of course, RNA splicing events or a separate promoter may sometimes place the messages of overlapping genes on different RNA's, e.g., some SV40 genes[30-32]

also the B promotor in øX174[24]. Still in many other cases the two messages
occur in one RNA chain.[29,33,34] In Table 4, we show some examples of
extensively overlapping genes. In each case the DNA from a closely related
virus or mitochondrion has also been sequenced. This permits to determine

TABLE 4. OVERLAPPING GENES FROM VARIOUS SOURCES

Origin of nucleic acid		Gene overlap
MS2	UUCCGACUGCGAGCUUAUUGUUAAGGCAAUGCAAGGUCUCCUAAAAGAUGGAAACCC	Coat-lysis
øX174	AAAAAGUCAGAUAUGGACCUUGCUGCUAAAGGUCUAGGGAGCUAAAGAAUGGAACAAC	A-B
G4	AAAAAAUCAGAUAUUGACAUGGCCGUAAAAGGCCUAGGGAAUAAAGAAUGGAACAAU	A-B
øX174	UGCUACUGACCGCUCUCGUGCUCGUCGCUGCGUUGAGGCUUGCGUUUAUGGUACGCU	D-E
G4	UGCUACCGACCGUUCACGCGCUCGCCGUGCUAUCGAGGCUUGCGUAUAUGGAACACU	D-E
Human mt.	CCAAAGCCCAUAAAAAUAAAAAAUUAUAACAAACCCUGAGAACCAAAAUGAACGAAAA	URF A6L
Bovine mt.	CUGACACCAACAAAAAUAUUAAAACAAAACACCCCUUGAGAAACAAAAUGAACGAAAA	
Mouse mt.	CCAAAAUCACUAACAACCAUAAAAGUAAAAACCCCUUGAGAAUUAAAAUGAACGAAAA	ATPase 6
Adeno 12	UAUUGCUUUUUUGGCAACCAUAUUGGAUAAAUGGAGCGAGA	19K-54K
Adeno 5	UGUUGCUUUUUUGAGUUUUAUAAAGGAUAAAUGGAGCGAAG	21K-55K

Phage øX174 and G4 sequences are from ref. 24 and 25. Mitochondrial DNA is
from 26, 27 and 28 and adeno virus 5 and 12 from ref. 29. 19K, 54K, 21K and
55K are proteins coded for by the Elb mRNA in the respective viruses. Stop
and start codons are underlined. The underlined start codon marks the beginning of the downstream gene.

whether out of phase stop codons, preceding the gene overlap have been
conserved. We note that all overlaps, except one, show such a conserved out
of phase stop codon. This indicates that expression of the downstream gene
can, in principle, be controlled as described in this paper.[44] In the A-B
gene overlap of øX174 and G4 the stop codon is conserved, although the codon
sequence that forms it calls for different amino acids in the two phages.
The D gene of phage øl74 has an out of phase UGA codon just before the
start of the overlapping E gene, which seems not to be conserved in the
closely related phage G4. Here, the corresponding sequence reads CGA. We

wonder whether this sequence diversion reflects an error in a DNA coping event or a different control mechanism.

SUMMARY AND DISCUSSION

In this paper we have presented the evidence that has led to a specific proposal for the control of expression of the L gene of bacteriophage MS2. In summary our conclusions are: 1. The L gene in native MS2RNA is not directly accessible to ribosomes. 2. The L gene is expressed by a fraction of ribosomes that after starting translation at the coat gene changes its reading frame at the appropriate region to terminate at either one of the ochre codons 1,652 or 1,672. This termination event elicits initiation of protein synthesis on the L message. The ribosomal phase shift is only necessary to reach the stop codon.

It must be noted that what we have basically shown is that every time termination at the indicated stop codons takes place, the L gene is expressed. We deduce that, when during phage infection lysis protein is synthesized, this is set off by termination. It is indeed difficult to prove that other ways to achieve lysis protein synthesis do not exist. Fortunately there is strong evidence that ribosomal fidelity and cell lysis are related. It has been shown that low error hosts will multiply MS2 phage, but will not release the progeny. This indicates that the L protein is not synthesized.[35] The group III RNA phage Q_β provides us with a striking parallel. This phage exploits the noise in translation to occasionally insert tryptophan at the UGA stop codon of its major coat protein gene. This nonsense suppression produces a small amount of carboxy extended coat protein, which is essential to phage growth.[36] Indeed, Q_β does not grow on streptomycin resistant, i.e., low error strains.[37,38] It appears that at least some forms of life depend on the inaccuracy of the translational machinery.

We do not know which type of frameshift (+1 or -1) is responsible for expression of the gene in vivo. In vitro studies have indicated a -1 phase shift in the region of interest. A frameshift polypeptide terminated at position 1,652 was found.[39] We also do not know the codons where the frameshift occurs. It seems to require a larger region than just a codon and its neighboring 5' and 3' nucleotides. We have noted before the possible role of the run of U residues around position 1620.[11,40] Extensive runs of either A or U residues may be observed close to most overlap regions shown in Table 4. On the other hand it has been argued that the probability of a frameshift increases after a missense error has been made.[41]

We suppose the L gene is a late addition to the phage's genomic
repertoire. The product and chance of occasional misreading have probably
become optimized by the reproductive advantage of carrying a lysis function.[43]
The position of the L gene in the phage genome is enigmatic. It is possible
to imagine the need of having this potentially dangerous protein under
translational control of the major coat protein. This setting ensures it
to be a late product, synthesized only when infection is well underway.
It is difficult to assess the necessity of overlap with the synthetase and
the intercistronic region.

ACKNOWLEDGEMENTS

We are grateful to Dr. Fiers and Dr. Remaut for support and advice. One
of us (J.V.D.) thanks Dr. Noller and Dr. Schleich for hospitality in their
laboratories at the University of California, Santa Cruz, during his leave
of absence from the University of Leiden. Mrs. Bakker-Steeneveld and Mr.
Overbeek are acknowledged for excellent technical assistance, R. S. Haxo
for correcting this manuscript, and B. Berkhout for help in the experiments.
R.K. was supported by SON.

REFERENCES
1. Horiuchi, K. in RNA phage (ed. Zinder, N. D.) 29-50 (Cold Spring Harbor
    Monogr. Ser. 1975).
2. Jeppesen, P. G. N., Argetsinger-Steitz, J., Gesteland, R. F. and Spahr,
    P. F. (1970) Nature 226 , 230-237.
3. Model, P., Webster, R. E. and Zinder, N. D. (1979) Cell 18, 235-246.
4. Atkins, J. F., Steitz, J. A., Anderson, C. W. and Model, P. (1979)
    Cell 18, 247-256.
5. Beremand, M. N. and Blumenthal, T. (1979) Cell 18, 257-266.
6. Fiers, W., Contreras, R., Duerinck, F., Haegeman, G., Iserentant, D.,
    Merregaert, J., Min Jou, W., Molemans, F., Raeymakers, A., van den
    Berge, A., Volckaert, G., and Ysebaert, M. (1976) Nature 260, 500-507.
7. Robertson, H. D. in RNA phages (ed. Zinder, N. D.) 113-145 (Cold
    Spring Harbor Monogr. Ser. 1975).
8. Zinder, N. D. and Lyons, L. B. (1968) Science 159, 84-86.
9. Devos, R., van Emmelo, J., Contreras, R. and Fiers, W. (1979) J. Molec.
    Biol. 128, 595-619.
10. Remaut, E., Stanssens, P. and Fiers, W. (1981) Gene 15, 81-93.
11. Kastelein, R. A., Remaut, E., Fiers, W. and van Duin, J. (1982) Nature
    295, 35-41.
12. Min Jou, H., Haegeman, G., Ysebaert, M. and Fiers, W. (1972) Nature 237,
    82-88.
13. Sugyama, T., Stone, H. and Nakada, D. (1969) J. Molec. Biol. 42, 97.
14. Steege, D. A. (1977(Proc. Nat. Acad. Sci. USA 74, 4163-4167.
15. Napoli, C., Gold, L. and Swebelius Singer, B. (1981) J. Molec. Biol.
    149, 433-449.
16. Oppenheim, D. S. and Yanofsky, C. (1980) Genetics 95, 785-795.
17. Rioji, M., Berland, R. and Kaji, A. (1981) Proc. Nat. Acad. Sci. 78,
    5973-5977.
18. Model, P. (1982) Nature 295, 15.

19. Martin, J. and Webster, R. E. (1975) J. Biol. Chem. 250, 8132-8139.
20. van Duin, J., Overbeek, G. P. and Backendorf, C. (1980) Eur. J. Biochem. 110, 593-597.
20. Hirashima, A. and Kaji, A. (1973) J. Biol. Chem. 248, 7580-7587.
21. van der Laken, K., Bakker-Steeneveld, H., Berkhout, B. and van Knippenberg, P. H. (1980) Eur. J. Bioche. 104, 19-23.
22. Jay, E., Seth, A. K. and Jay, G. (1980) J. Biol. Chem. 255, 3809-3812.
23. Zipori, B., Bosch, L., van Dieyen, G., and van der Hofstad, G. A. J. (1978) Eur. J. Biochem. 92, 225-233.
24. Sanger, F., Air, G. M., Barrel, B. G., Brown, N. L., Coulson, P. R., Fiddes, J. C., Hutchinson, C. A., Slocombe, P. M. and Smith, M. (1977) Nature 265, 687-695.
25. Godson, G. N., Barrell, B. G., Staden, R. and Fiddes, J. (1978) Nature 276, 236-247.
26. Anderson, S., Bankier, A. T., Barrell, B. G., de Bruijn, H. M. L., Coulson, A. R., Drouin, J., Eperon, J. C., Nierlich, D. P., Roe, B. A., Sanger, F., Schreier, P. H., Smith, A. J. H., Steden, R., and Young, O. G. (1981) Nature 290, 457-465.
27. Anderson, S., de Bruyn, H. M. L., Coulson, P. R., Eperon, J. C., Sanger, F. and Young, D. G. (1982) J. Mol. Biol., in press.
28. Bibb, M. J., van Etten, R. A., Wright, C. T., Welberg, M. W. and Clayton, D. A. (1981) Cell 26, 167-180.
29. Bos, J. L., Polder, L. J., Bernards, S., Schrier, P. J., van den Elsen, P. J., van der Eb, A. J. and van Ormondt, H. (1981) Cell 27, 121-131.
30. Fiers, W., Contreras, R., Haegeman, G., Rogiers, R., van de Voorde, A., van Heuverswyn, H., van Herreweghe, J., Volckaert, G. and Ysebaert, M. (1978) Nature 273, 113-120.
31. Reddy, V. B., Thimmappaya, B., Dher, R., Subramanian, K. N., Zin, B. S., Pan, J., Ghosh, P. K., Celma, M. L., and Weismann, S. M. (1978) Science 200, 494-502
32. Ziff, E. B. (1980) Nature 287, 491-499.
33. Montoya, J., Ojela, D. and Attardi, G. (1981) Nature 290, 465-470.
34. Ojela, D., Montoya, J. and Attardi, G. (1981) Nature 290, 470-474.
35. De Mars Cody, J. and Conway, T. W. (1981) J. Virol. 37, 813-820.
36. Hofstetter, H., Monstein, H. J. and Weissman, C. (1974) Experientia 30, 687.
37. Engelberg-Kulka, H., Dekel, L. and Israeli-Reches, M. (1977) J. Virol. 21, 1-6.
38. Gorini, L. in Ribosomes (eds. Nomura, M., Tissieres, A. and Lengyel, P.) 791-803 (Cold Spring Harbor Monograph Ser. 1974)
39. Atkins, J. F., Gesteland, R. F., Reid, B. R. and Anderson, C. W. (1979) Cell 18, 1119-1131.
40. Fox, T. D. and Weiss-Brummer, B. (1980) Nature 288, 60-63.
41. Kurland, C. G. in Ribosomes (eds. G. Chambliss, G. R. Craven, J. Davies, K. Davis, L. Kahan and M. Nomura) University Park Press (1979)
42. Maxam, A. and Gilbert, W. (1977) Proc. Nat. Acad. Sci. 74, 560-564.
43. Zinder, N. D. (1980) Interviology 13, 257-270.
44. Synthesis of the 55 k resp. 54 k proteins of adenovirus 5 and 12, however, does not appear coupled to translation of the upstream 21 k and 19 K messenger regions (H. van Ormondt, personal communication).

SECTION III: STRUCTURE AND EUKARYOTIC TRANSLATION

PROTEIN SYNTHESIS FOR CELL ARCHITECTURE

SHELDON PENMAN
Massachusetts Institute of Technology, 77 Massachusetts Avenue,
Cambridge, Masssachusetts 02139

Introduction

Considerable temerity is required to question the basic paradigm of one's

science. In full knowledge that the exercise rarely does one any good, I will

suggest that the paradigm underlying the biochemistry of protein synthesis,

developed for bacterial systems, is wearing thin with regard to higher

organisms. There is nothing incorrect in the studies of metazoan protein cell

synthesis and, indeed, the sophistication of present experimental programs often

is breathtaking. However, some of the fundamental qualities of metazoan

biology such as tissue patterns and architecture, have barely been touched.

Perhaps this is not suprising since the historical antecedents of protein

synthesis research focused on the enzymes of intermediary metabolism in bacteria

and the major questions were: what protein coding sequences are expressed and to

what degree?   The organization of the protein products or the architecture of

the synthesizing organism were, for the most part, of no particular concern.

It would seem that an understanding of how protin synthesis functions to

produce tissue structure is required to study the two most profound yet ellusive

problems in modern biology, i.e. development and evolution. Examining the

proteins themselves may not take us much farther than we are today. There is

suprisingly little difference in the proteins and the biochemical pathways of

the lowly Cnidaria and the lofty Vertebrata. What is different is form, pattern

and architecture. As Wilson has pointed out most cogently (King and Wilson,

1975) the often rapid evolution of species seems unrelated to the glacial rate

of change in proteins. Since protein coding sequences probably occupy less than
one percent of, for example, the mammalian genome it would seem incumbent upon
us to seek the information content in the rest of the DNA that directs the
organization of structural proteins and thus distinguishes us from the jelly
fish.

What percisely is the lacuna in our present paradigm? I suggest that we
suffer from a mistatement of the results of Beatle and Tatum. "One gene, one
enzyme" has always been troublesome since their experiments never really showed
any such correspondence. Alice Fulton, has put the problem most succinctly:
What Beatle and Tatum actually showed was "One enzyme, one gene". This is
clearly a very different statement and is easier to reconsile with the actuality
that most of the genes in the genetics of higher organisms deal with
deformations of form and pattern rather than lesions in biochemical pathways.
Much of genomic information is concerned not so much with what proteins are made
but how they are organized.

It may be all very well to point to apparent limitations of conventional
biochemistry but is it possible to suggest an alternative course? I believe
that the outlines of future studies in cell biology are begining to emerge.
They are suggested by new approaches that reveal more deeply the determinents of
cell architecture and the associated macromolecular metabolism. The
biochemistry of fiber systems is one such area. I will briefly describe another
here. It consists of studies using whole mount electron microscopy which view
cell structure in three dimensions, extractive procedures that separate cells
into their several architectural components, and experiments which suggest
regulation of genome expression by cell configuration.

Three Dimensional Cell Structure

For all of its tremendous power, the thin section technique has afforded us
a clouded picture of the cell interior; a picture seemingly supported by the
current techniques of cell fractionation and biochemical analysis. There are

two major limitations to the thin section; one is quite obvious. A thin section offers us only a two-dimensional slice of three-dimensional objects. Although everyone knows that sectioned material is but a poor representation of reality, few are really aware of how profoundly different the three-dimensional object may be. Serial section reconstruction requires patience of heroic proportions and there are but a few practioners. Their efforts have shown us how suprising the three-dimensional reality can be (a discussion of reconstruction from sections is offered in Ham and Cormack's classic text "Histology", 8th ed., 1979).

The other major limitation of the embedded thin section is more subtle and more serious. It has to do with the masking of important protein filaments by the electron scattering of the embedding plastic itself. Porter noted this problem with the advent of dense epoxy embedding materials; his words of caution were, alas, unheeded and forgotten (Pease and Porter, 1981).

We now know that the seemingly amorphous appearance of cell cytoplasm and nucleoplasm outside of organelle boundaries is due, in large part, to the failure to image the architectural elements poorly stained by the heavy metals necessary in conventional electron microscopic procedures. The true extent of this problem became obvious from comparing images of architectural elements of the cells seen in whole mounts with thin section electron microscopy. Architectural proteins, clearly visable in whole mount, disappear, often completely, in epoxy-embedded thin sections.

Our own work, which was for a long time concerned with nucleic acid biochemistry and cell fractionation, wandered or stumbled into the area of three-dimensional electron microscopy by a circuitous and unanticipated route. As a consequence of this history, we have concentrated our studies on the biochemistry and morphology of the filament network and surface lamina remaining after extraction with non-ionic detergents.

Our studies of cell architecture began with the seemingly unrelated question of the role of poly A, the curious homopolymer terminating most, though not all, mRNA molecules. We already knew that the mRNA of membrane bound polyribosomes is bound to the endoplasmic reticulum and that its poly (A) segment serves as one of the linking elements (Milcarek and Penman, 1974). Were the so-called "free" polyribosomes bound to some previously undetected structure and could poly (A) play a similar role in the binding of these messages? This question has now been answered in the affirmative (Fey and Penman, in preparation) and its pursuit revealed the presence of the skeletal network and the apparently obligatory binding of message to this framework for translation (Lenk et al., 1977; Lenk and Penman, 1979; Cervera, et al., 1981). We were thus led to study the nuclear and cytoplasmic networks and to reflect on why metazoan cells should be thus organized.

The Basic Extraction Procedure

These studies depend critically upon the method that separates the soluble from the structural proteins and then the separation of structural networks of the cytoplasm from the nucleus. Only in this way can the cell architecture both be viewed microscopically and characterized biochemically. A simple procedure is common to most of the experiments described here. Cultured cells are exposed to a sufficiently strong, non-ionic detergent, e.g. Triton x-100, in a suitably designed buffer. Most lipids are solubilized, together with about two-thirds of cellular proteins; these constitute the "soluble" fraction. Most of the architectural components of the cell remain as an intact entity which retains much of the morphology of the living cell. The complex cytoplasmic network with the polyribosomes attached are now readily visible with transmission electron microscopy. The detailed morphology and protein composition of the extracted structures very much depend on the conditions of extraction. The particular nonionic detergent appears relatively unimportant, provided that it is sufficiently lipophilic to effect a complete removal of lipids. Triton X-100 and NP-40 are especially suitable. Some of the various Brijs, Tweens, and Spans can serve, although their rate of lipid solubilization is rather slow. Most important is the ionic composition and pH of the extraction medium.

Our most recent extraction-buffer formulation, designed to optimize the preservation of preexisting structure and, at the same time, effect the complete removal of soluble components, is shown in Table 1. Although the concentrations of the constituents are not critical, large departures from these values result in a much reduced skeletal structure. The cation ion concentrations are close to the activities expected in the living cell interior. A pH of 6.8 appears optimum for preserving filament integrity (Solomon et al., 1979), and 300 mM sucrose has been found empirically to improve the retention of known structure-bound elements. Although the extraction medium appears very hypertonic, this does not result in osmotic stress on cell structure, since, in the presence of Triton, plasma-membrane integrity is breached within seconds and the cell interior rapidly equilibrates with the external medium. The length of extraction must be adequate for soluble components to diffuse completely from the cell interior. We have determined that 3 minutes is adequate for flat, fibroblastic cells, whereas thicker cells, such as those in epithelial sheets, require 10 minutes. Too brief an extraction results in thicker filaments that probably result from adventitious fixation of soluble proteins to skeletal fibers.

Whole mounts of the detergent-extracted cells have proven to be uniquely suited for visualizing cell architecture (Brown et al., 1976; Webster et al., 1978; Fulton et al., 1980; Fulton et al., 1981; Schliwa and Blerkom, 1981; Schliwa, 1982). Such whole mounts can afford a far more realistic view than the conventional thin section. In the absence of embedding plastic, required for most thin sections and soluble proteins, removed by extraction, the electron scattering

Table 1. Extraction Buffer

Constituent	Concentration
NaCl or KCl	100 mM
$MgCl_2$	3 mM
PIPES (pH 6.8)	10 mM
Sucrose	300 mM
PMSF	1 mM
EGTA	1 mM
Triton X-100	0.5%

by the architectural proteins forms clear images without heavy metal stains. Removal of the soluble proteins with detergent leaves a well defined skeletal structure and avoids, for now, questions of the organization of the soluble proteins in the cell. Most importantly, the significance of viewing the cell architecture through its entire thickness can not be over emphasized. Only in this way can the three-dimensional anastomosing networks be seen: these appear merely as specks or short fibers in thin sections (Wolosewick and Porter, 1979).

Networks of the Cytoplasm

The view of cell structure that has emerged from these studies using whole mounts and extractive procedures is summarized schematically in Figure 1. The extracted cell structure is shown here with the three dimensional anastanosing network with bound polyribosomes filling the cytoplasmic space. The surface is a protein

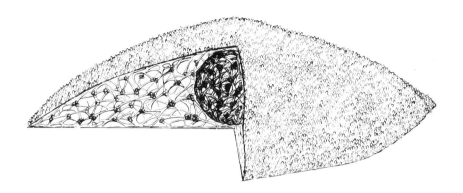

Figure 1:   Cell architecture visualized in the Triton extracted whole
mount.   This highly simplified sketch shows the networks
of the cytoplasm and of the nucleus after the removal of
chromatin. The surface lamina is the protein sheet
remaining after the removal of plasma membrane lipids.
Polyribosomes are shown bound to the cytoplasmic filaments
while in the nucleus the dark structures are hnRNP. The
actual networks are much more regionally heterogenesis.

lamina that remains after the removal of lipids by the non-ionic

detergent. The nucleus is shown as it appears after chromatin is

removed and it also has a space filling three dimensional network,

bounded by the nuclear lamina. The nuclear structure is distinct in

organization from the networks of the cytoplasm.

A few typical micrographs of extracted fibroblasts in whole mount

preparations are shown in Figure 2. A chicken embryo fibroblast is

shown at low magnification in Figure 2A.   Near the edge of the cell

is the fine filamentous network rich in microfilaments, characteristic

of motile regions of the cell. Closer to the nucleus, the network is

organized differently; the filaments are thicker, heterogeneous, and

densely studded with polyribosomes. At higher magnification the

heterogeneity of the fibers and the complexity of their inter-

connections becomes apparent as shown in the fibroblasts in 2B and 2C

and the aortic endothelial cell in 2D. There is a considerable range

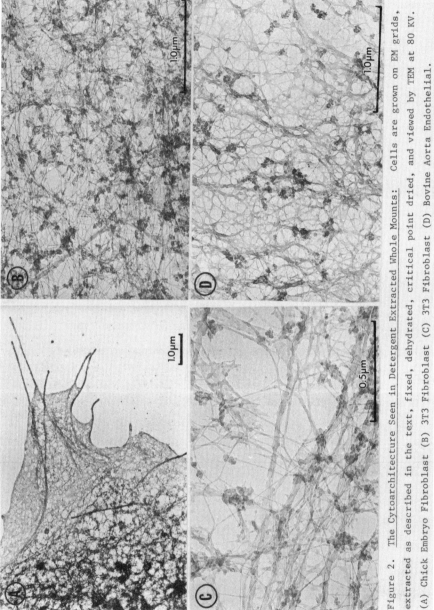

Figure 2. <u>The Cytoarchitecture Seen in Detergent Extracted Whole Mounts:</u> Cells are grown on EM grids, extracted as described in the text, fixed, dehydrated, critical point dried, and viewed by TEM at 80 KV. (A) Chick Embryo Fibroblast (B) 3T3 Fibroblast (C) 3T3 Fibroblast (D) Bovine Aorta Endothelial.

of filament size: many are 5-6 nanometers in diameter and may well be
microfilaments, but many are considerably larger. Of particular
interest are the numerous structures whose diameter changes
significantly throughout their length. These are probably more
complicated than simple homopolymers and are of as yet unknown
composition and function. The polyribosomes are bound to these
networks and, as Wolosewick and Porter (1976) observed in intact
cells, are usually found at the nexus of several filaments.

Polyribosomes appear completely bound to the architectural
structures remaining after Triton (Lenk et al., 1977; Fulton et al.,
1980; Cervera et al., 1980) When the polyribosomes are disassembled
in vivo by drugs or heat shock, only the ribosomes become soluble.
The mRNA remains bound to the cells architecture, or cytostructure,
presumably via the proteins of mRNP. Many different experiments
indicate that message binding to the skeleton fibers is not
adventitious (Fulton, et al., 1980; Cervera, et al., 1980). However,
the most compelling is cytological which shows that polyribosomes are
not free in the living cell. In flat, well spread cells,
polyribosomes are clustered in perinuclear regions and hence are not
freely diffusing. Figure 3A shows polyribosomes visualized by
ethidium fluorescence in an intact cell. The distribution of
cytoplasmic protein is shown in Figure 3B by the fluorescence of
Thiolyte, a general protein stain.

The observation that all polyribosomes were bound to some such
cellular structures came as something of a surprise after years of
classifying polyribosomes as "free" if they were not bound to

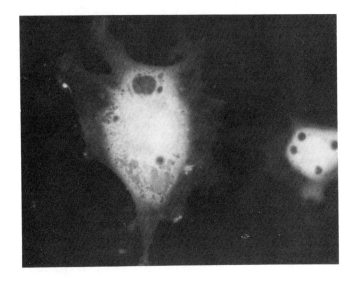

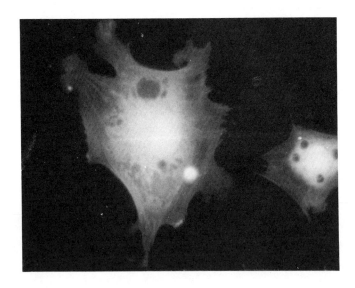

Figure 3:   Polyribosome distribution of intact bovine aorta
            endothelial cell. A. Polyribosomes stained with ethidime
            bromide. B. Total protein stained with Thiolyte.

membranes. The vigorous procedures developed in previous experiments
to effect a complete seperation of cytoplasm from the nucleus actually
served to disrupt the then unknown cytoplasmic network and thus yield
apparently freely sedimenting polyribosomes. These procedures gave
rise to a false impression of a fluid or gel-like cytoplasm which
conventional electron microscopy did not dispel. In contrast, the
comparatively gentle fractionation described here separates soluble
proteins from architectural proteins leaving the latter associated
with the nucleus. When this separation is made, polyribosomes are
completely retained on the cytoskeletal framework (CSK) (Figure 4).
This cytostructure can then be selectively solubilized and completely
separated from the nucleus by the use of a different detergent.

Perhaps the most profound implication of bound polyribosomes has
yet to be fully elucidated. In a recent paper, Alice Fulton has shown
that proteins destined for the cytoarchitecture associate with
structure at the time of synthesis and, indeed, while still nascent on
polyribosomes (Fulton et al., 1980). Thus it appears that for a
large class of cell structure, the common concepts of protein assembly
from solution are not applicable. We have been able to show with a
model system, the assembly of the nucleocapsid of VSV, that, in this
case at least, structural proteins become organized while associated
with the cytostructure and do not go through a soluble stage. The
fascinating possibility is raised that where proteins are made maybe
as important as which proteins are produced. Certainly, the regional
heterogeneity of the complex structures we see in cell cytoplasm would
suggest that mechanisms more sophisticated and subtle than in T even
phage assembly are operative.

254

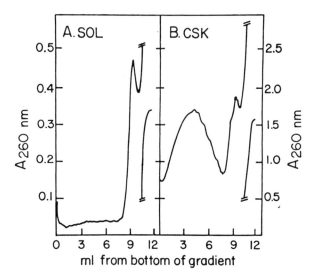

Figure 4:   Partition of polyribosomes between the soluble and
             cytoskeleton fractions of HeLa cells. The preparation was
             from cells infected with vescular stomatitis virus.
             Exactly the same results are obtained from uninfected
             cells. The Triton extract (A) is sedimented on a sucrose
             density gradient and the optical density monitored, The
             cytoskeleton (B) was solubilized and separated from the
             nucleus using a double-detergent procedure and the
             polyribosomes analyzed on the same sucrose gradient as in
             A. (from Cervera et al., 1981).

The Innermost Cellular Network: The Nuclear Matrix and Nuclear

Lamina

The cytoplasmic network terminates at the nuclear lamina, which

in turn encloses the nuclear interior. The nucleus has presented

formidable obstacles to structural studies. The density of chromatin

obstructs examination of nuclear architecture and its connections to

the cytoplasm unless drastic measures are adopted. The existence of

some form of nuclear matrix would seem obvious a priori from the

non-random distribution of chromatin. Furthermore, the co-ordinated

movements of chromatin into chromosomes indicate the existence of both scaffolding and contractile structures. Indeed, vigorous extraction procedures can remove chromatin and still leave a recognizable nuclear structure which has been partially characterized by biochemical measurements and by thin section microscopy. (Aaronson and Blobel, 1974; Berezney and Coffey, 1974; Cook et al., 1976; Keller and Riley, 1976; Herman et al., 1978; Adolph, 1980; Miller et al., 1978, Gerace and Blobel, 1980, McReady, et al., 1980, Puvion and Bernhard, 1975). This matrix has been shown to be the site of chromatin DNA replication (Vogelstein, et al., 1980; Berezney and Buchholtz, 1981), hnRNA attachment (Herman, et al., 1978; Miller, et al, 1978) as well as including the nuclear lamina and nuclear pores (Aronson and Blobel, 1974; Gerace and Blobel, 1980). However, even more than in the case of cytoplasmic architecture, the three-dimensional anastomosing network of nuclear filaments is very poorly represented in thin section and can be best visualized using the whole mount technique.

We have developed a relatively gentle procedure to prepare the nuclear filament network or nuclear matrix for electron microscopy as a whole mount (Capco et al., 1982). We do not isolate the nuclei since separation of the nucleus from the cytostructure requires harsh procedures which distort nuclear shape and result in a partial collapse of the filament network. Rather, we prepare the nuclear matrix still bound to the cytoplasmic architecture, or cytoskeleton. These structures are then viewed as cell whole mounts. Whole mounts of the nuclear matrix of a 3T3 mouse fibroblast is shown in Figure 5.

Collective Architecture and Biochemistry of Cells in Tissue; Breakdown in Malignancy.

The images of complex architectural networks composing the cytoplasm and nucleus of the cell and the first glimpse of their biochemical implications, lead to the obvious but difficult question;

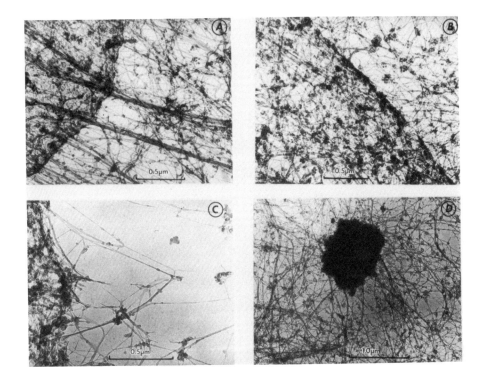

Figure 5:   The nuclear matrix viewed by whole mount electron
microscopy. 3T3 cells were grown on electron microscopy
grids and extracted with a high salt buffer and DNAase.
The resulting cytoskeleton and nuclear matrix are viewed
in transmission electron microscopy. 5A and 5B show the
nuclear matrix embedded in the remnant cytoskeleton.
Figure 5C is a high magnification view of the nuclear
lamina. The focal plane of the microscope was placed
inside the nuclear matrix for Figure 5D. The dark object
is a nucleous (from Capco et al., 1981).

why is the cell interior constructed this way? Some general

principles are apparent. The cellular networks are space filling

structures possessing considerable mechanical strength. However, they

are constructed subject to the constraint that the fibers can

withstand compression and tension but not bending movements or shear.

Such structures are well known in other contexts. The geodesic dome

is a simple example. There are many others which suggest, even more,

the networks seen in cells (Pugh, 1976). In one important regard the

cellular networks are unlike anything seen in human scale

architecture. They are dynamic structures with interconnections that

change with time and filaments that may undergo cycles of

polymerization. Much of cellular network behavior is accountable only

by its dynamism such as chromatin movement at mitosis and the

saltatory movement of organelles observed in time lapse

cinematography. The frozen images we see here undoubtedly reflect an

organization capable of intense and rapid movements. However, the

full significance of the structural networks become explicable only

when cells considered, not as independent entities, but as part of a

larger tissue organization.

Many aspects of cell behavior which appear incidental or are

unnoticed when cells are studied as individual entities, become of

paramount importance in the formation of the organized structure of

tissue. We have studied two of these. The first is the response of

gene activity to cell configuration or shape. This ability of the

genome to respond to architecture would seem necessary to regulate the

complex patterns adopted by the collective cell organization. The

second property is the propensity of the structural fibers in the

skeletal frameworks of adjacent cells to organize in response to their neighbors: a quality found only in cells competent to form organized tissue.

The shape dependence of cell behavior, examplified by the anchorage-dependence of fibroblasts, is more than a laboratory curiosity. We have studied fibroblasts as a model system for shape modulation of gene expression for many years (Benecke, et al., 1978; Benecke, et al., 1980; Ben-Ze'ev et al., 1980). Placing anchorage-dependent fibroblast into suspension culture leads to an extensive shut down of all major macromolecular processes and a elaborate program of recovery ensues upon replatting cells on a solid substrate. We determined that there are two principle signals when such a cell is placed in suspension. The loss of surface contact results in the shut down of cytoplasmic protein synthesis and this can be restored by the cells simply touching a suitable solid surface. However, nuclear metabolism responds to cell shape and is normal only when cells are almost completely spread. An adequate review of these findings would be too lengthy to present here and, very likely, the fibroblast represents but a simple version of the much more complex responses to shape change found in tissue forming cells such as those described below.

One important finding, which we will summarize, is the progressive loss of biochemical reponses to shape change as cells become less well growth regulated or more malignant in their phenotype. Folkman and coworkers made the seminal suggestion that malignant transformation is accompanied by and cassually related to the loss of shape regulation of cell growth (Folkman and Greenspan, 1975; Folkman and Moscona, 1978). We have shown that the decrease in growth regulation is accompanied by a progressive, stepwise loss of response in the basic macromolecular synthetic and processing systems (Wittelsberger, et al., 1981). We examined a series of mouse fibroblasts with varying degrees of growth control ranging from the well regulated primary diploid fibroblast to the completely unregulated, anchorage independent, doubly transformed mouse fibroblast line SVPy 3T3. Intermediate in growth regulating properties were the established lines 3T3, 3T6 and an HDP 3T6 line, developed by ourselves to be marginally anchorage dependent. The response of these cell lines to suspension culture is summarized in Figure 6. The cell types are in ascending order with the best

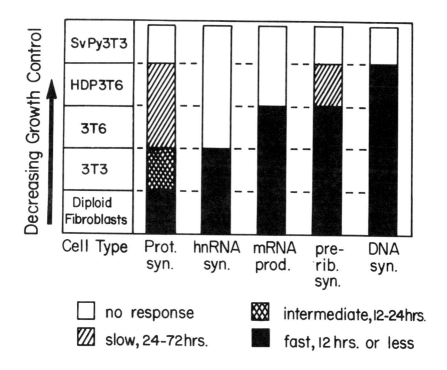

Figure 6:   Shutdown of major macromolecular processes in response to suspension culture (from Wittelsberber et al., 1981).

regulated at the bottom and the least regulated at the top. The loss
of responsiveness to architectural changes is apparent although the
stepwise nature of this loss was a suprise.

The second property of tissue forming cells, the organization of
their skeletal fibers in response to adjacent cells and other external
surfaces has been studied in model tissues such as the MDCK line of
canine kidney cells.

The organization of skeletal networks of tissue forming cells is
profoundly responsive to the exterior contacts of the cell. We have
been studying model tissues formed by the well differentiated cell
line, MDCK, derived from canine kidney. These cells grow as a highly
ordered monolayer colonies which closely resemble the epithelial cell
structure observed in vitro. On electron microscope grids cell
colonies are extracted and viewed as whole mounts much as single
cells. A low magnification view of such an epithelial cell sheet is
shown in Figure 7. The continuity pattern of cytoplasmic filament
organization across the cell border is apparent.

The organization of cell skeletons coherent with adjacent cell
structure is actually a well known phenomenon although its
significance has perhaps not been fully appreciated. Tight junctions
and especially desmasomes represent structures in which adjacent
participating cells must organize colinear filaments which terminate
at the same place on the intercell boundary. Actin filaments in
cardiac muscle and in epithelial cells appear to be anchored at the
cell surface in register with filaments in adjacent cells. Indeed,
the coherent organization of the skeletons in a model tissue such as
the kidney epithelium shown here, suggests that we might view tissue
not as collection of individual cells but, rather, as a organized
entity; in a sense we can view tissue as syncitium or macroskeleton

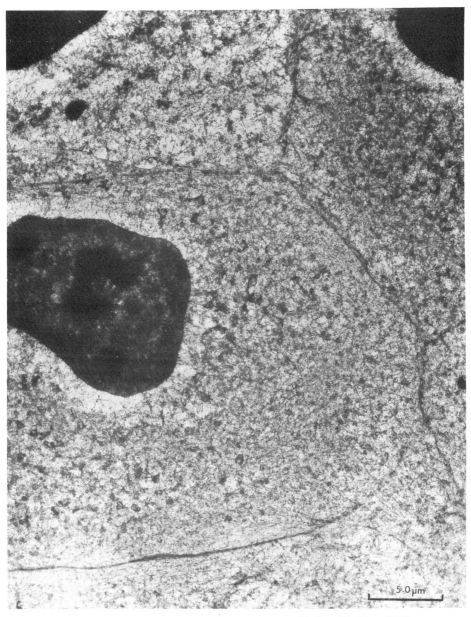

Figure 7:   Whole mount micrograph of the epithelial cell line MDCK.
Cells were grown as colonies on an electron microscope
grid.   The cells were extracted with Triton, fixed with
glutaraldehyde and osmium tetraoxide, dehydrated, critical
point dried and viewed by transmission electron
microscopy.

with nuclei embedded throughout. Whether this view of tissue as a supracellular skeleton organization is accurate will depend upon its conceptual utility. The long-range structural ordering of cytoskeletal structural elements in tissue forming cells takes on much greater significance in view of the ability of cell architecture to affect genomic activity.

Conclusion

The elaborate cell architecture shown here together with structurally related signals which function at the level of the genome offer an entree into the wide variety of biological phenomenon which have proven elusive to the approaches of traditional biochemistry. Intercell signalling, tissue organization and pattern formation in development can be seen as the operation of the genome in setting and controlling architecture. Only time will establish the validity of this structure based approach to cell and tissue behavior. Certainly, the promise seems clear.

REFERENCES

Aaronson, R.P. and G. Blobel (1974). On the attachment of the nuclear pore complex. J. Cell Biol. 62 746-754.

Adolph, K.W. (1980). Organization of chromosomes in HeLa cells: Isolation of histone depleted nuclei and nuclear scaffolds. J. Cell Sci. 42 291-304.

Benecke, B.J., A. Ben-Ze've, and S. Penman (1978). The control of messenger RNA production, translation, and turnover in suspended and reattached anchorage-dependent fibroblasts. Cell 14 931-939.

Benecke, B.J., A. Ben-Ze've, and S. Penman (1980). The regulation of RNA synthesis in suspended and reattached 3T6 fibroblasts. J. Cell Physiol. 103 247-254.

Ben-Ze've, A., A. Duerr, F. Soloman, and S. Penman (1979). The outer boundary of the cytoskeleton: a lamina derived from plasma membrane proteins. Cell 17 859-865.

Ben-Ze've, A., S. Farmer, and S. Penman (1980). Protein synthesis requires cell-surface contact while nuclear events respond to cell shape in anchorage-dependent fibroblasts. Cell 21 365-372.

Berezney, R. and D.S. Coffey (1974). Identification of a nuclear protein matrix. Biochem. Biophys. Res. Commun. 60 1410-1417.

Berezney, R. and L.A. Buchholtz (1981). Dynamic association of replicating DNA fragments with the nuclear matrix of regenerating liver. Exp. Cell Res. 132 1-13.

Branton, D., C.M. Cohen, and J. Tyler (1981). Interaction of cytoskeletal proteins on the human erythrocyte membrane. Cell 24 24-32.

Brown, S., W. Levinson, and J. Spudich (1976). Cytoskeletal elements of chick embryo fibroblasts revealed by detergent extraction. J. Supramol. Structure 5 119-130.

Bryant, S., V. French, P.J. Bryant (1981). Distal regeneration and symmetry. Science 212 993-1002.

Capco, D.G., K. Wan, and S. Penman (1982). The nuclear matrix: three dimensional architecture and protein composition. Cell 29, 847-858.

Cervera, M., G. Dreyfuss, and S. Penman (1980). Messenger RNA is translated when associated with the cytoskeletal framework in normal and VSV infected HeLa cells. Cell 23 113-120.

Cook, P.R., I.A. Brazell, and E. Jost (1976). Characterization of nuclear structures containing superhelical DNA. J. Cell Sci. 22 303-324.

Folkman, J. and H.P. Greenspan (1975). Influence of geometry on control of cell growth. Biochem. Biophys. Acta 417 211-236.

Folkman, J. and A. Moscona (1978). Role of cell shape in growth control. Nature 273 345-349.

Fulton, A.B., K. Wan, and S. Penman (1980). The spatial distribution of polysomes in 3T3 cells and the associated assembly of proteins into the skeletal framework. Cell 20 849-857.

Fulton, A.B., J. Prives, S.R. Farmer, and S. Penman (1981). Developmental reorganization of the skeletal framework and its surface lamina in fusing muscle cells. J. Cell Biol. 91 103-112.

Gerace, L. and G. Blobel (1980). The nuclear envelope lamina is reversibly depolymerized during mitosis. Cell 19 277-287.

Ham, A.W. and Cormack, D.H. (1979). Histology, 8th ed. J.B. Lippincott, Philadelphia, PA.

Herman, R., L. Weymouth, and S. Penman (1978). Heterogenous nuclear RNA-protein fibers in chromatin-depleted nuclei. J. Cell Biol. 78 663-674.

Keller, J.M. and D.E. Riley (1976). Nuclear ghosts: a nonmembranous structural component of mammalian cell nuclei. Science 193 399-401.

King, M.-C. and A.C. Wilson (1975). Evolution at two levels in humans and chimpanzees. Science 188 107.

Lenk, R., L. Ransom, Y. Kauffman, and S. Penman (1977). A cytoskeletal structure with associated polyribosomes obtained from HeLa cells. Cell 10 67-78.

Lenk, R. and S. Penman (1979). The cytoskeletal framework and poliovirus metabolism. Cell 16 289-301.

McCready, S.J., J. Godwin, D.W. Mason, I.A. Brazell, and P.R. Cook (1980). DNA is replicated at the nuclear cage. J. Cell Sci. 46 365-386.

Milcarek, C. and S. Penman (1974). Membrane-bound polyribosomes in HeLa cells: association of poly(A) with membranes. J. Mol. Biol. 89 327.

Miller, T.E., C.-Y. Huang, and O.A. Pago (1978). Rat liver skeleton and ribonucleoprotein complexes containing hnRNA. J. Cell Biol. 76 675-691.

Pease, D.C. and Porter, K.R. (1981). Electron microscopy and ultramicrotomy. J. Cell Biol. 91 287s-292s.

Pugh, A. (1976). Introduction to Tensegrity. University of California Press, Berkely.

Puvion, E. and W. Bernhard (1975). Ribonucleoprotein components in liver cell nuclei as visualized by cryoultramicrotomy. J. Cell Biol. 67 200-214.

Schliwa, M. and Blerkom, J. van. (1981) Structural interactions of cytoskeletal components. J. Cell Biol. 90 222-235.

Schliwa, M. (1982). Action of cytochalasin D on cytoskeletal network. J. Cell Biol. 92 79-91.

Solomon, F., M. Magendantz, and A. Salzman (1979). Identification with cytoplasmic microtubules of one of the coassembling microtubule associated proteins. Cell 18 431-438.

Vogelstein, B., D.M. Pardoll, and D.S. Coffey (1980). Supercoiled loops and eucaryotic DNA replication. Cell 22 79-85.

Webster, R.S., D. Henderson, M. Osborn, and K. Weber (1978). Three-dimensional electron microscopical visualization of the cytoskeleton of animal cells: immunoferritin identification of actin- and tubilin-containing structures. Proc. Natl. Acad. Sci. USA 75 5511-5515.

Wittelsberger, S.C., K. Kleene, and S. Penman (1981). Progressive loss of shape-responsive metabolic controls in cells with increasing transformed phenotype. Cell 24 859-866.

Wolosewick, J.J. and K.R. Porter (1976). Stereo high-voltage electron microscopy of whole cells of the human diploid line, WI-38. Am. J. Anat. 147 303-324.

Wolosewick, J.J. and K.R. Porter (1979). Microtrabecular lattice of the cytoplasmic ground substance, artifact or reality. J. Cell Biol. 82 114-139.

MESSENGER RNA CONFORMATION AND THE REGULATION OF PROTEIN SYNTHESIS

JOHN N. VOURNAKIS AND CALVIN P. H. VARY

Department of Biology, Syracuse University, 130 College Place, Syracuse,

New York, USA

INTRODUCTION

Importance of messenger RNA structure

A considerable literature exists that documents evidence for the relevance of secondary structure in messenger RNA function. The early work of
Lodish[1], Steitz[2] and Fiers, et. al.[3] strongly suggests that secondary structure acts to regulate protein synthesis in the bacteriophage f2, R17 and MS2
systems, respectively. Studies from Roth's laboratory on the regulation of
the histidine operon in salmonella[4] which combines genetic and translation
analysis, the work of Yanofsky and co-workers on the E. coli tryptophan
operon[5], and the model of enteric bacteria[6], all invoke the existence of
specific secondary structures in the respective mRNAs as a key regulatory
feature. Isertant and Fiers[7] used a series of recombinant plasmids containing the cro gene of λ phage to show a correlation between the level of expression of the cro protein and secondary structure models of the 5'-regions
of the mRNAs coded for by the respective plasmids. They argue that initiation of translation in prokaryotes is mainly influenced by the position of
the initiator AUG in the secondary structure, i.e. whether or not the AUG is
base-paired is a more important factor than the accessibility of the Shine
and Dalgarno sequence for ribosome binding[8]. They also argue that the
5'-region must be folded into a specific ribosome binding site following
transcription, prior to initiation.

Considerable recent work in eukaryotic systems is relevant to the role
of secondary structure in mRNA function. The observation initially by Protzel
and Morris[9], and more recently by Lizardi, et al.[10] that protein synthesis
can proceed in a discontinuous fashion in vivo, and the studies of Vary and
Morris[10] of the accumulation patterns of rabbit α and β globin nascent chain
polypeptides in vitro, suggest that ribosome transit is discontinuous along
the mRNA.

Work on ribosome binding sites in BMV mRNA[12], ovalbumin mRNA[13] alfalfa
mosaic virus mRNA[14], and globin mRNA[15] all invoke base-pairing schemes to

rationalize data on the size and sequence of 5'-region nucleotides protected by 40S and 80S ribosome initiation complexes. Kozak[16] has demonstrated that disruption of secondary structure by chemical means can alter the formation of 40S ribosomal initiation complexes in reovirus mRNA. Theoretical models for the translation of mRNA always involve secondary structure as a parameter.[17] Translation of mRNA can be affected by factors that affect RNA structure such as ionic conditions[18]. Direct measurements by physical techniques[19-22], and enzymatic means[23-24] demonstrate that eukaryotic mRNA has considerable specific secondary structure.

A recent publication by Rosa[26] describes experiments in which the enzymatic mapping technique developed in this laboratory[27,28] was used to study the structures of the protein synthesis initiator region in the three T7 late mRNA species IIIb, IV and V. Although it is known, as mentioned earlier, that prokaryotic ribosomes locate the correct initiator through base-pairing between the 3'-end of 16S rRNA and a region preceding the AUG[8], those factors that influence the translational efficiency of a mRNA are unclear. Minor changes in sequence around the initiator region seem to have a dramatic effect on the level of expression of proteins in prokaryotes[29,30]. It is plausible that secondary and tertiary structure changes caused by alterations in sequence can affect ribosome binding and thereby translational efficiency. The conclusions reached by Rosa[26] are consistent with this notion. In addition, secondary-structure models proposed by Rosa for ribosome binding regions in the three T7 mRNAs, which were arrived at by combining nuclease mapping data and computer-assisted thermodynamic secondary structure predictions using the strategy of Pavlakis et al.[24], indicate that the initiator AUG may in some cases be base-paired (mRNAs IIIb and IV) and in other cases unpaired (mRNA V). The relative efficiency of ribosome binding may, therefore, involve the extent to which stable hairpin regions must be disrupted (melted-out) during ribosome loading at the initiation site. A similar conclusion is reached by Pinck et al.[14] for the binding of ribosomes to alfalfa mosaic virus mRNA 3.

It is suggested, therefore, that knowledge of the structure of messenger RNA molecules may be key in attempting to begin to understand the subtleties of the regulation and modulation of protein synthesis in both prokaryotic and eukaryotic systems.

Methods for RNA Structure Analysis.

The development of rapid methods for DNA sequence analysis[31] resulted in

a general strategy that could be applied to RNA sequence and structure analysis[27]. Both enzymatic[27,28] and chemical[32] techniques for structure mapping
were developed that involve the partial cleavage of specifically end-labeled
RNAs followed by high resolution acrylamide gel electrophoresis and autoradiography to identify nucleotides that are either base-paired or unpaired.
The enzymatic technique has the advantages that both single-strand specific
(S1, T1, N. crassa and mung-bean nucleases, etc.) and base-pair specific
(cobra venom RNase) enzymatic probes are available and have been developed
as effective tools[28], and that either 5'- or 3'-end labeled RNA can be studied.
It has the major disadvantage that the enzyme probes are large molecules and
are unable, often, to penetrate regions of dense conformational packing
within RNA molecules. Therefore, enzymatic structure-mapping provides data
on the most exposed regions only. The enzymes used do not have significant
base-specificity, so that all four major nucleotides can be studied, although
the detailed structural specificities of certain of the enzymes are not
known. The chemical technique has the great advantage that the probes are
small and can penetrate throughout, thus providing nearly complete information
on base-pairing, base-stacking, and on shielding of reactive sites by tertiary structure, bound ions or bound proteins[32,33]. The major disadvantages
are that the method is limited to only three of the four major nucleotide
bases (A, G and C), and that it is applicable only to 3'-end labeled RNA.
The second limitation is particularly serious for proposed studies of
poly(A)-containing mRNA, requiring that an effective method for specifically
deadenylylating mRNA molecules be available.

Computer Aided Studies of RNA Secondary Structure

Computer methods for predicting RNA secondary structure have been developed by Pipas and McMahon[34], Studnicka et al.[35], and Zuker and Stiegler[36],
all of which impose thermodynamic criteria so as to obtain an optimal structure. These methods have limitations, and cannot generate an accurate structure prediction using sequence information alone[36,37]. A combination of
chemical reactivity, nuclease susceptibility and phylogenetic sequence
comparison data can be used to predict the structure of an RNA, as demonstrated by the impressive results of Noller and Woese[38] on E. coli 16S rRNA.
Comparing Zuker and Stiegler's computer-generated prediction of the 5'-end of
16S rRNA[36] to the Noller and Woese[38] consensus structure shows considerable
differences, even though Zuker and Stiegler take into account some experimental data. It is clear that there is no general method available for

unambiguously generating a RNA structure from sequence information alone. A new approach that combines thermodynamic and computer graphics based interactive methods has recently been described[37]. This method facilitates the input of experimentally derived structural data, and phylogenetic information. It has the potential for providing accurate secondary structures, but is not yet fully developed.

METHODS

Removal of m^7G from 5' terminus of Rabbit globin mRNAs

m^7G "Cap" removal was accomplished as previously described[24,39] 10-20 μg of mixed α and β Rabbit globin mRNAs in 10 μl of buffer containing 50 mM NaOAc pH 6.0 and 10 mM 2-mercaptoethanol and 5 units of tobacco alkaline pyrophosphatase (TAP) was incubated for 2 hr at 37°C. After the TAP incubation, enough 0.5 M Tris-HCl pH 8.5 was added to give a final concentration of 75 mM. Calf alkaline phosphatase was added to a final concentration of 0.1 Unit/ml. After incubation for 1 hr at 37°C the reaction mixture was immediately deproteinized with a 1:1 phenol/chloroform mixture and the mRNA precipitated with ethanol.

5' End group labeling of mRNA with ^{32}P.

Decapped and dephosphorylated mRNA was labeled with ^{32}P using $\gamma-^{32}P$-ATP (2-9000 Ci/mmol) and T_4 induced polynucleotide kinase[24,28,39]. Up to 0.4 A_{260} units of mRNA was labeled in a 10 μl reaction mixture containing 25 mM Tris HCl pH 8.5, 10 mM $MgCl_2$, 10 mM DTT, 2-4 nmols of $\gamma-^{32}P$-ATP and 2-4 units of T_4 polynucleotide kinase. After incubation for 30-60 min. at 37°C. The reaction mixture was brought to a final concentration of 25 mM EDTA, 8 M urea, 0.025% bromphenol blue and 0.05% xylene cyanol and heated to 50°C for 2 minutes prior to loading on a preparative polyacrylamide gel.

Preparative 2-dimensional polyacrylamide gel electrophoresis of 5'[^{32}P] labeled rabbit α and β globin mRNA.

5' labeled rabbit α and β globin mRNAs were electrophoresed in the first dimension on a 3.5% polyacrylamide gel containing 7M urea[24,28] at room temperature and 6-800 v until the xylene cyanol marker dye had migrated 37-40 cm. The partially resolved α andβglobin mRNA bands were located by autoradiography and a 1 cm x 4 cm band containing both mRNAs was cut out and transferred to a second set of gel plates oriented at 90' to the direction of electrophoresis. A 7% polyacrylamide 7M urea gel in the same buffer system was poured around the 1st dimension gel slice and allowed to polymerize for

10-15 hr. Following cooling of the gel to 4-5°C for electrophoresis was
performed for 24 hours at 5 watts. Radiolabeled spots corresponding to α and β
globin mRNA were located by autoradiography and excised from the gel. Elution of the labeled globin mRNA was performed in a volume of 1.5 ml containing 0.3M NH_4Ac pH 5.6 and 1 mM EDTA for 12 hr at 37°C. The elution solution
was isolated by centrifugation of the gel buffer suspension through silanized
glass wool in a 3 ml plastic syringe. To this solution was added 40 μg
carrier mixed tRNAs and the total RNA was isolated by addition of 3 volumes
of 95% ethanol followed by chilling to -80°C and centrifugation at 14,000 rpm
for 30 min. The pellet was redissolved in 100 μl of 0.3M NH_4Ac, 1 mM EDTA,
placed in 1.5 ml polypropylene tube along with 300 μl 95% EtOH and stored at
-20°C for use.

Deadenylylation of Rabbit α and β globin mRNAs

Ribonuclease H directed by one of several oligodeoxyribonucleotides was
used to selectively deadenylate mixed rabbit α and β globin mRNA [22,41].
100-300 pmols of globin mRNA was dissolved in 20 μl distilled deionized
sterile water. To this solution was added 3000 pmols of oligodeoxyribo-
nucleotide in 10 μl H_2O. It was heated to 75°C for 2 minutes. Sufficient 1M
KCl was then added to give a final potassium concentration of 80 mM. Following an additional minute of of incubation at 75°C the solution was cooled in
a thermal reservoir of sufficient capacity to bring the sample to 37°C in
approximately 30 minutes. To this mixture was added one tenth volume of
RNase H-Buffer 2[22] resulting in final concentration of 50 mM Tris-HCl pH 8.3,
25 mM $MgCl_2$, 0.08M KCl, 0.5 mg/ml BSA, and 0.001M DTT. Following further
incubation of the mixture at 25°C for 10 minutes, 1-1.5 units of calf thymus
RNase H was added and the reaction was allowed to proceed for 90 minutes at
33°C for the dT_8dGdC and dT_{11-17} oligodeoxynucleotides. Reactions directed
by dT_4dG were conducted at 15°C for 20 hr with 8-10 units of RNase H enzyme.

Isolation of Deadenylylated mRNA.

Purification of mRNA and removal of oligomer was accomplished by passage
of the reaction mixture over a 1.5 ml Sephadex G-75 column equilibrated in
0.3 M NaAc pH 5.6. Fractions corresponding to the void volume were pooled
and the RNA precipitated with 3 volumes of EtOH.

3'end labeling of deadenylylated globin mRNA with 5'[^{32}P]pCp.

Deadenylylated globin mRNA was labeled with 5'[^{32}P]pCp using T_4 induced
RNA ligase. 100 pmoles of deadenylated globin mRNA was dissolved in 20 μl of
buffer containing 10% dimethyl sulfoxide 50 mM Hepes pH 7.5, 10 mM $MgCl_2$,
5 mM DTT, 5 μM ATP, 100-200 pmoles of 5'[^{32}P] pCp and 400 units/ml T_4 RNA

ligase. The reaction was allowed to proceed for 24 to 48 hr at 4°C and incorporation was monitored by polyethyleneimine cellulose chromatography in 0.75M phosphate buffer pH 6.0. 3' end labeled $\alpha + \beta$ globin mRNA were purified as described for 5' labeled mRNA above.

Determination of 3' and 5' end heterogeneity

The terminal heterogeneity of 5' and 3' end labeled RNA was determined by total digestion of labeled RNA with T_1 ribonuclease followed by polyacrylamide gel electrophoresis and of the products and detection of labeled bands by autoradiography. Reactions containing 10 mM Tris-HCl pH 8.0, 1 mM EDTA, 3-5000 cpm (Cerenkov) of 5' or 3' end labeled mRNA and 0.01 units of T_1 ribonuclease in a volume of 10 µl was incubated for 60 min at 55°C. To the reaction was added an equal volume of sample buffer giving a final concentration of 8M urea, 0.025% bromphenol blue and 0.025% xylene cyanol. Electrophoresis was performed on a 20% polyacrylamide gel (.15 x 20x40 cm) in 1X Tris borate-EDTA buffer at 6-800 volts until the bromphenol blue had migrated 50% of the length of the gel, followed by detection of digestion products by autoradiography.

Partial digestion of end labeled α and β globin mRNA

Structural analysis of 5' end labeled α and β globin mRNA using the structure specific enzymes S_1 nuclease and cobra venom ribonuclease essentially as described previously[27,28], except that RNA samples were occasionally heated to 60°C in the absence of salts and slowly cooled in the presence of salts to 37°C.

Computer aided analysis of mRNA structural data

Data was recorded and hypothetical RNA models produced by several computer assisted methods as described in detail elsewhere[37].

RESULTS AND DISCUSSION

Preparation and purification of 5' labeled α and β globin mRNA

The methods described provide highly labeled α and β globin mRNA with little degradation. Due to the heterogeneity in the lengths of α and β globin mRNA molecules, polyacrylamide gel electrophoresis in one dimension does not achieve adequate separation of the mRNA species present. By taking advantage of a change in the relative mobility of the α and β globin mRNA molecules (α globin mRNA migrates faster in the first dimension, β globin mRNA faster in the second), an adequate separation of labeled mRNA species is achieved[40]. T1 RNase digestion of α and β globin mRNA shows that the prominent digestion products correspond to the seven nucleotide 5' α globin mRNA T_1 fragment and

271

the nine nucleotide long 5' β globin T₁ fragment, respectively (data not
included).

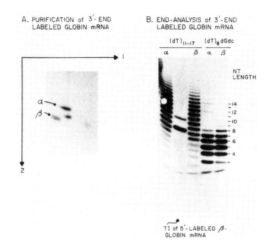

A. PURIFICATION of 3'- END
LABELED GLOBIN mRNA

B. END-ANALYSIS of 3'-END
LABELED GLOBIN mRNA

(dT)₁₁₋₁₇   (dT)₈dGdc
α   β   α   β

NT
LENGTH

- 14
- 12
- 10
- 8
- 6
- 4

T1 of 5'-LABELED β-
GLOBIN mRNA

Fig. 1. Rabbit globin mRNAs were deadenylylated by the action of calf thymus
ribonuclease H directed to the mRNA-poly(A) junction using the primers (dT)₁₁₋₁₇
(dT)₈-dG-dC. Following hybridization and deadenylylation the purified RNA
was labelled at its 3'-terminus with T₄ RNA ligase and 5'[³²P]pCp as described
above. A. Two-dimensional gel electrophoretic purification of 3'-end labelled
α- and β globin mRNAs was performed as described in Methods B. End-analysis
of 3'-end-labelled products was achieved by a complete T₁ RNAase digestion
(see Methods) followed by electrophoresis of the digestion products on a 20%
polyacrylamide/8.3M-urea gel. The size distribution of labelled T₁ oligonucleotides reveals the distance and frequency of cleavages made by RNAase H
relative to the terminal G residue in either the α- or β-globin mRNA.

Deadenylylation of Rabbit α and β globin mRNAs.

The presence of deadenylylated globin mRNAs is immediately evident in
the homogeneity of labeled bands in the first and second dimensions of preparative gel electrophoresis as seen in Figure 1. Direct analysis of the
deadenylylation reaction is accomplished by a total T₁ digestion pattern of
3' labeled mRNA molecules. Fig. 1 shows the oligoA distribution at the 3'
termini of α and β globin mRNA molecules digested with RNase H directed by
dT₁₁₋₁₇ and dT₈dGdC. It is evident that the specificity of the cleavage is
enhanced when the smaller oligodeoxynucleotide is used. It is also evident
that a limiting cleavage pattern is obtained in which several sites are
present and serve as acceptors for RNA ligase labeling with pCp. Since the

size of the T_1 released oligonucleotides is known the limiting site sizes obtained with dT_{11-17} and dT_8dGdC can be determined[40]. These data indicate that heterogeneity in RNase H catalyzed cleavage of the poly (A) sequence is due to the size of the oligonucleotide used to direct the cleavage. The reason for this is not entirely clear, since the oligonucleotides may be expected to bind anywhere along the length of the adenylate tract and may participate in site-site exclusion caused by adjacent molecules of oligo-deoxyribonucleotides overlapping to some extent. It is interesting to note that in neither case do cleavages occur closer than five nucleotides from the T_1 nuclease cleavage point. This may be due to failure of RNase H to cleave near misaligned DNA-RNA junctions or may reflect in addition the minimum site size requirements of the enzyme. It should be noted that no detectable random fragmentation of α or β globin mRNA molecules occurred under these conditions.

RNase H cleavage of globin mRNA molecules directed by dT_4dG

In order to further reduce the 3' end heterogeneity a smaller oligonucleotide primer, dT_4dG, was used. Using the same conditions as for the dT_8dGdC or dT_{11-17} primed reactions, only slight hydrolysis of the mRNA molecules occurred with dT_4dG[40]. This situation was not improved with increased salt or decreased temperature, longer incubation at increased enzyme concentration and at lowered temperature resulted in the digestion pattern observed in Figure 2[40]. T_1 digestion of the bands in Figure 2 following end labeling with pCp indicated a considerable decrease in heterogeneity such that most bands from the fragmented mRNA profile yield predominantly 1 or 2 bands on total T_1 digestion (Vary, C. P. H. and Vournakis, J. H., in preparation). This method provides, therefore, starting material for sequencing and structure mapping polyadenylated mRNA molecules from the 3' end inward.

Structural analysis of 5'-end labeled β globin mRNA.

Figure 3 shows a typical example of results obtained by partial hydrolysis of the β globin mRNA. Data for the 5' proximal region of the β globin message generated by the single strand specific S_1 nuclease and the base-pair specific cobra venom ribonuclease (CVR)[28] are shown. A complete enzymatic analysis of α and β globin mRNAs has been obtained recently (Vournakis, J. N. and Vary, C. P. H., in preparation). Of particular interest is the region surrounding the α and β globin mRNA initiator AUG sequences. It can be seen in Fig. 3 that the β initiator AUG lies in a S_1 susceptible region of the molecule and is symmetrically flanked by CVR sensitive areas indicating that this AUG may be exposed in a hairpin or bulge structure in the

3.5% POLYACRYLAMIDE

dT4dG

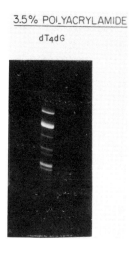

Fig. 2. Deadenylylation of rabbit globin mRNAs using RNase H directed by $(dT)_4$-dG. Discreet α- and β-globin mRNA fragments were generated by $(dT)_4$ dG directed cleavage of rabbit α- and β-globin mRNAs with RNAase H. These fragments are resolved by electrophoresis on a 3.5% polyacrylamide/8M urea gel as described in Methods. The first dimension of the usual two-dimensional electrophoretic technique used for polyadenylated mRNAs is sufficient to prepare homogeneous RNAase H fragments for further characterization.

β globin mRNA molecule. The corresponding region of the α globin mRNA molecule shows that the AUG is also exposed but not to such an extent as in the case of the β globin mRNA, since S_1 hits here only lightly[24]. The following trends have been observed to date:

1. Both molecules have regions of strong preference for S_1 nuclease and other regions which are susceptible to CVR.

2. Some regions of both α and β globin molecules appear to be masked from both enzymes possibly indicating a globular or buried conformation.

3. So-called control sites seem to lie in or very near single stranded exposed regions as determined by S_1 sensitivity and relative CVR nuclease insensitivity. These regions include the α and β globin mRNA splice points for the first and second intervening sequences, the α and β initiator AUG codons, the terminator codons, and the poly(A) addition signal AAUAAA.

It is tempting to suggest that presumptive regulatory sequences need to maintain exposed single stranded configurations for their various functions in mRNA expression and processing. Indeed recent models for the interaction of small nuclear RNAs with splice junctions in mRNA precursors[42] would

274

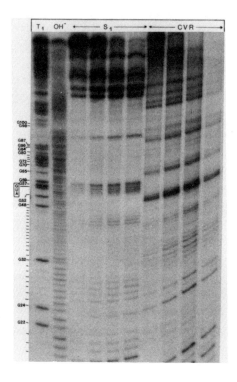

Fig. 3. This figure displays the CVR and S_1 nuclease susceptibility of the
5' region of the rabbit β-globin mRNA illustrating the structural disposition
of the AUG initiator codon. The region corresponding to the   AUG initiator
codon is highly susceptible to the action of single-strand-specific S_1
nuclease. Cobra venom ribonuclease specific for base paired phosphodiester
bonds yield no fragments within this region but gives evidence of base pairing symmetrically flanking the AUG initiation codon. The apparent increased
accessibility of the AUG codon may be due to its location in the loop of a
hairpin or bulge structure in the mRNA.

suggest that the regions near the splice junctions be single stranded before

and most likely after splicing has taken place.

Secondary Structure Models.

Several computer assisted techniques, mentioned earlier[34,37], exist for

predicting secondary structure of RNA from primary sequence, enzymatic cleav-

age and thermodynamic data. Models of rabbit β-globin mRNA obtained by three

different computer methods are published elsewhere[37,40]. In some cases no thermodynamic energy calculations were used, and the models were constructed to optimize agreement with nuclease susceptibility data[37]. The structure prediction in Figure 4 was obtained using the method of Pavlakis et. al.[24], which is a modification of previously described thermodynamic calculations[34,35]. The prediction was made by generating a minimum free-energy structure for 200 base sequences, starting at the 5'-end and moving in 50 base steps to the 3'-end of the molecule. No nuclease digestion data was used as input in generating this structure, the S1 data having been added later. Nuclease data were added to the model after the minimum free-energy structure was obtained. The Zuker RNA2 approach[37] is designed to feature short-range hairpin structures. In general, nuclease susceptibility data are in good agreement with the Zuker RNA2 prediction[40]. It is important to note that surprisingly dissimilar structures are generated by the three different computer methods[37,40].

Ribosome Transit and Secondary Structure.

Experiments to determine the size distribution of nascent chain oligopeptides in a reticulocyte in-vitro translation system were performed by Vary and Morris[11]. An example of the results of such a study is published[40] for rabbit β-globin mRNA programmed protein synthesis. These data demonstrate that the oligopeptide β-globin nascent chain size profile is non-uniform. Such a result is likely to be generated by ribosome pausing which may reflect mRNA structure. Figure 4 shows a correlation of a secondary structure model and the nascent chain data. It is indeed surprising that the presumptive ribosome pause sites occur at the 5'-side of stable local hairpin structures. This correlation suggests that ribosome pausing due to secondary structure within mRNA coding regions may be a subtle protein synthesis regulatory feature.

Conclusions and Future Directions

In vitro enzymatic digestions of rabbit α and β globin mRNAs have revealed the presence of extensive secondary structure in these molecules. These studies are points of departure in the study of mRNA structure and its influence on functions such as translation initiation, elongation and termination.

The overall pattern of secondary structures, the exposure of control sequences and functional regions such as the initiator codons, intervening splice point junctions and polyadenylation sequences have mechanistic relevance and provide a basis for comparison of the dispositions of these regions in different mRNA molecules. Certain problems must be addressed in the study

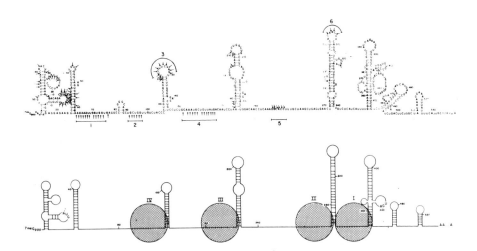

Fig. 4. Secondary structure model of rabbit β-globin mRNA (upper) and a schematic diagram (lower) of that structure indicating the apparent positions of ribosome accumulation as determined from the data in ref.40. Arrows indicate S_1 and/or T_1 susceptible regions of 5'-^{32}P-end labelled β-globin mRNA. Six such regions (labelled 1-6) are indicated in the upper figure. Shaded circular areas in the lower diagram are numbered I-IV and correspond to persistent ribosome positions that generate the peaks I-IV in ref. 40. The positions of the ribosomes were calculated from the apparent molecular weights of the predominant nascent oligopeptide size classes shown in ref.40.

of mRNA structure by these methods. These include the assumption that structure is not lost during isolation procedures which subject the RNA to denaturation. Though evidence exists to support this contention the only satisfactory answer must await the development of methods which probe the structure of mRNA molecules in their native state. Methods have been developed to probe RNA in ribonucleoprotein complexes. These involve the use of small chemical probes[32,33] which can penetrate ribonucleoprotein particles such as ribosomes and ribosomal subunits and react with the RNA in a structure specific manner. Unfortunately these methods, derived from chemical sequencing technology,[32] rely on the availability of 3' labeled RNA. This possibility is precluded in the case of most eukaryotic mRNAs due to the presence of a polyadenylic acid tract of heterogenous length. We illustrate, above, the use of RNase H directed to the mRNA poly(A)-mRNA junction with the primer dT_4dG to produce mRNA which is specifically deadenylylated. This approach provides a general method for production of mRNA which can be 3'-labeled and sequenced enzymatically or chemically, thus providing a doubling of the length range for mRNA sequence analysis. In addition, this methodology allows sequencing and structure mapping methodologies with chemical probes to be brought to bear on problems of mRNA structure in-situ in ribonucleoprotein complexes. This is possible since the base or structure specific modification reaction of the RNA is separate from the cleavage reaction. Therefore, RNA may be modified in vivo with structure specific probes such as DMS or DEPC, extracted, end labeled following deadenylylation, and then cleaved at the sites of modification by treatment with aniline and sodium borohydride. An alternative and complimentary scheme would employ 3' labeled mRNA obtained following deadenylylation to probe the structure of mRNA involved in 40S preinitiation complexes, 80S initiation complexes and polysomes. Since 3'-labeled mRNA possesses an intact cap structure unlike 5'-labeled mRNA, it also provides an opportunity to measure the rate at which labeled mRNA progresses through the stages of preinitiation and initiation complex formation and into polysomes, relative to a standard competing mRNA under mRNA limiting conditions. This will provide a functional correlate which will be useful in relating the structural disposition of the initiator AUG codon and surrounding sequences to a functional parameter such as the relative efficiency of incorporation of labeled mRNA into preinitiation and initiation complexes.

The proposal by Kozak[43] that initiation of protein synthesis in eukaryotic cells proceeds by a scanning mechanism, whereby the 40S ribosomal subunit attaches at the 5'-end and migrates to the correct AUG initiator codon, postulates that a specific, universal, initiator sequence exists that includes the AUG. According to the scanning model, secondary structure in the 5'-non-coding region would be inhibitory to initiation, since it could retard the movement of the 40S subunit. Recently, Sonenberg[44,45] has suggested that cap binding proteins required for intiation may be acting to unwind the 5'-end structure of capped eukaryotic mRNAs, thus making it possible for 40S subunits to "scan" to the AUG. This is illustrated schematically in Fig. 5. Thus, 5'-end structure could act to modulate protein synthetic rates at the initiation step such that mRNAs have less stable 5'-end structure would initiate with a faster rate. Structural studies with cap binding protein mRNA complexes of the sort mentioned earlier may yield some insight on these issues.

278

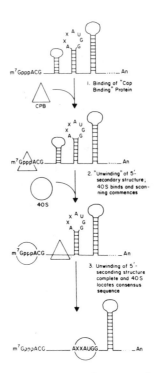

Fig. 5. A schematic diagram suggesting that secondary structure in the 5'-non coding region of eukaryotic mRNA may modulate the rate of initiation of protein synthesis.

ACKNOWLEDGEMENTS

This work was supported by grants from N.I.H. (GM 22280) and N.S.F. (PCM 8004620).

REFERENCES

1. Lodish, H. P.   (1970) J. Mol. Biol. 50, 689-702.
2. Steitz, J. A.   (1969) Nature 224, 957-967.
3. Fiers, et al.   (1976) Nature 260, 500-507.
4. Johnston. H. M., Barnes, W. M., Chumley, F. G., Bossi, L. and Roth, J.R.   (1980) Proc. Nat. Acad. Sci. NSA 77, 508-512.
5. Oxender, D. L., Zurawski, G. and Yanofsky, C.   (1979) Proc. Nat. Acad. Sci. USA 76, 5524-5528.
6. Keller, E. B. and Calvo, J. B.   (1979) Proc. Nat. Acad. Sci.   USA 76, 6186-6190.
7. Isertant, D. and Fiers, W.   (1980) Gene 9, 1-12.
8. Shine, J. and Dalgarno, L.   (1974) Proc. Nat. Acad. Sci.   USA 71, 1342-1346.
9. Protzel, A. and Morris, A. J.   (1974) J. Biol. Chem. 249, 4594-4600.

10. Lizardi, P. M., Mahdavi, V., Shields, D. and Candelas, G. (1979) Proc. Nat. Acad. Sci. USA 76, 6211-6215.
11. Vary, C. P. H. and Morris, A. J. unpublished observations; and Vary, C.P.H. Ph.D. Dissertation.
12. Alquist, P., Dasgupta, R., Shih, D. S., Zimmern, D. and Kaesburg, P. (1979) Nature 281, 277-282.
13. Schroeder, H. W., Liarkos, C. D., Gupta, R. C., Randerath, K. and O'Malley, B. W. (1979) Biochemistry 18, 5798-5808.
14. Pinck, M., Fritsch, C., Ravelonandro, M., Thivent, C. and Pinck, L. (1981) Nucl. Acids Res. 9, 1087-1100.
15. Legon, S., Model, P. and Robertson, H. D. (1977) Proc. Nat. Acad. Sci. USA 74, 2692-2696.
16. Kozak, M. (1980) Cell 19, 79.
17. Heijne, G., Nilsson, L. and Blomberg, C. (1978) Eur. J. Biochem. 92, 397-402.
18. Bergmann, J. E. and Lodish, H. F. (1979) J. Biol. Chem. 254, 459-468.
19. Holder, J. W. and Lingrel, J. B. (1975) Biochemistry 14, 4209-4215.
20. Van, N. T., et al. (1977) Biochemistry 16, 4090-4100.
21. Kallenbach, N. R., Brentani, M. M. and Brentani, R. R. (1979) Biopolymers 18, 1515-1531.
22. Vournakis, et al. (1976) Prog. Nucl. Acid Res. Mol. Biol. 19, 233-252.
23. Flashner, M. and Vournakis, J. N. (1977) Nucl. Acids Res. 4, 2307-2319.
24. Pavlakis, G. N., et al. (1980) Cell 19, 91-102.
25. Lodish, H. F. and Jacobsen, M. (1972) J. Biol. Chem. 247, 3622.
26. Rosa, M. D. (1981) J. Mol. Biol. 147, 55-71.
27. Wurst, R., Vournakis, J. and Maxam, A. (1978) Biochemistry 17, 4493-4499.
28. Vournakis, J. N., et al. (1981) in Gene Amplification and Analysis, vol. II: Analysis of Nucleic Acids by Enzymatic Methods, (Chirikjian, J. and Papas, T., eds.), Elsevier-North Holland, Inc., pp 267-298.
29. Dunn, J. J., Buzash-Pollert, E. and Studier, F. W. (1978) Proc. Nat. Acad. Sci. 75, 27412745.
30. Roberts, T. M., Kacich, R. and Ptashne, M. (1979) Proc. Nat. Acad. Sci. 76, 760-764.
31. Maxam, A. and Gilbert, W. A. (1977) Proc. Nat. Acad. Sci. 74, 560-564.
32. Peattie, D. A. and Gilbert, W. A. (1980) Proc. Nat. Acad. Sci. 77, 4679-4682,
33. Peattie, D. A. and Herr, W. (1981) Proc. Nat. Acad. Sci. 78, 2273-2277.
34. Pipas, J. M., and McMahon, J. E. (1975) Proc. Nat. Acad. Sci. 72, 2017-2021.
35. Studnicka, G. M., Rahn, G. M., Cummings, I. W., and Salser, W. A. (1978) Nucl. Acids Res. 5, 3365-3387.
36. Zuker, M. and Stiegler, P. (1981) Nucl. Acids Res. 9, 133.
37. Auron, P. E., Rindone, W. F., Vary, C. P. N., Celantano, J., and Vournakis, J. (1982) Nucl. Acids Res 10, 403-419.
38. Noller, H. F. and Woese, C. R. (1981) Science 212, 403-411.
39. Lockard, R., Rieser, L. and Vournakis, J. N. (1981) in Gene Amplification and Analysis, vol. II; Analysis of Nucleic Acids by Enzymatic Methods (J. Chirikjian and T. Papas, eds.) Elsevier-North Holland, Inc., pp 229-251.
40. Vary, C. P. H. and Vournakis, J. N. (1982) Biochem. Soc. Symp. 47, in press.
41. Stavrianopoulos, J. G. and Chargaff, E. (1978) Proc. Nat. Acad. Sci. USA 75, 4140-4144.
42. Lerner, M. R., Boyle, J. A., Mount, S. M., Wolin, S. L. and Steitz, J. A. (1980) Nature 283, 220-224.

43. Kozak, M. (1981) Nucl. Acids Res. $\underline{9}$, 5234-5252.
44. Sonenberg, N. (1981) Nucl. Acids Res. $\underline{9}$, 1644-1657.
45. Sonenberg, N., Guertin, D., Cleveland, D. and Trachsel, H. (1981)
        Cell $\underline{27}$, 563-572.

FUNCTION AND TOPOGRAPHY OF COMPONENTS OF INITIATION COMPLEXES OF EUKARYOTIC TRANSLATION

HEINZ BIELKA[+], PETER WESTERMANN[+], ODD NYGÅRD[++], JOACHIM STAHL[+],
ULRICH-AXEL BOMMER[+], FRANZ NOLL[+], GUDRUN LUTSCH[+] and BURKHARDT
GROSS[+]
[+]Institute of Molecular Biology, GDR Academy of Sciences,
GDR-1115 Berlin-Buch; [++]The Wenner-Gren Institute, Department
of Cell Physiology, University of Stockholm, S-113 45 Stockholm,
Sweden

INTRODUCTION

The understanding of the mechanism(s) by which protein synthesis is channeled in time and space at the ribosomal level needs
the elucidation of the arrangement of functional sites on the
ribosome and their interactions with nonribosomal components of
the translation system. In this respect, much less is known about
eukaryotic ribosomes in comparison to prokaryotic ones.

We have studied by means of different techniques macromolecular components of the ternary initiation complex and of the small
subunit of eukaryotic ribosomes (rat liver) involved in the organization of the so-called P site on which Met-tRNA$_f$, eIF-2, eIF-3,
and mRNA are joined together noncovalently in specific sequential
orders of events during initiation. This level of the initiation
process of protein biosynthesis, which is preferentially characterized by initiation factor mediated codon-anticodon (mRNA-tRNA)
interactions seems to be also of regulatory significance for the
entire translation process in eukaryotes.

RESULTS

1. Neighborhood of macromolecular components within ternary and quaternary initiation complexes

Cross-linking

Cross-linking has been successfully applied for the determination of the topography of proteins in ribosomal particles of
E.coli. We have used this approach to determine neighboring
components in the ternary (eIF-2 x Met-tRNA$_f$ x GMPPCP) and quaternary (eIF-2 x Met-tRNA$_f$ x GMPPCP x 40S ribosomal subunit)
initiation complexes from rat liver using the following bifunc-

tional reagents (Table 1): Dimethyl-5,6-dihydroxy-4,7-dioxo-3,8-diazadecanbisimidate (DBI), Methyl-5-(p-azidophenyl)-4,5-dithiapentanimidate (APTPI), Methyl-p-azidobenzoylaminoacetimidate (ABAI), and 1,2:3,4-Diepoxybutane (DEB)[1].

TABLE 1

BIFUNCTIONAL REAGENTS USED FOR CROSS-LINKING

Reagent		Distance of reactive groups Å
DEB	1,2:3,4-Diepoxybutane $H_2C-HC-CH-CH_2$	4
ABAI	Methyl-p-azidobenzoylaminoacetimidate $N_3-Ph-CO-NH-CH_2-C\!\!<^{NH}_{OCH_3}$	10
APTPI	Methyl-5-(p-azidophenyl)-4,5-dithiapentanimidate $N_3-Ph-S-S-CH_2-CH_2-C\!\!<^{NH}_{OCH_3}$	12
DBI	Dimethyl-5,6-dihydroxy-4,7-dioxo-3,8-diazadecanbisimidate $^{HN}_{H_3CO}\!\!>\!C-CH_2-NH-CHOH-CHOH-NH-CH_2-C\!\!<^{NH}_{OCH_3}$	14

Cross-linked dimers were separated and analyzed by various chromatographic and electrophoretic procedures and identified by one- and/or two-dimensional polyacrylamide gel electrophoresis as described in detail[2-5].

In experiments with the ternary initiation complex (see Table 2) it could be shown[2] that Met-tRNA$_f$ can be cross-linked to all three subunits (α, β, γ) of eIF-2, when APTPI or ABAI were used (distance of the reactive groups 12 and 10 Å, respectively). However, when DEB (reactive range 4 Å) was applied, only the ß-subunit of eIF-2 was attached to initiator-tRNA. From this finding it is concluded that the ß subunit of eIF-2 is the Met-tRNA$_f$ binding component within the ternary initiation complex[2].

The same result was obtained when quaternary initiation complexes were studied with DEB[3]; here also, Met-tRNA$_f$ was found covalently linked to eIF-2β (see Table 2).

TABLE 2

PROTEINS AND RNA OF THE SMALL SUBUNIT OF RAT LIVER RIBOSOMES AND NONRIBOSOMAL COMPONENTS (eIF-2 AND Met-tRNA$_f$) WHICH CAN BE CROSS-LINKED BY VARIOUS BIFUNCTIONAL REAGENTS

Substrates subjected to reaction	Cross-linking reagent	Cross-linked species Ribosomal components	Nonribosomal components
eIF-2•GMPPCP•Met-tRNA$_f$	ABAI		
	APTPI		eIF-2α,-ß, γ Met-tRNA$_f$
	DEB		eIF-2β Met-tRNA$_f$
eIF-2•GMPPCP	DEB		eIF-2β Met-tRNA$_f$
•Met-tRNA$_f$•40S subunit	DEB	Proteins S3a, S6	
	ABAI		Met-tRNA$_f$
	DBI	Proteins S3, S3a, S15 S3b, S6, S7, S10, S13, S19, S23/24)	eIF-2 (α and/or γ?)
	APTPI		
	ABAI	18S rRNA	eIF-2α and γ
	DEB		
	ABAI	Proteins S3, S3a, S6,	
	DEB	S8, S11, S16/18, S23/24, S25 with 18S rRNA	

In these studies it could furthermore be shown by using DEB and ABAI that in the quaternary initiation complex Met-tRNA$_f$ is also in close contact with ribosomal proteins S3a and S6[3] (nomenclature of proteins according to[4]).

In experiments concerning the binding site for eIF-2 on the 40S subunit of rat liver ribosomes, we could at first prove that mainly ribosomal proteins S3, S3a, and S15 (also smaller amounts S3b, S6, S7, S10, S13, S19 and S23/24) can be cross-linked with eIF-2[5].

In further studies with APTPI, ABAI, and DEB eIF-2 was found to be cross-linked also to 18S rRNA especially by the α- and the γ-subunit[6] (see also Table 2).

Application of antibodies

Further attempts to analyze proteins of the small subunit of rat liver ribosomes involved in interactions with components of the ternary initiation complex have been made by the application of antibodies against single ribosomal proteins. These antibodies were used to study their effect on the binding of the ternary initiation complex (with [3H]Met-tRNA$_f$) to 40S ribosomal subunits (for details see[7]). It was found that antibodies against ribosomal proteins S3, S6, and S13 are most efficient in blocking the quaternary complex formation. Antibodies against other S-proteins (e.g. anti-S5, -S7, -S9) have weaker inhibitory effects which might be due to the location of the corresponding proteins either only partially at the binding region or in its neighbourhood. Antibodies against protein S21 as well as the IgG fraction from non-immune antisera and antibodies against rat albumin are without any influence.

From these studies it seems justified to conclude that proteins S3, S6 and S13 are involved in interactions with components of the ternary initiation complex, which is in good agreement with cross-linking data described before (S3a has not yet been studied by using antibodies).

Affinity labeling

In studies with 80S and 40S ribosomal particles from rat liver, the 4-(N-2-chloroethyl-N-methylamino)-benzaldehyde [14C]acetal derivative of octauridylate ((pU)$_8$-RCl) containing the alkylating group in the 3'-position[8] and a heptauridylate derivative bearing the same reactive group via a phosphoamide bond in the 5'-position (ClR-(pU)$_7$)[9] were applied as mRNA analogue.

Binding of these reagents to ribosomal particles was inhibited by poly(U) and increased by addition of uncharged tRNA.

In experiments with the 3'-substituted (pU)$_8$-RCl, nearly exclusively proteins S3/3a, and to a much lesser extent S26, were found associated with the ^{14}C-label[8]. On the other hand, in studies with the 5'-substituted ClR-(pU)$_7$ label, protein S26 was the main target for the 5'-modified affinity probe and proteins S3/3a were labeled very weakly only[9].

In this context the work of Terao and Ogata[10] should briefly be mentioned. These authors found protein S6 cross-linked to poly(U) following ultraviolet irradiation.

Thus it follows that proteins S3, S3a, (S6) and S26 are located at or at least near the mRNA binding site on the small ribosomal subunit and that the mRNA seems to migrate from proteins S3/3a toward S26 during translocation.

2. Topography of proteins in the small ribosomal subunit
Cross-linking

This technique has been applied so far with eukaryotic ribosomes by Ogata's group in Niigata[11,12], by Tolan and Traut in Davis[13], and by Westermann et al. in our laboratory in Berlin-Buch[14].

In Table 3 those protein dimers are listed which have been found conformably by at least two of the three groups.

The fact that only a partial agreement exists in the results of these groups may be due to different ribosome species and/or different reagents used.

Among the cross-linked proteins, especially S3, S3a, S6, S13, S15 are of interest, because data about their function (see before) or their localization as analyzed by immune electron microscopy (see the following section) are available, which will be discussed later.

TABLE 3

CROSS-LINKED PROTEINS IN SMALL SUBUNITS OF EUKARYOTIC RIBOSOMES

Davis-group[13] (Rabbit reticulocytes)	Niigata-group[11,12] (Rat liver)	Berlin-Buch group[14+] (Rat liver)
S 2 - 3		S 2 - 3
S 3 - 3a		S 3 - 3a
	S 3/3a - 11	S 3/3a - 11
S 3a - 14	S 3/3a - 14	
	S 3b - 11	S 3b - 11
S 4 - 6	S 4 - 6	S 4 - 6
S 4 - 24	S 4 - 23/24	S 4 - 23/24
	S 5 - 7	S 5 - 7
S 5 - 13		S 5 - 13
S 5 - 25	S 5 - 25	S 5 - 25
	S 7 - 18	S 7 - 18
S 8 - 11	S 8 - 11	
	S16 - 19	S16 - 19
	S23 - 24	S23 - 24

+and unpublished results

Immune electron microscopy

This technique has been used so far successfully for the topo-
graphical analysis of proteins in eukaryotic ribosomes in our
group only.

We have mapped up to now antigenic domains of 10 proteins on
the small ribosomal subunit reacting with antibodies against the
corresponding pure proteins of rat liver ribosomes[7,15] (and unpublished results), namely S2, S3, S3a, S5, S6, S7, S9, S17, and
S21. The main results are demonstrated in the schematic drawing
of Fig. 1.

From the data summarized in Figure 1, it follows that the majority of the antibody binding sites is localized on the head
and the body region very close to the cleft between these two
parts of the small subunit and on the so-called protuberance
(comparable to the platform of E.coli ribosomes). Most of the

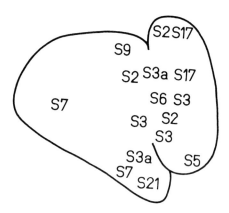

Fig. 1. Map of proteins on the small subunit of rat liver ribosomes.

proteins have two, and some have even three different antibody binding sites. This finding points to a more or less elongated in situ structure of at least some of the small ribosomal sub-unit proteins, which is in agreement with data obtained by hydro-dynamic measurement of isolated ribosomal proteins[16].

CONCLUSIONS

1. From the cross-linking data in Tables 2 and 3 and the re-sults obtained by immune electron microscopy (Fig. 1), it can be stated that the following proteins are arranged in more or less close   proximity   within the small ribosomal subunit: S2, S3, S3a, S4, S5, S6, S7, S11, S13, S15, S16, S19, S23, and S24.

2. Among these proteins, S3, S3a, S6, S13, and S15 were found by different techniques to be involved in interactions with com-ponents of the ternary initiation complex (eIF-2, Met-tRNA$_f$) and S3/3a and S6 (according to Terao and Ogata[10]) also in the bind-ing of mRNA (poly(U)).

3. Some of the proteins involved in interactions with poly(U), Met-tRNA$_f$ and eIF-2 (at least S3, S3a, S6) seem to be closely associated also with the 18S rRNA, since these proteins can be

cross-linked to the RNA within the small ribosomal subunit. With the exception of protein S3a, which could be covalently attached to the oxidized 3'-end of 18S rRNA by Svoboda and McConkey[17], the position of the other ribosomal proteins along the 18S rRNA chain is still unknown. Nevertheless it is tempting to conclude that also the 18S rRNA molecule with at least parts of its polynucleotide chain contributes to the organization of various functional domains on the ribosome. This assumption is supported by the finding that the α- and the γ-subunit of eIF-2 can be cross-linked to 18S rRNA.

4. The cross-linking studies, the antibody blocking experiments and the results of the affinity labeling studies clearly show that functional units are formed by more than one ribosomal protein only and that some of these proteins are involved at the same time in different reactions. These findings suggest that ribosomal functions are very probably organized by cooperative interactions of different ribosomal components arranged in close neighbourhood, which might be the molecular basis for channeling the entire translation process in the ribosomal structure.

5. In Figure 2 data are summarized in a tentative model about possible functions of components of the small ribosomal subunit and their spatial arrangement.

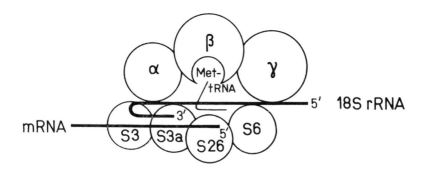

Fig. 2. Model of functional and topological data of proteins and RNAs of the quaternary initiation complex.

6. In the immune electron microscopic studies, antigenic determinants of proteins S3, S3a, and S6 were preferentially found in the neck region between the head and the body and, furthermore, on the protuberance (see Fig. 1). Therefore, it is tempting to assume that this site on the small ribosomal subunit is the structural domain at which essential steps of the initiation process are organized.

7. The location of protein S6 at the binding site for Met-tRNA$_f$ and eIF-2 is of special interest in so far as it is the only protein of the small subunit phosphorylatable in vivo, depending on conditions which also alter protein synthesis. This fact raises the question as to a possible regulatory function of S6 phosphorylation for the translation process. Preliminary studies have revealed that the ternary initiation complex binds equally well to 40S ribosomal subunits with unphosphorylated and phosphorylated S6[18,19]. On the other hand, differences in in vitro-interactions of poly(U) with 40S subunits of rat liver ribosomes with low and highly phosphorylated protein S6 have been described[20]. Thus, it seems worthwhile to further study this problem.

REFERENCES

1. Nygård, O., Westermann, P., and Hultin, T. (1981) Acta Chem. Scand. 35, 57-59.
2. Nygård, O., Westermann, P., and Hultin, T. (1980) FEBS-Lett. 113, 125-128.
3. Westermann, P., Nygård, O., and Bielka, H. (1981) Nucleic Acids Res. 9, 2387-2396.
4. McConkey, E. H., Bielka, H., Gordon, J., Lastick, S. M., Lin, A., Ogata, K., Reboud, J.-P., Traugh, J. A., Traut, R. R., Warner, J. R., Welfle, H., and Wool, I. G. (1979) Molec. gen. Genet. 169, 1-6.
5. Westermann, P., Heumann, W., Bommer, U.-A., Bielka, H., Nygård, O., and Hultin, T. (1979) FEBS-Lett. 97, 101-104.
6. Westermann, P., Nygård, O., and Bielka, H. (1980) Nucleic Acids Res. 8, 3065-3071.
7. Bommer, U.-A., Noll, F., Lutsch, G., and Bielka, H. (1980) FEBS-Lett. 111, 171-174.
8. Stahl, J., and Kobets, N. D. (1981) FEBS-Lett. 123, 269-272.
9. Stahl, J., and Kobets, N. D. (1982) FEBS-Lett., in press.
10. Terao, K., and Ogata, K. (1979) J. Biochem. 86, 605-617.
11. Terao, K., Uchiumi, T., Kobayashi, Y., and Ogata, K. (1980) Biochim. Biophys. Acta 621, 72-82.

12. Uchiumi, T., Terao, K., and Ogata, K. (1981) J. Biochem. 90, 185-193.
13. Tolan, D. R., and Traut, R. R. (1981) J. Biol. Chem. 256, 10129-10136.
14. Westermann, P., Gross, B., and Bielka, H. (1980) Acta Biol. Med. Germ. 39, 1147-1152.
15. Lutsch, G., Noll, F., Theise, H., Enzmann, G., and Bielka, H. (1979) Molec. gen. Genet. 176, 281-291.
16. Behlke, J., Theise, H., Noll, F., and Bielka, H. (1979) FEBS-Lett. 106, 223-225.
17. Svoboda, A. J., and McConkey, E. H. (1978) Biochem. Biophys. Res. Commun. 81, 1145-1152.
18. Bommer, U.-A., Bielka, H., Henske, A., and Kärgel, H.-J. (1981) Acta Biol. Med. Germ. 40, 1105-1110.
19. Leader, D. P., Thomas, A., and Voorma, O. (1981) Biochim. Biophys. Acta 656, 69-75.
20. Gressner, A. M., and Van der Leur, E. (1980) Biochim. Biophys. Acta 608, 459-468.

Published 1982 by Elsevier Science Publishing Co., Inc.
Marianne Grunberg-Manago and Brian Safer, editors
INTERACTION OF TRANSLATIONAL AND TRANSCRIPTIONAL
CONTROLS IN THE REGULATION OF GENE EXPRESSION

THE PROTEIN TRANSLOCATION MACHINERY OF THE ENDOPLASMIC RETICULUM

PETER WALTER, REID GILMORE, MATTHIAS MULLER AND GUNTER BLOBEL
The Rockefeller University, New York, NY

Substantial experimental data has recently been provided on the

cotranslational translocation of proteins across and integration into

the endoplasmic reticulum. So far, two components have been purified

from dog pancreas and shown to be required for this translocation pro-

cess.

One of these is the so-called Signal Recognition Particle (SRP), an

11S ribonucleoprotein (1). SRP consists of six non-identical polypep-

tide chains (M_r = 72,000, 68,000, 54,000, 19,000, 14,000 and 9,000 dal-

tons) (2) and one molecule of 7S RNA (1). The RNA has been identified by

partial sequence analysis (1) to be the previously described (3) and re-

cently sequenced (4.5) small cytoplasmic 7SL RNA (7S RNA, ScL). Both,

RNA and protein and required for SRP's activity. In dog pancreas at

physiological salt concentration (150mM potassium ions) the bulk of SRP

appears to be about equally distributed between a membrane-bound and a

free or ribosome/polysome-associated form (6).

The other component, termed SRP-receptor (7), is a protein of

72,000 daltons (8,9) that has been purified from detergent-solubilized

microsomal membranes by SRP-affinity chromatography (9). The SRP-

receptor is an integral membrane protein of the endoplasmic reticulum.

It consists of a large cytoplasmic domain of 60,000 daltons (10) that

can be severed from the membrane in an intact form by treatment with a

variety of proteases and can be added back to the proteolysed membranes

to reconstitute activity (7,11,12).

The function of these components in the protein translocation pro-

cess was deduced from in vitro reconstituted assay systems. Using such assays, SRP was found to function in decoding the information contained in the signal peptide of nascent secretory (13,14,15) lysosomal (16), and membrane (17) proteins to the effect that it mediates the specific attachment of the translating ribosome to the microsomal membrane (18). In the absence of microsomal membraines, SRP specifically arrests the elongation of secretory protein synthesis in vitro (13) just after the signal peptide has emerged from the ribosome, thereby preventing the completion of pre-secretory proteins (many of which could be potentially harmful to the cell) (19) in the cytoplasmic compartment. Upon interaction of these arrested ribosomes with a specific integral membrane protein, the SRP-receptor (7,20), on the microsomal membrane, this elongation arrest is released and the nascent chain is translocated across (19) or – as in the case of integral membrane proteins – integrated into (17) the lipid bilayer.

Both, SRP as well as the mode of cotranslational protein translocation seem to be highly conserved through evolution (15,21). SRP therefore appears to be an intergral and indispensable component of the protein synthesis machinery of living cells assuring the correct topogenesis of a specific subset of proteins. Considering its structural features and its intimate (although most likely transient) functional association with ribosomes, it could almost be regarded as a "third ribosomal subunit" functioning as the adapter between the cytoplasmic translation and the membrane-bound protein translocation machinery.

1. Walter, P. and Blobel, G. submitted to Nature (London) (1982).
2. Walter, P. and Blobel, G. Proc. Natl. Acad. Sci. USA 77, 7112-7116, (1980).
3. Zieve, G. and Penman, S. Cell 8, 19-31, (1976).
4. Ullu, E., Murphy, S. and Melli, M. Cell 29, 195-201, (1982).
5. Li, W.Y., Reddy, R., Henning, D., Epstein, P. and Bush, H. J. Biol. Chem. 257, 5136-5142, (1982).
6. Walter, P. and Blobel, G. in preparation.
7. Gilmore, R., Blobel, G. and Walter, P. submitted to J. Cell Biol. (1982).
8. Meyer, D.I., Louvard, D. and Dobberstein, B. J. Cell Biol. 92, 579-583, (1982).
9. Gilmore, R., Walter, P. and Blobel, G. submitted to J. Cell Biol. (1982).
10. Meyer, D.I. and Dobberstein, B. J. Cell Biol. 87, 503-508, (1980).
11. Walter, P., Jackson, R.C., Marcus, M.M., Lingappa, V.R. and Blobel, G. Proc. Natl. Acad. Sci. USA 76, 1795-1799, (1979).
12. Meyer, D.I. and Dobberstein, B. J. Cell Biol. 87, 498-502, (1980).
13. Walter, P., Ibrahimi, I. and Blobel, G. J. Cell Biol. 91, 545-550, (1981).
14. Stoffel, W., Blobel, G. and Walter, P. Eur. J. Biochem. 120, 519-522, (1981).
15. Mueller, M., Ibrahimi, I., Chang, C.N., Walter, P. and Blobel, G. J. Biol. Chem. in press (1982).
16. Erickson, A.H., Walter, P. and Blobel, G. in preparation.
17. Anderson, D.J., Walter, P. and Blobel, G. J. Cell Biol. 93, 501-506, (1982).
18. Walter, P. and Blobel, G. J. Cell Biol. 91, 551-556, (1981).
19. Walter, P. and Blobel, G. J. Cell Biol. 91, 557-561, (1981).
20. Meyer, D.I., Krase, E. and Dobberstein, B. Nature (London) in press (1982).
21. Talmadge, K., Stahl, S. and Gilbert, W. Proc. Natl. Acad. Sci. USA 77, 3369-3373, (1980).

SECTION IV: REGULATION OF INITIATION FACTOR ACTIVITY

THE CONTROL OF THE RATE OF PROTEIN SYNTHESIS INITIATION

HARRY O. VOORMA AND HANS AMESZ
Department of Molecular Cell Biology, Molecular Biology section, State University of Utrecht, Transitorium 3, Padualaan 8, 3584 CH Utrecht, The Netherlands

INTRODUCTION

One of the earliest observations in hemin-deficient lysates indicated that the level at which the inhibition of protein synthesis is executed is the formation of the 40S preinitiation complex[1]. Preceeding the disaggregation of polysomes one sees a strong decline in the amount of these 40S preinitiation complexes. For a long time the mechanism of the inhibition of protein synthesis by hemin-deficiency could not be elucidated since one was unable to obtain inhibition of initiation in model assay system employing purified initiation factors and ribosomal subunits.[2,3] The bottleneck in obtaining reliable data was the failure to design an assay system in which the initiation factors are used in a catalytic way.

Recent developments, i.e. the discovery of a new factor, formerly designated anti-HRI but recently renamed eukaryotic regulatory or recycling factor, eRF, provided new possibilities and made available a cell-free system which initiates at high rate and acts catalytically with respect to factors.[4,5] In essence eRF shows a strong resemblance to eIF-2-stimulating protein SP and restoring factor RF[6] and to Co-eIF-2C.[7]

RESULTS

Mode of action of eRF. The effect of eRF on the formation of 40S preinitiation complexes is depicted in Fig. 1, which result shows that the transfer of Met-tRNA to the 40S subunit is strongly enhanced in the presence of eRF and reaches a level which is similar to that of the ternary complex formation. It means that at low input concentrations of eIF-2 when no other components in the assay system are limiting, the transfer of Met-tRNA is complete, 100%, whereas in the absence of this factor, only about 20% is bound into a stable form. From Fig. 2 it is obvious that the stimulation of Met-tRNA binding at the 40S level also results in a stimulation of the methionyl-puromycin formation when the assay is carried out with 80S ribosomes instead of 40S subunits and initiation factor eIF-5 is added as well. The important result to emphasize at this

point is that in this assay system each pmole of eIF-2 is capable to catalyze
the formation of about 6 pmoles of methionyl-puromycin, which means a catalytic
use of eIF-2 and thus a recycling system. The levelling off at higher inputs of
eIF-2 is due to a limited availability of Met-tRNA.

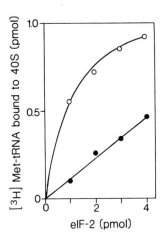 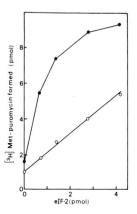

Fig. 1. The effect of eRF on 40S preinitiation complex formation. The total radioactivity on the 40S preinitiation complexes was determined in sucrose gradients. (●) minus eRF, (o)
plus eRF.

Fig. 2. The effect of eRF on the formation of methionyl-puromycin. Methionyl-puromycin formation was determined
after incubation for 60 min followed
by extraction with ethyl acetate. (o)
minus eRF, (●) plus eRF.

The data obtained so far bear some resemblance to results obtained with prokaryotic EF-Tu and EF-Ts as well as with eukaryotic EF-1. Therefore it was of
interest to establish the subunit composition of this factor, the more so as it
occurs in two forms, free eRF and complexed with eIF-2. In eRF we distinguished
4 subunits of M_r 80,000, 51,000, 35,000 and 27,000 whereas in the complex the
three subunits of eIF-2 of M_r 51,000, 48,000 and 33,000 were found besides those
four already mentioned. In most gels no clear separation between subunit 3
(51,000) and the β-unit of eIF-2 (51,000) is found.

Complex formation of eIF-2 and eRF. In recent studies we have focussed the
attention on the subunit composition of the eIF-2.eRF complex since the complex
may be the basis of the observed high rate of initiation. The complex formation
can be studied with glycerol gradients because eIF-2 sediments between 4-6S,
free eRF at about 10S and eIF-2.eRF at the 17S position. In our experiments we

have employed ^{14}C-labeled eIF-2 together with non-labeled eRF and established the conditions for complex formation between both factors.

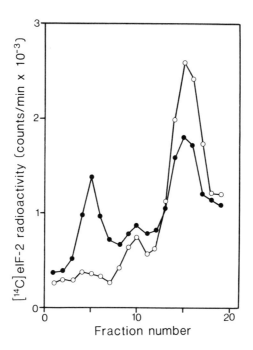

Fig. 3. Complex formation between ^{14}C-labeled eIF-2 and eRF. Analysis was performed on isokinetic 15-40% glycerol gradients. (o) incubation at 100 mM KCl, (●) incubation at 500 mM KCl, followed by 100 mM KCl.

An initial study revealed that dissociation of the complex occurred at a high salt concentration. Thereafter we found that the reverse of the reaction could be accomplished by a preincubation with high salt, this step being a prerequisite as is shown in Fig. 3. Since the radioactivity is found primarily at a too heavy position in the gradient we assume that a dimer of the complex has been formed rather than an extreme conformational change has taken place that may explain the high sedimentation value as well.

A similar experiment has been carried out in Fig. 4. showing that the exchange reaction requires similar conditions. In panel A one sees the result of an experiment in which the complex eIF-2.eRF has been incubated with ^{14}C-labeled eIF-2 after the dissociation step at 500 mM KCl has taken place. It is obvious that free eIF-2 can replace the eIF-2 moiety present in the complex. Another

300

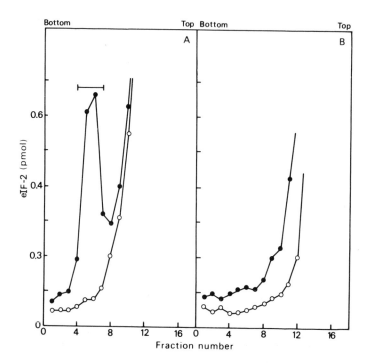

Fig. 4. Exchange of $[^{14}C]eIF-2$ and $[^{32}P]$-eIF-2 with eIF-2 in the complex eIF-2.
eRF. eIF-2.eRF was incubated with $[^{14}C]$-eIF-2, panel A and with $[^{32}P]$-eIF-2,
phosphorylated by HRI, panel B. (●) eIF-2.eRF present, (o) control without eIF-2
.eRF.

interesting observation is the failure of eIF-2 phosphorylated by the hemin-
regulated inhibitor, HRI, to displace non-phosphorylated eIF-2. It appears that
the phosphorylation of the α-subunit of eIF-2 impairs the complex formation.

In order to study the effect of phosphorylation by HRI to a greater extent
eIF-2 was incubated with HRI, with ATP, with both HRI and ATP and without addition, whereafter complex formation was analysed on gradients.

The results given in Fig. 5 clearly show that only upon incubation in the
presence of both HRI and ATP, a conditional requirement for phosphorylation of
the α-subunit of eIF-2, a strong inhibition of complex formation occurs, whereas
in both separate incubations as compared to the control no inhibition takes
place.

Since complex formation should occur under more physiological conditions than
the ones employed here, we continued our research and established that also un-

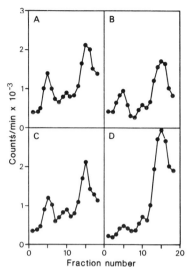

Fig. 5. Complex formation between eIF-2 and eRF, influence of incubation of HRI and ATP. (A) no addition, (B) and (C) incubation with either HRI or ATP, prior to complex formation, (D) incubation with both HRI and ATP prior to complex formation.

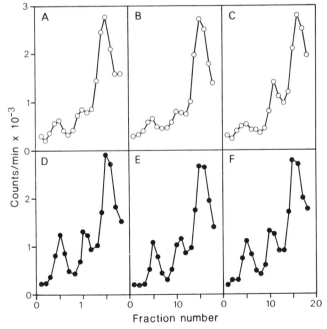

Fig. 6. Complex formation in absence and presence of DTT. (A) 0.4 mM GTP, (B) 0.4 mM GTP, 20 mM Hepes, (C) 0.4 mM GTP, 1 mM ATP, (D, E and F) as A, B and C with the addition of 1 mM DTT.

der conditions as to those in protein synthesis experiments complex formation can be demonstrated.

Further studies revealed that ATP, GTP and an energy-regenerating system were only capable to give marginal complex formation, whereas dithiothreitol, DTT, proved to be the most important component of the reaction mixture (see Fig. 6).

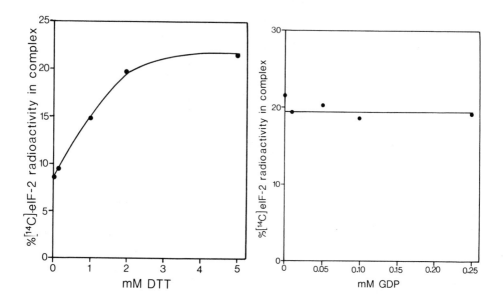

Fig. 7. Dependence of DTT in complex formation. Total radioactivity in the complex eIF-2.eRF is expressed as percentage of total radioactivity recovered from fractions of the gradient.

Fig. 8. Effect of GDP on complex formation. ^{14}C-labeled eIF-2 and eRF were incubated with an increasing concentration of GDP, followed by gradient analyses for complex formation.

In Fig. 7 the dependence on DTT of the complex formation has been given. At a concentration of 2 mM DTT one reaches a maximum, indicating that reducing conditions are favourable for complex formation. Another question to be raised is whether GDP is capable to inhibit complex formation because it is known that eIF-2 displays a very high affinity for GDP. In Fig. 8 an experiment is shown, which result has been obtained in the absence of creatine phosphate and creatine kinase. It is obvious that it does not give any indication about the impairment of complex formation by GDP.

This result should be borne in mind when one hypothesizes about regulation models of initiation of protein synthesis.

The following question to be answered is whether all three subunits take part in the complex formation and which is to be expected from the results of the exchange experiment.

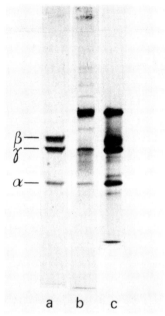

Fig. 9. The dependence on the β-subunit for complex formation. $[^{14}C]$-eIF-2 was incubated with a four-fold excess of eRF. The proteins present at the position of the complex and at the top of the gradient were concentrated and analysed by SDS-polyacrylamide gelelectrophoresis followed by autoradiography. a, complex, b, top of gradient and c, starting material.

In Fig. 9 the result is given of an experiment in which ^{14}C-labeled eIF-2 has been incubated with a four-fold excess of eRF in the presence of 2 mM DTT, 100 mM KCl and 1 mM Mg-acetate, followed by glycerol gradient analysis. Three gels are given, lane a: the complex $[^{14}C]$eIF-2.eRF, lane b: material remaining at the top of the gradient, lane c: the starting material showing three subunits of eIF-2 (and a contaminating band of 65k). The result is surprising since one sees near-stoichiometric amounts of the α, β and γ-subunit in the complex whereas at the top of the gradient one can detect only the α and γ subunit of eIF-2. The conclusion from this experiment seems justified that the β-subunit of eIF-2 is needed for complex formation.

In trying to summarize this part one can say that for complex formation between eIF-2 and eRF the prerequisites are: 1, the β-subunit of eIF-2; 2, the α-subunit of eIF-2 not being phosphorylated by HRI; 3, reducing conditions of 2 mM dithiothreitol appears to be favourable over the high salt treatment, whereas one should bear in mind that GDP does not interfere with complex forma- tion.

The results so far presented exclude that a ternary complex comprising merely eIF-2, Met-tRNA and GTP plays an important role in the formation of 40S preini- tiation complexes because of the exceedingly efficient transfer of Met-tRNA from the eIF-2.Met-tRNA.GTP.eRF quaternary complex. Since this transfer is in vitro even five à six times more efficient than the transfer from a ternary complex, in which formation eRF does not play a stimulating role, one should ask at this point which subunits are being bound into the 40S at the moment that Met-tRNA binds. To this end Met-tRNA binding experiments to 40S were performed with ei- ther ^{14}C-labeled eIF-2 or with ^{14}C-labeled complex. The 40S preinitiation com- plexes were analysed on sucrose gradients, whereafter from two regions, i.e. 40S position and the top region proteins were precipitated and run on SDS-polyacryl- amide gels followed by autoradiography. The experiments with ^{14}C-labeled eIF-2 and eIF-2.eRF complex are shown in Fig. 10, panels A-C and D-F respectively. In the panels the radioactivity profiles are given, whereas in the insets the auto- radiographs of the gels are shown.

In all cases the presence of the α- and γ-subunit of eIF-2 is established on 40S, the presence of the subunits of eRF can be wholly excluded, whereas the presence of the β-subunit is at least doubtful. In any case the ratio of β over both α- and γ seems much less than 1. Because the β-subunit is also missing in the experiments with labeled complex it is assumed that in these set of experi- ments the β-subunit remains bound into the eRF-moiety. Definite proof for this assumption is lacking, because eRF purified from the postribosomal supernatant does not contain sufficient β subunit to allow association of eIF-2 comprised of only α- and γ-subunits (see Fig. 9).

The effect of HRI on eRF-mediated stimulation. The effect of eRF on eIF-2- mediated reactions suggested that complex formation between both may be the ba- sis for the increase in Met-tRNA binding to 40S and 80S ribosomes. Already de- monstrated in Fig. 4 is the failure of eIF-2, phosphorylated by HRI upon incub- ation with ATP, to associate with eRF. The result suggested that HRI can bring down the eIF-2-mediated Met-tRNA binding to the level of binding obtained with- out eRF. Such an experiment is depicted in Fig. 11.

The assay system is maximally stimulated by eRF which is reached with about 1 μg of eRF added. Then in the presence of this amount of eRF HRI is added with increasing quantity. It is obvious that HRI brings down the initiation level to

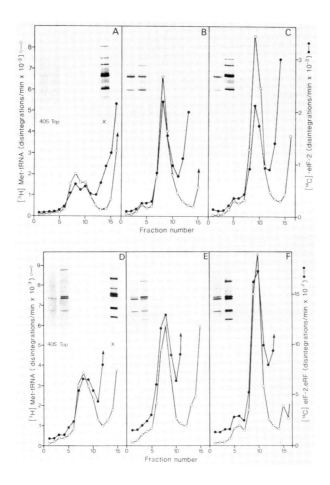

Fig. 10. The presence of subunits of eIF-2 and eIF-2 and eIF-2.eRF on 40S pre-initiation complexes. panels A, B, C: experiment with ^{14}C-eIF-2. panels D, E, F: experiment with ^{14}C-eIF-2.eRF. Insets: autoradiogram of gels showing the presence of proteins taken from 40S position and top of gradient. x: starting material.

the low level that is obtained without eRF. The next question to be raised is whether eIF-2 in the complex of eIF-2.eRF is protected against phosphorylation. A similar experiment as the previous one has been carried out but instead of eIF-2 plus and minus eRF an experiment was included with complex. The result can

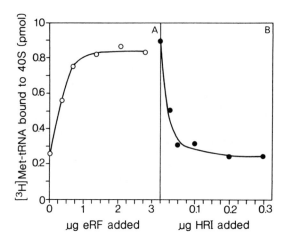

Fig. 11. Complete elimination of stimulatory activity of eRF by HRI. 40S initiation complexes were formed with increasing amounts of eRF (panel A). At an optimal concentration of 1 µg of eRF increasing quantity of HRI has been added, followed by analyses of 40S preinitiation complex (panel B).

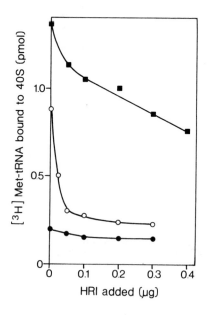

Fig. 12. Comparison of the effect of HRI on the formation of 40S initiation complexes. (●) in presence of only eIF-2, (o) in presence of eIF-2 and eRF, (■) in presence of the complex eIF-2.eRF.

best be explained by the assumption that a certain degree of protection occurs since the inhibition curve proceeds with a slope less steep than the one with separate eIF-2 and eRF, which result is depicted in Fig. 12. Furthermore it should be stressed that HRI is unable to exert any influence on the 40S initiation complex formation carried out with eIF-2 alone.

Is eRF a common factor in other eukaryotes? eRF isolated from rabbit reticulocytes is the key factor in the understanding of inhibition of protein synthesis mediated by HRI that has been activated by hemin-deficiency. By means of eRF a high rate of initiation is achieved if conditions for complex formation are met. Following phosphorylation and maybe also oxidation the complex formation is impaired, resulting in a low initiation rate. So eRF is the control switch that decides over the rate of initiation. Such an important regulating protein may be common to many cells. To check this assumption a neuroblastoma cell lysate was treated in a similar way as was done for the purification of eRF from reticulocytes[4]. Indeed a similar activity was found, also capable to restore protein synthesis in hemin-deficient lysates, to stimulate the methionyl-puromycin reaction, whereas it failed to give stimulation with phosphorylated eIF-2.

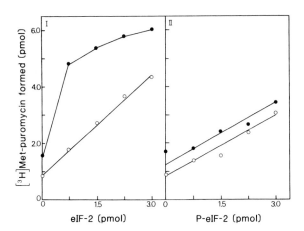

Fig. 13. The effect of "eRF" from neuroblastoma cells. Panel I, stimulation by eRF from neuroblastoma cells. (o) without eRF, (●) plus eRF. Panel II, lack of stimulation by eRF from neuroblastoma cells when phosphorylated eIF-2 is employed. Note that all other components are derived from reticulocytes.

The results are shown in Fig. 13, panel I showing the stimulating effect of the neuroblastoma eRF, (closed circles) whereas in panel II is shown that the factor failed to stimulate the reaction when phosphorylated eIF-2 was employed. Further characterization of this factor is in progress.

DISCUSSION

The eukaryotic regulatory / recycling factor, eRF, possesses interesting properties which makes it a key factor in the control of the rate of initiation. When complexed with eIF-2 it displays a high activity with respect to stimulation of Met-tRNA binding to 40S, which stimulation is also found on 80S initiation complex formation. Impairment of complex formation between eIF-2 and eRF inevitably results in a decrease in the initiation rate. Conditions which favour complex formation are a non-phosphorylated and reduced form of eIF-2, whereas GDP does not inhibit the formation, achieving a high rate of initiation.

When eIF-2 is phosphorylated by HRI or is oxidized no complex is further possible and the initiation rate is low. Fig. 14 shows a possible mode of action of eRF from which a number of hypotheses can be proposed.

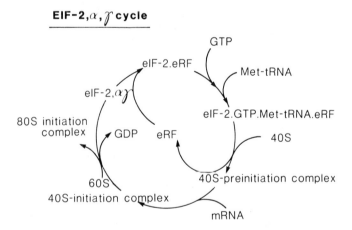

Fig. 14. General scheme for mode of action of eRF.

One possibility is that an eIF-2.GDP complex is generated upon formation of the 80S initiation complex. An EF-Tu.EF-Ts-type of action should then come into gear to reactivate the factor and made the factor recycle. Another question to be answered is as to whether all three subunits of eIF-2 recycle after 80S initiation complex formation, whereas the four subunits of eRF do so after completion of the 40S preinitiation complex. Reason for this assumption is the prerequisite for β-subunit for complex formation. At present we can say that this factor decides, depending of the state of eIF-2 whether a high or low rate is achieved. That GDP does not inhibit complex formation may be a positive indication for the existence of an eIF-2.GDP complex.

REFERENCES

1. Clemens, M.J., Henshaw, E.C., Rahaminoff, H. and London, I.M. (1973) Proc. Natl. Acad. Sci. USA, 71, 2946-2950.
2. Benne, R., Salimans, M., Goumans, H., Amesz, H. and Voorma, H.O. (1980) Eur. J. Biochem. 104, 501-509.
3. Safer, B., Jagus, R. and Crouch, D. (1980) J. Biol. Chem. 225, 6913-6917.
4. Amesz, H., Goumans, H., Haubrich-Morree, T., Voorma, H.O. and Benne, R. (1979) Eur. J. Biochem. 98, 513-520.
5. Goumans, H., Amesz, H., Voorma, H.O. and Benne, R. submitted.
6. Ochoa, S. (1981) Eur. J. Cell Biol. 26, 212-216.
7. Das, A., Bagchi, M., Roy, R., Ghosh-Dastidar, P. and Gupta, N.K. (1982) Biochem. Biophys. Res. Commun. 104, 89-98.

ACKNOWLEDGEMENT

Part of this work has been carried out in cooperation with Hans Goumans, René Paulussen and Rob Benne. We thank them for their suggestions for the final concept of this chapter.

Published 1982 by Elsevier Science Publishing Co., Inc.
Marianne Grunberg-Manago and Brian Safer, editors
INTERACTION OF TRANSLATIONAL AND TRANSCRIPTIONAL
CONTROLS IN THE REGULATION OF GENE EXPRESSION

THE MECHANISM OF TRANSLATIONAL INHIBITION IN HEMIN-DEFICIENT LYSATES

BRIAN SAFER, ROSEMARY JAGUS, ANDREJ KONIECZNY, AND DEBORAH CROUCH
Section on Protein Biosynthesis, Laboratory of Molecular Hematology,
National Heart, Lung, and Blood Institute, NIH, Bethesda, Maryland 20205

INTRODUCTION

In reticulocyte lysate supplemented with 20-40 μM hemin, the rate of translation closely approximates that of the intact cell. In the absence of hemin, however, control rates of protein synthesis are maintained for 4-6 min, and are followed by a $\geq$ 90 percent inhibition of translational activity.[1,2] Since 1) the onset of translational inhibition is accompanied by activation of a cAMP-independent kinase specific for the α-subunit of eIF-2, 2) the addition of exogenous eIF-2 to inhibited hemin-deficient lysate restores translation, and 3) inhibition of eIF-2 dependent Met-tRNA$_i$ binding to initiating $43S_N$ ribosomal subunits is observed, it was thought that eIF-2 phosphorylation was directly responsible for the inhibition of protein synthesis initiation[3] (see reviews 4,5). Evidence for a direct effect of phosphorylation on eIF-2 function was also obtained by several laboratories.[6-9] In this and other groups, however, eIF-2α phosphorylation did not inhibit either its Met-tRNA$_i$ binding activity or its ability to rescue translational activity in hemin-deficient lysates.[10-12] The major effect of eIF-2 phosphorylation appeared to be a conversion in its utilization from a highly catalytic mode to one in which eIF-2 was recycled inefficiently. The requirement for an efficient eIF-2 recycling process arises from: 1) its small pool size (30-50 pmol/ml lysate) relative to the rate of protein synthesis initiation (200 pmol initiations/min/ml) in reticulocyte lysate;[12] 2) inactivation of Met-tRNA$_i$ binding activity by binding of GDP;[13] eIF-2 is thought to be released as eIF-2.GDP upon 60S ribosomal subunit joining to the 48S preinitiation complex; and 3) an unfavorable K_D^{GDP} for GTP/GDP exchange under physiologic conditions.[14] Since an initial binding of GTP by eIF-2 increases the rate of Met-tRNA$_i$ binding 20-30 fold[15] and is required for ternary complex formation, failure to exchange bound GDP for GTP would inhibit catalytic utilization of eIF-2.

Recently, a factor which promotes the catalytic utilization of eIF-2 in hemin-deficient lysates has been purified to near homogeneity in several

laboratories and partially purified by others.* Our laboratory and those of
Voorma and Ochoa now appear to be in substantial agreement that a major function of this factor is to implement the exchange of bound GDP for GTP, required
for catalytic function of eIF-2 (for review see 5). While several lines of
evidence suggested that phosphorylation of eIF-2α inhibits the RF-mediated exchange of guanine nucleotides and inhibits catalytic function of eIF-2 in model
assays and reticulocyte lysate, the low extent of eIF-2α phosphorylation in
lysates remained problematic.[19,20] Although phosphorylation of eIF-2α
appeared to prevent RF.eIF-2 association, it was not clear why the remaining
nonphosphorylated eIF-2 (70%) was unable to interact normally with RF. A
reasonable explanation for this discrepancy appears to be that when examined
under physiologic conditions in the presence of guanine nucleotides, phosphorylation of eIF-2α actually increases the stability of the RF.eIF-2 complex; this
inhibits the apparent function of this complex to mediate guanine nucleotide
exchange. In addition, by sequestering the smaller RF pool into a nonfunc-
tional stable complex, interaction of RF and the larger nonphosphorylated eIF-2
pool is thus prevented.

RESULTS

Purification and Subunit Composition of the Recycling Factor Complex RF.eIF-2

At physiologic salt concentrations, eIF-2 exists in two states. The first
to be characterized was an α, β, and γ subunit+ trimer which exists either
free or bound to ribosomal preinitiation complexes.[4] By observing procedures
which avoid fractionation at high salt concentrations, eIF-2 could also be
isolated as part of a large polypeptide complex we designate RF.eIF-2. The
final purification of RF.eIF-2 by glycerol gradient centrifugation is shown in
Figure 1A. The RF.eIF-2 complex can be dissociated into free RF and eIF-2 by
glycerol gradient centrifugation at 500 mM KCl (Figure 1B). The subunit
compositions of RF.eIF-2 and RF are shown to the right of these panels in
Figure 1. RF.eIF-2 consists of 8 subunits, three of which are the α, β, and γ
subunits of eIF-2. The RF component consists of 5 distinct polypeptides,

---

* The RF.eIF-2 complex appears to be identical in subunit composition to
eRF (formerly designated anti-HRI) first isolated by Voorma and his col-
leagues[16] (see Voorma, this volume) and to eIF-2-SP (ESP) from Ochoa and
Siekierka[17], and functionally to the partially purified Co-eIF-2C of Gupta
and his associates[18] (see Gupta, this volume).

+ The α, β, and γ subunits of eIF-2 are defined by the apparent M_r, pI
and kinase specificity: α (38,000; 5.1; eIF-2α kinase), β (56,000; 5.2; Casein
kinase II), and γ (52,000; 8.9; no known phosphorylation).[21]

313

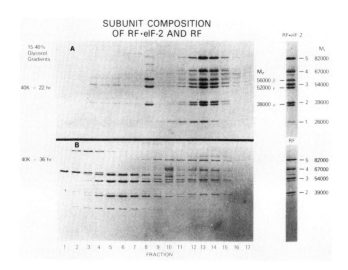

FIG. 1. RF.eIF-2 was purified from rabbit reticulocyte lysate by phospho-cellulose and DEAE chromatography, followed by glycerol gradient centrifuga-tion. The polypeptide composition of gradient fractions, resolved by SDS PAGE, are shown. A. Final purification of the RF.eIF-2 complex by glycerol gradient centrifugation. Two μg of free eIF-2 are added to fraction 8 to identify eIF-2 subunits in RF.eIF-2. B. An aliquot of the RF.eIF-2 pool was fractionated into free eIF-2 and RF by centrifugation at 500 mM KCl. The subunit composition of the RF.eIF-2 and free RF pools are indicated to the right of their respective glycerol gradient fraction.

M_r= 82,000, 67,000, 54,000, 39,000, and 26,000. Partial dissociation of the of the 26 kilodalton subunit from RF at high salt concentrations is observed. RF.eIF-2 sediments at 12S on linear 15-50% glycerol gradients, in agreement with an apparent M_r = 393,000.

Effect of RF on Translation on Hemin-Deficient Lysates

Addition of RF.eIF-2, RF or eIF-2 to hemin-deficient lysates maintains high translational activity (Figure 2). In the absence of any additions, biphasic kinetics of translational inhibition are observed. For each pmol RF.eIF-2 or RF added, 30 to 40 pmol globin are synthesized. While eIF-2 also rescues translational activity, much larger amounts of protein are required and catalytic function is not observed. Only 1 to 2 pmol globin are translated

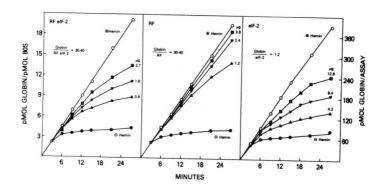

FIG. 2. Rescue of translation by RF·eIF-2, RF, and eIF-2. Increasing
amounts of RF·eIF-2 (left), RF (center), and eIF-2 (right) were added to rabbit
reticulocyte lysates not supplemented with exogenous hemin. Globin synthesis
was measured by incorporation of [^{14}C]valine. The ratio of pmol of globin
synthesized (above that in unsupplemented hemin deficient lysate) to the pmol
of exogenous factor added, calculated after 20 min incubation, is based on the
M_r of RF·eIF-2=390,000, RF=268,000 and eIF-2=122,000.

per pmol eIF-2 added. These data indicate that the ability of RF·eIF-2 to
maintain translational activity does not result from the Met-tRNA$_i$ binding
activity of its eIF-2 component, but rather may be a direct result of its RF
moiety. Essentially identical results were obtained when exogenous RF·eIF-2,
RF and eIF-2 were added (at 12 min) to inhibited hemin-deficient lysates.

The ability of RF to maintain translational activity in hemin-deficient
lysates appears to be related to its effect on Met-tRNA$_i$ binding by eIF-2 at
physiologic concentrations of GTP and GDP (Table I). In the absence of GDP,
130 μM GTP promotes the rapid and efficient binding of Met-tRNA$_i$ by eIF-2.
Based on a M_r = 122,000, 70 percent of eIF-2 is active and phosphorylation of
the α subunit has no effect. In the presence of RF·eIF-2, the increased Met-
tRNA$_i$ binding only reflects the additional eIF-2 moiety of RF·eIF-2. RF
itself does not bind Met-tRNA$_i$. At physiologic concentrations of GTP and
GDP, however, Met-tRNA$_i$ binding is decreased by 80 percent. This reflects
the inhibition of Met-tRNA$_i$ binding by GDP and the smaller dissociation

TABLE I

EFFECT OF RF AND GUANINE NUCLEOTIDES
ON MET-tRNA$_i$ BINDING BY elF-2

ADDITIONS	MET-tRNA$_i$ BOUND/1.5µg		MET-tRNA$_i$/eIF-2	
	eIF-2	eIF-2(αP)	eIF-2	eIF-2(αP)
130µM GTP	9.0	8.6	0.7	0.7
125µM GTP 5µM GDP	1.8	1.9	0.2	0.2
130µM GTP 0.4µg RF·eIF-2	9.8	9.9	0.8	0.8
125µM GTP 5µM GDP 0.4µg RF·eIF-2	6.7	2.3	0.6	0.2

1.5 µg eIF-2 or eIF-2 (αP) were incubated in a 50 µl reaction containing 50 mM KCl, 10 mM pH 7.1 HEPES, 2 mM DTT, 15 pmol [^{14}C]Met-tRNA$_i$, 1 mM MgCl$_2$, and guanine nucleotides and RF.eIF-2 as indicated. Binding was initiated by the addition of eIF-2. Following incubation for 2 min at 30°, bound Met-tRNA$_i$ was analyzed by nitrocellulose filtration.[15]

constant of eIF-2 for GDP in comparison to GTP (K_D^{GDP}=3 x 10^{-8}M; K_D^{GTP} $\approx$ x 10^{-6}M)[14]. At physiologic [GTP]/[GDP], however, RF.eIF-2 or RF stimulate Met-tRNA$_i$ binding catalytically. One pmol RF.eIF-2 or RF promotes the binding of an additional 4.9 pmol Met-tRNA$_i$. Phosphorylation of the α subunit of eIF-2, however, prevents this stimulation by RF of Met-tRNA$_i$ binding when both GTP and GDP are present.

Effect of RF on Guanine Nucleotide Exchange

The ability of RF to stimulate Met-tRNA$_i$ binding at physiologic concentrations of guanine nucleotides appears to result from its effect on GDP binding to eIF-2. The K_D^{GDP} of free eIF-2 was obtained by Scatchard analysis (Figure 3A). In agreement with the study of Walton and

Gill,[14] a dissociation constant for eIF-2.GDP of 3.9×10^{-8} is obtained. At physiologic concentrations of GDP and GTP, eIF-2 would largely

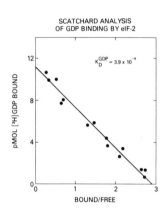

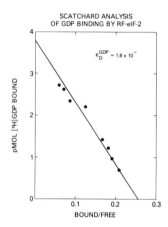

FIG. 3. The K_D^{GDP} of free eIF-2 and the RF.EIF-2 complex were obtained by Scatchard analysis of $8\text{-}[^3H]GDP$ binding to nitrocellulose filters as described in Table I except that 4 µg eIF-2 and 6 µg RF.eIF-2 were used and Met-tRNA$_i$ was omitted.

exist as an inactive eIF-2.GDP complex. The effect of the association of eIF-2 with RF, however, is to increase the dissociation constant for GDP 6-fold ($K_D^{GDP} = 1.8 \times 10^{-7}$ for RF.eIF-2; Figure 3B). The association of eIF-2 with RF may therefore serve as a GTP:GDP exchange mechanism analogous to the prokaryotic EF-Tu/EF-Ts system.[5,22] This requirement for RF to promote GTP:GDP exchange necessary for catalytic function of eIF-2 during protein synthesis can be bypassed by increasing the GTP/GDP ratio (Figure 4). In hemin-deficient lysates, control rates of protein synthesis are immediately restored by the addition of 1 mM GTP.Mg. GTP has no effect on the rate of translation in hemin-supplemented lysate. This immediate activation of translation is not the result of eIF-2 dephosphorylation, which does not occur to any significant extent until several minutes have elapsed (data not shown).

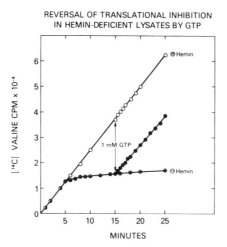

REVERSAL OF TRANSLATIONAL INHIBITION
IN HEMIN-DEFICIENT LYSATES BY GTP

FIG. 4. Reticulocyte lysates were incubated, as previously described, in the absence or presence (25 μM) of hemin. After 15 min incubation, 1 mM GTP.Mg was added to both lysates, and the incubation was continued for 10 min. An immediate restoration of translation to control rates is produced in the inhibited hemin-deficient lysate, which is not preceded by the dephosphorylation of eIF-2(αP).

Effect of eIF-2α Phosphorylation on RF.eIF-2 Association

Two mechanisms for translation inhibition in hemin-deficient lysates can be proposed based on the requirement for RF.eIF-2 association to achieve guanine nucleotide exchange and thus allow catalytic utilization of eIF-2. In the first, RF.eIF-2 association would be directly inhibited by eIF-2α phosphorylation. In the second, RF would be sequestered into an inactive pool by eIF-2(αP). Evidence that eIF-2α phosphorylation inhibits RF.eIF-2 association is shown in Figure 5. In panel A, a Coomassie blue stain of glycerol gradient fractions analyzed by SDS PAGE shows the RF.eIF-2 complex sedimenting at 12S. No free eIF-2 is seen. If RF.eIF-2 is first incubated with eIF-2α kinase and [^{32}P-γ-ATP], however, the RF.eIF-2 complex is extensively dissociated to free eIF-2 (7S) and RF (Figure 5B). In Figure 5C,

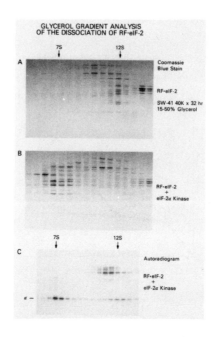

GLYCEROL GRADIENT ANALYSIS
OF THE DISSOCIATION OF RF·eIF-2

FIG. 5. A. RF.eIF-2 sediments as a 12S complex. B. Dissociation of RF.EIF-2
to free eIF-2 and RF (approximately 6S and 9S, respectively) by incubation with
γ-[32P]ATP and eIF-2α kinase. C. Autoradiogram of B. Glycerol gradient
fractions were analyzed by SDS PAGE in 15% gels. 40 μg of RF.eIF-2 purified
through the first glycerol gradient step were used in each analysis. Samples
were fractionated on 15-50% glycerol gradients containing 100 mM KCl, 2 mM
MgCl2, 10 mM pH 7.1 HEPES and 2 mM DTT for 32 hr at 40,000 rpm in a SW-41
rotor.

autoradiography confirms that the α subunit of released eIF-2 is extensively
phosphorylated. In contrast, phosphorylation of free eIF-2β does not inhibit
its rapid exchange with RF-bound eIF-2 (Figure 6). In panel A, sedimentation
of the RF.eIF-2 complex is shown. In panel B, the addition of [32P]eIF-2α
and [32P]eIF-2β does not alter the extent of RF.eIF-2 association. In
panel C, however, only [32P]eIF-2β is able to exchange with eIF-2 bound in
the RF.eIF-2 complex. [32P]eIF-2α remains as free eIF-2 sedimenting at 7S.
A major problem with this model, however, is the low extent of eIF-2α phosphorylation in hemin-deficient lysates. Although translation in hemin-deficient
lysates is inhibited by ≥ 90 percent, 70 percent of the total eIF-2 pool is not

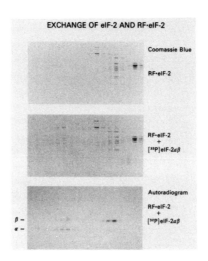

FIG. 6. A. Control RF.eIF-2. B. RF.eIF-2 incubated with 4 µg [^{32}P]-
eIF-2(αP) and [^{32}P]eIF-2(βP) for 4 min at 30°. C. Autoradiogram of B.
Samples were analyzed as described in Figure 5.

phosphorylated on the α subunit (19,20). The question posed, therefore, is
why this eIF-2 can not interact normally with RF to maintain nearly normal
rates of translation.

Guanine Nucleotides Stabilize the RF.eIF-2(αP) Complex Following eIF-2α Phosphorylation

Support for the second model of translational inhibition involving sequestration of a limited RF pool was obtained in the following set of experiments.
Since RF.eIF-2 interaction seemed to be involved primarily in the exchange of
guanine nucleotides, the effect of eIF-2α phosphorylation on RF.eIF-2 association was re-examined at physiologic concentrations and ratios of GTP and GDP.
Figure 7 shows that when the simultaneous effects of eIF-2α phosphorylation and
guanine nucleotides are examined, the opposite effect of eIF-2α phosphorylation
on RF.eIF-2 association is seen. In Figure 7A, phosphorylation of eIF-2α in
the absence of guanine nucleotides prevents exchange of free eIF-2 into
RF.eIF-2. Similar results are obtained when RF.eIF-2 is incubated with equal

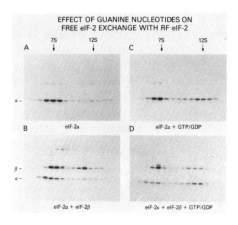

EFFECT OF GUANINE NUCLEOTIDES ON
FREE eIF-2 EXCHANGE WITH RF·eIF-2

FIG. 7. Exchange of eIF-2(αP) and eIF-2(βP) into RF.eIF-2. In the absence of guanine nucleotides, A. [^{32}P]eIF-2(αP) exchanges poorly into nonlabeled RF.eIF-2 and B. [^{32}P]eIF-2(βP), but not [^{32}P]eIF-2(αP), readily exchanges into RF.eIF-2. However, in the presence of 130 μM GTP and 6 μM GDP, [^{32}P]eIF-2(αP) readily exchanges into RF.eIF-2 (C) and (D) exchanges in preference to [^{32}P]eIF-2(βP). 10 μg RF.eIF-2 were incubated with 10 μg [^{32}P]eIF-2(βP) and/or [^{32}P]eIF-2(αP), with or without guanine nucleotides in 100 μl buffer containing 100 mM KCl, 2 mM MgCl$_2$, 10 mM pH 7.1 HEPES, and 2 mM DTT for 4 min at 30°. The distribution of [^{32}P] was analyzed by autoradiography of glycerol gradient fractions following SDS PAGE. Free eIF-2 and RF.eIF-2 sediment at 7S and 12S, respectively.

amounts of eIF-2(αP) and eIF-2(βP). Only eIF-2(βP) can exchange with RF-bound eIF-2 (Figure 7B). In the presence of GTP (130 μM) and GDP (5 μM), however, eIF-2(αP) can readily exchange into the RF.eIF-2 complex (Figure 7C). Moreover, Figure 7D shows that when RF.eIF-2 is incubated with equal amounts of eIF-2(αP) and eIF-2(βP), the exchange of eIF-2(αP) into RF.eIF-2 predominates. The RF.eIF-2(αP) complex, therefore, appears to be more stable than the RF.eIF-2(βP) complex in the presence of physiologic guanine nucleotide concentrations. These results are in direct contrast to those obtained when the effect of eIF-2α phosphorylation is examined in the absence of guanine nucleotides (Figure 5; 23, see also the results of Voorma et al., this volume).

Relative Size of the eIF-2 and RF Pools

Since phosphorylation of eIF-2α occurs only to a limited extent in hemin-deficient reticulocyte lysate (20-30 percent), it has been difficult to explain the large effect that phosphorylation has on the translational initiation process. Since eIF-2(αP) forms a more stable complex with RF than eIF-2 in the presence of physiologic guanine nucleotide concentrations, it was important to determine the relative sizes of the RF and eIF-2 pools. This was done by fractionating the total eIF-2 and RF.eIF-2 pool obtained by phosphocellulose and DEAE chromatography on 15-50% glycerol gradients. Figure 8A is a

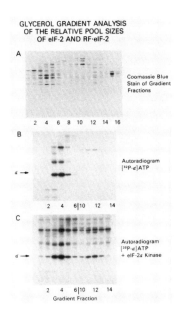

FIG.8. The relative sizes of the free eIF-2 and the RF.eIF-2 complex was determined by glycerol gradient fractionation of the DEAE RF.eIF-2/eIF-2 pool. A. Coomassie blue stain of glycerol gradient fractions resolved by SDS PAGE. B. Autoradiogram of A following the addition of [γ-32P]ATP and eIF-2α kinase to each gradient fraction.

Coomassie blue stain of the glycerol gradient fractions showing the free eIF-2 pool (peak fraction = 4) and the RF.eIF-2 complex (peak fraction = 12). Panel B is an autoradiogram of A, following the addition of [γ-32P]ATP to each gradient fraction. Phosphorylation of free eIF-2α is seen, since these

322

fractions also contain eIF-2α kinase (top band, fraction 4). The distribution
of eIF-2α throughout the gradient can be obtained, however, when each fraction
is supplemented with [γ-32P]ATP and eIF-2α kinase. The apparent RF:eIF-2
ratio from this data is approximately 1:6, by quantitation of alkali-labile
32P released from the α bands of the free and RF-complexed eIF-2.
Essentially identical results are obtained by ELISA analysis of Western blots
using sheep anti-rabbit eIF-2 serum (R. Jagus, personal communication). Since
eIF-2(αP) forms a more stable complex with RF than eIF-2 having a nonphosphorylated α subunit, the smaller pool of RF could be largely sequestered, even
though the extent of eIF-2α phosphorylation is limited (20-30 percent) in
hemin-deficient lysates.[19,20]

DISCUSSION

Figure 9 is a model of the mechanism of translational inhibition produced
in hemin-deficient reticulocyte lysates. The first regulated step during

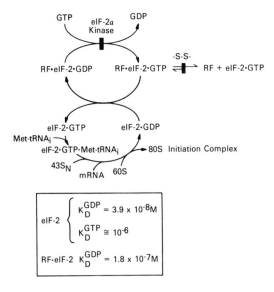

FIG. 9. The mechanism of translational inhibition in hemin-deficient reticulocyte lysate.

assembly of the 80S initiation complex is formation of the eIF-2.GTP.Met-tRNA$_i$ ternary complex. The initial binding of GTP to eIF-2 increases the rate of Met-tRNA$_i$ binding 20-fold. The ternary complex then binds to the 43S$_N$ ribosomal subunit to form the 43S preinitiation complex. Phosphorylation of eIF-2α has no effect on either ternary complex formation or its binding to initiating 43S$_N$. Following mRNA binding to form the 48S preinitiation complex, joining of the 60S ribosomal subunit is promoted by the hydrolysis of GTP in the bound ternary complex. This effects the release of bound initiation factor, including eIF-2 which, because of its high affinity for GDP (K_D^{GDP} = 3.9 x 10^{-8}M), is thought to be released as an inactive complex. Reactivation of eIF-2.GDP to active eIF-2.GTP by nonenzymatic exchange can not occur at the rate required for protein synthesis initiation, since the K_D^{GTP} of eIF-2 is at least one order of magnitude higher than the K_D^{GDP}. Reactivation can occur, however, if the concentration of GTP is experimentally increased to allow nonenzymatic exchange (Figure 4).

The recycling factor RF appears to fulfill the requirement for an enzymatic exchange mechanism. Exchange of free GTP for bound GDP is accomplished in an analogous manner to the exchange promoted by the prokaryotic factor EF-Ts, that is, the K_D^{GDP} of eIF-2 is increased 6-fold following its association with RF. The K_D^{GDP} of RF.eIF-2 is 1.8 x 10^{-7}. At physiologic guanine nucleotide concentrations (GTP = 130 μM, GDP = 5-10μM), GTP can now readily displace GDP to form a RF.eIF-2.GTP complex. The binary complex eIF-2.GTP is then displaced from RF.eIF-2.GTP by eIF-2.GDP to complete the exchange cycle. In hemin-deficient lysates, the partial phosphorylation of eIF-2α interferes with this process by forming a stable RF.eIF-2(αP).GDP complex that is unable to effect a rapid exchange of guanine nucleotides. However, at the present time, the molecular basis of this inhibition by phosphorylation is not known. One possibility is that the K_D^{GDP} of the RF.eIF-2(αP).GDP complex is not increased significantly over that of RF.eIF-2.GDP. This would result in the sequestration of RF into an inactive RF.eIF-2(αP).GDP complex.

This mechanism could explain a major inconsistency in mechanisms proposed previously for inhibition of protein synthesis initiation by eIF-2 phosphorylation. The original proposals for RF activity were based on either a dual pathway for Met-tRNA$_i$ binding to 43S$_N$, one involving free eIF-2 and a second more efficient pathway utilizing the RF.eIF-2 complex,[11,16] or a requirement of RF for guanine nucleotide exchange.[23] Inhibition of translation in hemin-deficient lysates in both models would result from the effect of eIF-2α

phosphorylation to prevent RF/eIF-2 association. The major problem with these proposals, however, was how to accommodate the low extent of eIF-2α phosphorylation (20-30 percent) in hemin-deficient lysates. The question raised was, why doesn't the remaining 70-80 percent of eIF-2 that is non-phosphorylated interact normally with RF? In addition, why should small amounts of RF rescue translational activity in a highly catalytic manner? Both of these inconsistencies can be accommodated by our proposed mechanism in which eIF-2(αP) sequesters the RF pool into a less active complex. Although the percent of eIF-2 phosphorylated by eIF-2 kinase is low, the smaller pool size of RF in comparison to that of eIF-2 (10-20 percent) would be surpassed by that of eIF-2(αP). Any increase in the RF pool size to a size greater than that of eIF-2(αP) would allow catalytic function of the nonphosphorylated eIF-2 pool to be resumed. The precise modulation of eIF-2 phosphorylation by highly active kinase and phosphatase activities and the effect of this on RF/eIF-2 association can therefore rapidly alter the size of the active eIF-2 pool by a series of metabolic events characteristic of other regulatory cascade systems.[24]

REFERENCES

1. Zucker, W.V. and Shulman, H.M. (1968) Proc. Natl. Acad. Sci. 59, 582-586.
2. Adamson, S.D., Herbert, E. and Godchaux, W. (1968) Arch. Biochem. Biophys. 125, 671-676.
3. Farrell, P.J., Balkow, K., Hurt, T., Jackson, R.J. and Trachsel, H. (1977) Cell 11, 187-200.
4. Jagus, R., Anderson, W.F. and Safer, B. (1981) Prog. Nucleic Acid Res. and Mol. Biol. 25, 127-185.
5. Safer, B. and Jagus, R. (1981) Biochimie 63, 709-717.
6. de Haro, C. and Ochoa, S. (1978) Proc. Natl. Acad. Sci. 75, 2713-2716.
7. Das, A., Ralston, R.O., Grace, M., Roy, R., Ghosh-Dastidar, P., Das, H.K., Yaghmai, B., Palmieri, S. and Gupta, N.K. (1979) Proc. Natl. Acad. Sci. 76, 5076-5079.
8. Kramer, G., Henderson, A.B., Grankowski, N. and Hardesty, B. (1979) in Ribosomes. Structure, Function, and Genetics, G. Chambliss, G.R. Craven, J. Davies, K. Davis, K. Kahan and M. Nomura, ed., University Park Press, Baltimore, pp. 825-846.
9. Ranu, R.S. and London, I.M. (1979) Proc. Natl. Acad. Sci. USA 76, 1079-1083.
10. Safer, B., Peterson, D. and Merrick, W.C. (1977) in Proceedings of the International Symposium on Translation of Synthetic and Natural Polynucleotides, A.B. Legocki, ed. Poznan, Poland, pp. 24-31.
11. Voorma, H.)., Goumans, H., Amesz, H. and Benne, R. (1982) in Current Topics in Cellular Regulation, B.L. Horecker and E.R. Stadtman, ed. (in press).
12. Benne, R., Amesz, H., Hershey, J.W.B. and Voorma, H.O. (1979) J. Biol. Chem. 254, 3201-3205.

13. Safer, B., Kemper, W. and Jagus, R. (1979) J. Biol. Chem. 254, 8091-8094.
14. Walton, G.M. and Gill, G.N. (1976) Biochim. Biophys. Acta 447, 11-19.
15. Safer, B., Adams, S.L., Anderson, W.F. and Merrick, W.C. (1975) J. Biol. Chem. 250, 9076-9082.
16. Amesz, H., Goumans, H., Haubrich-Morree, T., Voorma, H.O. and Benne, R. (1979) Eur. J. Biochem. 98, 513-520.
17. Siekierka, J., Mitsui, K.I. and Ochoa, S. (1981) Proc. Natl. Acad. Sci. 78, 220-223.
18. Das, A., Bagchi, M., Roy, R., Ghosh-Dastidar, P. and Gupta, N.K. (1982) Biochem. Biophys. Res. Commun. 104, 89-98.
19. Safer, B., Jagus, R. and Crouch, D. (1981) in Cold Spring Harbor Conference on Cell Proliferation, Volume 8: Protein Phosphorylation, New York, pp. 979-998.
20. Jagus, R., Crouch, D., Konieczny, A. and Safer, B. (1982) in Current Topics in Cellular Regulation, B.L. Horecker and E.R. Stadtman, ed. (in press).
21. Lloyd, M.A., Osborne, J.C., Safer, B., Powell, G.M. and Merrick W.C. (1980) J. Biol. Chem. 255, 1189-1194.
22. Kaziro, Y. (1978) Biochim. Biophys. Acta 505, 95-127.
23. Siekierka, J., Mauser, L., and Ochoa, S. (1982) Proc. Natl. Acad. Sci. 79, 2537-2540.
24. Stadtman, E.R. and Chock, P.B. (1978) in Current Topics in Cellular Regulation, Horecker, B.L. and Stadtman, E.R. eds., Academic Press, New York, pp. 53-95.

ACKNOWLEDGMENTS

We would like to recognize the skill and patience of Mrs. Kirsten Cook and Mrs. Eve Church in the preparation and editing of this manuscript.

THE CONTROL OF POLYPEPTIDE CHAIN INITIATION BY PHOSPHORYLATION

OF THE α SUBUNIT OF THE INITIATION FACTOR eIF-2

JOHN J. SIEKIERKA, LJUBICA MAUSER AND SEVERO OCHOA
The Roche Institute of Molecular Biology, Nutley, New Jersey, USA

INTRODUCTION

Protein synthesis in rabbit reticulocyte lysates, of which
greater than 90% is globin, is dependent upon adequate levels of
heme. In the presence of heme, protein synthesis will proceed at
rates close to those found in vivo. In the absence of heme, the
rate of protein synthesis is normal initially, but decreases to
about 10% of the control rate after about 5 minutes. This inhibition, due to heme deficiency, is the result of the activation of a
latent inhibitor of protein synthesis called the heme-controlled
inhibitor (HCI). A second inhibitor of protein synthesis, distinct
from HCI, is also found in reticulocytes. This inhibitor is activated by low concentrations of double-stranded RNA. In cells
other than reticulocytes, the double-stranded RNA activated inhibitor (DAI) is induced by interferon pretreatment and may play a
role in establishing the antiviral state. HCI and DAI are cAMP-
independent protein kinases which phosphorylate the smallest of
three subunits of the initiation factor eIF-2, the α subunit. The
only known substrate for HCI is the α subunit of eIF-2, while DAI
can also phosphorylate histone.[1] Though HCI and DAI are distinct
proteins, both appear to phosphorylate the same site(s) on the α
subunit of eIF-2.[2] It is the specific phosphorylation of the α
subunit of eIF-2 by HCI or DAI that renders it inactive in catalyzing polypeptide chain initiation. This paper describes the mechanism by which phosphorylation of the α subunit of eIF-2 interferes with its function in protein synthesis.

Role of eIF-2 in Polypeptide Chain Initiation and the Requirement for a New Initiation Factor, eIF-2·SP. The first step in
polypeptide chain initiation is the formation of a ternary complex
between the initiation factor eIF-2, GTP, and the initiator methi-
onyl transfer RNA (Met-tRNA$_i$) (for reviews see, 3,4). The formation of a ternary complex is preceded by binary complex formation
between eIF-2 and GTP.[5] The ternary complex subsequently binds to

a 40S ribosomal subunit forming a 40S initiation complex. A new factor, first described by this laboratory, was shown to greatly stimulate the formation of ternary complex. This factor was referred to as ESP for eIF-2 stimulating protein.[6] Factors similar or identical to ESP were subsequently described by other laboratories and were shown to prevent Mg^{+2} inhibition of ternary complex formation.[7,8] This observation has been confirmed for ESP.[9] Recently, we have isolated an apparently homogeneous complex of eIF-2 and ESP from the postribosomal supernatant of rabbit reticulocyte lysate.[10] We currently refer to this complex as eIF-2·SP where SP is the stimulatory factor. The apparent molecular weight of this complex is 450,000 as determined by sucrose density gradient centrifugation. SDS-polyacrylamide gel electrophoresis reveals that, in addition to containing the α, β, and γ subunits of eIF-2 (38, 52, and 64 kilodaltons, respectively), the complex also consists of polypeptides at 80, 65, 57, 40, and 32 kilodaltons which make up the SP portion (Fig. 1)

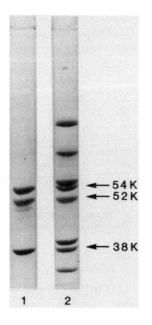

Fig. 1. SDS-polyacrylamide gel electrophoresis of eIF-2 (lane 1 and eIF-2·SP (lane 2). Each factor (7.8 μg) was subjected to electrophoresis in 10% acrylamide containing 0.26% N',N'-methylenebisacrylamide, for 2.5 hr at 6 mA per tube. Gels were stained with Coomassie blue. The arrows indicate the position and molecular weight (X 10^3) of the three eIF-2 subunits.

Ternary complex formation with our eIF-2 preparations is strongly inhibited by Mg^{+2}, and this inhibition is at the level of guanosine nucleotide binding to eIF-2 (Fig. 2A). Addition of catalytic amounts of eIF-2·SP to the Mg^{+2} inhibited ternary (GTP·eIF-2· Met-tRNA$_i$) or binary (GDP·eIF-2) complex assays, restores complex formation to the levels observed in the absence of Mg^{+2} (Fig. 2B).[11,12]

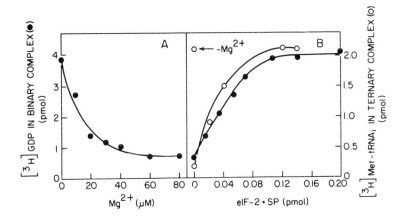

Fig. 2. (A) Inhibition of eIF-2·[^{3}H]GDP binary complex formation by Mg^{+2}. Five pmol of eIF-2 (30% pure) was used. (B) Relief of Mg^{+2} inhibition of binary (●) and ternary (O) complex formation by eIF-2·SP. Ternary complex was assayed with 5 pmol of eIF-2 (30% pure), 3 pmol [^{3}H]Met-tRNA$_i$ (62,000 cpm/pmol) and 0.5 mM Mg(OAc)$_2$. Binary complex was assayed at 0.1 mM Mg(OAc)$_2$. ←, Ternary complex formation in the absence of Mg^{+2} and eIF-2·SP

The formation of a 40S initiation complex has an absolute requirement for Mg^{+2}. Therefore, it appeared that eIF-2·SP may function by allowing the formation of ternary complex in the presence of the optimal concentrations of Mg^{+2} required for this step. No effect of eIF-2·SP on the transfer of ternary complex to the 40S ribosomal subunit has been observed.[13]

The Control of the Catalytic Function of eIF-2·SP by α Phosphorylation of eIF-2. Early work in this laboratory demonstrated that HCI and DAI inhibited ternary complex stimulation by crude ESP and that this inhibition required ATP.[1,15] By measuring

complex formation between eIF-2 and [³H]GDP, it can be demonstrated that phosphorylation of the α subunit of eIF-2 with HCI and
ATP, inhibits eIF-2·SP catalysis of binary complex formation in the
presence of Mg^{+2}.[11,12] The inhibition is specifically related to
α subunit phosphorylation since phosphorylation of the β subunit
with reticulocyte casein kinase is without effect.[15] In the absence of Mg^{+2}, the rate and extent of binary complex formation
with α or β phosphorylated eIF-2 is identical.[12] Phosphorylation
of the α subunit of eIF-2, therefore, does not affect its activity
per se, but rather, it prevents its interaction with eIF-2·SP in
the presence of Mg^{+2}.

MATERIALS AND METHODS

The preparation of rabbit reticulocyte lysates and eIF-2 has
been described.[10] Unless otherwise noted, the eIF-2 used in these
studies was between 80-90% pure. Amounts are given as pure eIF-2.
The preparation of eIF-2·SP from the reticulocyte lysate supernatant has also been described.[10] For the preparation of eIF-2[³H]GDP
samples (3 ml) containing: 20 mM Tris-HCl, pH 7.5; 100 mM KCl;
2 mM β mercaptoethanol; 20 µM [³H]GDP (Amersham, 3,300 cpm/pmol);
and 1.2 nmol of CM-350 eIF-2,[15] were incubated for 10 min at 30°C,
cooled in an ice bath and made 1.0 mM in $Mg(OAc)_2$. The ratio of
[³H]GDP to eIF-2 (mol/mol), as determined by retention of the complex on nitrocellulose filters, was 1. The samples were applied
to a 0.7 X 2.0 cm Sepharose 6B-heparin column previously equilibrated with buffer A (20 mM Tris-HCl, pH 7.5, 1.0 mM $Mg(OAc)_2$,
1.0 mM $Mg(OAc)_2$, 2.0 mM β mercaptoethanol, and 5% (vol/vol) glycerol) containing 100 mM KCl. The column was washed with this
buffer until all noncomplexed [³H]GDP was eluted. The eIF-2·[³H]
GDP complex was eluted with buffer A containing 400 mM KCl. The
eIF-2 complex (containing 80% pure eIF-2 as determined by SDS-
polyacrylamide gel electrophoresis) was stored in liquid nitrogen.
For the preparation of eIF-2 phosphorylated in the β subunit
[eIF-2(β³²P)], samples (2 ml) containing: 10 mM Tris-HCl, pH 7.5;
50 mM KCl; 4 mM $Mg(OAc)_2$; 1.0 mM DTT; 0.1 mM [γ³²P]ATP (4,000 cpm/
pmol); 1.9 nmol eIF-2; and 200 µg of rabbit reticulocyte casein
kinase,[15] were incubated for 20 min at 30°C. The phosphorylated
eIF-2 was isolated as described for eIF-2·[³H]GDP using Sepharose

6B-heparin chromatography. The phosphorylation of eIF-2 in the α subunit [eIF-2($α^{32}$P)] was conducted in a similar manner, but in the absence of KCl, and with 124 µg of partially purified heme-control-led translational inhibitor (HCI) prepared as described.[10] Both phosphorylated eIF-2 preparations contained approximately 1 mol of [^{32}P] per mol of eIF-2. GDP-free GTP and [^{35}S]Met-tRNA$_i$ were pre-pared as described. Protein was determined by the Bradford proce-dure with bovine serum albumin as the standard.[16]

The exchange of GDP or GTP with eIF-2 bound [^{3}H]GDP was assayed in reaction mixtures (50 µl) containing: 20 mM Tris-HCl, pH 7.5; 100 mM KCl; 1.0 mM Mg(OAc)$_2$; 1-2 pmol eIF-2·[^{3}H]GDP; and the indi-cated amounts of eIF-2·SP and either GDP or GTP. Reaction mixtures were incubated for 1 min at 30°C. Bound radioactivity was assayed by filtration on nitrocellulose membranes. After incubation, sam-ples were chilled in an ice bath, and immediately before assaying, diluted with 1.0 ml of ice-cold wash buffer (20 mM Tris-HCl, pH 7.5; 100 mM KCl, 1.0 mM Mg(OAc)$_2$, and 2.0 mM β mercaptoethanol). The samples were filtered on nitrocellulose membranes and washed with 10 ml of wash buffer. The filters were transferred to glass scintillation vials and 1.0 ml of methylcellosolve (Fisher) was added. After shaking for 5 min to dissolve the filters, 10 ml of Hydrofluor (National Diagnostics) was added and the samples counted in a liquid scintillation counter.

For the exchange of free [^{32}P]-labeled eIF-2 with SP-bound eIF-2, samples (250 µl) contained: 20 mM Tris-HCl, pH 7.5; 100 mM KCl; 0.8 mM Mg(OAc)$_2$; 25 µg of creatine kinase; and the indicated amounts of eIF-2($β^{32}$P) at 2,800 cpm/pmol or eIF-2($α^{32}$P) at 3,700 cpm/pmol and eIF-2·SP as indicated. Samples were incubated for 2-5 min at 30°C in plastic tubes. Creatine kinase was added to prevent losses of eIF-2 through adsorption to the tube walls.[17] After cooling in ice, 200 µl aliquots were layered onto a 4 ml, 10-30% linear sucrose gradient in 20 mM Tris-HCl, pH 7.5; 100 mM KCl; 0.8 mM Mg(OAc)$_2$; 1.0 mM DTT and centrifuged for 5 hr at 54,000 rpm in a Beckman SW 56 rotor. Gradients were fractionated with an Isco model 640 density gradient fractionator. Each fraction (0.3 ml) was mixed with an equal volume of water and 10 ml of Ready-Solv HP (Beckman) added. Samples were counted by liquid scintillation counting.

RESULTS

Catalysis of GDP-GTP Exchange by eIF-2 SP. Further studies on the role of Mg^{+2} in binary and ternary complex inhibition have indicated that in the presence of Mg^{+2}, the eIF-2·GDP binary complex is greatly stabilized as judged by the inability of excess GDP or GTP to compete with eIF-2-bound GDP.[12] Figure 3A shows that eIF-2-bound [³H]GDP cannot be exchanged with unlabeled free GDP in the presence of 1.0 mM Mg^{+2} unless eIF-2·SP is present. GTP can also

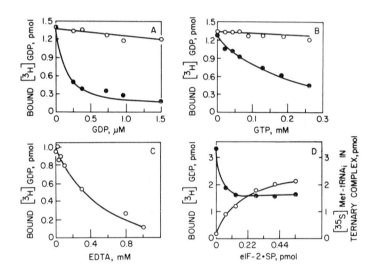

Fig. 3. Exchange of unlabeled GDP (A) or GTP (B) with eIF-2-bound [³H]GDP as a function of the GDP or GTP concentration in the presence of 1.0 mM Mg(OAc)₂ and in the absence (O) or presence (●) of 0.66 pmol of eIF-2·SP. (C) GDP exchange in the absence of eIF-2·SP upon removal of Mg^{+2} by EDTA. (D) Release of eIF-2-bound [³H]GDP (●) and ternary complex formation (O) at various concentrations of eIF-2·SP. Present in the ternary complex assay was 1.0 mM Mg(OAc)₂ 22 μM GTP and 4 pmol of [³⁵S]Met-tRNA_i (11,400 cpm/pmol).

replace eIF-2-bound [³H]GDP in the presence of eIF-2·SP, under identical conditions. (Fig. 3B). As indicated in figure 3C, removal of Mg^{+2} by chelation with EDTA destabilizes the eIF-2·[³H]GDP complex and permits the exchange of bound GDP for free GDP. From these data, it can be estimated that about a 1,000-fold higher concentration of GTP than GDP is required for the displacement of [³H]GDP from eIF-2 in the presence of eIF-2·SP and Mg^{+2}. When the exchange reaction is coupled to ternary complex formation, much lower concentrations of GTP are effective in replacing eIF-2-bound [^{3}H]GDP. The resulting eIF-2·GTP binary complex readily binds [^{35}S]Met-tRNA$_i$ to form a ternary complex. As shown in figure 3D, in the presence of Mg^{+2} and excess GTP, addition of eIF-2·SP decreases the amount of eIF-2·[^{3}H]GDP binary complex while increasing the level of GTP·eIF-2·[^{35}S]Met-tRNA$_i$ ternary complex. Due to the high affinity of GDP for eIF-2 (estimated to be about 100-fold greater than GTP),[14] conversion of the 40S (GTP·eIF-2·Met-tRNA$_i$· 40S) to the 80S (Met-tRNA$_i$·80S·mRNA) initiation complex, which is accompanied by GTP hydrolysis, probably releases eIF-2 as an eIF-2·GDP complex. Since only the eIF-2·GTP binary complex is capable of forming a ternary complex, some mechanism must be operative for the efficient removal of eIF-2-bound GDP in the presence of Mg^{+2}. It appears that eIF-2·SP may fulfill this requirement.

Exchange of Free eIF-2 with eIF-2 complexed to SP. The ability of eIF-2·SP to function catalytically in promoting ternary complex formation via the GDP-GTP exchange reaction suggests that the eIF-2 complexed to SP must be able to exchange with free eIF-2. Figure 4A shows that free eIF-2, labeled in the β subunit with reticulocyte casein kinase and [γ^{32}P]ATP [eIF-2(β^{32}P)], readily exchanges with unlabeled eIF-2 complexed to SP. The exchange is very rapid (going to completion in less than 1 minute at 0°C) and occurs in the presence or absence of Mg^{+2}. The eIF-2 exchange reaction is inhibited by prior treatment of eIF-2·SP with N-ethylmaleimide (Fig. 4B), a treatment which also inhibits the catalytic activity of eIF-2·SP in stimulating ternary complex formation. In addition, the eIF-2 exchange only occurs at the optimum pH for stimulation of ternary complex formation by eIF-2·SP (Fig. 4C and 4D). These experiments suggest that the eIF-2 exchange is directly related to the catalytic action of eIF-2·SP in ternary complex formation.

Phosphorylation of the α subunit of eIF-2 and its effect on the catalytic activity of eIF-2·SP. Both the GDP-GTP exchange and the eIF-2 exchange reactions are inhibited when the free eIF-2 present in these assays is phosphorylated in the α subunit by the heme-controlled inhibitor of protein synthesis (HCI). As indicated previously and in figure 5, eIF-2·SP catalyzes the exchange of

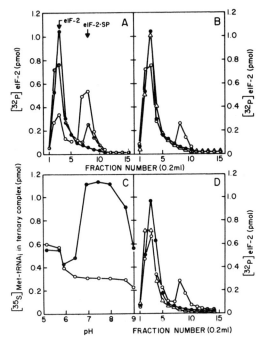

Fig.4. Exchange of free eIF-2 with SP-bound eIF-2. eIF-2(β^{32}P) was incubated with eIF-2·SP for 5 min at 30°C and the distribution of radioactivity between free and bound eIF-2 was analyzed by sucrose density centrifugation. (A) ●, Control, eIF-2(β^{32}P) alone (6.6 pmol); ◉, eIF-2(β^{32}P) and eIF-2·SP (8.8 pmol); O, eIF-2(β^{32}P) and eIF-2·SP (13.8 pmol). (B) ●, Control, eIF-2(β^{32}P) (6.6 pmol) alone; O, eIF-2(β^{32}P) and eIF-2·SP (8.8 pmol); Δ, eIF-2(β^{32}P) and N-ethylmaleimide treated eIF-2·SP (8.8 pmol).(C) pH dependence of eIF-2·SP catalyzed ternary complex formation. The eIF-2 used was about 30% pure (6 pmol) and 3 pmol of [^{35}S]Met-tRNA$_i$ (95,400 cpm/pmol) was used. O, no eIF-2·SP; ●, 0.2 pmol of eIF-2·SP. (D) Effect of pH on exchange. ●, Control, eIF-2(β^{32}P)(6.6 pmol) at pH 7.5; O, eIF-2(β^{32}P) and eIF-2·SP (8.8 pmol) at pH 7.5; Δ eIF-2 (β^{32}P) and eIF-2.SP at pH 6.5.

eIF-2-bound [^{3}H]GDP for GTP. The resulting eIF-2·GTP binary complex binds [^{35}S]Met-tRNA$_i$ to form a GTP·eIF-2·[^{35}S]Met-tRNA$_i$ ternary complex. Substitution of α phosphorylated eIF-2 [eIF-2(αP)] for eIF-2 in these experiments shows that while eIF-2(αP) is able to bind [^{3}H]GDP, this GDP cannot be exchanged for GTP in the presence of Mg^{+2} and eIF-2·SP (Fig. 5). Since [^{35}S]Met-tRNA$_i$ cannot bind to the eIF-2·[^{3}H]GDP complex, no ternary complex forms. It appears that the stimulation of [^{3}H]GDP binding to eIF-2 by eIF-2·SP and the guanosine nucleotide exchange reaction are related

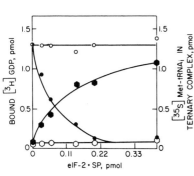

Fig.5. Effect of eIF-2·SP on ternary complex formation from eIF-2·[^{3}H]GDP or eIF-2(αP)·[^{3}H]GDP, and [^{35}S]Met-tRNA$_i$. Release of [^{3}H]GDP and ternary complex formation with [^{35}S]Met-tRNA$_i$ were assayed simultaneously as in Fig. 3D. Present in the assay were 1.0 mM Mg(OAc)$_2$, 65 μM GTP, 4 pmol of [^{35}S]Met-tRNA$_i$ (8,400 cpm/pmol). eIF-2·[^{3}H]GDP was phosphorylated in its α subunit with HCl and [γ^{32}P]ATP and reisolated with Sepharose-heparin chromatography. ● and ⬢, eIF-2·[^{3}H]GDP; ○ and ⬡, eIF-2(αP)·[^{3}H]GDP. ● and ○, [^{3}H] radioactivity; ⬢ and ⬡ [^{35}S] radioactivity

events. Preliminary results from our laboratory suggest that the eIF-2 used in our assays is isolated from the lysate as an eIF-2· GDP complex (Manne,V., Siekierka,J. and Ochoa,S.unpublished observations). It appears, therefore, that the stimulation of binary complex formation by eIF-2·SP is really a reflection of the exchange of eIF-2-bound GDP for [^{3}H]GDP in the presence of Mg^{+2}. Phosphorylation of the α subunit of eIF-2 prevents this exchange from occurring with no increase in the level of eIF-2-bound [^{3}H]GDP.

It is demonstrated in figure 6, that while eIF-2 phosphorylated in the β subunit [eIF-2(βP)] readily exchanges with eIF-2 bound to SP, eIF-2(αP) cannot. These results demonstrate that phosphorylation of the α subunit of eIF-2 directly blocks its interaction with SP.

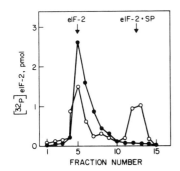

Fig.6. Failure of eIF-2(αP) to exchange with SP-bound eIF-2. Samples (200 μl) contained 14 pmol of eIF-2 phosphorylated with [γ^{32}P]ATP in either the α or β subunit and other additions were incubated for 2 min at 30°C and analyzed by sucrose density gradient centrifugation. ○, eIF-2 (βP) control; ●, eIF-2(αP).

These experiments support the view that phosphorylation of the α subunit of eIF-2 inhibits polypeptide chain initiation by preventing the interaction of eIF-2 with SP, a step required for the removal of eIF-2-bound GDP.

DISCUSSION

The results presented in this paper suggest that eIF-2·SP catalyzes the exchange of eIF-2-bound GDP for GTP during protein synthesis in the reticulocyte (Fig. 7A). This reaction bears a close

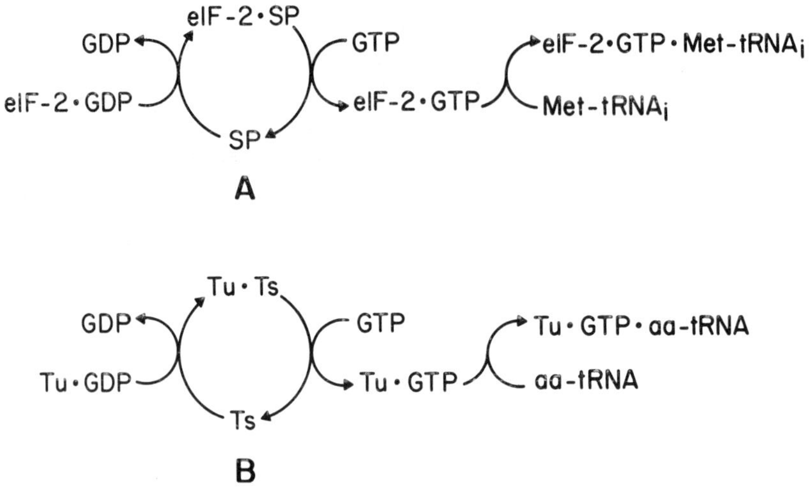

Fig. 7. The analogy between the initial steps of eukaryotic chain
initiation (A) and prokaryotic chain elongation (B).

resemblance to the exchange of prokaryotic EF-Tu-bound GDP with
GTP catalyzed by EF-Ts during prokaryotic polypeptide chain elongation (for a review, see 18). The similarity between the eIF-2·SP
complex in eukaryotic chain initiation (Fig. 7A) and the EF-Tu·EF-
Ts complex in prokaryotic chain elongation (Fig. 7B) is apparent.

In the presence of Mg^{+2}, eIF-2, like EF-Tu, has at least a 100
fold greater affinity for GDP than for GTP.[18] After each round of
initiation (or elongation) GTP is hydrolyzed with concomitant release of eIF-2·GDP (or EF-Tu·GDP) complex. We propose that eIF-2·
SP, like EF-Ts, catalyzes the exchange of eIF-2-bound GDP for GTP,
allowing the formation of ternary complex (GTP·eIF-2·Met-tRNA$_i$).
Under conditions where the α subunit becomes phosphorylated (e.g.
heme deficiency), the interaction between eIF-2 and SP is blocked
and the exchange of GDP with GTP prevented. Under these conditions, eIF-2 cannot participate in additional rounds of polypeptide
chain initiation, and protein synthesis is inhibited. Models similar to this one have been considered before[19,20] and a similar

mechanism has recently been suggested for Ascites system.[21]

REFERENCES

1. Grosfeld, H. and Ochoa, S. (1980) Proc. Nat. Acad. Sci. 77, 6526-6530.
2. Ernst, V., Levin, D.H., Leroux, A. and London, I.M. (1980) Proc. Nat. Acad. Sci. 77, 1286-1290.
3. Ochoa, S. and deHaro, C. (1979) Annu. Rev. Biochem. 48, 549-580.
4. Jagus, R., Anderson, W.F. and Safer, B. (1980) Prog. Nucleic Acid Res. Mol. Biol. 25, 127-185.
5. Safer, B., Adams, S.L., Anderson, W.F. and Merrick, W.C. (1975) J. Biol. Chem. 250, 9076-9082.
6. deHaro, C., Datta, A. and Ochoa, S. (1978) Proc. Nat. Acad. Sci. 75, 243-247.
7. Ranu, R.S. and London, I.M. (1979) Proc. Nat. Acad. Sci. 76, 1079-1083.
8. Das, A., Ralston, R.O., Grace, M., Roy, R., Ghosh-Dastidar, P., Das, H.K., Yaghmai, B., Palmieri, S. and Gupta, N.K. (1979) Proc. Nat. Acad. Sci. 76, 5076-5079.
9. Ochoa, S., Siekierka, J., deHaro, C. and Grosfeld, H. (1981) in Protein Phosphorylation, Cold Spring Harbor Conferences on Cell Proliferation, eds. Rosen, O.M. and Krebs, E.G. Cold Spring Laboratory, Cold Spring Harbor, N.Y. 8, 931-940.
10. Siekierka, J., Mitsui, K.-I. and Ochoa, S. (1981) Proc. Nat. Acad. Sci. 78, 220223.
11. Siekierka, J. and Ochoa, S. in Symposium on Hemoglobin Biosynthesis, ed. Goldwasser, E. Elsevier-North Holland, Amsterdam, in press.
12. Siekierka, J., Mauser, L. and Ochoa,S. (1982) Proc. Nat. Acad. Sci. 79, 2537-2540.
13. Siekierka, J., Datta, A., Mauser, L. and Ochoa, S. (1982) J. Biol. Chem. 257, 4162-4165.
14. Walton, G. M. and Gill, G.N. (1975) Biochim. Biophys. Acta 390, 231-245.
15. deHaro, C. and Ochoa, S. (1979) Proc. Nat. Acad. Sci. 76, 1741-1745.
16. Bradford, M.M. (1976) Anal. Biochem. 72, 248-254.
17. Benne, R., Amesz, H., Hershey, J.W.B. and Voorma,H.O. (1979) J. Biol. Chem. 254, 3201-3205.
18. Miller, D.L. and Weissbach, H. (1977) in Molecular Mechanisms of Protein Biosynthesis, eds. Weissbach, H. and Pestka, S. Academic Press, N.Y., pps. 323-373.
19. Cherbas, L. and London, I.M. (1976) Proc. Nat. Acad. Sci.73, 3506-3510.
20. Hunt, T. (1979) in Miami Winter Symposium: From Gene to Protein, eds. Russel, T.R., Brew, K., Schultz, J. and Harber, H. Academic Press, N.Y. 16, 229-243.
21. Clemens, M.J., Pain, V.M., Wong, S.T. and Henshaw, E.C. (1982) Nature (London) 296, 93-95.

PROTEIN SYNTHESIS INITIATION AND ITS REGULATION IN RETICULOCYTE LYSATES BY
eIF-2 AND eIF-2-ANCILLARY PROTEIN FACTORS

NABA K. GUPTA, MICHAEL GRACE, AMBICA C. BANERJEE, AND MILAN K. BAGCHI
Department of Chemistry, The University of Nebraska, Lincoln, Nebraska 68588-0304, USA

1. INTRODUCTION

The first step in peptide chain initiation in mammalian cells is the formation of a ternary complex between a eukaryotic peptide chain initiation factor, eIF-2, Met-tRNA$_f$ and GTP; Met-tRNA$_f$·eIF-2·GTP.

Recent work done in our laboratory and elsewhere has indicated that this ternary complex formation by eIF-2 and its proper functioning during peptide chain initiation in reticulocyte lysates is regulated by a complex mechanism involving several eIF-2-ancillary protein factors which we term Co-eIF-2A, Co-eIF-2B, Co-eIF-2C and RF, and also by eIF-2 kinases, such as HRI (heme-regulated protein synthesis inhibitor) and dsI (double stranded RNA-activated protein synthesis inhibitor). For recent reviews on the roles of eIF-2 and eIF-2-ancillary protein factors in regulation of protein synthesis initiation in reticulocyte lysates, see Refs. 1-5.

In this article, we will: (1) Briefly review our previous reports on the characteristics of Co-eIF-2A, Co-eIF-2B and Co-eIF-2C, and their roles in overall protein synthesis initiation, and also in Met-tRNA$_f$· 40S initiation complex formation. (2) Describe our recent results on the characteristics of RF and its mechanism of reversal of protein synthesis inhibition in heme-deficient reticulocyte lysates.

2. CHARACTERISTICS OF Co-eIF-2A, Co-eIF-2B AND Co-eIF-2C: A REVIEW.

2.1. Purification Scheme. Figure 1 shows a scheme for purification of eIF-2, Co-eIF-2A, Co-eIF-2B, Co-eIF-2C and also RF from rabbit reticulocyte lysates. eIF-2, Co-eIF-2A, Co-eIF-2B, and Co-eIF-2C activities are purified from the 0.5 M KCl wash of the reticulocyte ribosomes whereas the RF activity is purified from the cell supernatant. The bulk of Co-eIF-2A activity is not adsorbed onto a DEAE-cellulose column and so is separated from eIF-2, Co-eIF-2B and Co-eIF-2C activities at an early step in purification.[6,7] Co-eIF-2A

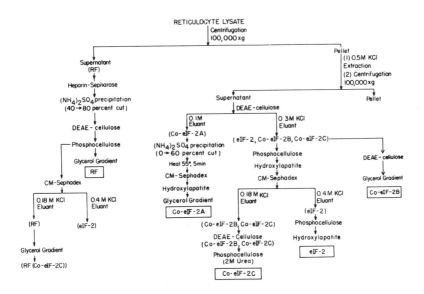

Fig. 1: Purification scheme for eIF-2 and eIF-2-ancillary factors.

activity has been further purified to homogeneity by heating, CM-Sephadex and hydroxylapatite chromatography, and glycerol density gradient centrifugation.[7]

eIF-2 activity co-purifies with Co-eIF-2B and Co-eIF-2C activities and is only separated from the latter activities at the CM-Sephadex chromatographic step. The Co-eIF-2B and Co-eIF-2C activities are eluted with 0.18 M KCl whereas eIF-2 activity is eluted with a buffer containing 0.4 M KCl. The eIF-2 activity obtained after the CM-Sephadex chromatographic step has been further purified to homogeneity using phosphocellulose and hydroxylapatite chromatography.[8] Separation of Co-eIF-2B activity (still containing Co-eIF-2C activity) from eIF-2 can be achieved by DEAE-cellulose chromatography using a KCl gradient.[9,10]

The CM-Sephadex purified Co-eIF-2B and Co-eIF-2C activities remain associated with a high molecular weight protein complex (approximate mol. wt. 500,000 daltons) composed of many polypeptides.[11] This protein complex also contains significant amounts of Co-eIF-2A activity. Apparently, all three eIF-2-ancillary factor activities remain associated in a protein complex. As mentioned above, the bulk of the Co-eIF-2A activity remains in the free form and this activity has been purified to homogeneity. The precise nature of the polypeptide component(s) responsible for Co-eIF-2B and Co-eIF-2C activities is not clear. These two activities may reside in a single polypeptide, and different domains of the polypeptide may be responsible for two different activities. Alternatively, these two activities may be associated with different polypeptide component(s) of the protein complex. It has not yet been possible to fully resolve these two activities. Using a phosphocellulose chromatographic procedure in the presence of 2 M urea, it has been possible to obtain a Co-eIF-2C preparation completely devoid of Co-eIF-2B activity.[11] Presumably, under the purification conditions, the Co-eIF-2B activity is unstable and is lost. However, attempts to detect any change in the polypeptide pattern in the Co-eIF-2C preparations, before and after the above phosphocellulose chromatographic step, using SDS-polyacrylamide gel electrophoresis, have not been successful.

The characteristics of the eIF-2-ancillary factors are described below. The factors have been described in order according to their roles in reaction sequences during peptide chain initiation.

2.2. <u>Co-eIF-2C</u>. Co-eIF-2C is a high molecular weight (mol. wt. 500,000 daltons) protein complex and is composed of many polypeptides.[11-21] In partial reactions, Co-eIF-2C stimulates ternary complex formation by eIF-2 in the presence of Mg^{2+} and such stimulation of eIF-2 activity is strongly inhibited by HRI and ATP.[11-18]

The characteristic effects of Co-eIF-2C on Met-tRNA$_f$ binding to eIF-2 are shown in Table 1. Homogeneous eIF-2 preparations bind efficiently to Met-tRNA$_f$ in the absence of Mg^{2+} but the presence of Mg^{2+} during the initial complex formation drastically reduces binding of eIF-2 to Met-tRNA$_f$. Addition of Co-eIF-2C in the absence of Mg^{2+} produced some stimulation of Met-tRNA$_f$ binding to eIF-2. The extent of this stimulation varies with different eIF-2 preparations. However, addition of Co-eIF-2C almost completely reversed Mg^{2+} inhibition of ternary complex formation by eIF-2. The presence of Mg^{2+} also similarly reduced Met-tRNA$_f$ binding to eIF-2α(P). As shown here, this Mg^{2+} inhibition of Met-tRNA$_f$ binding to eIF-2α(P) was not relieved by Co-eIF-2C. The eIF-2α(P) used in this experiment was prepared by phosphorylating eIF-2 using HRI and ATP, and was freed from HRI and ATP by phosphocellulose chromatography.

The above observations suggest that in the presence of Mg^{2+}, eIF-2 changes into an inactive form and requires Co-eIF-2C for restoration to the active

TABLE 1

EFFECTS OF ADDITION OF Co-eIF-2C ON Met-tRNA$_f$ BINDING TO eIF-2.

Addition	[^{35}S]Met-tRNA$_f$ Bound, pmol	
	-Mg^{2+}	+Mg^{2+}
eIF-2	1.70	0.60
eIF-2 + Co-eIF-2C	2.03	1.70
eIF-2α(P)	1.54	0.57
eIF-2α(P) + Co-eIF-2C	1.48	0.59

Standard Millipore filtration assay conditions for Met-tRNA$_f$ binding to eIF-2 were used. Where indicated, Co-eIF-2C and Mg^{2+} (1 mM) were added. Data obtained from Das et al.[15]

conformation necessary for Met-tRNA$_f$ binding. Co-eIF-2C does not recognize eIF-2α(P) and thus fails to permit Met-tRNA$_f$ binding to eIF-2α(P) in the presence of Mg^{2+}. This lack of recognition of eIF-2α(P) by Co-eIF-2C could possibly be the cause of protein synthesis inhibition by HRI in reticulocyte lysates.

As noted above, significant but variable stimulation of Met-tRNA$_f$ binding to eIF-2 by Co-eIF-2C preparation was also observed in the absence of Mg^{2+}. This stimulation was more apparent when aged eIF-2 preparations were used. The results of our preliminary experiments suggest that a part of this stimulation of Met-tRNA$_f$ binding to eIF-2 may possibly be due to the presence of Co-eIF-2A activity in Co-eIF-2C preparations. On the other hand, Co-eIF-2C activity itself may be responsible for the variable stimulation of Met-tRNA$_f$ binding observed with different eIF-2 preparations. A possible explanation for this observation is as follows: eIF-2 can exist in either active or inactive form and the conversion of the active into inactive form can be promoted by several agents including Mg^{2+} and aging. Co-eIF-2C activity is responsible for conversion of eIF-2 into an active conformation. In the absence of Mg^{2+}, the Co-eIF-2C stimulation of Met-tRNA$_f$ binding to eIF-2 may possibly be due to restoration of the aged and inactive eIF-2 molecules to the active conformation.

Recently Clemens et al.[20] and Siekierka et al.[21] have proposed a mechanism for Mg^{2+} inhibition of eIF-2 activity and Co-eIF-2C reversal of Mg^{2+} inhibition. According to this mechanism, eIF-2 is purified as eIF-2·GDP. In the absence of Mg^{2+}, GDP is easily exchangeable with exogenously added GDP and also can be easily displaced by GTP to form the eIF-2·GTP needed for ternary complex formation. However, in the presence of Mg^{2+}, GDP remains tightly bound to eIF-2 and prevents eIF-2 binding to GTP and Met-tRNA$_f$. The function of Co-eIF-2C is to permit replacement of GDP by GTP. A likely explanation consistent with our proposed mechanism is that Co-eIF-2C restores the active conformation of eIF-2 and in this conformation eIF-2 bound GDP can be replaced by GTP.

As mentioned in Section 2.1, Co-eIF-2C activity is purified as a high molecular protein complex composed of many polypeptides. Such Co-eIF-2C preparations also contain Co-eIF-2A and Co-eIF-2B activities. Co-eIF-2C activity can be freed from Co-eIF-2B activity using a phosphocellulose chromatographic procedure and an elution buffer containing 2 M urea. The final Co-eIF-2C preparation showed a single protein band under nondenaturing conditions. In the presence of SDS, the gel picture showed several prominent polypeptide bands (approximate mol. wt.: 100,000, 67,000; 53,000; 45,000 and 38,000) and numerous faint bands.

The requirement of Co-eIF-2C in overall protein synthesis has not yet been unequivocally demonstrated. The specific polypeptide component(s) responsible for Co-eIF-2C activity has not been identified and antibodies against the active polypeptide component(s) have not been prepared. There is some indirect evidence that Co-eIF-2C activity is involved in the overall protein synthesis activity of the system. (i) Co-eIF-2C activity stimulates Met-tRNA$_f$·40S·AUG initiation complex formation[11] (see Section 2.5). (ii) Co-eIF-2C activity is inhibited by protein synthesis inhibitors such as HRI and dsI,[11-17] and (iii) active RF preparations contain Co-eIF-2C activity which is not inhibited by HRI plus ATP.[19,22]

2.3. Co-eIF-2A. Co-eIF-2A is a low molecular weight, heat-stable protein factor.[6,7] Co-eIF-2A activity from rabbit reticulocytes has been purified to homogeneity. The molecular weight of Co-eIF-2A in homogeneous preparations is approximately 25,000 daltons.[7] Co-eIF-2A-like activity has been reported to be present in widely divergent eukaryotic cells such as mouse ascites tumor cells,[24] wheat germ[25-27] and Artemia salina.[28-30] Two different forms of Co-eIF-2A-like activities have been reported to be present in wheat germ[26,27] and in Artemia salina.[29,30]

Co-eIF-2A stimulates Met-tRNA$_f$ binding to eIF-2 by 2-3-fold and also stabilizes the complex. The results presented in Table 2 show the characteristic effects of addition of Co-eIF-2A on Met-tRNA$_f$ binding to eIF-2.

As shown in Exp. 1, Co-eIF-2A stimulated Met-tRNA$_f$ binding to eIF-2 and also to eIF-2α(P) by 2-3-fold. Co-eIF-2A can recognize both eIF-2 and eIF-2α(P) with similar efficiency. The results presented in Exp. 2, Table 2, indicate that Co-eIF-2A also stabilizes the ternary complex. Addition of physiologically im-

344

TABLE 2

EFFECTS OF ADDITION OF Co-eIF-2A ON Met-tRNA$_f$ BINDING TO eIF-2.

Additions	[^{35}S]Met-tRNA$_f$ Bound, pmol	
	-Co-eIF-2A	+Co-eIF-2A
Experiment 1.		
eIF-2	0.50	1.25
eIF-2α(P)	0.45	1.20
Experiment 2.		
eIF-2	0.50	1.2
eIF-2 + Globin mRNA (1.5 μg)	0.10	1.1
eIF-2 + Hemin (50 μM)	0.06	0.9
eIF-2 + Aurintricarboxylic acid (30 μM)	0.10	1.0

Standard Millipore filtration assay conditions for Met-tRNA$_f$ binding to eIF-2 were used. Data obtained from Das et al.[15] and Roy et al.[31]

portant compounds such as mRNA's and hemin, and also aurintricarboxylic acid, drastically reduced Met-tRNA$_f$ binding to eIF-2. Such inhibition of ternary complex (Met-tRNA$_f$·eIF-2·GTP) formation by eukaryotic mRNA's, has been reported previously by other laboratories.[32,33] As shown in Exp. 2, Table 2, the ternary complexes formed in the presence of Co-eIF-2A were almost fully resistant to globin mRNA, hemin and aurintricarboxylic acid. The above observations suggest that the ternary complex does not exist in the free form because it is too easily degraded in the presence of some normal cell constituents such as mRNA's and hemin. The binding of Co-eIF-2A to ternary complex may stabilize it under physiological conditions.

The requirement of Co-eIF-2A for protein synthesis in reticulocyte lysates was demonstrated using antibodies prepared against homogeneous Co-eIF-2A preparations.[34] Addition of anti-Co-eIF-2A drastically inhibited protein synthesis in reticulocyte lysates and such inhibition was significantly relieved by preincubation of anti-Co-eIF-2A specifically with homogeneous Co-eIF-2A. Importantly, addition of homogeneous eIF-2 even at high concentrations did not relieve protein synthesis inhibition by anti-Co-eIF-2A. The specific requirement for Co-eIF-2A for restoration of protein synthesis activity in anti-Co-eIF-2A- treated lysates strongly suggests that Co-eIF-2A is an integral part of the protein synthesis machinery.

Ghosh-Dastidar et al. studied the molecular mechanism of interaction of Co-eIF-2A with eIF-2 and other protein factors, such as Co-eIF-2B and Co-eIF-2C, using a fluorescence polarization technique.[35] For these studies, Co-eIF-2A was fluorescently labelled by dansylation and the changes in fluorescence polarization were measured in the presence of different protein factors, Met-tRNA$_f$, GTP and Mg^{2+}. These studies showed that Co-eIF-2A interacts specifically with the ternary complex and does not interact with free eIF-2, Co-eIF-2C, Met-tRNA$_f$ or GTP, or with eIF-2 plus either Met-tRNA$_f$ or GTP; the fluorescence polarization value of dansyl Co-eIF-2A increased significantly only when full complements of the ternary complex were present.

Based on the above observation, it has been proposed that the first step in peptide chain initiation is the formation of a ternary complex Met-tRNA$_f$·eIF-2·GTP. This ternary complex formation in the presence of Mg^{2+} requires Co-eIF-2C. The next step involves interaction of Co-eIF-2A with the ternary complex and formation of a stable quaternary complex, Met-tRNA$_f$·eIF-2·GTP Co-eIF-2A.

2.4. Co-eIF-2B. As mentioned earlier, the high molecular weight protein complex which contains Co-eIF-2C activity also, in most cases, contains another associated eIF-2-ancillary factor activity, namely Co-eIF-2B. In partial reactions, Co-eIF-2B actively promotes dissociation of the preformed ternary complex in the presence of high Mg^{2+} (5 mM) and low temperature (0°C). The characteristic effects of Co-eIF-2B on eIF-2 activity are summarized in Table 3.

TABLE 3

EFFECTS OF ADDITION OF Co-eIF-2B ON Met-tRNA$_f$ BINDING TO eIF-2.

Additions	[^{35}S]Met-tRNA$_f$ Bound, pmol	
	Stage I −Mg^{2+}	Stage II +5 mM Mg^{2+}
eIF-2	1.35	1.60
eIF-2 + Co-eIF-2B	1.70	0.52
eIF-2α(P)	1.66	1.50
eIF-2α(P) + Co-eIF-2B	1.55	1.45

Standard two-stage method for assay of Co-eIF-2B activity was used. Data obtained from Das et al.[15]

The reaction was carried out in two stages. In Stage I, the ternary complex was formed with homogeneous eIF-2 in the presence or absence of Co-eIF-2B. The reaction mixture was then mixed with Mg^{2+} (5 mM, final concentration) and incubated at ice bath temperature for 5 mins (Stage II). The ternary complex retained after Stage II incubation was assayed by standard Millipore filtration. As shown here (Table 3), the ternary complex was formed with homogeneous eIF-2 in the absence of Mg^{2+} and the preformed complex was completely resistant to Mg^{2+}, added in Stage II. The Co-eIF-2B preparation used in this experiment contained very little eIF-2 activity and did not significantly alter Met-tRNA$_f$ binding to eIF-2 in the absence of Mg^{2+}. However, when both eIF-2 and Co-eIF-2B were present in Stage I incubation, the ternary complex was extensively disso-ciated upon addition of Mg^{2+} in Stage II. A ternary complex was also formed with eIF-2α(P) and the complex was similarly resistant to further additions of Mg^{2+}. However, the complex formed using eIF-2α(P) in the presence of Co-eIF-2B was not dissociated during the second stage incubation with Mg^{2+}. These results show that Co-eIF-2B does not recognize eIF-2α(P).

The precise role of Co-eIF-2B activity in protein synthesis, and also the significance of the ternary complex dissociation by Co-eIF-2B in the overall peptide chain initiation process, is not clear. Presumably under physiological conditions of protein synthesis, and at physiological Mg^{2+} concentrations, Co-eIF-2B does not cause enough dissociation of the ternary complex to signifi-cantly impede peptide chain initiation; addition of Co-eIF-2B does not inhibit peptide chain initiation or overall protein synthesis in reticulocyte lysates. As noted above, the Co-eIF-2B activity is inhibited by HRI and ATP. This observation suggests that Co-eIF-2B activity is possibly involved in peptide chain initiation.

2.5. <u>Requirements for Met-tRNA$_f$·40S Initiation Complex Formation.</u> Previous reports indicated that the initiator tRNA binding to 40S ribosomes in crude reticulocyte lysates[36] and also in the fractionated system[37] precedes mRNA binding. Staehelin and coworkers reported that eIF-2 alone promotes Met-tRNA$_f$ binding to 40S ribosomes in the absence of mRNA.[37] Several years ago, we reported that Met-tRNA$_f$ binding to 40S ribosomes required, in addition to eIF-2, another high molecular weight protein complex which presumably contained Co-eIF-2B and Co-eIF-2C activities and that such binding was almost entirely dependent on added AUG-codon.[9,38] The requirement of AUG codon for formation of a stable Met-tRNA$_f$·40S initiation complex has now been reported from several laboratories.[25,29,30,39,40]

As noted in Section 2.1, we have now developed a purification procedure for Co-eIF-2C which results in complete removal of Co-eIF-2B activity from the final Co-eIF-2C preparation. Such Co-eIF-2C preparation, however, contains signifi-cant amounts of Co-eIF-2A activity. We have now used this more purified Co-eIF-2C preparation to investigate the precise factor requirements for Met-tRNA$_f$·40S·AUG complex formation. As shown in Fig. 2, the purified Co-eIF-2C preparation strongly stimulated Met-tRNA$_f$ binding to 40S ribosomes in the presence of eIF-2. Such Met-tRNA$_f$ binding to 40S ribosomes was entirely dependent on added AUG codon (data not shown). Also, Co-eIF-2C stimulated Met-tRNA$_f$ binding to 40S ribosomes was almost completely inhibited by HRI plus ATP, mimicking a physiological response observed in reticulocyte lysates.[41]

The above experiments thus establish the minimum factor requirements, namely eIF-2 and Co-eIF-2C, for formation of a stable Met-tRNA$_f$·40S·AUG initiation complex. We also examined the effects of the addition of Co-eIF-2A and Co-eIF-2B on eIF-2 and Co-eIF-2C promoted Met-tRNA$_f$·40S·AUG complex formation. As noted above, purified Co-eIF-2C preparation also contains significant amounts of Co-eIF-2A activity. In our experiments, addition of a partially

Fig. 2: Requirements and characteristics of Met-tRNA$_f$·40S·AUG complex formation. Standard sucrose density gradient centrifugation assay methods were used. Where indicated, eIF-2, Co-eIF-2C and HRI were added. Data obtained from Das et al.[11]

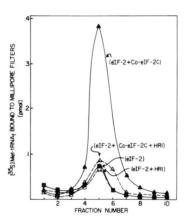

purified Co-eIF-2A preparation did not cause any significant increase in eIF-2 and Co-eIF-2C promoted Met-tRNA$_f$·40S·AUG complex formation. We have not yet been able to obtain Co-eIF-2B preparation completely free from Co-eIF-2C activity. However, we have found that addition of a factor preparation enriched in Co-eIF-2B activity consistently gives 30-50% stimulation of Met-tRNA$_f$·40S·AUG complex formation in the presence of saturating amounts of Co-eIF-2C. This observation suggests that the Co-eIF-2B and Co-eIF-2C activities are distinct activities although they may not be produced by separate polypeptides.

Recently, Siekierka et al.[40] have reported that eIF-2 and a Co-eIF-2C factor preparation, which they call eIF-2·SP and which also contains Co-eIF-2B activity, are both required for Met-tRNA$_f$ binding to <u>Artemia salina</u> 40S ribosomes in the presence of Mg^{2+} and AUG codon. These authors concluded, however, that the Co-eIF-2C (eIF-2·SP) factor requirement is entirely due to its reversal of Mg^{2+} inhibition of ternary complex formation. Co-eIF-2C had no significant effect when the ternary complex was preformed in the absence of Mg^{2+} and was then used for Met-tRNA$_f$·40S·AUG complex formation. These results are apparently different from results of similar experiments done with reticulocyte 40S ribosomes reported previously from our laboratory.[8] We have observed that addition of a partially purified Co-eIF-2C preparation significantly enhances (2-3-fold) Met-tRNA$_f$ transfer to 40S ribosomes from Met-tRNA$_f$·eIF-2·GTP complex preformed in the absence of Mg^{2+}. Our results suggest that other factor activities, possibly including Co-eIF-2A and Co-eIF-2B activities, are also required for Met-tRNA$_f$·40S·AUG initiation complex formation. The differences observed by Siekierka et al.[40] could possibly be due to the use of heterologous 40S ribosomes, namely <u>Artemia salina</u> ribosomes, in their experiments. As will be discussed in a later section, there is no clear evidence that the lower eukaryotic organisms contain Co-eIF-2B, Co-eIF-2C and RF-like activities. Also, unlike reticulocyte eIF-2, the eIF-2 activities from wheat germ and <u>Artemia salina</u> are not significantly inhibited by Mg^{2+}. It is conceivable that the <u>Artemia salina</u> 40S ribosomes as used by Siekierka et al.[40] do not recognize reticulocyte Co-eIF-2C activity.

The factor requirements for Met-tRNA$_f$·40S initiation complex formation with physiological mRNA's are less clear. The eIF-2 and Co-eIF-2C preparations which efficiently promote Met-tRNA$_f$·40S·AUG initiation complex formation were completely inactive in promoting Met-tRNA$_f$·40S initiation complex formation in the presence of a physiological mRNA such as globin mRNA. The results of our recent experiments suggest that Met-tRNA$_f$·40S initiation complex formation requires, in addition to eIF-2 and Co-eIF-2C, a relatively crude Co-eIF-2A preparation (after heat step, see Fig. 1).[5,42] The precise nature of the active component in this factor preparation is not clear at present. The requirement for Co-eIF-2A for Met-tRNA$_f$·40S initiation complex formation with physiological mRNA is consistent with our previous observation that Co-eIF-2A is required for formation of a stable Met-tRNA$_f$·eIF-2·GTP complex in the presence of physiological mRNA's.

2.6. <u>A Proposed Mechanism for Met-tRNA$_f$·40S Complex Formation</u>. Based on the results of our experiments and others, we propose a provisional model for Met-tRNA$_f$·40S initiation complex formation in reticulocyte lysates (Fig. 3). According to this mechanism, the first step in peptide chain initiation is the formation of the ternary complex, Met-tRNA$_f$·eIF-2·GTP. This ternary complex formation in the the presence of Mg^{2+} requires Co-eIF-2C. Co-eIF-2A then binds to the ternary complex and forms a stable quaternary complex Met-tRNA$_f$·eIF-2·GTP·Co-eIF-2A. This complex then transfers Met-tRNA$_f$ to 40S ribosomes and may require one or more additional factors for such transfer in the presence of physiological mRNA's. The role of Co-eIF-2B in peptide chain initiation is not apparent. This factor activity might be involved in release and recycling of eIF-2 and Co-eIF-2A after Met-tRNA$_f$·eIF-2·GTP·Co-eIF-2A is bound to 40S ribosomes.

$$\text{Met-tRNA}_f + \text{eIF-2} + \text{GTP}$$
$$\xrightarrow[\text{Mg}^{2+}]{\text{Co-eIF-2C}} \text{Met-tRNA}_f\text{·eIF-2·GTP} \qquad (i)$$

$$\text{Met-tRNA}_f\text{·eIF-2·GTP} + \text{Co-eIF-2A}$$
$$\longrightarrow \text{Met-tRNA}_f\text{·eIF-2·GTP·Co-eIF-2A} \qquad (ii)$$

$$[\text{Met-tRNA}_f\text{·eIF-2·GTP·Co-eIF-2A} + 40\text{S} + \text{mRNA}$$
$$\xrightarrow[\text{Other Factor(s)}]{\text{Co-eIF-2B}} \text{Met-tRNA}_f\text{·40S·mRNA}] \qquad (iii)$$

Fig. 3: A provisional mechanism for Met-tRNA$_f$·40S·mRNA complex formation.

Protein synthesis inhibitors such as HRI and dsI phosphorylate eIF-2 and the eIF-2α(P) thus formed is not recognized by Co-eIF-2C (and Co-eIF-2B) and is thus inactive in peptide chain initiation.

It is possible that the above regulatory mechanism of protein synthesis initiation involving Co-eIF-2B, Co-eIF-2C and eIF-2 kinases may operate only in mammalian cells. Several laboratories have reported the presence of eIF-2-ancillary factor activities in lower eukaryotic organisms such as wheat germ[25-27] and Artemia salina.[28-30] However, the characteristics of the eIF-2-ancillary factors reported from these laboratories more closely resemble those described for reticulocyte Co-eIF-2A. Although these organisms contain eIF-2 activities, they are unlike reticulocyte eIF-2 in that the eIF-2 activities present in wheat germ,[27] and Artemia salina[29,30] are not significantly inhibited by Mg^{2+}. Also, there has been no indication that these cells contain

Co-eIF-2B, Co-eIF-2C and RF-like activities nor is protein synthesis in these cells inhibited by eIF-2 kinases.

It may be that the complex regulation involving eIF-2 kinases and eIF-2-ancillary protein factors, Co-eIF-2B, Co-eIF-2C and also RF, is the consequence of evolution and is unique for mammalian cells.

3. RF

3.1. <u>Characteristics of RF</u>. Gross[43,44] and Ranu and London[45] reported the presence of a protein factor, RF, in reticulocyte cell supernatant which reverses protein synthesis inhibition in heme-deficient reticulocyte lysates. This RF activity has been extensively purifed by Amsez et al.,[46] Ralston et al.,[22] Siekierka et al.[19] and Safer and Jagus[2]. Amsez et al. and Ralston et al.,[26] have independently reported that purified RF preparation does not dephosphorylate eIF-2, whose phosphorylation has been catalyzed by HRI, nor prevent phosphorylation of eIF-2 by HRI. Ralston et al.[22] provided evidence that RF acts at the level of ternary complex formation and can reverse HRI inhibition of ternary complex formation. In partial reactions, RF stimulated ternary complex formation with eIF-2α(P) in the presence of 1 mM Mg^{2+}.

We have now further purified RF activity (Fig. 1).[23] Purified RF preparation contains significant amounts of eIF-2, Co-eIF-2A, Co-eIF-2B and Co-eIF-2C activities, but the Co-eIF-2B and Co-eIF-2C activities in the RF preparation are completely resistant to HRI and ATP.

The precise mechanism of the altered recognition of Co-eIF-2B and Co-eIF-2C activities in the RF preparation is not clear. We investigated the possibility that the active RF preparation contains additional factor(s) which renders Co-eIF-2B and Co-eIF-2C activities insensitive to HRI and ATP. For these studies, we fractionated the phosphocellulose purified RF preparation (Fraction V (Fig. 1)) further and examined the characteristics of the component factor activities from fractionated RF preparations. Upon further fractionation using a CM-Sephadex chromatographic procedure, the RF preparation (Fraction VI) was almost completely freed from eIF-2 activity. Such RF preparations, however, still contained Co-eIF-2A, Co-eIF-2B, and Co-eIF-2C activities. As shown in Fig. 4A and B, both Fraction V (phosphocellulose purified, RF(PC)) and Fraction VI (CM-Sephadex purified, RF(CM)) RF preparations efficiently stimulated protein synthesis in heme-deficient reticulocyte lysates. At higher concentrations, both factor preparations increased the initial rate of protein synthesis over that observed in the presence of hemin, and in both cases protein synthesis continued over a long period. However, we have observed that the Fraction VI RF preparation, which is devoid of eIF-2 activity, is significantly less stable than the Fraction V RF preparation, and rapidly loses RF activity upon storage. This loss in RF activity is more pronounced at high dilution. Conceivably, the eIF-2 stabilizes the RF protein complex. Upon further fractionation of Fraction VI RF preparation using density gradient centrifugation, the only remaining RF activity was found in a single symmetrical peak (∿ 21 S) coincident with Co-eIF-2C activity.[5] However, such "RF" preparation was very inefficient in stimulating protein synthesis in heme-deficient reticulocyte lysates, and the protein synthesis rate declined rapidly after a short period (Fig. 4C). The characteristics of the protein synthesis stimulation observed with this glycerol density gradient fractionated RF preparation resembled those observed with a purified Co-eIF-2C preparation from ribosomal high salt wash (Fig. 4D).

Fig. 4: Protein synthesis inhibition reversal activities of RF fractions (Fraction V (RF(PC)), Fraction VI (RF(CM)), Fraction VII (glycerol gradient RF)) and Co-eIF-2C. Protein synthesis in reticulocyte lysates was assayed using the standard procedure described previously.[22] Protein concentrations (μg) used in different experiments were: **Fig. A:** RF(PC) 3 (▲), 6 (■), 9 (●); **Fig. B:** RF(CM) 3 (▲), 6 (■), 9 (●); **Fig. C:** (Glycerol gradient RF), 1 (▲), 2 (■), 3 (○), 4 (●); **Fig. D:** Co-eIF-2C 8 (▲), 16 (■), 24 (●). Protein synthesis in the presence and absence of optimal concentrations of hemin (20 μM) is shown (dotted lines) for comparison. Data obtained from Grace et al.[23]

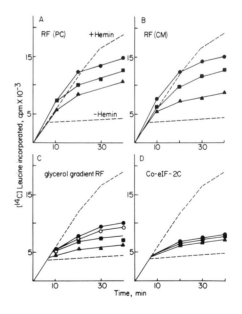

We analyzed the characteristics of Co-eIF-2C activities in Fraction V to VII RF preparations in the presence and absence of HRI and ATP (Fig. 5). As mentioned earlier, the phosphocellulose purified RF preparation (Fraction V) contained eIF-2 and Co-eIF-2C activities, and efficiently formed ternary complexes in the presence of Mg^{2+}. Such complex formation was completely insensitive to HRI and ATP at different concentrations of the fraction tested (Fig. 5A). Upon further fractionation, using a CM-Sephadex chromatographic procedure, the RF preparation (Fraction VI) can be freed from eIF-2 activity. This preparation, however, retained the Co-eIF-2C activity. As shown in Fig. 5B, the Co-eIF-2C activity in this RF preparation was partially inhibited by HRI and ATP at low RF concentration, and progressively regained HRI-insensitivity with increasing RF concentration. After glycerol gradient centrifugation, the RF preparation (Fraction VII) regained almost complete sensitivity to HRI and ATP, and HRI inhibition of ternary complex formation could not be reversed by increasing

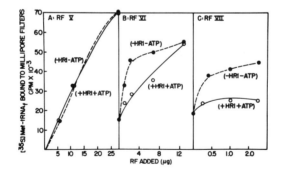

Fig. 5: HRI Sensitivity of Co-eIF-2C activities in RF fractions. Standard Millipore filtration assay conditions in the presence of 1 mM Mg^{2+} were used. Different concentrations of the RF fractions were used as indicated. Fractions VI and VII RF preparations were assayed in the presence of 1 μg exogenous eIF-2. Data obtained from Grace et al.[23]

concentrations of RF (Fig. 5C). These observations strongly suggest that a factor(s) present in active RF preparations (Fraction V) renders Co-eIF-2C activity insensitive to HRI and ATP, and this factor is progressively removed by further fractionation using CM-Sephadex chromatography and glycerol density gradient centrifugation.

To investigate the possible presence of some polypeptide component(s) in active RF preparations which renders the Co-eIF-2C activity insensitive to HRI and ATP, we examined the polypeptide components in Fractions V and VI RF preparations. For this experiment, we incubated the RF preparations, Fractions V and VI, and for comparison a partially purified eIF-2 preparation, in the presence of HRI and $[\gamma\text{-}^{32}P]$ATP, and analyzed the polypeptide components in each fraction using SDS-polyacrylamide gel electrophoresis and autoradiography (Fig. 6).

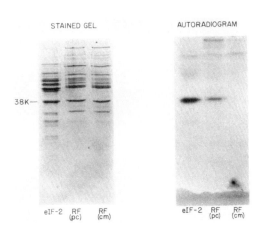

Fig. 6: Phosphorylation of eIF-2 and the RF Fractions V and VI. Partially purified eIF-2 and RF preparations, Fractions V and VI were phosphorylated in the presence of HRI and $[\gamma\text{-}^{32}P]$ATP and were then analyzed by NaDodSO$_4$ polyacrylamide gel electrophoresis and autoradiography. Data obtained from Grace et al.[23]

Both RF preparations (Fractions V and VI) showed numerous polypeptide bands, including several prominent polypeptide bands of approximate molecular weights 100K, 67K, 53K, 43K, 38K and 25K. As reported previously, several of these polypeptide bands are also present in a highly purified Co-eIF-2C preparation from ribosomal salt wash. As shown in Fig. 6, both Fractions V and VI RF preparations contained significant amounts of 38K polypeptide component(s) which moved similarly to the 38K polypeptide component of eIF-2. However, whereas the 38K polypeptide component in eIF-2 and in Fraction V RF preparation was extensively phosphorylated by HRI and $[\gamma-^{32}P]$ATP, the 38K polypeptide component in Fraction VI RF preparation remained almost completely unphosphorylated. As noted previously, Fraction V RF preparation contained significant amounts of eIF-2 activity (specific activity (pmol Met-tRNA$_f$ bound per mg protein); 400) while the Fraction VI RF preparation was almost completely devoid of eIF-2 activity (specific activity, 27).

The 38K polypeptide components in eIF-2 and Fractions V and VI RF preparations were further analyzed by two-dimensional gel electrophoresis (Fig. 7). The 38K polypeptide in eIF-2 moved as a single component with a mobility corresponding to a pK value of approximately 5 (Fig. 7A), and this polypeptide was extensively phosphorylated by HRI and $[\gamma-^{32}P]$ATP (Fig. 7D). Fractions V and VI RF preparations contained two 38K polypeptide components, and one 38K polypeptide component had a pI similar to that of the 38K polypeptide of eIF-2. The intensities of these 38K spots in both Fractions V and VI RF preparations appeared comparable. However, whereas this 38K polypeptide component in Fraction V preparation (Fig. 7E) was extensively ^{32}P labelled, the corresponding 38K polypeptide component in Fraction VI RF preparation (Fig. 7F) remained almost completely unlabelled.

The above results indicate that the Fraction VI RF preparation, which has negligible eIF-2 activity, contains the 38K polypeptide (α-subunit) of eIF-2 and this 38K polypeptide is not phosphorylated by HRI and ATP. The Fraction V RF preparation appears to contain two closely spaced 38K polypeptide components (Fig. 7B), possibly corresponding to the unphosphorylated α-subunit of eIF-2 and phosphorylated 38K subunit from contaminating eIF-2. It is conceivable that the nonphosphorylated 38K polypeptide component represents 38K-subunit as in the Fraction VI preparation, and this form of the polypeptide is not phosphorylated by HRI and ATP.

3.2. <u>Mechanism of RF Action</u>. The RF preparation which actively reverses protein synthesis inhibition in heme-deficient reticulocyte lysates also contains Co-eIF-2B and Co-eIF-2C activities, and the Co-eIF-2B and Co-eIF-2C ac-

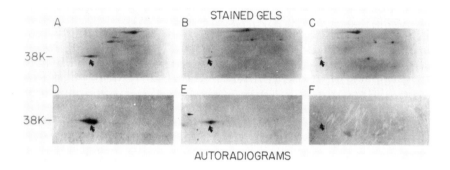

Fig. 7: Two-dimensional gel electrophoresis and autoradiography of partially purified eIF-2 and partially purified RF fractions (Fractions V and VI). Partially purified eIF-2 and the RF preparations, Fractions V and VI, were phosphorylated using HRI and $[\gamma-^{32}P]$ATP and were then analyzed by two dimensional gel electrophoresis. Segments of the stained gels and the corresponding autoradiograms around the 38K polypeptide regions are shown here. A and D: eIF-2; B and E: RF Fraction V; C and F: RF Fraction VI. Data obtained from Grace et al.[23]

tivities in RF preparation are completely resistant to HRI and ATP. However, the mechanisms of RF action and the altered recognition of Co-eIF-2C in RF preparation are not clear. Upon further fractionation of Fraction V RF preparation, the Co-eIF-2C activity in the RF preparation regains sensitivity to HRI and ATP. This observation suggests that some factor(s) present in the Fraction V RF preparation renders the Co-eIF-2C activity resistant to HRI and ATP, and this factor(s) is progressively removed during purification. The nature of this putative factor(s) is not clear at present. No clear difference in polypeptide composition can be seen in comparing Fractions V and VI RF preparations. Another observation discussed in this paper is that an RF preparation (Fraction VI) which is almost completely devoid of eIF-2 activity and which can reverse protein synthesis inhibition in heme-deficient reticulocyte lysates is enriched in a 38K polypeptide component indistinguishable from the α-subunit of eIF-2 on two-dimensional gel electrophoresis. Also, this 38K polypeptide component in

Fraction VI RF preparation cannot be phosphorylated by HRI and ATP. This observation indicates that this active RF preparation contains the α-subunit of eIF-2, and the α-subunit, in this form, is not phosphorylated by HRI and ATP.

The significance of the presence of the α-subunit of eIF-2 in the RF preparation may be related to the basic mechanism of RF action. A possible explanation for RF action is that RF promotes replacement of the phosphorylated α-subunit in eIF-2 with the nonphosphorylated α-subunit from the RF, and the nonphosphorylated eIF-2 thus reformed is recognized by Co-eIF-2B and Co-eIF-2C and is active in protein synthesis initiation. It is conceivable that this replacement reaction is promoted by a postulated "exchange factor" present in active RF preparations (Fraction V), and this factor is progressively lost during further fractionation of RF using CM-Sephadex chromatography and glycerol density gradient centrifugation with resultant loss of HRI resistance of Co-eIF-2C activity. The proposed model is depicted diagrammatically as follows:

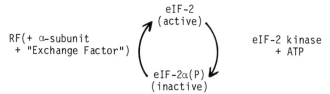

According to the proposed model, eIF-2α(P), formed by phosphorylation of eIF-2 by HRI and ATP, is not recognized by Co-eIF-2B and Co-eIF-2C, and is inactive in protein synthesis initiation. RF reverses protein synthesis inhibition in heme-deficient reticulocyte lysates by replacing the phosphorylated α-subunit in eIF-2 with the nonphosphorylated α-subunit present in RF, and the reformed eIF-2 is recognized by Co-eIF-2B and Co-eIF-2C, and is active in protein synthesis initiation.

Finally, it should be noted that the present observation of the presence of a pool of α-subunit of eIF-2 in RF preparations protected from phosphorylation by HRI and ATP, may provide a rationale for recent reports from several laboratories regarding the lack of correlation of the extent of phosphorylation of eIF-2 and the degree of protein synthesis inhibition observed in the heme-deficient reticulocyte lysates.[2,47,48] These reports indicate that only 25 to 30% of eIF-2 becomes phosphorylated under conditions of almost total protein synthesis inhibition in heme-deficient reticulocyte lysates. The presence of a pool of α-subunit of eIF-2 in reticulocyte lysates, which is not phosphorylated by HRI and ATP, could explain this lack of correlation between the extent of phosphorylation of α-subunits and the degree of protein synthesis inhibiton in heme-deficient reticulocyte lysates. Our present results suggest that only eIF-2 bound α-subunit is phosphorylated by HRI and ATP.

ACKNOWLEDGMENTS

This research work was supported by the United Public Health Service Research Grants GM 22079 and 18796, by a University of Nebraska-Lincoln Research Council Grant, and NIH Biomedical Research Grant RR 07055. The authors are grateful to Dr. George A. Vidaver for critical reading of the manuscript.

REFERENCES
1. Austin, S.A., and Clemens, M.J. (1980) FEBS Letters, 110, 1-6.
2. Safer, B., and Jagus, R. (1981) Biochimie, 63, 709-717.
3. Gupta, N.K. (1982) in Protein Biosynthesis in Eukaryotes, Bercoff, R. ed., Plenum Publishing Co., New York, pp 419-440.
4. Gupta, N.K. (1982) in Current Topics in Cellular Regulation, Horecker, B.L., and Stadtman, E.R., ed., Academic Press, New York, Vol. 21, pp 1-33.
5. Gupta, N.K., Bagchi, M., Das, A., Ghosh-Dastidar, P., Grace, M., Nasrin, N., Ralston, R., and Roy, R. in The Regulation of Hemoglobin Biosynthesis, Goldwasser, E., ed., Elsevier Press, Amsterdam (in press).
6. Dasgupta, A., Majumdar, A., George, A.D., and Gupta, N.K. (1976) Biochem. Biophys. Res. Commun., 71, 1234-1241.
7. Dasgupta, A., Das, A., Roy, R., Ralston, R., Majumdar, A., and Gupta, N.K. (1978) J. Biol. Chem., 253, 6054-6059.
8. Das, A., Bagchi, M., Ghosh-Dastidar, P., and Gupta, N.K. (1982) J. Biol. Chem., 257, 1282-1288.
9. Majumdar, A., Roy, R., Das, A., Dasgupta, A., and Gupta, N.K. (1977) Biochem. Biophys. Res. Commun., 78, 161-169.
10. Majumdar, A., Dasgupta, A., Chatterjee, B., Das, H.K., and Gupta, N.K, (1979) in Methods in Enzymology, Moldave, K., and Grossman, L., ed., Academic Press, New York, Vol. 60, pp. 35-52.
11. Das, A., Bagchi, M., Roy, R., Ghosh-Dastidar, P., and Gupta, N.K. (1982) Biochem. Biophys. Res. Commun., 104, 89-98.
12. deHaro, C., Datta, A., and Ochoa, S. (1978) Proc. Natl. Acad. Sci., U.S.A., 75, 243-247.
13. deHaro, C., and Ochoa, S. (1978) Proc. Natl. Acad. Sci., U.S.A., 75, 2713-2716.
14. Ranu, R.S., and London, I.M. (1979) Proc. Natl. Acad. Sci., U.S.A., 76, 1079-1083.
15. Das, A., Ralston, R.O., Grace, M., Roy, R., Ghosh-Dastidar, P., Das, H.K., Yaghmai, B., Palmieri, S., and Gupta, N.K. (1979) Proc. Natl. Acad. Sci., U.S.A., 76, 5076-5079.
16. Grosfeld, H., and Ochoa, S. (1980) Proc. Natl. Acad. Sci., U.S.A., 77, 6526-6530.
17. Ranu, R.S. (1980) Biochem. Biophys. Res. Commun., 97, 252-262.
18. Das, H.K., Das, A., Ghosh-Dastidar, P., Ralston, R.O., Yaghmai, B., Roy, R., and Gupta, N.K. (1981) J. Biol. Chem., 256, 6319-6323.
19. Siekierka, J., Mitsui, K., and Ochoa, S. (1981) Proc. Natl. Acad. Sci., U.S.A., 78, 220-223.

358

20. Clemens, M.J., Pain, V.M., Wong, S., and Henshaw, E.C. (1982) Nature, 296, 93-95.
21. Siekierka, J., Mauser, L., and Ochoa, S. (1982) Proc. Natl. Acad. Sci., U.S.A., 79, 2537-2540.
22. Ralston, R., Das, A., Grace, M., Das, H.K., and Gupta, N.K. (1979) Natl. Acad. Sci., U.S.A., 76, 5490-5494.
23. Grace, M., Ralston, R., Banerjee, A., and Gupta, N.K. Submitted for publication.
24. Reynolds, S., Dasgupta, A., Palmieri, S., Majumdar, A., and Gupta, N.K. (1977) Arch. Biochem. Biophys. 184, 325-335.
25. Treadwell, B.V., Mauser, L., and Robsinson, W.G., (1979) in Methods in Enzymology, Moldave, K., and Grossman, L., ed., Academic Press, New York, Vol. 60, pp. 181-193.
26. Osterhout, J.J., Phillips-Minton, J., and Ravel, J.M., (1979) Fed. Proc., 38, 327.
27. Lax, S.R., Osterhout, J.J., and Ravel, J.M., J. Biol. Chem. (in press).
28. Malathi, V.G., and Mazumdar, R., (1978) FEBS Letters, 86, 155-159.
29. MacRae, T., Houston, K.J., Woodley, C.L., and Wahba, A.J. (1979) Eur. J. Biochem., 100, 67-76.
30. Woodley, C.L., Roychowdhury, M., MacRae, T.H., Olson, K.W., and Wahba, A.J. (1981) Eur. J. Biochem., 117, 543-551.
31. Roy, R., Ghosh-Dastidar, P., Das, A., Yaghmai, B., and Gupta, N.K. (1981) J. Biol. Chem., 256, 4719-4722.
32. Kaempfer, R., Rosen, H., and Israeli, R. (1978) Proc. Natl. Acad. sci., U.S.A., 75, 650-655.
33. Barrieux, A., and Rosenfeld, M.G. (1978) J. Biol. Chem., 253, 6311-6314.
34. Ghosh-Dastidar, P., Yaghmai, B., Das, A., Das, H.K., and Gupta, N.K. (1980) J. Biol. Chem., 255, 365-368.
35. Ghosh-Dastidar, P., Giblin, D., Yaghmai, B., Das, A., Das, H.K., Parkhurst, L.J., and Gupta, N.K. (1980) J. Biol. Chem., 255, 3826-3829.
36. Darnbrough, C.H., Legon, S., Hunt, T., and Jackson, R.J. (1973) J. Mol. Biol., 76, 379-403.
37. Schreier, M.H., Erni, B., Staehelin, T. (1977) J. Mol. Biol., 116, 727-753.
38. Chaterjee, B., Dasgupta, A., Palmieri, S., and Gupta, N.K. (1976) J. Biol. Chem., 251, 6379-6387.
39. Stringer, E.A., Chaudhuri, A., Valenzuela, D., and Maitra, U.(1980) Proc. Natl. Acad. Sci., U.S.A., 77, 3356-3359.
40. Siekierka, J., Datta, A., and Ochoa, S. (1982) J. Biol. Chem., 257, 4162-4165.
41. Legon, S., Jackson, R.J., and Hunt, T. (1975) Nature, New Biol., 251, 150-152.
42. Bagchi, M., Das, A., Roy, R., Ghosh-Dastidar, P., Chakraborty, I., and Gupta, N.K. (1982) Fed. Proc., 41, 1040.
43. Gross, M., (1975) Biochem. Biophys. Res. Commun., 67, 1507-1515.
44. Gross, M., (1976) Biochim. Biophys. Acta, 447, 445-449.
45. Ranu, R.S., and London, I.M., (1977) Fed. Proc., 36, 868.
46. Amesz, H., Gouman, S.H., Haudrich-Morre, T., Voorma, H.O., and Benne, R., (1979) Eur. J. Biochem., 98, 513-520.
47. Farrell, P.J., Hunt, T., and Jackson, R.J. (1978) Eur. J. Biochem., 89, 517-521.
48. Leroux, A., and London, I.M. (1982) Proc. Natl. Acad. Sci., U.S.A., 79, 2147-2151.

INTERACTIONS OF CAP BINDING PROTEINS WITH EUKARYOTIC mRNAs

STANLEY M. TAHARA[+], MAUREEN A. MORGAN[+], JAMIE A. GRIFO[++], WILLIAM C. MERRICK[++]
AND AARON J. SHATKIN[+]
[+]Roche Institute of Molecular Biology, Nutley, New Jersey, USA;   [++]Department
of Biochemistry, Case Western Reserve University, Cleveland, Ohio, USA

INTRODUCTION--FUNCTIONAL IMPORTANCE OF CAPPING

A prominent structural feature of eukaryotic mRNAs is the 5'-terminal cap structure, m^7GpppN (Fig. 1). Addition of the cap occurs during or shortly after initiation of transcription. Consequently, caps are found on primary nuclear transcripts as well as mature cytoplasmic mRNAs. Other mRNA modifications include methylation of internal bases and 3'-terminal polyadenylation. Capping of mRNA apparently is universal among eukaryotes, and naturally uncapped cellular mRNAs have not been reported. In addition, most viruses that infect animals, plants and insects direct the synthesis of capped virus-specific mRNAs. The few known exceptions include animal picornaviruses (e.g. poliovirus) and Cowpea mosaic and Satellite tobacco necrosis viruses of plants (for reviews, see 1-3).

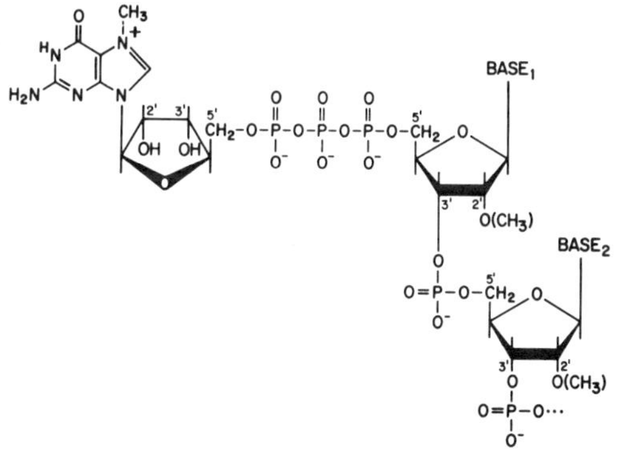

Fig. 1. Structure of mRNA 5'-terminal cap.

Transcription Initiation and Capping. Studies of whole cells and nuclear fractions indicate that cap formation probably occurs very close in time to the initiation of transcription.[4] In several viral transcription systems caps are present on short, nascent oligonucleotides corresponding to mRNA 5'-terminal sequences,[5-7] suggesting that capping may be intimately linked with the start of transcription in these systems. Tight coupling between the two processes is also implied from the positive allosteric effect of S-adenosylmethionine (SAM) and related compounds on transcription initiation by the RNA polymerase in purified insect cytoplasmic polyhedrosis virus (CPV).[7]

Influenza virus transcription involves another type of cap-related phenomenon.[8] Capped cellular RNAs are used as primers by the virion-associated polymerase for transcription of the viral genome. In an unusual mechanism of initiation, the cap and approximately 10-15 adjacent nucleotides are transferred from the unrelated primer RNA to the virus-specific transcriptase products. Capped mRNAs are considerably better primers for the influenza transcriptase than uncapped mRNAs, consistent with cap recognition by the virion enzyme complex. Recent findings demonstrate that one of the influenza structural P proteins has cap recognizing activity and is involved in initiation of primer-dependent influenza transcription. [9,10]

Caps and mRNA Stability. Capping increases mRNA stability. Capped reovirus mRNAs have a longer lifetime as compared to the corresponding uncapped viral transcripts after microinjection into Xenopus oocytes or incubation in cell extracts.[11] Similar results were obtained subsequently with CPV transcripts[12] and trout protamine mRNAs.[13] Translation of another cellular mRNA (rabbit globin) also was decreased ~10-fold by decapping prior to microinjection into oocytes.[14] Other studies with cell extracts and reovirus mRNAs that were radiolabeled at 5'-terminal as compared to internal sites indicate that caps stabilize mRNAs by protecting against 5'-exonucleolytic degradation (Y. Furuichi, unpub. results).

Facilitation of Translation by Caps. Involvement of the cap in translation is indicated by the inhibitory effect of m^7GMP and other cap analogs on binding of mRNA to ribosomes.[1,2,15] Removal of the cap by enzymatic or chemical procedures leads to a decrease in the ability of mRNA to form initiation complexes. The importance of the cap for initiation is further indicated by the finding that in some mRNAs it is part of the 40S ribosome binding site as measured by protection against RNase digestion.[16] Attachment of the ribosomal subunit at or near the mRNA 5'-end may involve interaction initially with the cap, followed by "scanning" of the ribosome along the mRNA until it encounters the correct (usually first) AUG codon for beginning peptide bond formation.[17]

Inhibition of initiation by cap analogs is attributed to a competition for some translational component which specifically recognizes the mRNA 5'-cap. In agreement with this idea, a "cap-binding protein" was detected in initiation factor preparations from rabbit reticulocytes by an affinity labeling procedure.[18] Reovirus mRNA containing 5'-terminal ^3H-methyl groups was oxidized with sodium periodate to form a 2', 3'-dialdehyde on the ribose moiety of the cap m^7G. The mRNA was reacted with ribosomal salt wash proteins to allow initiation factor(s) that interact with the cap to form Schiff bases between primary amino groups and the oxidized m^7G. After reduction with $NaBH_3CN$ to stabilize any Schiff bases and ribonuclease digestion to remove the RNA, a ^3H-labeled "cap-binding" protein of approximately 24,000 molecular weight was detected by sodium dodecyl sulfate (SDS)-polyacrylamide gel electrophoresis and fluorography. As a criterion of cap specificity, crosslinking was inhibited by m^7GDP and other cap analogs. The cap binding protein was purified from initiation factors by m^7GDP-Sepharose affinity chromatography[19] and shown subsequently to consist of a single polypeptide of apparent molecular weight ∿24,000 (CBP I).

Correlation of Cap Analog Effects on mRNA Translation and Cross-linking to Cap Binding Protein. A role of CBP I and the cap in the initiation of translation was clearly illustrated by comparing the effects on mRNA function of m^7GMP and a chemically modified cap analog, 7-methylguanosine 5'-phosphate methyl ester ($m^7GMP.me$). The latter compound was synthesized to test the putative importance of a "rigid" conformation of the cap resulting from an electrostatic interaction between the delocalized positive charge on the m^7G imidazole ring and the negatively charged phosphoryl moieties. Although a decrease in phosphate net charge by O-methylation would be expected to change this interaction, conformational analysis by solution NMR studies showed no significant differences in structure between m^7GMP and $m^7GMP.me$.[20] However, $m^7GMP.me$ was almost inactive as a cap analog and inhibited initiation complex formation only weakly as compared to m^7GMP.[20] Treatment of the methyl ester with snake venom phosphodiesterase removed the methyl moiety, and the resulting m^7GMP regained activity. Venom treatment followed by alkaline phosphatase digestion converted the $m^7GMP.me$ to the inactive m^7G nucleoside.

As a corollary to the ribosome binding experiments, the cap analogs were also compared for reversal of the stimulatory action of CBP I on Sindbis virus capped mRNA translation in HeLa cell extracts.[21] At a concentration of 0.25 mM, $m^7GMP.me$ had little or no effect while 0.1 mM m^7GMP decreased viral capsid synthesis by 90%.[20] To determine if the capacity to inhibit translation was correlated with an effect on cap binding protein interaction with mRNA, cap analogs were tested for inhibition of crosslinking. $m^7GMP.me$ diminished cross-linking by 11% at 0.25 mM while m^7GMP inhibited by ∿85%. These results taken together indicate that O-methylation of the phosphoryl moiety is sufficient to diminish cap analog activity of m^7GMP as measured by inhibition of mRNA binding to ribosomes, stimulation of capped mRNA translation in cell-free protein

362

synthesizing extracts and by crosslinking of cap-binding protein to the oxidized cap of mRNA. From comparisons of cap analog activity and NMR conformational analyses, it is apparent that in addition to 7-alkylation, the polyphosphate linkage of m^7G to the 5' end of mRNA is also important for functional interactions of the cap with translational components.

Cap Binding Protein and Host Shut-off in Poliovirus-infected HeLa Cells. The strong inhibition of cellular capped mRNA translation accompanying poliovirus infection of HeLa cells is not due to mRNA breakdown[22] and may be mediated by inactivation of cap binding protein. Loss of mRNA cap binding activity presumably would prevent initiation by cellular mRNAs and allow preferential utilization of the uncapped viral RNA. Early reports indicated that ability to restore capped mRNA translation in extracts of poliovirus-infected cells copurified with the 24,000 dalton cap binding protein.[23] However, this restoring activity was labile and identity to CBP I was based mainly on crosslinking to oxidized mRNA. Later studies demonstrated that CBP I could also be found in a larger complex with a molecular size in the range of 8-10S (CBP II).[24] The two forms of cap binding protein both actively stimulated capped mRNA translation in extracts of uninfected HeLa cells. However, in extracts prepared after poliovirus infection, capped mRNA translation was promoted by CBP II and not by CBP I. The activity of the CBP II complex was stable, unlike the restoring factor reported previously. The additional polypeptides in CBP II may directly confer stable restoring activity on CBP I. Alternatively, CBP II may provide other components required for expression of capped templates in poliovirus-infected cells. In either case, since CBP II rather than CBP I possesses stable restoring activity, we have carried out additional studies to try to define the functional forms of CBP.

MATERIALS AND METHODS

L-[^{35}S]methionine (specific activity = 1,000 Ci/mmol) and [^{3}H-methyl]S-adenosylmethionine (60 Ci/mmol) were from Amersham; cytidine 3',5'-bis[^{32}P]-phosphate (2900 Ci/mmol) was from New England Nuclear; ATP, inosine triphosphate (ITP), and dithiothreitol (DTT) were Calbiochem products; 5-bromouridine triphosphate (BrUTP) was from Sigma; and sucrose was Schwarz-Mann ultrapure grade. All other chemicals were of reagent grade or greater purity.

Cap binding proteins were isolated from rabbit reticulocyte lysates,[24] and eukaryotic initiation factors 4A and 4B were purified as described.[25,26]

Reovirus mRNAs were synthesized in vitro using viral cores. Inosine and 5-bromouridine substituted mRNAs were obtained by replacement of GTP and UTP, respectively, in transcription mixtures.[27,28] Periodate oxidation of [^{3}H]-

methyl-labeled reovirus mRNAs and crosslinking to proteins were performed as detailed previously,[18] except that 10 ug of bovine serum albumin were added as carrier prior to acetone precipitation of crosslinked complexes.

Protein synthesizing extracts were prepared from uninfected and poliovirus infected HeLa cells by the procedure of Rose et al.[29] Sindbis virus mRNA was isolated from virus-infected chicken embryo fibroblasts.[30] Translation products were analyzed by SDS-polyacrylamide gel electrophoresis in a modified Laemmli system followed by autoradiography.[24]

For mRNA filter binding studies, globin mRNA purified from rabbit reticulocyte ribosomes by oligo-dT cellulose chromatography and sucrose gradient centrifugation was 3'-end labeled by incubation with [^{32}P]pCp and T4-RNA ligase as described.[25] Labeled mRNA was sedimented in a sucrose gradient, and material sedimenting at 9S was pooled, made 0.3 M potassium acetate and precipitated by addition of 2-1/2 volumes of absolute ethanol. Analysis of the radiolabeled mRNA by agarose gel electrophoresis under denaturing conditions yielded a single band in the position of full-length globin mRNA. Binding of mRNA to factors was assayed in a total volume of 50 µl containing 20 mM Tris-HCl (pH 7.5), 90 mM KCl, 5 mM MgCl$_2$, 1 mM dithiothreitol (DTT), 0.15 mM phosphoenolpyruvate, 0.2 I.U. rabbit muscle pyruvate kinase and 20,000 cpm of [^{32}P]-labeled globin mRNA (0.15 pmol). Purified initiation factors and other components were added as indicated. Reactions were terminated by addition of ice-cold buffer containing 20 mM Tris-HCl (pH 7.5), 100 mM KCl, 0.1 mM EDTA, 1 mM DTT, 2.5 mM magnesium acetate and 10% glycerol, and samples were collected on nitrocellulose filters. Reaction tubes were rinsed twice with buffer and applied to the filters which were dried and counted by liquid scintillation spectrometry.

RESULTS

Composition and Activity of Two Forms of Cap Binding Protein. CBP II containing stable restoring activity was obtained previously as a large complex in the 8-10S size range by preparative sucrose gradient centrifugation of an ammonium sulfate precipitate of ribosomal salt wash protein.[24] A direct size estimate of CBP II was made by assaying sucrose gradient fractions for stimulation of Sindbis virus capped mRNA translation in HeLa cell extracts. As shown in Fig. 2, CBP II measured by incorporation of ^{35}S-methionine into Sindbis virus capsid sedimented in a peak position corresponding to a protein with a size of 7S. Activity in pooled fractions 7-15 was further analyzed by chromatography on m^{7}GDP-Sepharose. Fractions eluted from the affinity resin in m^{7}GDP consisted of major polypeptides of molecular weights ~225K, 55K, 48K and 24K as determined by

SDS-polyacrylamide gel electrophoresis (see Fig. 4 below).

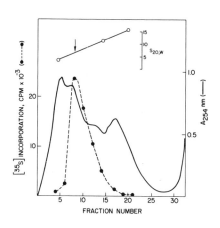

FRACTION NUMBER

Fig. 2. Sucrose gradient analysis of CBP II activity. The 0-40% ammonium sulfate fraction of rabbit reticulocyte ribosomal salt wash (16 mg) was sedimented for 24 hr at 40,000 rpm through a 12.5 ml, 15-30% linear sucrose gradient containing 0.5 M KCl.[24] Gradients were fractionated and simultaneously monitored for absorbance at 254 nm. Aliquots (3 μl) were taken from indicated fractions and assayed for stimulation of Sindbis virus mRNA transla-tion in HeLa cell extracts.[24] After incu-bation with [^{35}S]methionine for 60 min at 37° 5 μl samples were spotted on Whatman 3MM discs and processed as described pre-viously.[21] Sedimentation positions of the protein standards ovalbumin (3.7S), cata-lase (11.3S), and beta-galactosidase (15.9S) were determined in a parallel gradient.

Affinity purified cap binding proteins I and II were shown previously to stimulate Sindbis capsid synthesis in extracts of uninfected HeLa cells.[24] By contrast, in extracts prepared from poliovirus-infected HeLa cells only CBP II restored translation of Sindbis virus mRNA[24] or VSV capped mRNA (unpub. results). The stimulatory activities of CBP I and II on translation of a cellular mRNA, i.e. rabbit globin mRNA, were also compared. The results in Fig. 3 indicate that rabbit globin mRNA translation is stimulated by CBP II in uninfected and in infected cell extracts. As in the case of capped viral mRNA, globin mRNA translation was not restored in infected extracts by CBP I. However, unlike Sindbis virus mRNA, translation of the cellular mRNA was only slightly stimulated by CBP I in uninfected cell extracts. The results indicate that CBP II but not CBP I restores translation of capped cellular and viral mRNAs in extracts of poliovirus-infected HeLa cells. The weak stimulation by CBP I with globin as compared to Sindbis mRNA suggests that CBP may have capped mRNA discriminatory activity. In addition, translation of naturally uncapped EMC viral RNA was not stimulated by CBP I or II in either uninfected or polio-infected extracts (unpub. results).

The identity of the polypeptides in CBP II is of particular interest since these components apparently confer on CBP I the ability to restore capped mRNA translation. The possible presence of other initiation factors in CBP II was considered since different mRNAs may show different factor requirements for translation under limiting conditions.[31] Sindbis mRNA translation was not

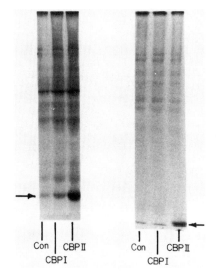

Fig. 3.   Differential activity of cap binding proteins in protein synthesis extracts prepared from uninfected and poliovirus infected HeLa cells.   Globin mRNA (1 ug/25 µl reaction) was translated in the presence of [^{35}S]methionine in uninfected (left) or infected (right) extracts.   Affinity purified cap binding proteins added to the incubation mixtures were: Con, none; CBP I (0.15 µg); and CBP II (0.14 µg). ^{35}S-Labeled proteins were analyzed by autoradiography after electrophoresis in 10% polyacrylamide gels under denaturing conditions as previously described.   Arrows denote position of globin.

restored in infected extracts by purified eIF-3 (prepared from ribosomal salt wash by 0-40% $(NH_4)_2SO_4$ fractionation and two successive centrifugations in sucrose gradients containing 0.1 M and 0.5 M KCl respectively, with fractions from the 13-18S region pooled according to polypeptide composition).   Factor eIF-2 (>90% purified[26]) did not restore translation in infected cell extracts and was only slightly stimulatory in uninfected cell extracts.   The specific activity of eIF-2 was 5-fold less than CBP II in uninfected extracts, consistent with a general rather than cap-specific stimulation.   Consideration of the molecular weights of other known factors suggested that CBP II may contain eIF-4A, although according to previous purification protocols[26] it fractionates in the 40-70% ammonium sulfate cut as compared to 0-40% used to precipitate CBP II.   However, by SDS-polyacrylamide gel electrophoresis the 48,000 molecular weight component of CBP II had the same mobility as purified eIF-4A (Fig. 4).   In addition, the 48,000 dalton component in CBP II crosslinked to oxidized mRNA in an ATP-Mg^{+2} dependent reaction (data not shown).   The presence of putative eIF-4A in the 0-40% ammonium sulfate cut could be due to association with a protein complex (CBP II) having different solubility characteristics.

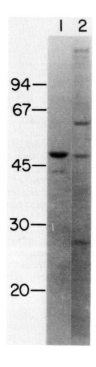

Fig. 4. Analysis of purified rabbit reticulocyte initiation factor eIF-4A and CBP II. Samples were analyzed by electrophoresis in a 10% polyacrylamide gel containing 0.1% SDS. Lane 1, eIF-4A (1.1 μg)[25] and lane 2, CBP II (1.2 μg)[24] were cut from the same gel and aligned according to the positions of the standard proteins which included soybean trypsin inhibitor (20,000), carbonic anhydrase (30,000), ovalbumin (45,000), bovine serum albumin (67,000) and phosphorylase (94,000).

ATP-stimulated Binding of Capped mRNA to Initiation Factors. Recent studies of eIF-4A function indicated that this factor in the presence of eIF-4B binds globin mRNA by an ATP-Mg^{2+}-dependent interaction as measured by mRNA retention on nitrocellulose filters.[25] Binding required both eIF-4A and 4B and was inhibited by m^7GMP (Table 1). Hydrolysis of the beta-gamma phosphate linkage of ATP was apparently required for mRNA binding as shown by the inability of AMPP(NH)P to substitute in the reaction. Recent experiments indicate that eIF-4A has an ATPase activity that is dependent on eIF-4B and RNA (J.A. Grifo and W.C. Merrick, unpub. results).

Purified eIF-4B usually contains CBP I. Therefore we considered whether the inhibition of ATP-Mg^{2+} stimulated mRNA binding by m^7GMP might be due to involvement of CBP I in this reaction. To test for factor interaction with mRNA, eIF-4A and -4B were incubated with oxidized reovirus mRNA under crosslinking conditions. In the presence of ATP-Mg^{2+} both eIF-4A and eIF-4B crosslinked to the radiolabeled, oxidized mRNA (Fig. 5). Crosslinking of eIF-4A and -4B (but not of CBP I that was present in the latter) required both ATP and Mg^{2+}. Furthermore, ATP hydrolysis was essential, and AMPP(NH)P could not replace ATP in the reaction (data not shown). In addition, the interaction was inhibited by the cap analog,

TABLE 1

EFFECT OF ATP ON eIF-4A AND -4B BINDING OF GLOBIN mRNA

Exp. No.	Additions	mRNA bound	ATP-stimulated binding
		(cpm)	(cpm)
1	eIF-4A	201	
	eIF-4A + ATP	226	25
	eIF-4B	687	
	eIF-4B + ATP	756	69
	eIF-4A + eIF-4B	978	
	eIF-4A + eIF-4B + ATP	5,168	4,190
	eIF-4A + eIF-4B + AMPP(NH)P	1,268	290
2	eIF-4A + eIF-4B	701	
	eIF-4A + eIF-4B + ATP	5,164	4,463
	eIF-4A + eIF-4B + ATP + m^{7}GMP	2,202	1,500

Binding assays were performed with eIF-4A (3 ug) and eIF-4B (2.5 ug) as described.[25] Nucleotides were present at 1 mM as indicated except 2 mM m^{7}GMP. Values shown have been corrected for background binding of 262 and 201 cpm in experiments 1 and 2, respectively.

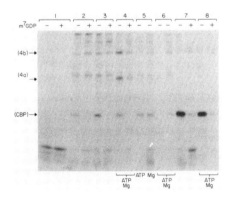

Fig. 5. Crosslinking of cap binding proteins to oxidized, 5'[^{3}H]methyl-labeled reovirus mRNA. Assays were performed as described in Materials and Methods. Proteins and amounts used were: eIF-4A (2.3 ug), eIF-4B (1.7 ug) and affinity purified CBP I$_{0,2}$ (1.7 ug). The fluorogram obtained after gel electrophoresis of crosslinked complexes is shown. Gel lanes contain eIF-4A (1 and 6-left), eIF-4B (2 and 6-right), eIF-4A + eIF-4B (3-5), and CBP I (7,8). Crosslinking was performed in the absence of ATP and Mg^{2+} (1-3,7), in the presence of 2 mM ATP and Mg-acetate (4,6 and 8), with 2 mM ATP alone (5-left) or 2 mM Mg^{2+} alone (5-right). "-" and "+" above each lane refer to incubations done in the absence or presence of 1 mM m^{7}GDP. Positions of eIF-4B (M_r = 80,000), eIF-4A (M_r = 48,000) and CBP I are based on migration positions determined after Coomassie Blue staining and are indicated by arrows.

m^7GDP. CBP II was also examined for ability to crosslink to oxidized mRNA. In addition to the 24,000 molecular weight band, the presence of ATP-Mg^{+2} specifically stimulated crosslinking of the 48,000 molecular weight polypeptide. On the basis of their similar ATP-dependent crosslinking activity and comigration in SDS-polyacrylamide gels, the 48,000 molecular weight component in CBP II may correspond to eIF-4A.

Effect of mRNA Secondary Structure on Protein Crosslinking to 5' Cap. ATP is required for correct binding of mRNA to ribosomes during initiation of protein synthesis. Experiments of Kozak indicate that ATP hydrolysis probably occurs during 40S ribosomal subunit scanning of the 5'-end of mRNA, and that ATP-dependent melting of mRNA secondary structure may be important for positioning ribosomes at the AUG start codon.[32] Sonenberg et al. have suggested that the components responsible for ATP-dependent RNA "unwinding" may be cap binding proteins.[33] We tested this hypothesis in the crosslinking assay by using oxidized reovirus mRNAs that contained different amounts of secondary structure. RNAs were synthesized with ITP in place of GTP or BrUTP instead of UTP. As shown previously,[27,28] I-substitution results in a loss of secondary structure, presumably due to the decreased stability of I:C as compared to G:C base pairs. Conversely, BrU-substitution increases secondary structure stability. The effect of these alterations in mRNA structure on ATP dependent crosslinking to eIF-4A and -4B was examined. As seen in Fig. 6, absence of mRNA secondary structure did not eliminate the ATP requirement for crosslinking to these factors. The trace level of CBP I which was present in eIF-4B also was crosslinked. Similar results were obtained with BrU-substituted mRNA. These findings suggest that eIF-4A and eIF-4B crosslinking to cap requires ATP for processes other than melting of 5'-proximal base-paired regions in the mRNA.

Possible Role of eIF-4A and eIF-4B in Restoring Activity. The observation that CBP II may contain eIF-4A led us to examine the effects of the purified factor on capped mRNA translation in polio-infected HeLa extracts. The eIF-4A preparation shown in Fig. 4 was tested for restoring activity in comparison with CBP I, CBP II and partially purified eIF-4B. Like CBP I, single addition of eIF-4A or eIF-4B to infected cell extracts did not restore Sindbis virus mRNA translation (Fig. 7 and Table 2). The combinations of eIF-4B with either eIF-4A or CBP I also were inactive. Additions of eIF-4A and CBP I or eIF-4A, CBP I and eIF-4B were 21-25% as effective as CBP II. The results suggest that the polypeptides of molecular weight greater than eIF-4A that are present in CBP II[24] are required for restoration.

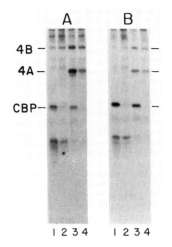

Fig. 6. Effect of mRNA secondary structure on crosslinking of cap binding proteins to oxidized reovirus mRNA. Crosslinking assays were performed as for Fig. 5 except that the oxidized mRNAs were I-substituted (panel A) or BrU-substituted (panel B). All incubations contained eIF-4A (1.7 µg) and eIF-4B (3 µg in panel A, 2.5 µg in panel B) and were done in the absence (lanes 1 and 2) or presence (lanes 3 and 4) of 1 mM ATP plus 3 mM Mg-acetate. Incubations shown in lanes 2 and 4 of both panels contained 1 mM m^{7}GDP, and those in lanes 1 and 3 were without the cap analog.

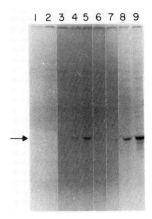

Fig. 7. Restoration of Sindbis capsid synthesis in polio-infected HeLa cell extract by addition of initiation factors and/or cap binding proteins. Sindbis virus mRNA was translated, and the [^{35}S]methionine-labeled products were analyzed by autoradiography after polyacrylamide gel electrophoresis as for Fig. 3. Arrow indicates position of the M_r = 33,000 capsid protein. Reaction mixtures of 25 µl contained ∿0.01 µg Sindbis mRNA and ∿1 µg of each purified initiation factor and/or cap binding protein as indicated in Table 2. For quantitation, also see Table 2.

TABLE 2

RESTORATION OF CAPPED mRNA TRANSLATION IN POLIO-INFECTED EXTRACTS

Lane No.	Protein added				Relative Activity
	4A	4B	CBP I	CBP II	
1	-	-	-	-	0
2	+				0
3		+			0
4			+		0.05
5	+		+		0.21
6	+	+			0
7		+	+		0.04
8	+	+	+		0.25
9				+	1

The numbered lanes of the autoradiogram in Fig. 7 were scanned densitometrically to quantitate the amounts of Sindbis capsid protein synthesized under different conditions. Activities are relative to capsid formed in the presence of CBP II (lane 9) which was taken as unity.

DISCUSSION

At least two translationally active forms of CBP (I and II) can be isolated from reticulocyte ribosomal salt wash. In solutions of low ionic strength (e.g. 0.1 M KCl), CBP I of molecular weight ~24,000 is associated with either eIF-3 or eIF-4B.[18,21] This association may have functional significance or be simply an artifact of isolation. In the presence of 0.5 M KCl, CBP I is not associated with either eIF-3 or eIF-4B but can be obtained in a fast sedimenting, discrete complex (CBP II) that markedly stimulates capped mRNA translation in vitro. The CBP II complex is stable to further purification by m^7GDP-Sepharose affinity chromatography and gel filtration, consistent with a specific association of its constituent polypeptides that include CBP I. The finding that CBP II (but not CBP I) restores capped mRNA translation in extracts prepared from poliovirus-infected HeLa cells suggests that it resembles the in vivo form of CBP. In agreement with this possibility, CBP II is 27-fold more active than CBP I in stimulating globin mRNA translation in uninfected HeLa cell extracts.

Inactivation of capped, host mRNA translation during poliovirus infection may involve disaggregation of a native CBP complex by a virus-specific process or a virus-induced host mechanism. Recent reports suggest that poliovirus infection leads to inactivation of CBP either directly[34] or by dissociation of CBP I from eIF-3.[35] A simple loss of CBP I from eIF-3 may be insufficient to explain the shut-off of host protein synthesis since reconstitution of eIF-3 with CBP I did not restore capped mRNA translation in poliovirus-infected extracts (unpub. results). Additional components apparently are involved in the shutoff phenomenon. In this respect the unidentified polypeptides of CBP II may be important, e.g. one or more of these components may be modified during virus infection, resulting in a switch from host to viral protein synthesis and a shift in the sedimentation position of CBP.

Interaction of mRNA with ribosomes is a complex event mediated by several initiation factors. One of the primary mRNA recognition events during initiation may occur via an interaction between cap binding protein (either CBP I or II) and the 5'-cap to form a protein-RNA complex. In addition, both eIF-4A and eIF-4B in the presence of ATP apparently bind to mRNA at or near the 5'-end, possibly displacing cap binding protein in the process. Results presented here indicate that the interaction of eIF-4A and eIF-4B (but not CBP I) with mRNA requires ATP hydrolysis. This requirement can be demonstrated by either filter binding of mRNA-protein complexes or by chemical crosslinking to the radiolabeled cap of oxidized mRNA. In the presence of ATP-Mg^{+2}, CBP I and both initiation factors eIF-4A and eIF-4B crosslinked similarly to caps in native and I-substituted mRNAs. From this observation and other filter binding results, it appears that coupling of γ-phosphate hydrolysis of ATP is required during mRNA-factor interaction for some process other than opening 5'-proximal structures in the mRNA.

Our studies suggest that the ∿50,000 and 80,000 molecular weight "cap binding proteins" previously detected[33] in crude initiation factor preparations by crosslinking in the presence of ATP-Mg^{+2} probably are identical to eIF-4A and eIF-4B, respectively. The suggestion that crosslinking of cap binding proteins to oxidized mRNA is an active process requiring ATP hydrolysis for melting 5'-proximal secondary structure seems unlikely from the results obtained with I-substituted mRNA which presumably have readily accessible caps. The newly identified interaction of eIF-4A and eIF-4B with the 5' end of capped mRNA may imply that the shut-off of capped mRNA translation by poliovirus also involves interaction of these proteins with CBP I/II. A more complete understanding of the shut-off mechanism induced by poliovirus awaits further studies on the partial reactions involved in the binding of mRNA to 40S initiation complexes.

REFERENCES

1. Shatkin, A.J. (1976) Cell 9, 645-653.
2. Banerjee, A.K. (1980) Microbiological Reviews 44, 175-205.
3. Darnell, J.E. (1979) Prog. Nucleic Acid Res. Mol. Biol. 22, 327-353.
4. Salditt-Georgieff, M., Harpold, M., Chen-Kiang, S. and Darnell, Jr., J.E. (1980) Cell 19, 69-78.
5. Yamakawa, M., Furuichi, Y., Nakashima, K., LaFiandra, A.J. and Shatkin, A.J. (1981) J. Biol. Chem. 256, 6507-6514.
6. Babich, A., Nevins, J.R. and Darnell, Jr., J.E. (1980) Nature 287, 246-248.

7. Furuichi, Y. (1981) J. Biol. Chem. 256, 483-493.
8. Krug, R.M. (1981) Current Topics in Microbiology and Immunology, Vol. 93, 125-149.
9. Ulmanen, I., Broni, B.A. and Krug, R.M. (1981) Proc. Nat. Acad. Sci. USA 78, 7355-7359.
10. Blaas, D., Patzelt, E. and Kuechler, E. (1982) Virology 116, 339-348.
11. Furuichi, Y., LaFiandra, A. and Shatkin, A.J. (1977) Nature (London) 266, 235-239.
12. Shimotohno, K., Kodama, Y.,Hashimoto, J. and Miura, K-I. (1977) Proc. Nat. Acad. Sci. USA 74, 2734-2738.
13. Gedamu, L. and Dixon, G.H. (1978) Biochem. Biophys. Res. Commun. 85, 114-124.
14. Lockard, R.E. and Lane, C. (1978) Nucleic Acids Res. 5, 3237-3247.
15. Filipowicz, W. (1978) FEBS Letters 96, 1-11.
16. Kozak, M. (1977) Nature 269, 390-394.
17. Kozak, M. (1981) Current Topics in Microbiology and Immunology 93, 81-123.
18. Sonenberg, N., Morgan, M.A., Merrick, W.C. and Shatkin, A.J. (1978) Proc. Nat. Acad. Sci. USA 75, 4843-4847.
19. Sonenberg, N., Rupprecht, K.M., Hecht, S.M. and Shatkin, A.J. (1979) Proc. Nat. Acad. Sci. USA 76, 4345-4349.
20. Darzynkiewicz, E., Antosiewicz, J., Ekiel, I., Morgan, M.A., Tahara, S.M. and Shatkin, A.J. (1981) J. Mol. Biol. 153, 451-458.
21. Sonenberg, N., Trachsel, H., Hecht, S. and Shatkin, A.J. (1980) Nature 285, 331-333.
22. Fernandez-Munoz, R. and Darnell, J.E. (1976) J. Virology 126, 719-726.
23. Trachsel, H., Sonenberg, N., Shatkin, A.J., Rose, J.K., Leong, K., Bergmann, J.E., Gordon, J. and Baltimore, D. (1980) Proc. Nat. Acad. Sci. USA 77, 770-774.
24. Tahara, S.M., Morgan, M.A. and Shatkin, A.J. (1981) J. Biol. Chem. 256, 7691-7694.
25. Grifo, J.A., Tahara, S.M., Leis, J.P., Morgan, M.A., Shatkin, A.J. and Merrick, W.C. (1982) J. Biol. Chem. 257, 5246-5252.
26. Merrick, W.C. (1979) in Methods in Enzymology "Nucleic Acids and Protein Synthesis", Moldave, K. and Grossman, L. eds., Vol. LX, Academic Press, New York, pp. 101-108.
27. Morgan, M.A. and Shatkin, A.J. (1980) Biochemistry 5960-5966.
28. Kozak, M. (1980) Cell 19, 79-90.
29. Rose, J.K., Trachsel, H., Leong, K. and Baltimore, D. (1978) Proc. Nat. Acad. Sci. USA 75, 2732-2736.
30. Cancedda, R. and Shatkin, A.J. (1979) Eur. J. Biochem. 94, 41-50.
31. Lodish, H.F. (1976) Ann. Rev. Biochem. 45, 39-72.
32. Kozak, M. (1980) Cell 22, 459-467.
33. Sonenberg, N., Guertin, D., Cleveland, D. and Trachsel, H. (1981) Cell 27, 563-572.
34. Lee, K.A.W. and Sonenberg, N. (1982) Proc. Nat. Acad. Sci. USA (in press).
35. Hansen, J., Etchison, D., Hershey, J.W.B. and Ehrenfeld, E. (1982) J. Virology 42, 200-207.

CAP FUNCTION AND REGULATION OF TRANSLATION DURING POLIOVIRUS INFECTION

NAHUM SONENBERG AND KEVIN A.W. LEE
Department of Biochemistry, McGill University, 3655 Drummond St., Montreal
Quebec, CANADA H3G 1Y6

INTRODUCTION

Cap Binding Proteins of Eukaryotic mRNAs

The cap structure, m^7GpppN, has been found at the 5' terminus of most eukaryotic mRNAs[1-3]. The only exceptions are some viral RNAs such as picornavirus RNAs[4-7] and a few plant viral RNAs[8-10].

It has been well documented that the cap structure facilitates the binding of mRNA to eukaryotic ribosomes (for a recent review see ref. 3), and subsequently suggested that specific protein factors recognize the cap-structure[11-13]. Although several studies, using nitrocellulose filter binding techniques, have suggested the involvement of initiation factors eIF-4B and eIF-2 (refs. 12 and 13 respectively) in cap structure recognition, these studies were inconclusive due to the absence of an unequivocal identification of the active component in these preparations and possible non-specific effects of cap analogues in these assays[14]. A more direct assay to identify polypeptides that can bind specifically to the eukaryotic mRNA cap structure has been developed involving covalent cross-linking of mRNA with an oxidized cap structure to polypeptides that bind near or at the cap[15]. Using the cross-linking technique, it has been demonstrated that a 24 kilodalton (Kd) polypeptide from rabbit reticulocytes interacts specifically with the cap structure[16], as indicated by cap analogue inhibition. This protein has been called the 24 Kd cap binding protein (24-CBP). A polypeptide of similar mobility on SDS/polyacrylamide gels and with the same cross-linking properties has also been detected in L, Ehrlich ascites[16], and HeLa cells[17-19] and in wheat-germ embryos (K.L., N.S. and A. Marcus unpublished observations). The 24-CBP has been isolated from rabbit reticulocytes[20-21] using a m^7GDP-Sepharose-4B column for affinity chromatography, and was shown to preferentially stimulate the translation of capped mRNAs in a HeLa cell free translation system[22].

In more recent experiments, it has been shown that in addition to the 24-CBP, other polypeptides having molecular masses of $\sim$ 28, 50 and 80 Kd can recognize and subsequently be cross-linked to the oxidized 5' cap structure of reovirus mRNA. Cross-linking of the higher molecular weight CBPs requires the presence of ATP-Mg^{++} in contrast to the 24-CBP, and non-hydrolyzable analogues of ATP can not substitute for ATP in the cross-linking reaction[23]. In a recent publication, Grifo et al.[24] demonstrated that eIF-4A and eIF-4B can be cross-linked specifically to the cap structure in an ATP-Mg^{++} dependent manner, suggesting that the 50 and 80 Kd CBPs may actually correspond to eIF-4A and eIF-4B respectively.

In an attempt to gain insight into the possible relationships between the different polypeptides that interact with the cap structure, monoclonal antibodies were raised against CBPs[25-27]. Two of the antibodies characterized were shown to interact on nitrocellulose blots with several proteins in ribosomal high salt wash and post-ribosomal supernatant of rabbit reticulocytes. Some of the immunoreactive polypeptides were isolated by immunoaffinity chromatography and compared by tryptic mapping. Polypeptides having molecular weights of 210, 160, 50 and 28 Kd were shown to share common peptides with the 24-CBP indicating a structural similarity between them[25]. In addition, these monoclonal antibodies were shown to inhibit the cross-linking of the higher molecular weight CBPs and the 24-CBP[23]. This result suggests that the different cross-linkable proteins are structurally related, may be by virtue of the presence of a cap binding domain and that they are all able to recognize the cap structure. Alternatively, the oxidized mRNA might interact with a complex consisting of the 24-CBP and the high molecular weight proteins.

Poliovirus Infection and the Shut-Off of Host Protein Synthesis

Infection of mammalian cells by picornaviruses elicits a complex response, a salient feature of which is the inhibition of host mRNA and protein synthesis[28]. The mechanism by which host protein synthesis is selectively inhibited is however, poorly understood.

During the past twenty years, the mechanism by which poliovirus exerts the shut-off of host protein synthesis in HeLa cells has been extensively investigated[29,30]. Poliovirus infection results in the disruption of host specific polysomes although host mRNA is not degraded[31,32] and there are no detectable changes in patterns of host mRNA capping, methylation or polyadenylylation[33]. Furthermore, Ehrenfeld and Lund have demonstrated that host mRNA extracted from infected cells remains functional in wheat germ cell-free translation system[34].

It has been shown that the inhibition of host protein synthesis occurs at the level of initiation[32] and subsequently that preparations of crude initiation factors (IF) from the infected cells can stimulate translation of polio-

virus RNA but not host mRNA, in HeLa cell extracts[35]. Rose et al[36] have used
vesicular stomatitis virus (VSV) mRNA as a model for host mRNA, since super-
infection of VSV infected HeLa cells by poliovirus results in inhibition of
VSV mRNA translation in an apparently identical manner to that of host mRNAs
[35,37]. They found that initiation factor eIF-4B was the only factor that
could restore translation of VSV mRNA in cell extracts from poliovirus-infected
cells and concluded that inactivation of this factor was responsible for the
inhibition of VSV and host protein synthesis[36]. However, a different study
using a reconstituted reticulocyte translation system indicated that eIF-3
activity purified from poliovirus-infected cells was severely impaired, thus
suggesting that this initiation factor or a co-purified component became inactivated as a consequence of poliovirus infection[38]. Later, it has been
shown that an apparently homogeneous 24 Kd protein isolated by a multi-step
procedure copurified with the ability to restore the capacity of poliovirus-
infected HeLa cell extracts to efficiently translate VSV mRNA (this activity
will be referred to as restoring activity)[39]. This polypeptide was found to
be identical to the 24-CBP isolated from rabbit reticulocyte IF by m^7GDP-
Sepharose 4B affinity chromatography[20,21] in a number of respects, including
migration on SDS/polyacrylamide gels, ability to specifically cross-link to
the oxidized 5' cap of reovirus mRNA and by two dimensional peptide mapping[39].
However, it has been observed that this purified restoring activity is very
labile[39]. The observation that 24-CBP copurified with eIF-3 and eIF-4B[16,39,
40] suggests that the effects ascribed to these factors might have been due to
the presence of 24-CBP and the shut-off phenomenon might actually be a result
of inactivation of the 24-CBP.

There is now evidence to suggest that functional 24-CBP is part of a
higher molecular weight complex. Firstly, it has been shown that 24-CBP
sediments as a ∿200 Kd protein complex in sucrose gradients[40]. Secondly,
Tahara et al[41] recently described an 8-10S protein complex purified by m^7GDP
chromatography from IF of rabbit reticulocytes, which contained several higher
molecular weight polypeptides in addition to the 24-CBP and possessed stable
restoring activity. This finding is consistent with the results described
above demonstrating the existence of several higher molecular weight polypeptides which are structurally related to the 24-CBP of rabbit reticulocytes
[25,26], and with the observation that some polypeptides with molecular weights
that are strikingly similar to those of the latter polypeptides can specifically recognize the cap structure, as determined by the cross-linking
technique[23].

In contrast to the almost ubiquitous nature of the cap structure at the 5' terminus of eukaryotic mRNAs, poliovirus RNA is not capped[4,5] and its translation initiation must therefore occur by a cap independent mechanism. Inactivation of CBPs would most likely result in the selective inhibition of host mRNA translation but allow translation of poliovirus RNA.

The nature of the viral dependent factor which effects the shut-off of host protein synthesis is unknown. It was originally shown that translation of viral mRNA is absolutely required for the shut off[42,43], indicating that a viral gene product is responsible, either directly or indirectly. However, there is no strong evidence to indicate the involvement of a specific viral protein, although studies with ts viral mutants in structural proteins, which are defective in shut-off, have implicated these protein(s) as effectors in the shut-off mechanism[44]. Rose et al. have reported that cell extracts from poliovirus-infected cells have activity which can slowly reduce the ability of uninfected extracts to translate capped mRNAs and can also inactivate the restoring activity present in eIF-4B preparations. More recently, Brown and Ehrenfeld have demonstrated that crude IF from poliovirus-infected cells can specifically restrict the translation of capped mRNAs in rabbit reticulocyte lysate[45]. This activity is heat labile, again indicating that it resides in a protein, which could either be virally coded or induced.

In this article we describe results obtained from experiments to determine the fate of cap binding proteins during poliovirus infection and discuss the implications of these results in terms of the shut-off phenomenon and the difference in mechanism of translation initiation of capped and naturally uncapped mRNAs in general.

RESULTS

Cell free extracts prepared from poliovirus-infected cells have been shown to have a reduced ability to translate capped mRNAs, whereas translation of naturally uncapped mRNAs is not impaired[17,36,41,46]. To find whether our cell-free extracts had these characteristics, we tested their ability to translate capped mRNA and naturally uncapped mRNA. Fig. 1 shows that extracts prepared from mock-infected cells were capable of translation both Sindbis (capped) and EMC (naturally uncapped) mRNAs, whereas translation of Sindbis mRNA was restricted in extracts from poliovirus-infected cells although these extracts translated EMC mRNA with the same efficiency as extracts from mock-infected cells. Crude IF preparations from mock-infected cells had very little effect on the translation of Sindbis mRNA in mock-infected cell extracts, indicating that the IF do not contain any active component missing in the cell

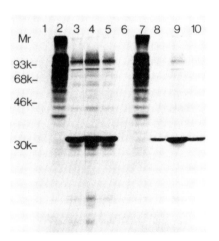

Fig. 1. Translation in poliovirus-infected and mock-
infected HeLa cell extracts and
the effect of initiation
factors. Cell free extracts
prepared from poliovirus-
infected or mock-infected
cells[36] were used for in vitro
protein synthesis as
described[19,36]. Products were
resolved on SDS/polyacrylamide
gels followed by fluoro-
graphy[16]. As a source of initiation factors we used a high-
salt ribosomal wash prepared
essentially as described[47]
with a few modifications[19].
Lanes 1-5, translation products in mock-infected HeLa
cell extracts. Lanes 6-10,
translation products in poliovirus-infected HeLa cell
extracts. The following
amounts of mRNA and IF were
added: lanes 1 and 6, no RNA;
2 and 7, 0.5μg EMC RNA[48]; 3
and 8, $\sim$ 1μg Sindbis mRNA[49]; 4
and 9, $\sim$ 1μg Sindbis mRNA plus
90 μg IF from mock-infected
cells; 5 and 10, $\sim$ 1μg Sindbis
mRNA plus 90 μg IF from poliovirus-infected cells.

extracts. However, IF preparations from mock-infected cells were able to
restore the ability of poliovirus-infected cell extracts to translate the
capped Sindbis mRNA (Fig. 1). In contrast, IF preparations from poliovirus-
infected cells did not possess restoring activity. These results are consistent with the reports of Rose et al.[36] and Hansen and Ehrenfeld[17],
indicating that a factor, crucial for the translation of capped mRNA, is inactivated in IF preparations from poliovirus-infected cells.

We wanted to test the hypothesis that the factor inactivated in poliovirus
infected cells is a cap binding protein. Therefore, we assayed IF preparations from mock-infected and infected cells for polypeptides with the ability
to cross-link specifically to the cap structure. Cross-linking was performed
in the presence of ATP-Mg^{++}, which is required for the cross-linking of CBPs
other than the 24-CBP. As shown in Fig. 2, oxidized reovirus mRNA can be
cross-linked to several polypeptides in crude IF from mock-infected cells.
Cross-linking of the 26, 28, 32, 50 and 80 Kd polypeptides was inhibited by
m^{7}GDP, indicating a cap specific interaction between these proteins and the

mRNA*. The same analysis of IF from poliovirus-infected cells showed a markedly reduced level of cross-linking of the different CBPs (Fig. 2). Hansen and Ehrenfeld[17] performed similar experiments and reported that the amount of 24-CBP in IF from poliovirus-infected HeLa cells is not reduced relative to preparations from mock-infected cells. One significant difference between the two studies is the absence of ATP-Mg^{++} in Hansen and Ehrenfeld's cross-linking reaction mixture[17]. Consequently, we performed cross-linking experiments in the absence of ATP-Mg^{++}, conditions which have been shown to allow cap specific cross-linking of the 24-CBP only. Comparison of the cross-linking of the 24-CBP from infected and mock-infected IF clearly shows that the cross-linking in infected preparations is markedly reduced as compared to mock-infected (Fig. 3). We believe that our results are not necessarily at variance with those of Hansen and Ehrenfeld[17], since careful examination of their data reveals that the extent of cross-linking of the 24-CBP is distinctly less in preparations from infected cells. We have consistently observed reduced levels of all CBPs in infected preparations, although when cross-linking was performed in the absence of ATP-Mg^{++}, the reduction in the amount of the 24-CBP from infected preparations was somewhat less compared to the reduction observed in the presence of ATP-Mg^{++} (compare Fig. 3 to Fig. 2).

The mechanism by which CBPs are inactivated during poliovirus infection is not known. It is possible that CBPs become modified or degraded in polio-virus-infected cells. One approach to elucidate the mode of inactivation of CBPs is to study the inactivation process in vitro and isolate the putative inactivating factor which could be either virally coded or induced. Rose et al.[36] have reported that cell-free extracts from poliovirus infected cells contain an activity that slowly reduces the ability of uninfected extracts to translate capped mRNAs in vitro. Similarly, Brown and Ehrenfeld[45] have demonstrated an activity in IF preparations from poliovirus-infected cells that specifically restricts the translation of capped mRNAs in reticulocyte lysate. Consequently, it was of interest to first determine whether IF preparations from infected cells could impair the cross-linking ability of the different CBPs and secondly to verify that these preparations could also restrict translation of capped mRNA in vitro.

---

*Note that in this and the following experiments the fastest migrating CBP is a 26 Kd polypeptide. It has identical mobility to the 24-CBP from rabbit reticulocytes and consequently we refer to it as the 24-CBP, in order to avoid confusion. The change in mobility from previous reports is probably due to changes in the gel system.

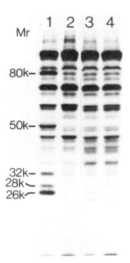

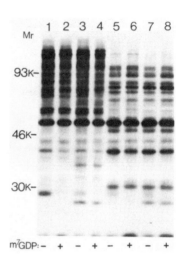

Fig. 2. Cross-linking pattern of CBPs from poliovirus-infected and mock-infected HeLa cells in the presence of ATP-Mg^{++}. Initiation factor preparations from mock-infected (92 µg) or poliovirus-infected cells (98 µg) were incubated with 0.7 µg (57,000 cpm) of [^{3}H] oxidized reovirus mRNA in an incubation mixture containing 1 mM ATP and 0.5 mM Mg(OAc)$_2$ as described[19], for 10 min at 30° in a final volume of 30 µl. After incubation, 3 µl of 0.2 M NaBH3CN was added and the mixture was left on ice for 3 hrs, followed by the addition of 3 µl of RNase A (5 mg/ml) and incubation for 30 minutes at 37°C to degrade the mRNA[15-17]. Cross-linked proteins were resolved in SDS/polyacrylamide gels, followed by treatment with PPO/DMSO and exposure to Kodak X-Omat XR-1 film at -70°[16]. Lanes 1 and 2, IF from mock-infected cells; lanes 3 and 4, IF from poliovirus-infected cells; lanes 2 and 4 contained 0.67 mM m^{7}GDP.

Fig. 3. Cross-linking pattern of CBPs from poliovirus-infected and mock-infected HeLa cells in the absence of ATP. Cross-linking was performed with 0.7 µg of [^{3}H] methyl oxidized reovirus mRNA ($\sim$ 57,000 cpm) in 10 mM Hepes buffer (pH 7.5) containing 45 mM KCl, 1 mM Mg(OAc)$_2$, 0.5 mM DTT, 70 µM PMSF as described in the legend to Fig. 2. Incubation mixtures contained 82 µg of 0-40% ammonium sulfate fraction of IF from mock-infected cells (lanes 1 and 2), 98 µg of 0-40% ammonium sulfate fraction of IF from infected cells (lanes 3 and 4), 93 µg of 40-70% ammonium sulfate fraction of IF from mock-infected cells (lanes 5 and 6), 90 µg of 40-70% ammonium sulfate fraction of IF from infected cells (lanes 7 and 8). m^{7}GDP was included at a concentration of 0.67 mM as indicated in the figure. Following cross-linking, samples were processed for SDS/polyacrylamide gel electrophoresis and fluorography as in the legend to Fig. 2.

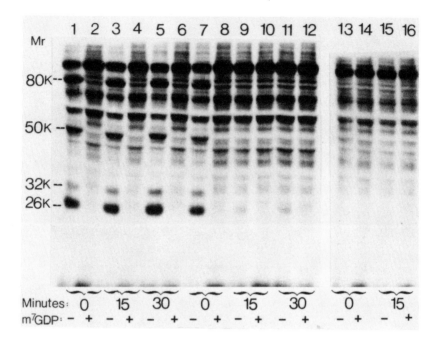

Fig. 4. Effect of mixing IF from mock-infected and poliovirus-infected cells on cross-linking of CBPs to mRNA. Initiation factors from mock-infected cells (92 µg) were incubated with IF from poliovirus-infected cells (36 µg) at 37° for the times indicated in the figure, prior to the addition of [^{3}H]methyl-labeled oxidized reovirus mRNA (0.5 µg, 42,000 cpm), under cross-linking conditions in the presence of 1 mM ATP as described in the legend to Fig. 2. As control experiments, IF from mock-infected and poliovirus-infected cells were preincubated separately. Each incubation was performed in the absence or presence of 0.67 mM m^{7}GDP. Following incubation, samples were processed for electrophoresis as described. Lanes 1-6, preincubation of IF from mock-infected cells; 7-12, preincubation of IF from mock-infected cells with IF from poliovirus-infected cells; 13-16, preincubation of IF from poliovirus-infected cells.

To determine whether IF preparations from infected cells have the ability to impair CBP function, we mixed IF from poliovirus-infected cells with IF from mock-infected cells and tested the cross-linking ability of the CBPs (Fig. 4). Simple mixing of IF from poliovirus-infected cells with IF from mock-infected cells did not diminish the cross-linking ability of the various CBPs (lane 7). However, preincubation of this mixture drastically diminished the ability of the CBPs to cross-link to the mRNA (lanes 9 and 11). These results indicate that IF from poliovirus-infected cells contain an activity that impairs the

ability of the various CBPs to recognize the cap structure of the mRNA and
would presumably effect a reduction in cellular protein synthesis. These
results also exclude the possibility that IF from poliovirus-infected cells
contain a preformed inhibitor of CBP function, since no effect could be
observed without preincubation.

To analyze the effect of IF from poliovirus-infected cells on translation
of capped mRNAs in mock-infected cell extracts, we preincubated the extracts
in presence or absence of IF prior to the translation of reovirus mRNA
(capped) or satellite tobacco necrosis virus (STNV) mRNA (uncapped). Fig. 5
shows that preincubation of the extract in the absence or presence of IF from
mock-infected cells or addition of IF from infected cells without preincubation had no significant effects on the translation of reovirus mRNA (lanes 3-
5). However preincubation in the presence of IF from infected cells dramatically inhibited translation (lane 6). In contrast to the distinct inactivation of reovirus mRNA translation, there was no detectable inhibition of
STNV translation in extracts from mock-infected cells that were preincubated
with IF from poliovirus-infected cells (compare lane 9 to lane 8). In
addition, IF from poliovirus infected cells could also neutralize the
restoring activity of IF from mock-infected cells upon preincubation (data
not shown).

The limited amount of sequence data available for capped mRNAs has
enabled computer aided predictions of thermodynamically stable secondary
structures of these mRNAs[51-53]. It has been demonstrated that mRNAs in which
inosine is substituted for guanosine with a consequent reduction in secondary
structure, are less dependent on the cap structure and ATP for initiation
complex formation than native mRNAs[54-56]. In addition, it has been shown that
monoclonal antibodies directed against cap binding proteins can inhibit
initiation complex formation with native but not with inosine substituted
mRNA[26]. Taken together, these results suggest that mRNA with reduced
secondary structure is less dependent on the cap structure and CBPs for
initiation complex formation. These observations and the finding that cross-
linking of CBPs (except for the 24-CBP) is dependent on the presence of hydro-
lyzable ATP-Mg^{++}[23], prompted the suggestion that CBPs might function by de-
stabilizing or melting the secondary structure of capped mRNAs to enable the
binding of 40S ribosomal subunits in an energy dependent process[23,26]. If
this hypothesis is correct, one would expect to find that ribosome binding to
capped mRNAs has a greater dependence on cap binding proteins at higher K$^+$
concentrations, since these conditions would favor more stable secondary

382

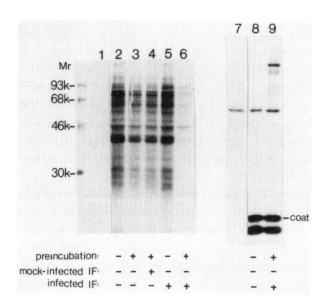

Fig. 5. Effect of preincubation of HeLa cell extracts with IF from mock-infected or poliovirus-infected cells on protein synthesis. Micrococcal nuclease treated HeLa cell extracts[22,36] prepared from poliovirus-infected or mock-infected cells were preincubated at 37° with the indicated amounts of IF from poliovirus-infected or mock-infected cells in 25 µl incubation mixtures containing the components required for protein synthesis except for 35S-methionine and mRNA. At the times indicated, mRNA and 35S-methionine were added and incubation was continued for 60 minutes at 37°. Translation in mock-infected HeLa cell extracts with no added RNA (lanes 1 and 7), 2 µg of reovirus mRNA (lanes 2-6) or 1 µg of STNV RNA (lanes 8 and 9). Initiation factors (20 µg) were added where indicated in the figure and preincubation time was: lane 3, 25 min.; 4, 25 min.; 6, 12 min.; and 9, 20 min. The auto-radiograph for STNV translation products was exposed for longer than the one for reovirus translation products.

structure. Hence, if cap binding proteins are limiting in the cell free extracts, increasing K^+ concentrations would result in a decrease in mRNA binding. However, denatured mRNA should not be susceptible to variation in salt concentrations since its secondary structure is not greatly altered under these circumstances. The results shown in Fig. 6 are in accord with this rationale. Native reovirus mRNA binds efficiently to ribosomes in a wheat-

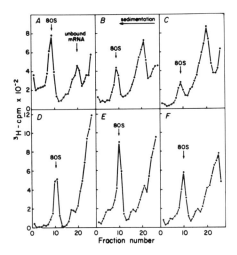

Fig. 6. Binding of native and inosine substituted reovirus mRNA to ribosomes
in mock-infected HeLa cell extracts as a function of K$^+$ concentration. [^{3}H]-
methyl labeled native reovirus mRNA (13,000 cpm) or inosine substituted reovirus mRNA (13,000 cpm) was incubated in 50 µl of a mock-infected HeLa extract
at 30° for 10 min. under conditions required for formation of translation
initiation complexes. Initiation complexes were analyzed in glycerol gradients
by centrifugation for 90 minutes at 48,000 rpm and 4° in a Beckman SW50.1
rotor[15,26,50]. Panels A-C, native mRNA and D-F, inosine substituted mRNA.
The final concentrations of KOAc (excluding 20 mM KCl contributed by the HeLa
cell extract) were as follows (mM): A - 45; B - 85; C - 125; D - 45; E - 100;
F - 180.

germ extract at 45 mM K$^+$ but the binding is reduced with increasing concentrations of the salt (from 50% of mRNA bound at 65 mM K$^+$ to 15% at 145 mM K$^+$).
This salt dependent reduction is not observed with inosine substituted reovirus mRNA (Fig. 6,panels D-F), but to the contrary, binding is stimulated by
two fold when K$^+$ concentrations are increased from 65 mM to 120 mM and then
decreases to the original value at higher concentrations of salt.

If CBPs are required only for the translation of mRNAs with secondary
structure and if CBPs become inactive during poliovirus infection then it is
predicted that inosine-substituted reovirus mRNA should still be able to form
initiation complexes in cell-extracts from poliovirus-infected cells. In Fig.
7 the binding pattern of native and relaxed reovirus mRNAs to ribosomes from
poliovirus-infected cells and mock-infected cells is shown. Native reovirus
mRNA cannot form initiation complexes in extracts from poliovirus-infected
cells, under all salt concentrations used (panels A-C), which is consistent
with the inability of these extracts to translate capped mRNAs. However,

384

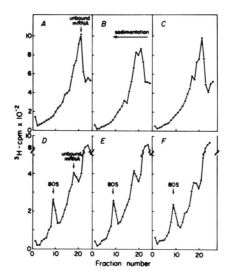

Fig. 7. Binding of native and inosine-substituted reovirus mRNA to ribosomes in poliovirus-infected HeLa cell extracts as a function of K+ concentration. [3H] methyl-labeled native reovirus mRNA (13,000 cpm) or inosine substituted mRNA (18,000 cpm) was incubated in 50 μl of a poliovirus-infected HeLa extract and initiation complex formation was analyzed as described in the legend to Fig. 6 and elsewhere[15,57]. Panels A-C, native mRNA; D-F, inosine substituted mRNA. The final concentrations of KOAc (excluding 20 mM KCl contributed by the HeLa cell extract) were as follows (mM): A and D - 70; B and E - 105; C and F - 145.

these extracts can still promote binding of inosine substituted mRNA, although the efficiency is reduced to some extent compared to that in extracts from mock-infected cells. These findings present strong evidence that the factors inactivated in poliovirus-infected cell extracts are required only for the translation of mRNAs with significant secondary structure. This idea is substantiated by evidence suggesting that the capped AMV 4 mRNA which does not contain a high degree of secondary structure in its 5' non-translated region[58] can be translated with similar efficiency in both poliovirus-infected and mock-infected cell extracts (data not shown).

DISCUSSION

Cap binding proteins with similar molecular masses to those of rabbit reticulocytes can be detected in IF preparations from HeLa cells. These poly-peptides are detected by their ability to specifically cross-link to the oxidized 5' terminus of reovirus mRNA. The cross-linking of five polypeptides (Mr = 26, 28, 32, 50 and 80 Kd) could be inhibited by m^{7}GDP, indicating that this cross-linking is cap specific. The interpretation of these results could be that all of these polypeptides are structurally and functionally related or alternatively that they exist as a complex and that binding of mRNA to this complex results in cross-linking to the different proteins of the complex. Previous data from experiments employing monoclonal antibodies have indicated structural similarities between the 24-CBP and the higher molecular weight CBPs of rabbit reticulocytes[25]. This common immunoreactivity could be readily explained by the presence of a structurally identical cap binding domain in these polypeptides, thus suggesting that they are all individually able to recognize the cap structure.

When cross-linking was performed with IF preparations from poliovirus-infected HeLa cells a sharp decrease in the amount of detectable CBPs was observed. It is noteworthy, however, that a residual amount of the 24-CBP was always found to be able to cross-link to mRNA after poliovirus infection. This residual fraction is also observed when CBPs become inactivated in vitro by incubation with IF from poliovirus-infected cells (Fig.4): even after longer times of incubation there was a residual fraction of the 24-CBP which could specifically cross-link to the cap structure. It is not clear how these results should be interpreted. However, it is possible that there exist two different populations of the 24-CBP, one being susceptible to regulation by the poliovirus-dependent inactivating factor, the other being resistant.

As described earlier, Hansen and Ehrenfeld[17] have reported that the cross-linking ability of the 24-CBP from poliovirus-infected cells is not grossly altered. In a more recent publication[18] they reported that the 24-CBP from poliovirus-infected cells cannot associate with eIF-3 under conditions in which it is normally found associated with this multi-subunit initiation factor. The conflicting evidence presented by our experiments and those of Hansen and Ehrenfeld[17] could reflect a quantitative difference in the residual amount of the 24-CBP which can still cross-link after poliovirus-infection. Whereas in our case this portion does not amount to more than 10% of the level found in IF preparations from mock-infected cells, in Hansen and Ehrenfeld's data, the residual amount of the 24-CBP seems to be at least 50% of the original. However, more rigorous experimental investigation is clearly required before any firm conclusions can be drawn.

Previous data have suggested that CBPs are required to destabilize the secondary structure of certain capped mRNAs in order to facilitate the binding of 40S ribosomal subunits[26]. Consequently, it was not surprising to find that the synthetic inosine substituted reovirus mRNA used in these studies was able to form initiation complexes in cell extracts from poliovirus-infected cells, in which the CBPs are apparently inactivated. In addition, we have evidence to indicate that a capped mRNA, AMV-4, which has very little potential for forming secondary structure at the 5' non-translated region[58], can be translated with similar efficiency in both infected and mock-infected cell extracts. These results support the idea that CBPs are required mainly to melt the secondary structure present in capped mRNAs[23,26] and that this function is inactivated in poliovirus-infected cells. The fact that capped mRNAs vary in their sensitivity to inhibition by cap analogues[1-3,26] and also in the degree to which their translation is inhibited in poliovirus-infected lysates (our data), indicates that the degree of cap dependence varies between different capped mRNAs. It would be of significance to determine whether this variation is correlated with the degree of secondary structure in the different mRNAs.

At the present time it is not clear exactly which structural features of naturally uncapped mRNAs are responsible for allowing their cap independent translation. One plausible model to explain the dispensability of the cap structure for translation of these mRNAs is that their 5' non-coding sequences are devoid of any extensive secondary structure. This is true for STNV RNA in which the 5' end can only form an energetically unstable hairpin loop ($\Delta G° = -0.26$ Kcal/mole nucleotide[59,60]). It is therefore clearly of importance to obtain computer generated predictions of the probable secondary structures at the 5' non-translated portions of the picornavirus RNAs.

Finally, it is of interest to note the findings reported by Jackson[61], that some naturally uncapped mRNAs are less dependent on ATP for initiation complex formation than are capped mRNAs. Since this is also the case for inosine substituted mRNA it is tempting to speculate that naturally uncapped mRNAs might be devoid of extensive secondary structure. These observations provide further evidence that the requirement for the cap structure and ATP are related aspects of translation initiation.

ACKNOWLEDGMENTS
We are grateful to D. Guertin for expert technical assistance and to A. LaFiandra, A.J. Shatkin, V. Stollar and J.M. Clark for viruses and viral RNAs. This research was supported by grants from the National Cancer Institute and Medical Research Council of CANADA. K.A.W.L. is a recipient of a predoctoral research fellowship from the Cancer Research Society (Montreal).

REFERENCES

1. Shatkin, A.J. (1976) Cell, 9, 645-653.
2. Filipowicz, W. (1978) FEBS Letters, 96, 1-11.
3. Banerjee, A. (1980) Bacteriol. Rev., 44, 175-205.
4. Hewlett, M.J., Rose, J.K., and Baltimore, D. (1976) Proc. Natl. Acad. Sci. USA, 73, 327-330.
5. Nomoto, A., Lee, Y.F., and Wimmer, E. (1976) Proc. Natl. Acad. Sci. USA, 73, 375-380.
6. Frisby, D., Eaton, M., and Fellner, P. (1976) Nucleic Acids Res., 3, 2771-2787.
7. Sangar, D.V., Rowlands, D.J., Harris, T.J.R., and Brown, F. (1977) Nature, 268, 648-650.
8. Horst, J., Fraenkel-Conrat, H., and Mandeles, S. (1971) Biochemistry, 10, 4748-4752.
9. Klootwijk, J., Klein, I., Zabel, P., and Van Kammen, A. (1977) Cell, 11, 73-82.
10. Ghosh, A., Dasgupta, R., Salerno-Rife, T., Rutgers, T., and Kaesberg, P. (1979) Nucleic Acids Res., 7, 2137-2146.
11. Filipowicz, W., Furuichi, Y., Sierra, J.M., Muthukrishnan, S., Shatkin, A.J. and Ochoa, S. (1976) Proc. Natl. Acad. Sci. USA, 73, 1559-1563.
12. Shafritz, D.A., Weinstein, J.A., Safer, B., Merrick, W.C., Weber, L.A., Hickey, E.D., and Baglioni, C. (1976) Nature, 261, 291-294.
13. Kaempfer, R., Rosen, H. and Israeli, R. (1978) Proc. Natl. Acad. Sci. USA, 75, 650-654.
14. Sonenberg, N., and Shatkin, A.J. (1978) J. Biol. Chem., 253, 6630-6632.
15. Sonenberg, N., and Shatkin, A.J. (1977) Proc. Natl. Acad. Sci. USA, 74, 4288-4292.
16. Sonenberg, N., Morgan, M.A., Merrick, W.C., and Shatkin, A.J. (1978) Proc. Natl. Acad. Sci. USA, 75, 4843-4847.
17. Hansen, J., and Ehrenfeld, E. (1981) J. Virol., 38, 438-445.
18. Hansen, J., Etchison, D., Hershey, J.W.B., and Ehrenfeld, E. (1982) J. Virol., 42, 200-207.
19. Lee, K.A.W., and Sonenberg, N. (1982) Proc. Natl. Acad. Sci. USA, 79, in press.
20. Sonenberg, N., Rupprecht, K.M., Hecht, S.M., and Shatkin, A.J. (1979) Proc. Natl. Acad. Sci. USA, 76, 4345-4349.
21. Rupprecht, K.M., Sonenberg, N., Shatkin, A.J., and Hecht, S.M. (1981) Biochemistry, 20, 6570-6577.
22. Sonenberg, N., Trachsel, H., Hecht, S.M., and Shatkin, A.J. (1980) Nature, 285, 331-333.
23. Sonenberg, N. (1981) Nucleic Acids Res., 9, 1643-1656.
24. Grifo, J.A., Tahara, S.M., Leis, J.P., Morgan, M.A., Shatkin, A.J. and Merrick, W.C. (1982) J. Biol. Chem., 257, in press.
25. Sonenberg, N., and Trachsel, H. (1982) in Current Topics in Cellular Regulation, Horecker, B.L. and Stadtman, E.R. eds., Academic Press, New York, Vol. 21, pp. 65-88.
26. Sonenberg, N., Guertin, D., Cleveland, D., and Trachsel, H. (1981) Cell, 27, 563-572.
27. Sonenberg, N., Skup, D., Trachsel, H., and Millward, S. (1981) J. Biol. Chem., 256, 4138-4141.
28. Bablanian, R. (1975) in Progress in Medical Virology, Melnick, J.L., ed., Karger, Basel, Vol. 19, pp. 40-83.
29. Baltimore, D. (1969) in the Biochemistry of Viruses, Levy, H.B., ed., Dekker, New York, pp. 101-176.
30. Lucas-Lenard, J.M. (1979) in The Molecular Biology of Picornaviruses, Perez-Bercoff, R., ed., Plenum Press, New York, pp. 78-99.

388

31. Leibowitz, R., and Penman, S. (1971) J. Virol., 8, 661-668.
32. Kaufman, Y., Goldstein, E., and Penman, S. (1976) Proc. Natl. Acad. Sci. USA, 73, 1834-1838.
33. Fernandez-Munoz, R., and Darnell, J.E. (1976) J. Virol., 18, 719-726.
34. Ehrenfeld, E., and Lund, H. (1977) Virology, 80, 297-308.
35. Helentjaris, T., and Ehrenfeld, E. (1978) J. Virol., 26, 510-521.
36. Rose, J.K., Trachsel, H., Leong, K., and Baltimore, D. (1978) Proc. Natl. Acad. Sci. USA, 75, 2732-2736.
37. Doyle, M., and Holland, J.J. (1972) J. Virol., 9, 22-28.
38. Helentjaris, T., Ehrenfeld, E., Brown-Luedi, M.L., and Hershey, J.W.B. (1979) J. Biol. Chem. 254, 10973-10978.
39. Trachsel, H., Sonenberg, N., Shatkin, A.J., Rose, J.K., Leong, K., Bergmann, J.E., Gordon, J., and Baltimore, D. (1980) Proc. Natl. Acad. Sci. USA, 77, 770-774.
40. Bergmann, J.E., Trachsel, H., Sonenberg, N., Shatkin, A.J., and Lodish, H.F. (1979) J. Biol. Chem., 254, 1440-1443.
41. Tahara, S.M., Morgan, M.A., and Shatkin, A.J. (1981) J. Biol. Chem., 256, 7691-7694.
42. Penman, S., and Summers, D. (1965) Virology, 27, 614-620.
43. Helentjaris, T., and Ehrenfeld, E. (1977) J. Virol., 21, 259-267.
44. Steiner-pryor, A., and Cooper, P.D. (1973) J. Gen. Virol., 21, 215-225.
45. Brown, B.A., and Ehrenfeld, E. (1980) Virology, 103, 327-339.
46. Bonatti, S., Sonenberg, N., Shatkin, A.J., and Cancedda, R., (1980) J. Biol. Chem., 255, 11473-11477.
47. Schreier, M.H., and Staehelin, T. (1973) J. Mol. Biol., 73, 329-349.
48. Eggen, K.L., and Shatkin, A.J. (1972) J. Virol., 9, 636-645.
49. Cancedda, R., and Shatkin, A.J. (1979) Eur. J. Biochem., 94, 41-50.
50. Kozak, M., and Shatkin, A.J. (1977) J. Biol. Chem., 252, 6895-6908.
51. Kozak, M., and Shatkin, A.J. (1977) J. Mol. Biol., 112, 75-96.
52. Pavlakis, G.N., Lockard, R.E., Vamvakopoulos, N., Rieser, L., RajBhandary, U.L., and Vournakis, J.N. (1980) Cell, 19, 91-102.
53. Nussinov, R. and Tinoco, I. Jr. (1981) J. Mol. Biol., 151, 519-533.
54. Kozak, M. (1980) Cell, 19, 79-80.
55. Kozak, M. (1980) Cell, 22, 459-467.
56. Morgan, M.A., and Shatkin, A.J. (1980) Biochemistry, 19, 5960-5966.
57. Both, G.W., Furuichi, Y., Muthukrishnan, S., and Shatkin, A.J. (1976) J. Mol. Biol., 104, 637-658.
58. Koper-Zwarthoff, E.C., Lockard, R.E., Alzner-DeWeerd, B., RajBhandary, U.L. and Bol, J.E. (1977) Proc. Natl. Acad. Sci. USA, 74, 5504-5508.
59. Leung, D.W., Browning, K.S., Heckman, J.E., RajBhandary, U.L., and Clark, J.M. Jr. (1979) Biochemistry, 18, 1361-1366.
60. Ysabaert, M., Van Emmelo, J., and Fiers, W., (1980) J. Mol. Biol., 143, 273-287.
61. Jackson, R.J. (1982) in Protein Biosynthesis in Eukaryotes, Perez-Bercoff, R., ed., Plenum Press, New York, pp. 363-418.

CAP BINDING PROTEIN RELATED POLYPEPTIDES IN CYTOPLASMIC
RIBONUCLEOPROTEIN PARTICLES OF CHICK EMBRYONIC MUSCLE

ASOK K. MUKHERJEE*, DIPAK CHAKRABARTY*, KEVIN A.W. LEE[+],
ANDRE DARVEAU[+], NAHUM SONENBERG[+], AND SATYAPRIYA SARKAR*[+]
*Department of Muscle Research, Boston Biomedical Research Institute, Boston,
Massachusetts 02114, U.S.A.; [+]Department of Biochemistry, McGill University,
Montreal, Quebec, Canada H3G1Y6; [+]Department of Neurology, Harvard University
Medical School, Boston, Massachusetts 02115, U.S.A.

INTRODUCTION

It is currently believed that a cap structure, $m^7G5'ppp5'N$, which is present
at the 5' terminus of all eukaryotic cellular mRNAs analyzed to date, is required for efficient translation of mRNAs and functions during the polypeptide
chain initiation process.[1,2] Several polypeptides, M_r 24×10^3, 28×10^3, 50×10^3
and 80×10^3, have been shown to recognize the cap structure (referred to as
cap binding proteins or CBP).[3-6] The M_r 24,000 CBP has been purified to apparent homogeneity and shows biological activity in in vitro systems.[4,5] Monoclonal antibodies directed against CBPs interacted with a protein, M_r 28×10^3
and several high molecular weight polypeptides.[7] Furthermore, the immuno-
crossreacting high molecular weight proteins and the M_r 24,000 CBP share common
peptides.[7]

In the cytoplasm of eukaryotic cells mRNAs are associated with proteins
to form messenger ribonucleoprotein (mRNP) particles.[8,9] Although it has
been suggested that the mRNA-associated proteins may be involved in the regulation of different aspects of mRNA metabolism,[9] their precise biological functions remain to be defined. Several components of the translation system,
such as elongation and initiation factors, have been reported to be present
among the protein components of cytoplasmic mRNP particles and the RNA-binding
proteins of rabbit reticulocytes and wheat embryos.[10,11]

We have previously reported the isolation and characterization of various
classes of cytoplasmic RNP particles from chick embryonic muscle. Among these
are a novel class of translation inhibitory 10 S RNP (iRNP) containing a 4 S
RNA (iRNA);[12-14] the 24-40 S free mRNP particles;[15,16] and the polysomal
mRNP.[8,17,18] Our results suggest that these particles may be involved in
translational regulation during myogenesis.[8,13,14] It was of interest to
study whether or not the presence of CBP-related polypeptides in different
classes of RNP particles could be correlated with possible models of translational control involving RNP particles. In this study using a monoclonal

antibody to purified CBP isolated from rabbit reticulocytes,[7] we have shown
that there is specific association of CBP-related polypeptides in certain
types of RNP particles, and some of the low mol. wt. CBP-related polypeptides
present in the RNP particles may arise due to proteolytic cleavage of an immunoreactive RNP-protein M_r 78x10[3].

MATERIALS AND METHODS

The 10 S iRNP containing 4 S iRNA, and the 20-40 S free mRNP particles
were isolated from the postpolysomal supernatant of 14-day old chick embryonic
leg and breast muscle by a combination of ultracentrifugation and sucrose
gradient fractionation as previously described.[12,13] The 10 S iRNP was further
purified by gel filtration on ultrogel ACA34.[13] The free mRNP particles were
further fractionated on oligo-dT-cellulose column into two fractions: poly(A)-
lacking unbound mRNP containing 8-20 S mRNAs which code for a subset of cellular
proteins, M_r 15,000-42,000;[15] and the bound poly(A)[+]mRNP fraction eluted with
low salt buffer at 45°C (thermal chromatography), as previously described.[17,18]
The polysomal mRNP was isolated from the 0.5 M KCl-sucrose washed EDTA-dissociated polysomes by oligo-dT-cellulose chromatography.[17,18] For the isolation
of poly(A)-protein segments of polysomal mRNP, the particles were digested
with T_1 and pancreatic RNase under conditions in which the poly(A)-protein
segments are resistant to nucleolytic cleavage.[8] After chromatography of
the digest on oligo-dT-cellulose column, the bound poly(A)-protein segments
were isolated by thermal chromatography.[17-19] The preparative procedures
for the isolation of different types of RNP particles used in this study and
their characterization have been previously described in detail.[8,12-15,17-19]

For the isolation of protein components of RNP, three methods were used.
The RNP particles were deproteinized by extraction with phenol/CHCl$_3$/isoamyl
alcohol and the proteins were precipitated with acetone from the phenol layer.[13]
Alternatively, the RNP particles were digested with RNase T_2, dialyzed against
10 mM Tris-HCl, pH 7.4, 1 mM EDTA, 5 mM MgCl$_2$ and 50 mM NaCl, and the material
was lyophilized. The proteins of 10 S iRNP were isolated by DEAE-cellulose
chromatography of the RNP under dissociating conditions, as previously
reported.[14] The proteins were analyzed by one and two-dimensional SDS-polyacrylamide gel electrophoresis (SDS-PAGE), as previously described.[13]

For the detection of immunoreactive polypeptides, the proteins following
SDS-electrophoresis were blotted to nitrocellulose sheet (Millipore) according
to the procedure of Towbin et al.[20] The blot was preincubated for 1 hr at
25°C with 2.5% bovine serum albumin and 5% horse serum in buffer A (10 mM

Tris, pH 7.4, 0.15 M NaCl). After washing with buffer A the blot was incubated
overnight with a solution containing purified anti-CBP monoclonal antibody[7]
at 1-2 μg/ml, 1% bovine serum albumin and 0.5% horse serum in buffer A. The
blot after washing with buffer A was reacted with horseradish peroxidase con-
jugated goat anti-mouse IGG fraction and the immunoreactive bands were stained
with a mixture of diaminobenzidine (1 mg/ml), imidazole (1 mg/ml) and 0.01%
H_2O_2.[20]

For peptide mapping, the protein bards were sliced from stained SDS-polyacrylamide gels, the protein was labeled with $Na^{125}I$ (0.4 mC_i per slice) and
digested with trypsin in gel slices.[21] The resulting peptides were eluted
from the gel and analyzed on Polygram-cellulose 300 thin layer chromatographic
plate. The first dimension was electrophoresis in pyridine/acetic acid/ace-
tone/water (1:2:8:40 v/v) at pH 4.4 for 75 min at 800 volts. The second dimension was chromatography in n-butanol/acetic acid/water/pyridine (25:3:12:10
v/v) for 5-6 hr.[21] The thin layer plates were exposed to Kodak XR-5 films
for autoradiography.

RESULTS AND DISCUSSION

The Coomassie blue-stained band patterns of proteins present in different
RNP particles of chick embryonic muscle, resolved by SDS-PAGE and the corresponding immunostaining patterns obtained after reaction with anti-CBP monoclonal antibody is shown in Fig. 1. Both 20-40 S free poly(A)⁻mRNP particles
(panel A, gel 1) and the 10 S translation inhibitory iRNP particles (panel
B, gel 1) contain a complex set of proteins (M_r 15,000-150,000), which is
in agreement with our previous report.[12-15] Among these multiple proteins,
a single major band, M_r 78,000, was found to react with anti-CBP antibody.
A few minor bands, mostly in the molecular weight range of 25,000-50,000,
were also detected by immunostaining in some preparations. The free poly(A)⁺
mRNP particles of chick embryonic muscle[8,18] also gave a major immunostained
band (M_r 78,000) and some minor bands ranging in size from 30,000 to 60,000
daltons (results not shown). The polysomal mRNP particles (panel C, gel 1),
in contrast to iRNP and the 20-40 S free poly(A)⁻mRNP, gave a relatively simple
protein pattern in which the size of the major bands ranged between 35,000
and 78,000 daltons.[8,18] None of these bands gave any detectable immunostaining
after reaction with anti-CBP antibody (panel C, gel 2). In order to confirm
the absence of CBP-related polypeptides in polysomal mRNP, we also tested
various preparations of polysomal mRNP isolated by different techniques such
as oligo-dT-cellulose chromatography of EDTA-dissociated polysomes and differential elution of the bound mRNP particles at $25^{\circ}C$ and $45^{\circ}C$ to yield subpopula-

tions of polysomal mRNP, as reported by us.[22] None of these preparations showed any immunostained band (results not shown here), although they all contained the poly(A)-bound protein (M_r 78,000), as revealed by Coomassie blue staining.

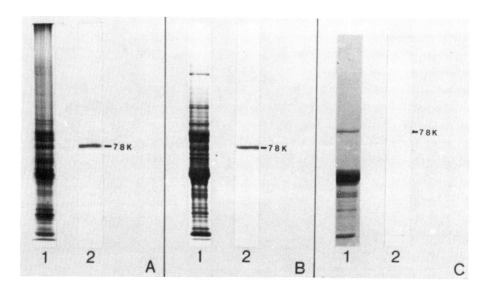

Fig. 1. Reaction of anti-CBP monoclonal antibody with different cytoplasmic RNP particles of chick embryonic muscle. For details see Materials and Methods. Panel A: 20-40 S free mRNP particle. Gel 1, Coomassie blue staining. Gel 2, immunostaining. Protein loaded, 85 µg. Gradient gels (10-15%) were used. Panel B: 10 S iRNP particle. Gel 1, Coomassie blue staining. Gel 2, immunostaining. Protein loaded, 95 µg. Gradient gels (10-15%) were used. Panel C: Polysomal mRNP. Gel 1, Coomassie blue staining. Gel 2, immunostaining. Protein loaded, 60 µg. Seven and one-half per cent gels were used.

The above-mentioned results suggest that the CBP-related protein (M_r 78,000) identified in free mRNP and the 10 S iRNP is distinct from the poly(A)-bound protein of 78,000 mol. wt., which is known to be present in polysomal mRNP particles of a wide variety of eukaryotic cells.[9,17,23] In order to test this view further, the 10-12 S poly(A)-protein segments were isolated by digestion of KCl-sucrose washed chick embryonic muscle polysomes with a mixture of pancreatic and T_1 RNase and chromatography of the digest on oligo-dT-cellulose column, followed by elution of the bound poly(A)-protein complex at 45°C with low salt buffer.[17,19] The nitrocellulose blot of the SDS-PAGE of the poly(A)-protein fragment was assayed for CBP-related proteins. In agreement with the results obtained with intact polysomal mRNP (Fig. 1, panel C), no detectable immunostaining was observed (results not shown), although the band of the poly(A)-associated protein (M_r 78,000) was visible by Coomassie blue staining.

When the protein components of 20-40 S free poly(A)⁻mRNP and iRNP were first isolated and then analyzed for CBP-related polypeptides, the results showed different patterns than those obtained with intact particles. Among the proteins of free mRNP particles the immunostained components corresponded to a major new band, M_r 43,000 and a much less intense band, M_r 78,000 (Fig. 2, panel A, gel 2). The proteins of 10 S iRNP showed a single immunoreacting component, M_r 29,000 (panel B, gel 3). The reciprocal relationship between the intensities of the 78,000 mol. wt. component and the lower mol. wt. new

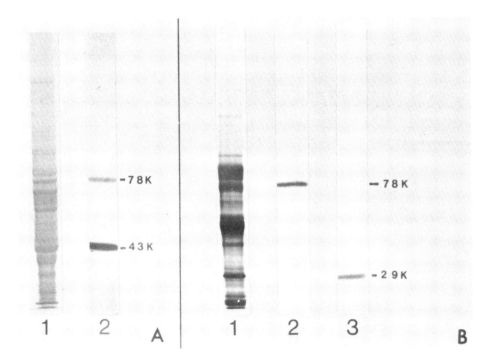

Fig. 2. Reaction of anti-CBP monoclonal antibody with proteins isolated from cytoplasmic RNP particles of chick embryonic muscle. For details see Materials and Methods and legend to Fig. 1. Panel A: Proteins isolated from 20-40 S free mRNP particle. Gel 1, Coomassie blue staining. Gel 2, immunostaining. Panel B: Proteins isolated from 10 S iRNP. Gel 1, Coomassie blue staining. Gel 2, immunostaining of proteins stored in 8 M urea. Gel 3, immunostaining of proteins stored in the absence of 8 M urea.

bands (Figs. 1 and 2) suggests that the M_r 43,000 and 29,000 bands may have resulted from cleavage of the M_r 78,000 polypeptide during the isolation and subsequent handling of the proteins of RNP particles. Interestingly, when proteins of 10 S iRNP were stored in 8 M urea, a single immunostained band of M_r 78,000, similar to the pattern obtained with intact particles (Fig. 1), was observed (Fig. 2, panel B, gel 2), suggesting that the proteolytic cleavage of 78,000 dalton CBP-related protein was inhibited under these conditions. Several CBP polypeptides (M_r 24,000, 28,000, 50,000 and 80,000) have been identified by crosslinking reticulocyte initiation factors to the cap structure.[6,7] Also, CBP-related proteins appear to be structurally related, as shown by the presence of common peptides.[4,7] Comparison of the immunostained band patterns (Figs. 1 and 2) indicates that specific classes of cytoplasmic RNP particles of chick embryonic muscle contain a family of immunoreactive CBP-related polypeptides similar in size to those found in other eukaryotic cells. The multiplicity of these proteins found in RNP particles may arise due to cleavage of a 78,000 dalton immunoreactive RNP protein.

In order to probe whether the CBP-related polypeptides found in free mRNP particles are correlated with protein synthesis, we attempted to determine the nature of the immunoreactive proteins in polysomes. The Coomassie blue staining patterns of proteins of polysomes washed with 0.25 M (low salt) or 0.5 M KCl (high salt) is shown in Fig. 3. The staining intensity of most proteins was similar, which is in agreement with the view that the majority of ribosomal proteins are not washed off with salt concentration up to 1 M. However, the staining intensity of several proteins is decreased in the preparation of 0.5 M KCl-washed polysomes, suggesting that these polysome-associated proteins are not integral ribosomal components. A single major immunostained protein, M_r 43,000, is found in both polysomal preparations (lanes 3 and 4) but its intensity is less in 0.5 M KCl-washed polysomes (lane 4). Since no immunostained band is detected in polysomal mRNP samples (Fig. 1), the M_r 43,000 CBP-related protein located in polysomes may have been lost during the isolation of polysomal mRNP. Recent work has shown that anti-CBP antibody reacts with a protein, M_r 50,000, located in reticulocyte polysome[7] and in the cytoskeleton fraction in BHK cell.[24] These results imply that the M_r 50,000 CBP-related protein may be involved in binding of mRNA to cytoskeleton, which is currently believed to be required for translation.[25] The M_r 43,000 CBP-related polypeptide which we have identified in chick embryonic muscle polysome (Fig. 3) is similar in size to the M_r 50,000 CBP-related protein reported in reticulocyte and BHK cells and may function in a similar manner.

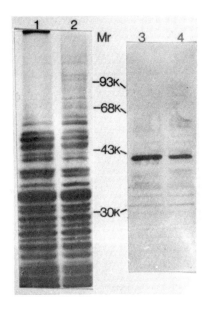

Fig. 3. SDS-PAGE patterns of chick embryonic muscle polysomes and the corresponding anti-CBP crossreacting immunostained band patterns. For details see Materials and Methods. Lane 1, Coomassie blue staining of SDS-gels of 0.25 M KCl-washed poly-somes. Lane 3, immunostaining of the corresponding nitrocellulose blot. Lane 2, Coomassie blue staining of SDS-gels of 0.5 M KCl-washed poly-somes. Lane 4, immunostaining of the corresponding nitrocellulose blot.

In order to gain more insight into the nature of the immunoreactive M_r 78,000 polypeptide identified in the 10 S iRNP and the 20-40 S free mRNP parti-cles, we have analyzed these polypeptides by two-dimensional gel electrophore-sis and peptide mapping. The proteins of 10 S iRNP and the 20-40 S free mRNP particles were resolved by two-dimensional gel electrophoresis.[13,26] The proteins were then blotted onto a nitrocellulose sheet and assayed for immuno-staining. The major band obtained in the case of 10 S iRNP and the 20-40 S free mRNP particles had a molecular weight of 78,000 and pI value of about 7.6 (Fig. 4, A and B). Peptide mapping using a sequential combination of electrophoresis and chromatography on thin layer plate[21] shows that the M_r 78,000 polypeptides present in 20-40 S free mRNP (Fig. 5A) and 10 S iRNP (Fig. 5B) share many common peptides (indicated by arrows). These results indicate that the immunoreactive M_r 78,000 polypeptides found in the two types of RNP particles are structurally related. An estimate of the extent of their homology will require complete sequence determination of these two polypeptides. The presence of CBP-related M_r 78,000 polypeptide in the 10 S iRNP is somewhat surprising, since both iRNA and iRNP are potent inhibitors of in vitro transla-tion of capped and uncapped mRNAs.[12-14] Determination of nucleotide sequence at the 5-terminus of iRNA, which is now in progress in our laboratory, will be helpful in understanding the interaction of CBP-related polypeptides with iRNA.

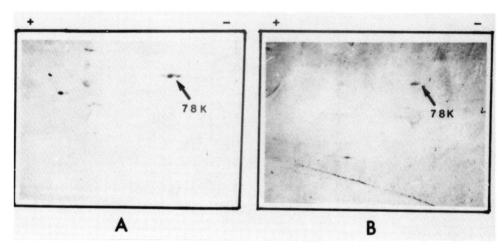

Fig. 4. Immunostained band patterns obtained with two-dimensional gel electrophoresis of cytoplasmic RNP particles of chick embryonic muscle. For details see Materials and Methods. Panel A: 20-40 S free mRNP particle. Panel B: 10 S iRNP particle.

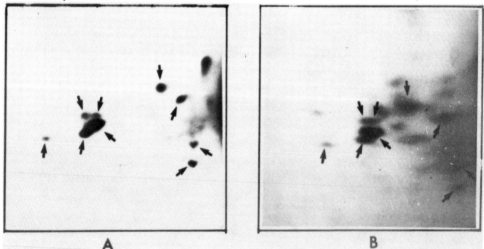

Fig. 5. Tryptic peptide patterns of the M_r 78,000 immunoreactive polypeptides isolated from cytoplasmic RNP particles of chick embryonic muscle. For details see Materials and Methods. Electrophoresis was from right to left (first dimension) and chromatography was from bottom to top (second dimension). The bottom right corner represents the origin. The autoradiograph is shown. Panel A: M_r 78,000 polypeptide from 20-40 S free mRNP particle. Panel B: M_r 78,000 polypeptide from 10 S iRNP particle.

In summary, the following features emerge from the results presented here. A family of immunoreactive polypeptides (M_r 78,000; 43,000 and 29,000) are identified in the free mRNP particles and the 10 S translation inhibitory RNP of chick embryonic muscle by reaction with anti-CBP antibody. Their molecular sizes are similar to those reported for the family of CBP-related proteins isolated from other eukaryotic cells.[6,7,24] The M_r 78,000 polypeptide appears to be the single immunoreactive species found in intact RNP. In contrast, the lower mol. wt. proteins (M_r 43,000 and 29,000) appear in preparations of the isolated proteins, presumably due to cleavage of the M_r 78,000 polypeptide. The immunoreactive 78,000 dalton protein identified in cytoplasmic RNP is distinct from the poly(A)-bound mRNP-protein of similar mol. wt., which is known to be present in a wide variety of eukaryotic cells.[9,17,23] The observation that polysomes contain a single immunoreactive protein, M_r 43,000, in contrast to the M_r 78,000 polypeptide present in free mRNP particles, suggests that the CBP-related proteins undergo a dynamic exchange during the transit of mRNA from the free mRNP pool to polysomes. Such a dynamic exchange for the majority, if not all, of the mRNA-associated proteins, has been previously postulated by us based on the subcellular localization and biological activity of eukaryotic mRNAs.[18,19,27]

Some of the results presented here are in agreement with those obtained by Drs. H. Trachsel and A. Vincent, who have independently identified CBP-related polypeptides in cytoplasmic mRNP particles of reticulocytes by reaction with anti-CBP antibody (personal communication).

ACKNOWLEDGMENTS

This work was supported by grants from the National Institutes of Health (AM 13238), the Muscular Dystrophy Association of America and Medical Research Council of Canada (No. 7214). We thank Dr. Hans Trachsel for his kind gift of the anti-CBP monoclonal antibody. We also thank Ms. Swantana Mukherjee and Ms. Denise Guertin for expert technical assistance.

REFERENCES

1. Shatkin, A.J. (1976) Cell 9, 645-653.
2. Banerjee, A. (1980) Bacteriol. Rev. 44, 175-204.
3. Sonenberg, N., Morgan, M.A., Merrick, W.C., and Shatkin, A.J. (1978) Proc. Nat. Acad. Sci. 75, 4843-4847.
4. Sonenberg, N., Rupprecht, K.M., Hecht, S.M., and Shatkin, A.J. (1979) Proc. Nat. Acad. Sci. 76, 4345-4349.
5. Sonenberg, N., Trachsel, H., Hecht, S.M., and Shatkin, A.J. (1980) Nature 285, 331-333.
6. Sonenberg, N. (1981) Nucleic Acds Res. 9, 1643-1656.

7. Sonenberg, N. and Trachsel, H. (1982) in Current Topics in Cellular Regulation, Horecker, B.L. and Stadman, E.R. ed., Vol. 21, Academic Press, New York, pp. 65-88.
8. Jain, S.K. and Sarkar, S. (1979) Biochemistry 18, 745-753.
9. Preobrazhensky, A.A. and Spirin, A.S. (1978) Prog. Nucleic Acid Res. Mol. Biol. 21, 1-38.
10. Vlasik, T.N., Ovchinnikov, L.P., Radjabov, Kh. M., and Spirin, A.S. (1978) FEBS Lett. 88, 18-20.
11. Ovchinnikov, L.P., Spirin, A.S., Erni, B., and Staehlin, T. (1978) FEBS Lett. 88, 21-26.
12. Mukherjee, A.K., Guha, C., and Sarkar, S. (1981) FEBS Lett. 127, 133-138.
13. Sarkar, S., Mukherjee, A.K., and Guha, C. (1981) J. Biol. Chem. 256, 5077-5086.
14. Mukherjee, A.K. and Sarkar, S. (1981) J. Biol. Chem. 256, 11301-11306.
15. Mukherjee, A.K. (1981) J. Cell Biol. 91, 366a.
16. Bag, J. and Sarkar, S. (1975) Biochemistry 14, 3800-3807.
17. Jain, S.K., Pluskal, M.G., and Sarkar, S. (1979) FEBS Lett. 97, 84-90.
18. Jain, S.K., Roy, R.K., Pluskal, M.G., Croall, D.E., Guha, C., and Sarkar, S. (1979) Mol. Biol. Rep. 5, 79-85.
19. Roy, R.K., Lau, A.S., Munro, H.N., Baliga, B.S., and Sarkar, S. (1979) Proc. Nat. Acad. Sci. 76, 1751-1756.
20. Towbin, H., Staehlin, T., and Gordon, J. (1979) Proc. Nat. Acad. Sci. 76, 4350-4354.
21. Elder, J.H., Pickett, R.A., II, Hampton, J., and Lerner, R.A. (1977) J. Biol. Chem. 252, 6510-6515.
22. Jain, S.K., Roy, R.K., Pluskal, M.G., and Sarkar, S. (1980) in Biomolecular Structure, Conformation, Function and Evolution, Vol. 2, Srinivasan, R. ed., Pergamon Press, Oxford, pp. 461-472.
23. Blobel, G. (1973) Proc. Nat. Acad. Sci. 70, 924-928.
24. Zumbe, A., Stahli, C., and Trachsel, H. (1982) Proc. Nat. Acad. Sci. 79, 2927-2931.
25. Cervera, M., Dreyfuss, G., and Penman, S. (1981) Cell 23, 113-120.
26. O'Farrell, P.H. (1975) J. Biol. Chem. 250, 4007-4021.
27. Roy, R.K., Sarkar, S., Guha, C., and Munro, H.N. (1981) in The Cell Nucleus, Vol. 9, Part B, Busch, H. ed., Academic Press, New York, pp. 289-308.

THE ROLE OF mRNA COMPETITION IN REGULATING TRANSLATION IN NORMAL FIBROBLASTS

WILLIAM E. WALDEN AND ROBERT E. THACH[+]
[+]Department of Biology, Washington University, St. Louis, Missouri 63130 USA

INTRODUCTION

The fact that different mRNAs may be translated at different rates in eukaryotic cells has been established for many years[1,2,3]. However, the molecular mechanisms which determine translation rates of individual mRNAs are not well understood. Even in general kinetic terms relatively little is known beyond the fact that initiation, and not elongation or termination, is the chief control point for the overall process[3]. An extremely useful kinetic model has been proposed by Lodish which describes the results of mRNA specificity in the initiation step[2]. This is based on two postulates: that different mRNAs bind to ribosomes with different rate constants, and that ribosomes close to the initiation site on the mRNA can interfere with subsequent initiation events. Shortly after the elaboration of this model various alternatives or supplements to it were proposed, to account for the same or related phenomena. One of these is the concept that mRNAs must compete for a limiting component of the initiation machinery in order to be translated, and that competitive inhibition of translation of one mRNA by other mRNAs may be an important factor in regulating their initiation rates. This was first proposed to account for the shutoff of cellular protein synthesis induced by encephalomyocarditis (EMC) virus infection[4]. It was postulated that prior to binding to the 40S initiation complex, a mRNA must first be recognized by a "message discriminatory initiation factor", so-called because it can bind to different mRNAs with different affinities. Also implicit were the assumptions that the molar concentrations of cellular mRNA and discriminatory factor should be roughly equivalent, and that the affinity of viral mRNA for the factor was much greater than that of the average host mRNA. Under these conditions, increasing levels of EMC viral mRNA would outcompete host mRNA in binding to the discriminatory factor, thereby reducing, and ultimately shutting off, host translation. We shall refer to this set of postulates as "the competition model", in order to emphasize that its unique feature is the competitive inhibition of translation of any given mRNA by other mRNAs.

Subsequent studies have confirmed and extended the competition model. The competitive advantage of EMC viral over cellular mRNA, and the significance of this fact for the shutoff of cellular translation, have been amply documented [5-9]. In addition, it is now clear that mRNA competition plays a central role in the replication of a number of other animal viruses, notably vesicular stomatitis and reo[10-12]. Kabat and Chappell[13] extended this concept to uninfected cells by showing that it could account for the regulation of α- and β-globin synthesis in rabbit reticulocytes; their model was suggested as a complement to that of Lodish[2], which did not deal with competition in the sense used here. Competition also appears to be involved in determining the effects of insulin deprivation[14] and density dependent growth inhibition[15-17] on the regulation of translation in fibroblasts and other cell types.

The precise identification of the message discriminatory factor (which has been abbreviated as "F", ref. 17), remains elusive. Early work, employing relatively crude initiation factor preparations, suggesting that eIF-4A and -4B were both involved in specific mRNA recognition[5-13]. This conclusion has recently been supported by the results of Grifo, et al., who have demonstrated the formation of an ATP-dependent complex between eIF-4A, eIF-4B and mRNA[18]. However, the work of Kaempfer, et al., has implicated eIF-2 in the mRNA recognition process[9,19]. In addition, it has been suggested that a 33,000 dalton component of eIF-3 may play this role in muscle cells[20]. This proliferation of candidates for F is confusing. In our view it stems from several problems, including a) lack of complete purity of factors employed, b) lack of a completely homologous in vitro system which faithfully reflects conditions in an in vivo correlate, c) lack of a sophisticated mathematical model suitable for quantitative data analysis, and d) possibly subtle but real differences between systems.

Messenger RNA competition has been analysed mathematically by a number of workers[13,17,21,22]. The most recent model, which includes all of the features described above, is that of Godefroy-Colburn and Thach[17]. This model deals with many aspects of the translational process, and hence employs a large number of parameters and constants. However, all but a very few of these can be assigned known values, hence the number of adjustable parameters available for modelling or "curve-fitting" is in fact relatively small. Moreover, within a certain limited range of conditions this model can be reduced to yield a formulation very similar to that of Lodish[2], the major difference being that a mRNA-factor complex (which is in equilibrium with limiting free factor and excess free mRNA), rather than mRNA per se, binds to the activated 40S ribosome[23].

As described above, the competition model has been very useful in explaining a number of cases of translational regulation in special circumstances, chiefly virus infection and cellular growth control. The most extensive documentation of this phenomena has been conducted in reovirus-infected mouse cells[12,17,24-26]. These studies showed that: a) reovirus infection of mouse SC-1 fibroblasts does not alter the basic kinetic parameters or specificity of the host translational apparatus; b) reovirus mRNAs compete in vivo with host mRNAs for a limiting, message discriminatory component (that is, F); c) the mole ratio of F to total host mRNA is slightly less than unity; d) most reovirus mRNAs have a relatively low affinity for F, hence they are outcompeted by host mRNAs and therefore initiated at an abnormally low rate; e) the modest inhibition of host translation in reovirus-infected SC-1 cells is due solely to competitive inhibition by viral mRNA; f) the relative competitive efficiencies of host and viral mRNAs can be reproduced in vitro, provided that appropriate ionic conditions are chosen; g) both in vivo and in vitro results can be modelled mathematically with a high degree of accuracy. These results provide the strongest evidence to date that mRNA competition can be an important regulatory phenomenon.

One point that deserves special comment is the fact that the mRNA competition model has developed so slowly. This has been due to three basic problems: a) Early work on competition was done in vitro [4,5,13], at a time when the extreme variability of such results with ionic conditions, activity of translation systems, etc., was also becoming apparent; thus it was entirely appropriate that any such results should be viewed with caution. However, in recent years the sophistication of in vitro systems has increased enormously to the point where in vitro results can be compared directly to those obtained in vivo [11,12,19,24,25]. Thus a greater degree of confidence can now be placed in such data. b) Early attempts to demonstrate mRNA competition in vivo were severely limited by lack of suitable technology. The first suggestive data was obtained with inhibitors of translation[6]. Once again a cautious interpretation was appropriate, due to the potential for unknown side effects of the drugs employed. However, subsequent studies employing a large number of inhibitors in several different systems have confirmed the validity of this approach[14,27,28]. Moreover, the major conclusions thereby obtained have been recently supported by studies on polysome sizes and translational yields for individual mRNAs[12]. In addition , it is becoming apparent that mutations in single ribosomal proteins are sufficient to render eukaryotic cells resistant to cycloheximide[29] or emetine[30], suggesting that these inhibitors are highly specific for protein synthesis and may not produce additional side effects. c) Early work was done in the absence of a rigorous mathematical model for competition. It was therefore difficult to rule out alternative explanations for subtle kinetic phenomena. However, the derivation and application of highly sophisticated models is now routine, so this final handicap would seem to have been eliminated[17,22]. Thus, at the present time, all three of these problems would seem to have been overcome.

While considerable progress has been made in our understanding of the competition phenomenon, it remains a relatively new and little studied concept. Many questions remain to be answered before the breadth of the phenomenon can be assessed. Among the most fundamental of these is whether competition occurs in normal, rapidly growing cells. While the results of mathematical modelling predict that this must be the case in SC-1 and vero fibroblasts[17], no direct evidence bearing on this point has been obtained. We have recently addressed this question using techniques previously developed for the analysis of translation kinetics in reovirus infected cells[12]. Preliminary results suggesting that competition does exist in growing SC-1 fibroblasts are presented below.

MATERIALS AND METHODS
Cells and Viruses

SC-1 cells, obtained from Dr. J. Hartley of the National Cancer Institute, were grown in monolayer cultures in McCoy 5A medium as previously described[31], except that the cultures were supplemented with 10% FCS. For cycloheximide experiments confluent monolayers were trypsinized and replated at a cell density of $2-7 \times 10^3$ cells/cm^2 in McCoy 5A medium containing 10% FCS. The cells were allowed to attach and grow in this medium for 24 hours at which time the medium was replaced with McCoy 5A containing 2% FCS. Experiments were performed 24 hours later when cell densities were approximately $1-3 \times 10^4$ cells/cm^2. The cell doubling time at this density is approximately 20 hours.

In Vivo Protein Synthesis Analysis

SC-1 cell monolayers in 35 mm cluster dishes (Costar) were preincubated at 37°C for 15 minutes in methionine free MEM-E containing 2% dialyzed FCS and cycloheximide (Sigma Chemical Co.) as indicated in RESULTS. The monolayers were washed once and labeled for 30 min at 37°C in 0.5 ml of this medium plus cycloheximide containing 25 μCi of [^{35}S]methionine (Amersham; 600-1400 Ci/mmole). Pulsed cells were then washed once with 2 ml of warm McCoy 5A medium containing 2% FCS. An additional 2 ml of this medium was added to the monolayer and incubation continued at 37°C for 15 min in order to chase all radioactivity into completed proteins. Labeled cells were washed twice with cold PBS and lysed with PBS containing 1% SDS. TCA was added to the lysates to a final concentration of 10%. The resulting precipitates were collected by centrifugation, washed twice with acetone, dried, and dissolved in sample buffer (20 mM Tris-HCl, pH 6.8, 2% SDS, 20% glycerol). Protein concentrations were determined by the method of Lowry et al.[32] Sample volumes were adjusted to give an equal protein concentration and 2-mercaptoethanol was added to a final concentration of 5%. After adding bromophenol blue, an aliquot containing approximately 2 µg of protein was analyzed by polyacrylamide gel electrophoresis as described further below.

Polysome Preparation

Polysomes were prepared as described by Walden, et al.[12] Cell cultures (approx. 1×10^8 cells per gradient) were refed 30 min prior to harvest with fresh growth medium. At harvest the growth medium was removed and the cells were rapidly cooled by adding ice cold polysome buffer (200 mM sucrose, 50 mM KCl, 5 mM Mg(OAC)$_2$, 20 mM Tris-HCl, pH 7.4, and 100 µg/ml cycloheximide) and submerging the culture vessel into an ice-salt bath (-10°C). All subsequent manipulations were done at 4°C. Cells were collected by scraping with a rubber policeman and pelleted by centrifugation at 800 x g for 5 min. Pelleted cells were washed once with polysome buffer. The cells were pelleted again and the packed cell volume measured. An equal volume of polysome buffer supplemented with 1 mM dithiothreitol and 0.1% diethyl pyrocarbonate (both from Sigma Chemical Co.) was added and the suspension transferred to a Dounce homogenizer. NP-40 (BDH Chemicals Ltd.) was added to a final concentration of 1%. The cells were disrupted with 7 strokes of a tight fitting pestle. Sodium deoxycholate (Calbiochem-Behring Corp.) was added to a final concentration of 1% and the nuclei were removed by centrifugation at 3,000 x g for 5 min. The post nuclear supernatant was layered onto a 25%-50% (w/v) sucrose gradient in 50 mM KCl, 5 mM Mg(OAC)$_2$ and 20 mM Tris-HCl, pH 7.4. The polysomes were resolved by centrifugation in a Beckman SW41 rotor at 36,000 RPM for 110 min at 4°C. Gradient fractions (1.5 ml) were collected and monitored for absorbance at 260 nm using a Gilford Recording Spectro-photometer. After dilution with an equal volume of TMK buffer (20 mM Tris-HCl, pH 7.4, 3 mM Mg(OAc)$_2$, and 100 mM KCl), each fraction was layered over a 30% (w/v) sucrose cushion in TMK buffer. The polysomes were pelleted in a Beckman Type 65 rotor at 40,000 RPM for 10 hours at 4°C. The resulting pellet was gently rinsed with TMK buffer and resuspended into 25 µl of the same buffer.

In Vitro Translation of Polysomal RNA

Polysomal RNA was phenol/chloroform extracted from 12.5 µl of polysome suspension, precipitated with ethanol and dissolved in sterile water. Equal

404

volumes of the RNA samples were translated in vitro using the fractionated
translation system described by Brendler et al.[24] from Krebs ascites tumor
cells. Each 25 µl assay contained 20 mM Tris·HCl, pH 7.4, 2.5 µg partially
hydrolysed mouse ribosomal RNA, 2 mM dithioerythritol (Sigma Chemical Co.),
1 mM ATP (Sigma Chemical Co.), 0.1 mM GTP (Sigma Chemical Co.), 0.6 mM CTP
(Sigma Chemical Co.), 4 µg creatine kinase (Boehringer-Mannheim Biochemicals),
0.1 A_{260} units MOPC 460 tumor tRNA, 1.5 µg spermidine (Calbiochem-Behring
Corp.), 10 mM creatine phosphate (Sigma Chemical Co.), 320 µM S-adenosyl-
homocysteine (Calbiochem-Behring Corp.), 40 µM 19 amino acids minus methionine
(Eastman Kodak Co.), 5 µCi ^{35}S-methionine (Amersham/Searle), 0.6 µM unlabeled
methionine (Eastman Kodak Co.,), 0.03 A_{260} units 40S ribosomes, 0.12 A_{260}
units 60S ribosomes, 25 µg to 40 µg ribosomal salt wash and 25 µg pH 5 fraction
proteins. Incubation of the protein synthesis assays was for 2 hours at 30°C.

Polyacrylamide Gel Electrophoresis

SDS-polyacrylamide gel electrophoresis was performed by the method of
Laemmli[33]. Samples to be analyzed were boiled for 5 min in a water bath.
Twenty-five microliter samples were layered onto a 1.6 mm thick slab gel
consisting of a 4% acrylamide stacking gel and a 5% to 12% acrylamide linear
gradient separating gel. Electrophoresis was started at 110 volts and the
voltage was increased to 125 volts when the dye had entered the separating
gel. The run was terminated after 5.5 to 6 hours, and the gel was fixed and
autoradiographed. The exposed film was scanned using a Joyce-Loebl double
beam scanning microdensitometer. The amount of incorporated radioactivity in
each protein band was determined by measuring the peaks from such scans. All
film analyzed had been exposed in the linear range.

RESULTS

The approach used to demonstrate mRNA competition in normal growing cells
was similar to that used previously for reovirus-infected cells[12]. To identify
mRNAs that have low affinities for discriminatory factor, cells were treated
briefly with low concentrations of cycloheximide, pulse labeled with
[^{35}S]methionine, and then chased with excess cold methionine (in order to
minimize post-translational processing effects) as described in MATERIALS AND
METHODS. Labeled proteins were then analyzed by SDS-PAGE and autoradiography,
as shown in Fig. 1. While labeling of the great majority of proteins is
reduced monotonically by cycloheximide addition, synthesis of a few (for
example, proteins C, D and Z in Fig. 1) is actually stimulated by low
concentrations of the inhibitor. Several of these were chosen for further

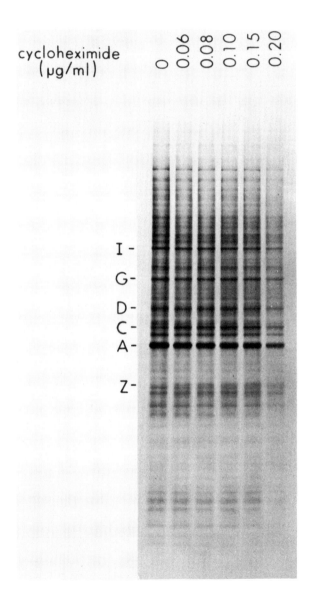

Fig. 1. Effect of cycloheximide on protein synthesis rates in growing SC-1 cells. Cells were treated with cycloheximide at the concentrations indicated, labeled with [^{35}S]methionine, "chased" with cold methionine, and analyzed by SDS-PAGE and autoradiography as described in MATERIALS AND METHODS.

study. Quantitation of the stimulatory effect of cycloheximide on protein C
synthesis is shown in Fig. 2A. A small but reproducible increase in the rate
of C synthesis is evident. This increase has been noted in reovirus infected
as well as in control cells, and is maximal at approximately 0.05 µg/ml
cycloheximide. At higher doses of cycloheximide inhibition of C synthesis
occurs, and the slope of this inhibition curve is identical to those for the
non-stimulated proteins represented by A (tentatively identified as actin).
Most proteins in the cell are of the latter type (e.g. proteins I and G in Fig.
1); thus the average response to cycloheximide, represented by the total TCA
precipitable radioactivity, is very similar to that of protein A. The
reproducibility of the stimulatory effect of cycloheximide on protein C

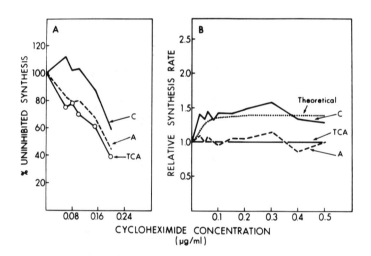

Fig. 2. Effect of cycloheximide on protein synthesis rates in growing SC-1
cells. Panel A: protein bands A and C in Fig. 1 were quantitated by
densitometry of the autoradiogram and total protein synthesis was quantitated
by TCA precipitation of aliquots from each sample prior to SDS-PAGE. Panel B:
comparison of results from four different experiments for synthesis of protein
C; a theoretical curve for the effect of cycloheximide on translation of a
minor (5% of total mRNA) message with a low (2% of normal) affinity for the
discriminatory factor calculated as in ref. 17 is also shown; both types of
data were normalized using total protein synthesis as 100% at each
cycloheximide concentration.

synthesis is indicated in Figure 2B, where the results from four different experiments are plotted (normalized to the TCA precipitable radioactivity data). These results are similar to those seen in reovirus-infected SC-1 cells, where synthesis of the most prominent viral proteins is enhanced by cycloheximide[12]. Moreover, the stimulation of host proteins in uninfected cells is quantitatively consistent with the mathematical model for mRNA competition previously developed[17], as shown in Fig. 2B.

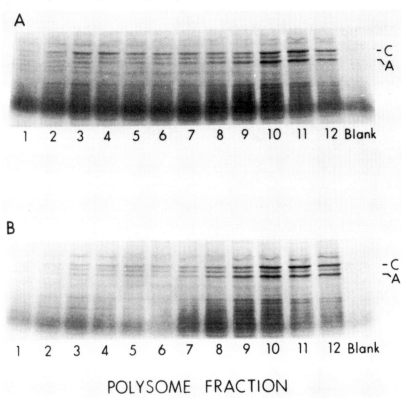

POLYSOME FRACTION

Fig. 3. Polysomal distribution of mRNAs for proteins A and C. Polysomes were prepared from SC-1 cells grown for 30 min in the absence (panel A) or presence (panel B) of 0.2 μg/ml cycloheximide, and then fractionated by sucrose gradient sedimentation. A representative portion of the RNA extracted from each gradient fraction was translated (at below half-saturating levels) in a fractionated cell-free system prepared from Krebs ascites cells. Translation products were analyzed by SDS-PAGE and autoradiography. The average polysome size (in ribosomes per mRNA) in each gradient fraction was: 0, fraction 1; 0, fraction 2; 0, fraction 3; 1, fraction 4; 1-2, fraction 5; 2-3, fraction 6; 3-4, fraction 7; 4-5, fraction 8; 6-7, fraction 9; 8-11, fraction 10; 12-18, fraction 11; very large polysomes, fraction 12.

408

These results suggest that the mRNAs encoding protein C and other cellular proteins are weak initiators and are outcompeted for a limiting initiation factor by stronger cellular mRNAs. Further evidence for this interpretation was obtained from an analysis of the polysomal distribution of the mRNA for C. Polysomes from SC-1 cells were prepared and fractionated by sucrose gradient sedimentation as described in MATERIALS AND METHODS. The mRNA for C was identified by translation of each fraction in a murine cell-free system[24], followed by SDS-PAGE and fluorography. As shown in Fig. 3A, the distribution of C mRNA is bimodal. Quantitation of this data (Fig. 4A) indicates that the two peaks of C mRNA occur at 8-12 and 0-1 ribosomes per message, respectively. That the polypeptides produced by these two classes of mRNA are identical was confirmed by the partial proteolysis technique of Cleveland et al.[34] as shown

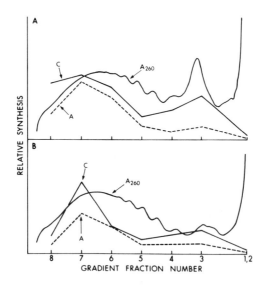

Fig. 4. Polysomal distribution of mRNAs for proteins A and C. Protein synthesis data from an experiment similar to that shown in Fig. 3 was quantitated by densitometry to indicate the presence of specific mRNAs in sucrose gradient fractions. The A_{260} profile of the polysomes from which fractions were taken for translation is indicated. Panel A: normal cells. Panel B: cells grown for 30 min in the presence of 0.2 µg/ml cycloheximide.

in Fig. 5. This curious result suggests that polypeptide C is encoded by two mRNAs. One of these initiates at a normal rate, since it contains a normal number of ribosomes (compare the distribution of A polysomes and C polysomes in Fig. 4A). The other is a weak initiator, as judged by the fact that it contains only one-eigth (or less) the normal number of ribosomes.

This data suggests that the average initiation rate on the mRNA(s) encoding C is usually low, which is consistent with its ability to be stimulated by cycloheximide.

The effect of a low dose of cycloheximide on these polysomal distributions bears out this interpretation. As shown in Fig. 3B and 4B, inclusion of 0.2 μg/ml cycloheximide in the cell culture media causes a shift of monosomes into

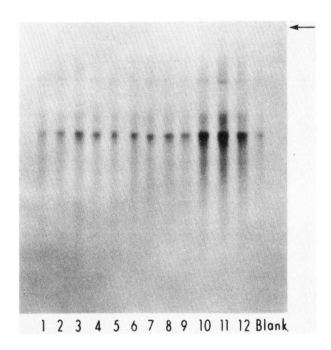

Fig. 5. Partial proteolytic mapping of protein C encoded by polysomal and monosomal mRNA. The C bands from the SDS-PAGE gel shown in Fig. 3A were excized, treated with S. aureus protease, and analyzed again by SDS-PAGE as described[34]. The arrow indicates the mobility of undigested protein C.

polysomes, while inhibiting the overall rate of protein synthesis by about 50%. This treatment drives most of the weakly initiating component of C mRNA into large polysomes, thus confirming that it is in fact a translatable message in vivo.

The curious distribution of the C mRNAs is by no means unique. A number of other proteins such as protein D, appear to be encoded by two classes of mRNA,

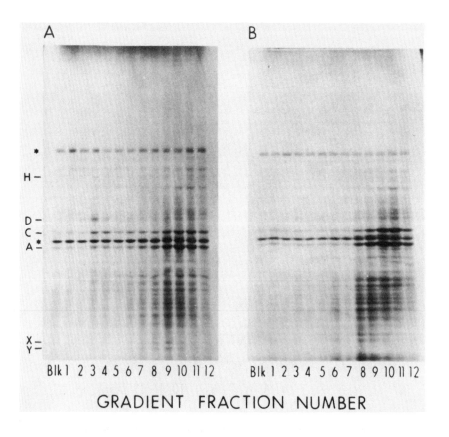

Fig. 6. Polysomal distribution of mRNAs encoding a wide range of proteins in the absence (panel A) or presence (panel B) of 0.2 μg/ml cycloheximide. Experimental procedures and lane designations were as described for Fig. 3.

one strongly and one weakly initiating (FIg. 6). Even the A protein exhibits a minor component of weakly initiating mRNA, which nevertheless can be driven into polysomes by low levels of cycloheximide. A similar observation has previously been made by Geoghegan, et al.[35]. This component is such a small fraction of the total A mRNA that it does not influence the normal cycloheximide inhibition curve of this protein (Fig. 2).

Still other patterns of mRNA translatability are seen. Some mRNAs exist only in very large polysomes, with no trace of a component at the top of the gradient (e.g., proteins X and Y in Fig. 3). Their position is unaffected by cycloheximide. These are presumed to be homogeneous, strongly initiating mRNAs. Conversely, a few mRNAs exist exclusively in small polysomes (or as free mRNP), and these can be driven into larger polysomes with cycloheximide (e.g., protein H in Fig. 6). Several of these correlate in size with proteins whose synthesis is stimulated by cycloheximide (Fig. 1), although identity remains to be rigorously established. In any event, these observations suggest that the intrinsic rate of initiation varies widely among the population of mRNAs in the normal cell.

In order to rigorously establish the identity of proteins produced in vivo and in vitro, another method of analysis which complements SDS-PAGE is required. For this purpose we have used isoelectric focussing. Corresponding pairs of bands from SDS-PAGE gels are readily excised and compared by this method. For example, in initial experiments the in vivo and in vitro versions of protein C, prepared both in the presence and absence of cycloheximide, were proven to have the same pI value, thereby confirming their identity (data not shown). Identification of other protein bands by this technique is in progress.

DISCUSSION

The results presented above suggest that mRNA competition exists in normal, rapidly growing SC-1 fibroblasts, and may be an important regulatory phenomenon. It is evident that this is of the same type as has been observed in reovirus-infected SC-1 cells[12]. This conclusion is in accord with predictions derived from mathematical modelling, which suggests that host mRNAs are in slight molar excess relative to the discriminatory initiation factor for which they must compete[17].

It is evident that according to the competition model, slight variations in the concentration of active discriminatory factor can have large effects on the translation of the weakly initiating mRNAs[17]. In addition, the synthesis of proteins encoded by a family of mRNAs consisting of strong and weak

412

initiators could be subtly modulated in response to changing conditions of factor concentration, etc. Whether such control mechanisms are actually used by cells to turn on or off the expression of some genes, or modulate that of others, remains to be determined.

Our results also suggest that the mRNAs for several proteins or groups of proteins, such as C and D, are heterogeneous with regard to initiation efficiency. The molecular basis for this heterogeneity is not known at present. In vitro translation experiments with both mRNA and mRNP preparations are in progress in order to elucidate this question. It will be of particular interest to determine whether the proteins encoded by the poorly initiating components are identical to those synthesized by the more rapidly initiating mRNAs (as appears to be the case for C) or represent distinct variants within a given family of genes.

Analysis of the distribution of mRNAs among large and small polysomes (or free mRNPs) suggests a wide diversity of intrinsic initiation rates. Some mRNAs, present only in very large polysomes, appear to have very high initiation rates, whereas others, ordinarily found in monosomes or free mRNP, are presumed to be very poor initiators. Similar observations have recently been reported by others[11,15,35,36].

ACKNOWLEDGEMENTS

This work was supported by a grant from the National Science Foundation (PCM-7911936). We are indebted to Mrs. Surekha Adya for expert technical assistance, and to Dr. Vincent Zenger for help in preparing the Figures.

REFERENCES

1. Lodish, H.F. and Jacobson, M. (1972) J. Biol. Chem. 247, 3622-3629.
2. Lodish. H.F. (1974) Nature, 251, 385-388.
3. Lodish, H.F. (1976) Ann. Rev. Biochem. 45, 39-72.
4. Lawrence, C. and Thach, R.E. (1974) J. Virol. 14, 598-610.
5. Golini, F., Thach, S.S., Birge, C.H., Safer, B., Merrick, W.C. and Thach, R.E. (1976) Proc. Natl. Acad. Sci. USA, 73, 3040-3044.
6. Jen, G., Birge, C.H. and Thach, R.E. (1978) J. Virol. 17, 640-647.
7. Svitkin, Y.V., Ginevskaya, V.A., Ugarova, T.Y. and Agol, V.I. (1978) Virology, 87, 199-203.
8. Abreu, S. and Lucas-Lenard, J. (1976) J. Virol. 18, 192-194.
9. Rosen, H., DiSegni, G. and Kaempfer, R. J. Biol. Chem. 257, 946-951.
10. McAllister, P.E. and Wagner, R.W. (1976) J. Virol. 18, 5257, 946-951.
11. Lodish, H.F. and Porter, M. (1980) J. Virol. 36, 719-733.
12. Walden, W.E., Godefroy-Colburn, T. and Thach, R.E. (1981) J. Biol. Chem. 256, 11739-11746.
13. Kabat, D. and Chappell, M.R. (1977) J. Biol. Chem. 252, 2684-2690.

14. Ignotz, G.G., Hokari, S., DePhilip, R.M., Tsukada, K. and Lieberman, I. (1981) Biochem. 20, 2550-2557.
15. Lee, G.T.-Y. and Engelhardt, D.L. (1979) J. MOl. Biol. 129, 221-233.
16. Sonnenshein, G.E. and Braverman, G. (1976) Biochem. 15, 5497-5501.
17. Godefroy-Colburn, T. and Thach, R.E. (1981) J. Biol. Chem. 256, 11762-11773.
18. Grifo, J., Leis, J., Shatkin, A.J. and Merrick, W.M. J. Biol. Chem. in press.
19. DiSegni, G., Rosen, H. and Kaempfer, R. (1979) Biochem. 18, 2847-2854.
20. Gette, W.R. and Heywood, S.M. (1979) J. Biol. Chem. 254, 9879-9885.
21. Lodish, H.F. and Froshauer, S. (1977) J. Biol. Chem. 252, 80804-8811.
22. Bergmann, J.E. and Lodish, H.G. (1979) J. Biol. Chem. 254, 11927-11937.
23. Thach, R.E., unpublished observation.
24. Brendler, T.G., Godefroy-Colburn, T., Carlill, R.D. and Thach, R.E. (1981) J. Biol. Chem. 256, 11747-11754.
25. Brendler, T.G., Godefroy-Colburn, T., Yu, S. and Thach, R.E. (1981) J. Biol. Chem. 256, 11755-11761.
26. Detjen, B.M., Walden, W.E. and Thach, R.E. J. Biol. Chem. submitted.
27. Yau, P.M.-P., Godefroy-Colburn, T., Birge, C.H., Ramabhadren, T.V. and Thach, R.E. (1978) J. Virol. 27, 648-658.
28. Ramabhadran, T.V. and Thach, R.E. (1980) J. Virol. 34, 293-296.
29. Stöcklein, W., Piepersberg, W. and Böck, A. (1981) FEBS Letters, 136, 265-268.
30. Madjar, J.-J., Nielson-Smith, K., Frahm, M. and Roufa, D.J. (1982) Proc. Natl. Acad. Sci. USA, 79, 1003-1007.
31. Ramabhadran, T.V., Hartley, J.W., Rowe, W.P., Godefroy-Colburn, T., Jhabvala, P.S. and Thach, R.E. (1979) J. Virol. 32, 123-130.
32. Lowry, O.H., Rosebrough, N.J., Farr, A.L. and Randall, R.J. (1951) J. Biol. Chem. 193, 265-275.
33. Laemmli, U.K. (1970) Nature 227, 680-685.
34. Cleveland, D.W., Fischer, S.G., Kirschner, M.W. and Laemmli, U.K. (1977) J. Biol. Chem. 252, 1102-1106.
35. Geoghegan, T., Cereghini, S. and Browerman, G. (1979) Proc. Natl. Acad. Sci. USA, 76, 5587-5591.
36. Croall, D.E. and Morrison, M.R. (1981) J. Mol. Biol. 140, 549-564.

SECTION V: TRANSLATIONAL REGULATION IN CELLS

MECHANISMS OF INTERFERON ACTION: CHARACTERISTICS OF RNase L, THE
$(2'-5')(A)_n$- DEPENDENT ENDORIBONUCLEASE

GEORGIA FLOYD-SMITH AND PETER LENGYEL
Dept. of Molecular Biophysics and Biochemistry, Yale University,
New Haven, CT 06511

INTRODUCTION

Interferons are a family of proteins, many if not all of which share some sequence homology[1-4]. They occur in vertebrates from fish to man and are biological regulators of cell function. Unless induced, their concentration is below the detectable level in most organisms and in cells in culture.

Human interferons are specified by at least 15 different interferon genes and can be classified into three antigenically distinct types: alpha, beta and gamma. Alpha and beta interferons are induced in a large variety of cells by various agents including certain viruses, bacteria and double-stranded RNA. Gamma interferon is induced in lymphoid cells by mitogens and by antigens to which the cells have been sensitized.[4,5] The first known intermediates in interferon induction are the messenger RNAs for interferons.[5] These are translated into the interferons which are secreted from the producing cells. The secreted interferons bind to surface receptors of cells[6] and affect in these numerous, seemingly diverse, biological phenomena.[1-3,5] Thus in cells treated with interferons, the replication of a large variety of viruses is inhibited. The treatment may also impair cell proliferation and cell motility and in lymphoid cells may alter various immunological processes including the antibody response, delayed hypersensitivity, natural killer cell recruitment, macrophage activation and the expression of several cell surface antigens.[7-9]

Most of these manifestations of interferon treatment depend on RNA and protein synthesis taking place in the cells after their exposure to interferons. This exposure causes an increase in the accumulation of several messenger RNAs and of the corresponding proteins. Among the proteins whose levels are increased in interferon-treated cells are several enzymes which serve as mediators of interferon action.[3] The identification of some of these enzymes was facilitated by studies on interferon-treated, virus infected cells. The aim of these studies was to establish the step (or steps) in the replication of particular viruses which is (or are) inhibited in cells treated with interferons. They revealed that viral RNA and protein accumulation are inhibited in interferon-treated cells.[10] These findings prompted a comparison of extracts from interferon-treated cells and control cells for enzyme activities affecting RNA and protein metabolism. The comparison resulted in the discovery of several such enzymes and enzyme systems which were present in higher levels in extracts from interferon-treated cells than in those from control cells.[3]

The best characterized of these is an endonuclease system, which if activated by double-stranded RNA, causes an accelerated cleavage of RNA.[11] The first enzyme in this system is $(2'-5')(A)_n$*synthetase. Upon activation by double-stranded RNA this generates from ATP $(2'-5')$ linked oligoadenylates.[12] The enzyme has been purified to homogeneity from both human and mouse cells.[13-15] The stoichiometry of the reaction it catalyzes is:

$$nATP \longrightarrow (2'-5')(A)_n + (n-1)pyrophosphate$$

---

*The abbreviations used are: $(2'-5')(A)_n$, $2'-5'$-linked $pppA(pA)_{n-1}$ where n is between 2 and about 15; $(2'-5')(A)_3[^{32}P]pCp$, $2'-5'$-linked $pppA(pA)_2$ to which $[^{32}P]$cytidine $5',3'$ diphosphate has been ligated; PMSF, phenylmethylsulfonylfluoride; RNase L, $(2',5')(A)_n$-dependent endoribonuclease; SDS, sodium dodecyl sulfate

It is notable that the equilibrium of this reaction is strongly in favor
of synthesis (over 97%) since it does not require "pulling" by the cleavage
of pyrophosphate. This seems to indicate that the free energy of (2'-5')
linked oligoadenylates is different from that of (3'-5') linked
oligoadenylates. Among the products of the enzyme di-, tri- and
tetraadenylates are the most abundant and the amount of longer
oligoadenylates diminishes with an increasing chain length.[12,14,16] At
present the only well established biochemical function of (2'-5') linked
oligoadenylates is the activation of a second enzyme of the endonuclease
system i.e. RNase L. If activated by (2'-5')(A)$_n$ RNase L acts as an
endonuclease.[11] Further characteristics of this enzyme are described in
subsequent sections. Interferon treatment also increases the level (at
least in some cell lines) of a phosphodiesterase cleaving (2'-5')(A)$_n$.[17]
Another enzyme induced by interferons is a protein kinase which if
activated by double-stranded RNA phosphorylates and thereby impairs the
activity of the peptide chain initiation factor eIF-2, thus inhibiting
protein synthesis.[18-20]

Other biochemical effects of interferons on cells include an alteration
in the methylation of viral and perhaps host messenger RNAs[21-23] and a
decrease in the proportion of unsaturated fatty acids in membrane
phospholipids resulting in a change in the rigidity of the cell
membrane.[24,25] The agents mediating these alterations and their role in
interferon action remain to be elucidated.

Here we present further data on RNase L together with a survey of our
present knowledge of this enzyme.

MATERIALS AND METHODS

Published procedures were followed for culturing mouse Ehrlich ascites
tumor cells and, if so indicated for treating the cells with mouse
interferon (specific activity 10^8 NIH standard reference units per mg

protein) for 24 hours, for preparing the low speed supernatant fraction (S30) from the extracts of the cells[26], for assaying the binding[27], and the UV irradiation promoted crosslinking of $(2'-5')(A)_3[^{32}P]pCp$ to proteins of the cell extract,[28] for preparing bacteriophage R17 RNA and for assaying RNase L activity.[26]   The RNase L DEAE-cellulose fraction was prepared as reported earlier.[26] The RNase L modified DEAE-cellulose fraction was prepared as the RNase L DEAE-cellulose fraction except that the last step in the procedure (i.e., the concentration of the pooled RNase L-active fractions by precipitation with 80% ammonium sulfate) was not performed. If so indicated the modified DEAE-cellulose fraction was further purified. For this purpose an aliquot (23 mg protein) from this fraction was dialyzed against buffer A (50 mM Tris Cl (pH 7.8), 10 mM 2-mercaptoethanol, 1 mM EDTA, 50 μM PMSF, 10% glycerol) containing 50 mM KCl and applied to a 10 ml phosphocellulose column equilibrated in the same buffer also containing 50 mM KCl. The column was developed with a linear KCl gradient (50-500 mM) in buffer A. The RNase L-active material (which was eluted at 200 mM KCl; 2 mg protein in 10 ml) was dialyzed against buffer B (10 mM Tris Cl (pH 8.2), 10 mM 2-mercaptoethanol, 1 mM EDTA, 50 μM PMSF, 10% glycerol) and applied to a 1 ml poly(A)-agarose (PL Biochemicals) column which had been equilibrated with buffer B. The column was developed with a linear KCl gradient (0-1 M) in buffer B. The RNase L-active material eluting at 200 mM KCl was supplemented with 1 mg/ml of bovine serum albumin, dialyzed against buffer B and stored at -70° (poly (A)-agarose fraction).

RESULTS

Discovery. RNase L was discovered in the course of comparing the rates of cleavage of reovirus mRNAs in extracts from interferon-treated cells with those in extracts from control cells.[29] A faster cleavage in the extracts from interferon-treated cells turned out to depend on the presence of (genomic) double-stranded RNA from reovirions contaminating the reovirus mRNA preparation. Subsequent studies revealed that: a) mRNA free of double-stranded RNA was cleaved at the same rate in the two extracts; b) added double-stranded RNA from reovirions or of poly(I).poly(C) accelerated the cleavage only in extracts from interferon-treated cells;[29] c) in addition to double-stranded RNA, ATP was also required for accelerating the cleavage and d) double-stranded RNA and ATP were needed only for the activation of an endoribonuclease system not for its action.[30]

The double-stranded RNA and ATP dependent RNA cleavage was found to involve two enzymes: 1) RNase L, a latent endoribonuclease and 2) the enzyme catalyzing the synthesis of the activators of RNase L, small thermostable compounds.[31] These activators turned out to be (2'-5') linked oligoadenylates.[31-34] This set of compounds was discovered originally as inhibitors of protein synthesis[16,35] Their formation from ATP is catalyzed by an enzyme ((2'-5')(A)$_n$ synthetase which is present in extracts from interferon-treated cells and requires activation by double-stranded RNA. Thus double-stranded RNA and ATP activate RNase L indirectly: Double-stranded RNA activates (2'-5')(A)$_n$ synthetase which converts ATP to (2'-5')(A)$_n$, the activators of RNase L. Moreover, the inhibition of protein synthesis by (2'-5')(A)$_n$ is an indirect consequence of RNA cleavage by RNase L which had been activated by (2'-5')(A)$_n$.

Assay. The most reliable assay of RNase L is based on a comparison of the rate of cleavage of labeled RNA (usually viral mRNA) in cell extracts without or with added (2'-5')(A)$_n$. The differential cleavage is followed either by sucrose gradient centrifugation[29] or gel electrophoresis of labeled RNA after incubation with the enzyme.[31]

The (2'-5')(A)$_n$ has to be the trimer or longer to activate RNase L from Ehrlich ascites tumor, L929 or HeLa cells and the tetramer or longer to activate the enzyme from rabbit reticulocyte lysates.[36] Furthermore to be able to activate RNase L, the oligoadenylates have to have triphosphate or diphosphate moieties at their 5' termini.[37]

Reversibility of the activation. The activation of RNase L by
$(2'-5')(A)_n$ is fully reversible: The addition of $(2'-5')(A)_n$ to the
reaction mixture including latent RNase L results in the activation of the
enzyme. The removal of $(2'-5')(A)_n$ from the activated enzyme (by
precipation with ammonium sulfate or by gel filtration) results in the
reversion of the enzyme to the latent state. Readdition of $(2'-5')(A)_n$ to
the now latent enzyme reactivates it.[26]

Binding of $(2'-5')(A)_n$. The following results indicate that the
activation of RNase L involves the binding of $(2'-5')(A)_n$ to the enzyme:
a)whereas free $(2'-5')(A)_n$ passes through nitrocellulose filter RNase
L preparations partially purified from Ehrlich ascites tumor cells can
retain $(2'-5')(A)_n$ on nitrocellulose filters. b) The agent retaining the
oligoadenylates copurifies with RNase L during both ion exchange
chromatography and gel filtration. 3) The treatment of RNase L with
N-ethylmaleimide abolishes its activatability by $(2'-5')(A)_n$ and its
ability to retain these compounds on nitrocellulose filters.

The binding of a labeled derivative of $(2'-5')(A)_n$, i.e.
$(2'-5')(A)_3[^{32}P]pCp$, by RNase L serves as the basis of a fast and
quantitative assay for RNase L and also for $(2'-5')(A)_n$.[38,39]

Crosslinking of $(2'-5')(A)_n[^{32}P]pCp$. The labeled derivative of
$(2'-5')(A)_n$ i.e. $(2'-5')(A)_3[^{32}P]pCp$ was covalently crosslinked to a
protein in cytoplasmic extracts from mouse Ehrlich ascites tumor cells by UV
irradiation[28] or after alkaline phosphatase treatment and periodate
oxidation of the compound followed by reduction.[40] The crosslinked
protein has an apparent molecular weight of 77,000 as determined by gel
electrophoresis in sodium dodecyl sulfate (Figure 1). It appears to be
identical with RNase L according to the following criteria:
Cochromatography on DEAE-cellulose and Sephacryl S300.

Fig. 1. Crosslinking of
(2'-5')(A)$_3$[^{32}P]pCp to
proteins from control and
interferon-treated Ehrlich
ascites tumor cells.
Aliquots of S30 fractions
(100 µg protein) from control
Ehrlich ascites tumor cells (C)
or cells treated with 1000 NIH
units/ml mouse interferon (I)
were incubated in the presence
of (2'-5')(A)$_3$[^{32}P]pCp, and
subsequently irradiated with
UV light (554 nm) for 1 hr
(36,000 J/m^2) as described
by Floyd-Smith et al.[28]
Aliquots (60 µl) were
analyzed by gel electro-
phoresis and autoradio-
graphy. Positions in the
electropherogram of the
size markers phosphorylase a
(92,500 daltons), bovine
serum albumin (68,000
daltons), ovalbumin (45,000)
daltons) and beta-lacto-
globulin (30,000 daltons) are
indicated. The scanning of the
radioautographs with a soft
laser scanning densitometer
(LKB) revealed that the intensity of the labeled
77,000 dalton band was
2.7 fold higher in track I
than in track C.

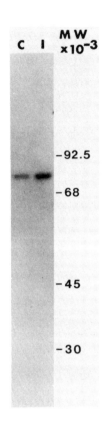

The crosslinking is oligonucleotide specific. It is inhibited by 10 nM

(2'-5')(A)$_n$ or 1 µM (2'-5')ApApA, i.e. compounds known to block even at low

concentrations the binding of (2'-5')(A)$_3$[^{32}P]pCp to RNase L. (3'-5')ApApA

inhibits only at 0.1 mM concentration and 1 mM ATP, 2'AMP or 3',5'pCp have no

effect.[28]

(2'-5')(A)$_3$[^{32}P]pCp is also crosslinked by UV irradiation to a protein

with a molecular weight of 78,000 (as determined by gel electrophoresis in

sodium dodecyl sulfate) in a cytoplasmic extract from human HeLa cells.

Furthermore (2'-5')(A)$_3$[^{32}P]pCp can also be crosslinked to protein(s) of

75,000 to 77,000 daltons in nuclear extracts from Ehrlich ascites tumor cells.

The broad, crosslinked protein band appears in some radioautographs as consisting of two adjacent protein bands of 75,000 and 77,000 daltons.[28] The existence in nuclear extracts of proteins to which $(2'-5')(A)_3[^{32}P]pCp$ can be crosslinked is consistent with reports on the occurrence of $(2'-5')(A)_3[^{32}P]pCp$ binding proteins in nuclei from HeLa cells.[41]

Size of RNase L. Recent studies revealed that the peaks of RNase L activity, $(2'-5')(A)_3[^{32}P]pCp$ binding activity and $(2'-5')(A)_3[^{32}P]pCp$ crosslinking activity coincide in the course of gel filtration on Sephacryl S300 and have an apparent molecular weight between 70,000 and 90,000. Since the protein to which $(2'-5')(A)_3[^{32}P]pCp$ can be crosslinked has an apparent molecular weight of 77,000 as determined by gel electrophoresis in sodium dodecyl sulfate, this suggests that the RNase L-active material eluting from the Sephacryl S300 column in our conditions is primarily a monomer.[28]

Earlier sizing studies on the RNase L from Ehrlich ascites tumor cells by gel filtration, in which much higher protein concentrations were used than in the recent studies, indicated a size of about 185,000 daltons for the enzyme.[26] To reconcile these seemingly contradictory results we compared the sedimentation patterns of RNase L at different protein concentrations by centrifugation through a glycerol gradient. RNase L (as detected by $(2'-5')(A)_3[^{32}P]pCp$ binding) sedimented in the form of two peaks with apparent molecular weights of 180,000 and 80,000 when sedimented at high protein concentration (6 mg/ml) and in the form of a single peak with an apparent molecular weight of 80,000 when sedimented at low protein concentration (0.5 mg/ml, Figure 2). These results indicate that the apparent molecular weight of RNase L (at least when tested after only partial purification) is affected by the protein concentration. Further studies are needed to establish whether RNase L occurs in vivo as a monomer, a multimer or bound to other proteins.

The sedimentation velocity of RNase L (tested at 0.5 mg protein/ml) is affected only very slightly, if at all, by the presence of

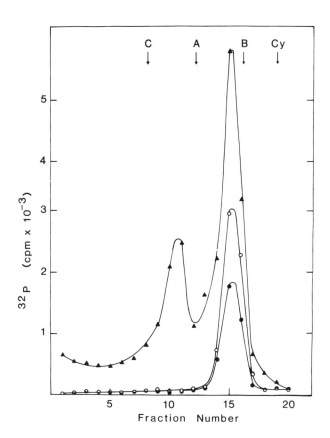

Fig. 2. Sedimentation patterns of RNase L in a glycerol gradient: Effects of protein concentration and of $(2'-5')(A)_n$. An aliquot (75 µg) protein in 150 µl) from the modified DEAE-cellulose fraction was applied on top of a 4.5 ml glycerol gradient (15-30% v/v) in buffer B containing 90 mM KCl (○). An identical aliquot was applied on top of a second glycerol gradient identical to the first one except that it was supplemented with 0.1 nM $(2'-5')$pppApApA and 0.05 nM $(2'-5')(A)_3[^{32}P]$pCp (●). An aliquot (600 µg protein in 100 µl) from the DEAE-cellulose fraction was applied on top of a third identical glycerol gradient without added $(2'-5')(A)_3$ (▲). The gradients were centrifuged at 200,000 g and 4° for 16 hr. 200 µl fractions were collected. 50 µl aliquots from each of the fractions of the two gradients without $(2'-5')(A)_3[^{32}P]$pCp were assayed for $(2'-5')(A)_3[^{32}P]$pCp binding activity. 50 µl aliquots from each fraction of the gradient which had been supplemented with $(2'-5')(A)_3[^{32}P]$pCp and $(2'-5')$pppApApA were filtered through nitrocellulose filters and the amount of label retained was determined. For further details see MATERIALS AND METHODS. The positions in the gradients of the sedimentation markers catalase (240,000 daltons) (C), aldolase (158,000 daltons)(A), bovine serum albumin (68,000 daltons)(B) and cytochrome c (12,500 daltons)(Cy) are indicated.

$(2'-5')(A)_3[^{32}P]pCp$ and $(2'-5')(A)_n$ in the gradient (Figure 2). This is

the case even if the $(2'-5')(A)_3$ is present throughout the gradient at a

concentration fully activating RNase L (i.e., 10 µM in adenosine monophosphate

equivalents). In this case the RNase L activity was assayed by following RNA

cleavage (not shown).

These results are in line with earlier findings indicating that the activa-

tion of RNase L by $(2'-5')(A)_n$ is not accompanied by a large change in size

or conformation. These and other data make it unlikely that the activation

should involve the release or binding of a subunit from RNase L.[26]

Inducibility by interferons. Using RNA cleavage as an assay for RNase L, it

was established earlier that the level of this enzyme is increased in Ehrlich

ascites tumor cells about 3 fold after treatment with 1000 units/ml of inter-

feron.[42] The effects of interferon treatment on the level of RNase L were re-

tested using the more quantitative $(2'-5')(A)_3[^{32}P]pCp$ binding and cross-

linking assays. The results obtained confirm the earlier finding.

The binding assay reveals a 2 fold increase in the level of RNase L upon

treatment of the cells with 500 units/ml of interferon and a 2.2 to 3.3 fold

increase upon treatment with 1000 units/ml of interferon(Table 1). The

crosslinking assay indicates a 2.7 fold increase in the level of the enzyme in

cells treated with 1000 units/ml of interferon (Figure 1).

Cleavage of RNA. Earlier studies revealed that: 1) Activated RNase L

cleaves from the four homopolyribonucleotides poly(A), poly(C), poly(G), and

poly(U) only poly(U)[43] and 2) in natural RNA's the enzyme cleaves pre-

ferentially at the 3' side of UA, UG and UU sequences.[43,44]

Here we present data indicating that the RNA fragments produced by cleavage

with activated RNase L have 3' phosphate and 5' hydroxide termini (Table 2).

The experiments which serve as the basis for this conclusion involve the

incubation first with activated RNase L (to cause only partial cleavage of the

RNA substrate) and subsequent incubation with either RNase T_2 or nuclease

TABLE 1.

INDUCTION OF RNase L IN EHRLICH ASCITES TUMOR CELLS BY INTERFERON

Experiment	Treatment of cells	$(2'-5')(A)_3[^{32}P]pCp$ bound (cpm)	Extent of induction
1	control	1190	3.3
	interferon (1000 U/ml)	3920	
2	control	1010	2.2
	interferon (1000 U/ml)	2210	
3	control	1002	2.0
	interferon (500 U/ml)	1990	

The binding of $(2'-5')(A)_3[^{32}P]pCp$ to aliquots (30 µg protein) of the S30 fractions from control   cells or in   cells which had been treated with interferon at the concentrations indicated was assayed by determining the amount of radioactivity retained on nitrocellulose filters according to published procedures. The extent of induction was calculated by dividing the number of cpms bound in the extracts from interferon-treated cells by that bound in the extracts from control cells. The results from three independent experiments are shown.

P_1 to obtain cleavage to mononucleotides and nucleosides.

The partial cleavage of the $[^3H]$-poly(U) with activated RNase L has no effect on the distribution of cleavage products generated by RNase T_2 (which degrades RNA to nucleoside 3' monophosphates[45]).

However the partial cleavage by activated RNase L changes the distribution of cleavage products generated by nuclease P_1 (which degrades RNA to nucleoside 5' phosphates and nucleoside 3' phosphates to nucleosides[46]). The partial cleavage with RNase L increases the percent of U and pUp and decreases the percent of pU (Table 2). (The smaller increase in the precent of pUp than in that of U (compare row 4 to row 3 in Table 2) might be a consequence of a) a slow cleavage by nuclease P_1 of pUp to pU and b) the production by nuclease P_1 of U from Up.) These results indicate that the cleavage of phosphodiester linkages by activated RNase L generates 3' phosphate and 5' hydroxyl termini.

The preferred RNase L cleavage sites in an RNA segment with a well-defined secondary structure (that is the 3' terminal region of bacteriophage R17 RNA[47]) are shown in Figure 3. The major RNA cleavage sites are at the

428

Fig. 3. Effect of RNA
secondary structure on RNase
L cleavage in the 3'
terminal region of
bacteriophage R17 RNA.
Bacteriophage R17 RNA
was labeled on its 3'-
terminal hydroxyl with
[^{32}P]pCp (2000 Ci/mmole,
New England Nuclear) and
T4 RNA ligase according
published procedures.[49]
The RNA (4 µg) was di-
gested in a 30 µl volume
with RNase L (12 µg/ml)
from the poly(A)-agarose
fraction and (2'-5')pppApApA
(1 µM in adenosine mono-
phosphate equivalents) in
25 mM Tris Cl (pH 8.2),
75 mM KCl, 5 mM Mg acetate
and 10 mM 2-mercaptoethanol
at 30° for 30 minutes.
For allowing the identification of the cleavage
sites further aliquots of
the 3' end labeled-R17 RNA
were partially digested with
RNase T$_1$ or alkali. The
digested samples were
analyzed by gel electro-
phoresis and autoradio-
graphy and the RNase L
cleavage sites were de-
termined by visual in-
spection. The nucleo-
tide sequence and secon-
dary structure of the 3'
terminal region of R17
RNA together with the
major (solid arrows) and
minor RNase L cleavage
sites (dashed arrows)
are shown.

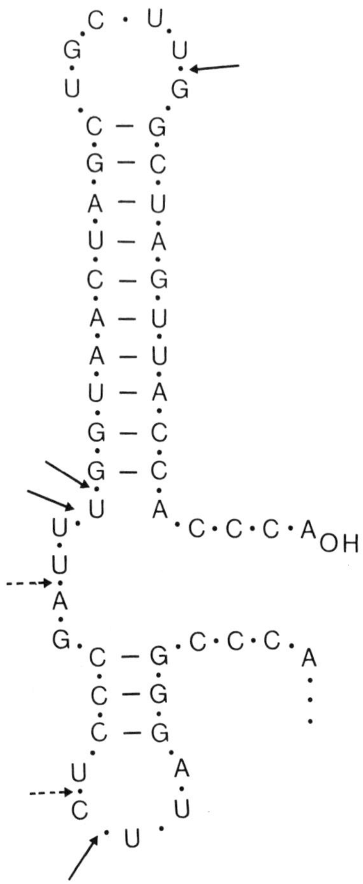

TABLE 2

MODE OF RNA CLEAVAGE BY RNase L. PRODUCTS FORMED WHEN POLY(U) IS FIRST
PARTIALLY CLEAVED WITH RNase L IN THE PRESENCE OF $(2'-5')(A)_n$ AND THEN
CLEAVED TO MONONUCLEOTIDES (OR NUCLEOSIDES) WITH RNase T_2 or Nuclease P_1.

First Incubation		Second Incubation		Products Formed (%)			
RNase L	$(2'-5')(A)_n$	RNase T_2	Nuclease P_1	pUp	Up	pU	U
+	-	+	-	0.13	99.43	-	0.44
+	+	+	-	0.12	99.41	-	0.47
+	-	-	+	0.23	-	99.31	0.46
+	+	-	+	0.66	-	96.76	2.58

The reaction mixtures (60 μl) for the first incubation (i.e. partial
cleavage by RNase L) contained: RNase L (1.4 μg protein from the poly(A)-
agarose fraction in 20 μl of buffer B), 6 μl of buffer 10X C (250 mM Tris Cl
(pH 8.2), 750 mM KCl, 100 mM 2-mercaptoethanol, 50 mM Mg acetate),
$[5-^3H]$poly(U) (2.7 μg, 674,000 cpm) without or with $(2'-5')$pppApApA (1 μM
in adenosine monophosphate equivalents) as indicated. They were incubated at
30° for 30 min. The reaction was stopped by the addition of sodium dodecyl
sulfate to a final concentration of 2% (v/v), unlabeled poly (U) to a final
concentration of 1 mg/ml and water to a final volume of 100 μl. The samples
were phenol extracted, ethanol precipitated and diluted to a concentration of
1 mg/ml of poly (U). In the second incubation 10 μl aliquots of the above
samples (containing 40,000 cpm of partially cleaved $[^3H]$poly(U) were
supplemented with either ribonuclease T_2 (0.5 units) and 1 μl of 200 mM
ammonium acetate (pH 4.5) and were incubated in a total volume of 12 μl at
37° for 3 hr or were supplemented with nuclease P_1 (3 units) and 1 μl of
400 mM ammonium acetate (pH 6.0) and were incubated in a total volume of 16
μl at 50° for 1 hr. The incubated reaction mixtures were analyzed by
chromatography on PEI cellulose plates using uridine 3',5' diphosphate,
uridine 3'phosphate, uridine 5' phosphate and uridine as markers. The
cleavage products comigrated with the markers and were scraped from the plate
and counted. The values shown represent the average % in 3 experiments.

3' side of UU sequences, a minor cleavage at the

3' side of UC, and a barely detectable one at the 3' side of GA. All these

cleavages occur in single-stranded loop structures and none of them in

complementary, hydrogen bonded structures. This is the case, though there are

several UU and UA sequences in the selfcomplementary regions of the RNA.

These results are in line with earlier data[48] indicating that RNase L, even

if activated, does not cleave double-stranded regions in RNA.

DISCUSSION

There are several problems concerning RNase L that remain to be solved.

These include the isolation of pure enzyme. Furthermore, it remains to be

established if the activation and/or action of RNase L involve(s) the cleavage of $(2'-5')(A)_n$. $(2'-5')(A)_n$ with extensively

modified 2' terminal ribose moieties can activate RNase L.[50] This might

indicate that the 2' terminus of $(2'-5')(A)_n$ does not have to be cleaved by

RNase L either. These results do not rule out the possibility, however, that

activated RNAse L might cleave unmodified $(2'-5')(A)_n$.

Findings consistent with a role of RNase L in mediating some of the

antiviral and/or cell growth inhibitory effects of interferons were reviewed

elsewhere.[11] One of these findings concerns a cell line (NIH 3T3 clone 1)

which is nonresponsive to some of the antiviral and anticellular activities

of interferons and at the same time is low in RNase L activity.[51] The

causal relationship between these deficiencies of this cell line remain to be

established.

A new aspect of the control of RNase L in virus-infected cells was

reported recently[40]: The loss or inactivation of RNase L in L, Ehrlich

ascites tumor and HeLa cells after infection with encephalomyocarditis virus.

This loss or inactivation does not take place in cells treated with

interferons. It remains to be seen whether or not the persistence of RNase L

activity in these encephalomyocarditis virus-infected cells which had been

pretreated with interferon is a consequence of the induction of RNase L by

interferons.

Though at present, RNase L is the only enzyme known to be affected by

$(2'-5')(A)_n$ it is possible that other enzymes affected by this agent will

be discovered in the future. Thus it may be pertinent to note that the

UV-promoted crosslinking of $(2'-5')(A)_3[^{32}P]pCp$ is not completely

restricted to RNase L. We find that a long exposure of the autoradiographs

from crosslinking experiments (in which $(2'-5')(A)_n$ has been used at

subnanomolar concentrations) reveals further minor bands of crosslinked

proteins: 46,000 dalton proteins in Ehrlich ascites tumor and HeLa cell S30

fractions and 61,000, 42,000 and 34,000 dalton proteins in Ehrlich ascites tumor cell nuclear fractions.[28] The amounts of these is however much smaller than that of the 77,000 dalton protein.

Furthermore, when $(2'-5)(A)_3[^{32}P]pCp$ is used at an about 1 nM concentration (i.e., a concentration at or above which it has been reported to exist in cells treated with interferon and infected with encephalomyocarditis virus)[38], see also [41] then in extracts of nuclei from interferon-treated Ehrlich ascites tumor cells even further proteins become crosslinked, including a 115,000 dalton protein (not shown). The significance of the crosslinking of $(2'-5')(A)_3[^{32}P]pCp$ to proteins other than RNase L will have to be established.

SUMMARY

RNase L is an endoribonuclease which if activated by $(2'-5')(A)_n$, cleaves single-stranded RNA primarily at the 3' side of UA, UG and UU sequences generating 3' phosphate and 5' hydroxyl terminated products. The activation of RNase L is reversible. The level of the enzyme is increased 2 to 3 fold in cells after treatment with interferons. A derivative of $(2'-5')(A)_n$, (i.e. $(2'-5')(A)_3[^{32}P]pCp$) can be crosslinked to RNase L by UV irradiation. RNase L has an apparent molecular weight of 77,000 as determined by gel electrophoresis in sodium dodecyl sulfate and about 80,000 as determined by gel filtration and glycerol gradient centrifugation. The activation of RNase L by $(2'-5')(A)_n$ does not seem to cause a significant change in the size of the enzyme. There are several findings indicating that RNase L is among the mediators of interferon action.

ACKNOWLEDGEMENTS

These studies were supported by NIH research grants AI-12320 and CA-16038 and a National Service Award to G. F.-S. We thank Drs. B. Jayaram and H. Schmidt for the purified mouse interferon preparations.

REFERENCES

1. Stewart II, W. E. (1979) The Interferon System. Springer-Verlag, New York, pp. 421.
2. Vilcek, J., Gresser, I., Merigan, T. C., eds (1980) Ann. N.Y. Acad. Sci. vol. 350, pp. 641.
3. Lengyel, P. (1982) Ann. Rev. Biochem. 51, 251-282.
4. Weissmann, C. (1981) Interferon 3, (Gresser, I. ed.) pp. 101-134 Academic Press, New York
5. DeMaeyer, E., DeMaeyer-Guignard, J. (1979) Comprehensive Virology 15, 205-284.
6. Aguet, M. (1980) Nature (London) 284, 459-461.
7. Sehgal, P. B., Pfeffer, L. M., Tamm, I. (1981) In Chemotherapy of Viral Infections, (ed. Came, P. E., Caliguri, L. A.) in press.
8. Taylor-Papadimitriou, J. (1980) Interferon 2, (Gresser. I., ed.) pp. 13-42, Academic Press, New York.
9. Gresser, I. (1977) Cell. Immunol. 34, 406-415.
10. Sen, G. C. (1982) Prog. Nucl. Acids Res. 27, 105-156.
11. Lengyel, P. (1981) Interferon 3 (Gresser, I. ed.) pp. 78-99, Academic Press, New York.
12. Ball, L. A. (1981) In The Enzymes,(ed. Boyer, P. D.), Academic Press, New York, in press.
13. Dougherty, J. P., Samanta, H., Farrell, P. J., Lengyel, P. (1980) J. Biol. Chem. 255, 3813-3816.
14. Samanta, H., Dougherty, J. P., Lengyel, P. (1980) J. Biol. Chem. 255, 9807-9813.
15. Yang, K., Samanta, H., Dougherty, J., Jayaram, B., Broeze, R. and Lengyel, P. (1981) J. Biol. Chem. 256, 9324-9328.
16. Kerr, I. M., Brown, R. E. (1978) Proc. Natl. Acad. Sci. USA 75, 256-260.
17. Schmidt, A., Chernajovsky, Y., Shulman, L., Federman, P., Berissi, H., Revel, M. (1979) Proc. Natl. Acad. Sci. USA 76, 4788-4792.
18. Lebleu, B., Sen, G. C., Shaila, S., Cabrer, B., Lengyel, P. (1976) Proc. Natl. Acad. Sci. USA 73, 3107-3111.
19. Zilberstein, A., Kimchi, A., Schmidt, A., Revel, M. (1978) Proc. Natl. Acad. Sci. USA 75, 4734-4738.
20. Roberts, W. K., Hovanessian, A., Brown, R. E., Clemens, M. J., Kerr, I. M. (1976) Nature (London) 264, 477-480.
21. Sen, G. C., Shaila, S., Lebleu, B., Brown, G. E., Desrosiers, R. C., Lengyel, P. (1977) J. Virol. 21, 69-83.
22. Kroath, H., Gross, H. J., Jungwirth, C., Bodo, G. (1978) Nucl. Acids Res. 5, 2441-2454.
23. deFerra, F., Baglioni, C. (1981) Virology 112, 426-435.
24. Chandrabose, K., Cuatrecasas, P., Pottathil, R. (1981) Biochem. Biophys. Res. Commun. 98, 661-668.
25. Apostolov, K., Barker, W. (1981) FEBS Lett. 126, 261-264.
26. Slattery, E., Ghosh, N., Samanta, H. and Lengyel, P. (1979) Proc. Natl. Acad. Sci. USA 76, 4778-4782.
27. Nilsen, T. W., Maroney, P. A., and Baglioni, C. (1981) J. Biol. Chem. 256, 7806-7811.
28. Floyd-Smith, G., Yoshie, O., and Lengyel, P. (1982) J. Biol. Chem. 257, in press.
29. Brown, G. E., Lebleu, B, Kawakita, M., Shaila, S., Sen, G. C., Lengyel, P. (1976) Biochem. Biophys. Res. Commun. 69, 114-122.

30. Sen, G. C., Lebleu, B., Brown, G. E., Kawakita, M., Slattery, E. and Lengyel, P. (1976) Nature (London) 267, 370-373.
31. Ratner, L., Wiegand, R., Farrell, P., Sen, G. C., Cabrer, B., Lengyel, P. (1978) Biochem. Biophys. Res. Commun. 81, 947-957.
32. Baglioni, C., Minks, M. A., Maroney, P. A. (1978) Nature (London) 273, 684-687.
33. Clemens, M. J. and Williams, B.R.G. (1978) Cell 13, 565-572.
34. Farrell, P. J., Sen, G. C., Dubois, M. F., Ratner, L., Slattery, E., Lengyel, P. (1978) Proc. Natl. acad. Sci. USA 75, 5893-5897.
35. Hovanessian, A. G., Brown, R. E., Kerr, I. M. (1977) Nature (London) 268, 537-540.
36. Williams, B.R.G., Golgher, R. R., Brown, R. E., Gilbert, C. S., Kerr, I. M. (1979) Nature (London) 282, 582-586.
37. Martin, E. M., Birdsall, N.J.M., Brown, R. E., Kerr, I.M. (1979) Eur. J. Biochem. 95, 295-307.
38. Knight, M., Cayley, P. J., Silverman, R. H., Wreschner, D. H., Gilbert, C. S., Brown, R. E. and Kerr, I. M. (1980) Nature (London) 288, 189-192.
39. Nilsen, T. W., Wood, D. L., and Baglioni, C. (1981) J. Biol. Chem. 256, 10751-10754.
40. Cayley, P. J., Knight, M. and Kerr, I. M. (1981) Biochem. Biophys. Res. Commun. 104, 376-382.
41. Nilsen, T. W., Wood, D. L. and Baglioni, C. (1982) J. Biol. Chem. 257, 1602-1605.
42. Ratner, L., Ph.D. Dissertation, Yale University, 1979.
43. Floyd-Smith, G., Slattery, E. and Lengyel, P. (1981) Science 212, 1030-1032.
44. Wreschner, D. H., McCauley, J. W., Skehel, J. J. and Kerr, I. M. (1981) Nature (London) 289, 414-417.
45. Uchida, T. and Egami, F. (1967) J. Biochemistry (Tokyo) 61, 44-53.
46. Fujimoto, M., Kuninaka, A. and Yoshino, H. (1974) Agr. Biol. Chem. 38, 785-790.
47. Cory, S., Adams, J. M., Spahr, P.-F. and Rensing, U. (1972) J. Mol. Biol. 63, 41-56.
48. Ratner, L., Sen, G. C., Brown, G. E., Lebleu, B., Kawakita, M., Cabrer, B., Slattery, E., Lengyel, P. (1977) Eur. J. Biochem. 79, 565-577.
49. Donis-Keller, H., Maxam, A. M. and Gilbert W. (1977) Nucl. Acids Res. 4, 2527-2538.
50. Torrence, P. F., Imai, J., Lesiak, K., Johnston, M. I., Jacobsen, H., Friedman, R. M., Sawai, H., and Safer, B. (1982) in UCLA Symposia on Molecular and Cellular Biology, Vol. 25 (eds. Merigan, T. C. and Friedman, R. M). Academic Press, New York.
51. Panet, A., Czarniecki, C. W., Falk, H. and Friedman, R. M. (1981) Virol. 114, 567-572.

EXPRESSION OF HUMAN FIBROBLAST AND HUMAN IMMUNE INTERFERON GENES IN HOMOLOGOUS AND HETEROLOGOUS CELLS.

ERIK REMAUT, RIK DERYNCK, RENE DEVOS, HILDE CHEROUTRE, JEAN CONTENT[+], ROLAND CONTRERAS, WIM DEGRAVE, DIRK GHEYSEN, PATRICK STANSSENS, JAN TAVERNIER, YOICHI TAYA, HUGO VAN HEUVERSWYN and WALTER FIERS
Laboratory of Molecular Biology, State University of Ghent, Ledeganckstraat 35, B-9000 Gent, (Belgium); [+]Institut Pasteur du Brabant, rue de Remorqueur 28, B-1040 Brussels, (Belgium)

INTRODUCTION

Interferons are secreted proteins that are produced by certain cell types of most vertebrates upon exposure to viruses or after treatment with specific inducers such as synthetic double-stranded RNA. They induce an antiviral state in their target cells and have also been shown to have immunomodulating and antiproliferative activities[1]. The latter property in particular prompted a great deal of research directed towards the potential clinical application of these potent molecules in the treatment of neoplasia. Human interferons can be classified by their biological and chemical properties into three major groups: leukocyte interferon (IFN-α), fibroblast interferon (IFN-β) and immune interferon (IFN-γ)[1]. Clinical trials have been carried out mainly with leukocyte interferon and to a lesser extent with fibroblast interferon. The poor availability of the immune interferon has so far precluded clinical trials.

With the particular intention of making available larger quantities of IFN-β and IFN-γ, we decided to clone these genes and study their expression in both a eukaryotic and prokaryotic expression vehicle.

CLONING OF HUMAN IFN-β AND HUMAN IFN-γ cDNA

The IFN-β interferon

IFN-β or fibroblast interferon is synthesized predominantly by fibroblast cells after viral infection. It is a hydrophobic, glycosylated protein that retains functional integrity after treatment at pH2[1]. In order to clone the coding sequence for human IFN-β we isolated mRNA from primed and superinduced VGS cells (a human diploid fibroblast cell line). This mRNA was partially purified by centrifugation in a sucrose-gradient containing 50% formamide. Fractions of the gradient were assayed for

IFN-β activity after injection of an aliquot into <u>Xenopus laevis</u> oocytes, which translated and secreted the IFN-β . The amounts of IFN-β produced were measured by a cytopathic effect-inhibition assay on human fibroblasts trisomic for chromosome 21 after challenge with vesicular stomatitis virus. The mRNA present in the most active fractions was pooled and converted into double-stranded DNA, tailed with poly dT and annealed into the poly dA-tailed <u>Pst</u> I site of pBR322. After transformation in HB101 a library of about 17000 transformants was obtained. These transformants were screened for the presence of IFN-β information by a procedure called group selection[2]. Plasmid DNA was isolated from groups of 50 clones and chemically linked to diazobenzyloxymethyl-paper. The paper discs were then used for hybridization with active IFN-β mRNA and both the hybridized and the unhybridized fractions were tested for biological activity, again by injection into <u>Xenopus laevis</u> oocytes and determination of cytopathic effect- inhibition. By this procedure several groups of 50 clones selectively retaining the human IFN-β mRNA were found. One such group was divided into individual clones and these were analysed as described. The first plasmid obtained was named pHFIF-1[2]. The plasmid contained a cDNA insert that was obviously too short to account for the full size of the IFN-β mRNA. Using pHFIF-1 DNA as a probe in colony hybridization, several other clones that contained IFN-β sequences were subsequently isolated. Sequence analysis of the cloned cDNA revealed that these clones contained a cDNA that was rearranged, presumably during the <u>in vitro</u> manipulations to arrive at the cDNA. This was possibly due to an unusual primary or secondary structure of the IFN-β mRNA[3]. Nevertheless, from the several clones obtained, it was possible to reconstruct the sequence of the IFN-β mRNA[2] (fig. 1). The nucleotide sequence revealed a single long reading frame accounting for 187 amino acid residues. The sequence of the first 13 N-terminal amino acid residues of authentic human IFN-β has been determined[4]. This sequence is colinear with amino acid residues 22 to 34 of the abovementioned reading frame. The first 21 residues therefore, most likely represent the signal peptide known to be present in secreted proteins . This sequence is presumably cleaved off during transmembrane migration of the nascent polypeptide. Taniguchi et al[6] and Goeddel et al[7] independently isolated a human IFN-β gene.

The IFN-γ interferon

IFN-γ is a type II interferon. It is acid-labile and it is produced by certain T-lymphocytes after treatment with mitogens such as Staphylococcal enterotoxin A. This protein seems to have a more pronounced antiproliferative effect on tumour cells than IFN-γ[8,9]. Very recently, Gray et al[10] reported the cloning of human IFN-γ cDNA and its expression

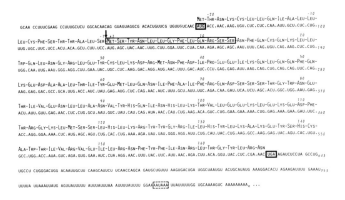

Fig. 1. Nucleotide sequence of human IFN-β mRNA and corresponding amino acid sequence. The boxed amino acid sequence indicates the amino-terminal sequence deduced from direct analysis of the native mature protein[4].

in monkey cells and in the bacterium <u>Escherichia</u> <u>coli</u>. We had independently initiated a search for a human IFN-γ gene. Cultures of lymphocytes obtained from a single human spleen were used as starting material. After stimulation with Staphylococcal enterotoxin A, IFN-γ activity was detectable in the medium. That this activity represented a genuine immune type interferon was suggested not only by the nature of the induction procedure used, but also by its insensitivity to anti-α or anti-β antisera[11].

The procedure used to clone the IFN-γ cDNA was quite similar to the procedure outlined above for IFN-β. The poly A$^+$-RNA fraction was isolated from induced spleen lymphocytes and fractionated on a sucrose gradient. The fractions that showed biological activity after injection into <u>Xenopus</u> oocytes were pooled and recentrifuged through a sucrose gradient in the presence of 50% formamide. The mRNA in the active fractions was converted to double-stranded DNA, tailed with poly dG and cloned into the filled-in poly dC tailed <u>Bam</u> HI site of pSV529 (see further). After annealing of the vector with an inserted fragment, the <u>Bam</u> HI sites are restored. A clone containing IFN-γ cDNA was searched for by the group selection technique, i.e. hybridization-elution and oocyte translation exactly as described for

IFN-β . Finally, three clones of almost full size were isolated. These were designated pHIIF-SV-γ1, γ5 and γ10[12]. The nucleotide sequence of these clones reveals a single open reading frame potentially coding for 166 amino acids (fig. 2). No data are available on the N-terminal amino acid sequence of authentic IFN-γ. However, it can be assumed that this protein, like most other secreted proteins, contains an N-terminal sequence which is processed away along with migration through the membrane. Delineation of this presumptive signal peptide remains a matter of speculation. At positions 18 to 21 the sequence Ser-Leu-Gly-Cys is found. The same sequence is found in a similar position in several IFN-α interferons[13, 14]. By analogy with the latter case one might suppose that cleavage of IFN-γ precursor also takes place between the Gly and the Cys residues. Based on this analogy Gray et al[10] constructed a recombinant plasmid in which the presumed mature IFN-γ sequence was linked to the initiating AUG of the attenuator of the E. coli trp operon[15]. The IFN-γ activity produced by bacterial cultures containing this recombinant plasmid accounted for 2.5 x 10[8] units per litre of culture.

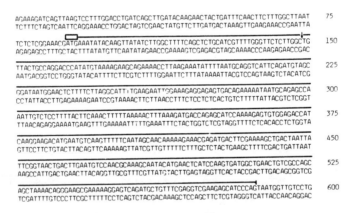

Fig. 2. Nucleotide sequence of part of the human IFN-γ cDNA. The coding region is indicated by a heavy line above the nucleotide sequence; the vertical arrow indicates the position where the presumed signal peptide is cleaved off.

ISOLATION OF GENOMIC DNA CODING FOR HUMAN IFN-β AND IFN-γ

The availability of plasmids containing coding regions for IFN-β or IFN-γ enabled us to probe human chromosomal DNA in search of the gene copies. The procedure used was identical for both types of interferons. The library of human genomic DNA fragments, cloned in derivatives of bacteriophage λ by Maniatis and co-workers[16], was probed with radiolabelled plasmid DNA from either pHFIF-21 or pHIIF-SV-γ1[17]. DNA fragments of human origin present in the identified recombinants were then transferred to plasmid vectors.

The IFN-β genomic DNA

Plasmid pgHFIF-4 contains a 1.9 Kb <u>Eco</u> RI fragment encompassing the entire coding region of IFN-β. As was previously shown for a human IFN-α gene[18], the IFN-β gene did not contain intervening sequences[19]. The nucleotide sequence of the chromosomal IFN-β DNA was in complete agreement with the previously established cDNA sequence and confirmed the absence of introns in the gene[20]. The sequence was extended some 280 bp upstream from the start of the mRNA. Several regions of homology with known upstream sequences of other eukaryotic mRNAs could be recognized. These include the Hogness-Goldberg box[21], here located at position -31, and the concensus sequence of Benoist et al[22], here located between positions -93 and -85. An interesting region of homology is found around position -110 (fig. 3). The genes for human IFN-α[18], human insulin[23], chicken ovalbumin[22] and chicken conalbumin[22,24], all contain a similar sequence also located at a rather long distance (-137 on average), from the transcriptional start. In

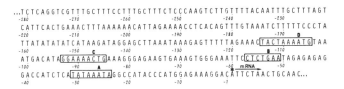

Fig. 3. Nucleotide sequence of the upstream flanking region of the chromosomal gene for IFN-β. The sequences showing homology with other upstream sequences are boxed. Their significance is discussed in the text.

contrast, this sequence was not apparent in viral genomes or in constitutively expressed genes. It is tempting to ascribe a role to this element in the induction process of the abovementioned genes.

The IFN-γ genomic DNA

When the human DNA library was screened with labelled cDNA of pHIIF-SV-γ0 (a nearly complete cDNA clone) two positives were scored out of 80.000 plaques. The recombinant phage designated CH$_4$A-γ2 was chosen for further study. This phage had acquired four Eco RI fragments of which two adjacent ones with a total length of 8 Kb hybridized to the radiolabelled cDNA probe. The size of this genomic fragment indicates that, unlike the genes for IFN-α and IFN-β, the human gene for IFN-γ does contain introns. Fine mapping of recloned fragments confirmed the presence of at least two introns[17].

The chimaeric phage was shown to contain a genuine and functionally complete IFN-γ gene by injection of the DNA into the nuclei of Xenopus laevis followed by assay of the IFN-activity released in the medium.

EXPRESSION OF IFN-β and IFN-γ GENES IN EUKARYOTIC SYSTEMS

Gheysen and Fiers[25], reported the construction of an SV40-derived expression vehicle suitable for expression studies in eukaryotic cells. Plasmid pSV529 is a chimaera consisting of an ampicillin resistance gene and an origin of replication derived from pBR322 on the one hand and 1.4 genome equivalents of SV40 on the other hand. The pBR322 part allows the plasmid to be propagated and selected in Escherichia coli; the SV40 information (T-antigen and origin of DNA replication) sustains replication to high copy numbers after transfection in kidney cells of the African Green Monkey. In pSV529 the gene for the major structural protein VP$_1$ has been deleted (between 0.945 and 0.145 SV40 map units) and a unique Bam HI insertion site was introduced at the junction point. It is predicted that foreign genes inserted into the Bam HI site of pSV529 will be transcribed under the control of the SV40 late promoter.

The IFN-β gene

Plasmid pPLaHFIF 67-1 contains the complete coding region for human IFN-β located on a single Bgl II fragment[26]. This fragment was ligated into Bam HI cleaved pSV529. The Bam HI and Bgl II enzymes produce identical staggered ends that can be ligated to each other[27]. Following transformation in HB101, two representative recombinant plasmids were recovered. Plasmid pSV529-HFIF1 contained the IFN-β gene in the sense orientation relative to the direction of transcription of the SV40 late promoter whereas plasmid pSV529-HFIF3 displayed the anti-sense configuration[25]. The IFN-β fragment present on these plasmids encompasses the entire coding region of the gene. The distal Bgl II site is located immediately after the terminating UGA codon. Preceding the 5'- coding end a small inverted region originating from the 3'-untranslated end was picked up in the original cloning steps[26]. The SV40 part of the plasmids contains the donor and acceptor sequences implicated in splicing of the precursor to the major late 16S mRNA[28, 29] and in addition the regions for polyadenylation and transcription- termination[30].

Plasmid DNA of pSV529-HFIF1 or pSV529-HFIF3 was introduced into cultured kidney cells from the African Green Monkey (cell line AP8) by a modification of the DEAE-dextran method[31, 25]. At various times after transfection, samples were withdrawn from the medium and their interferon titre determined on T21 human cells. Plasmid pSV529-HFIF1 directed the synthesis of up to 2×10^3 international units of IFN-β per ml of medium. This activity was completely neutralized by anti-IFN -antiserum. Under similar conditions pSV529-HFIF3 did not produce detectable levels of IFN-β interferon, indicating that the expression observed in pSV529-HFIF1 was under the control of the SV40 late promoter. IFN-β activity was barely detectable in the medium of transfected mouse or rat cells which are non-permissive for SV40.

In another experiment, the Bgl II fragment containing the coding region for IFN-β was tailed with poly dG at the filled in Bgl II sites and annealed to Bam HI-cleaved, filled-in and poly dC tailed pSV529 vector. A resulting plasmid, pSV529-HFIFG3, containing the IFN-β gene inserted in the sense orientation downstream from the SV40 late promoter was transfected into AP8 cells. This plasmid gave rise to IFN-β interferon titres which were about four-fold lower than the titres obtained with pSV529-HFIF1. The Bgl II fragment present in the former plasmids does not contain the region upstream of the transcriptional start point of IFN-β mRNA. In fact, the actual initiation site of the mRNA is also lacking in these clones. In the absence of such postulated control regions, it is not surprising that addition of the inducer poly rI:poly rC had no effect on the amount of IFN-β interferon produced.

We have made further modifications to the pSV529 vector to enable easy insertion of the 1.9 Kb Eco RI fragment from pgHFIF-4. The new recombinant includes some 280 nucleotides upstream from the starting site of the IFN-β mRNA, derived from the genomic clone and possibly containing the control functions of the gene. Both orientations of the genomic DNA insert with respect to the SV40 promoter were obtained. Treatment with poly rI:poly rC resulted in a 30-fold stimulation of the production of IFN-β , irrespective of the orientation of the insert. Induced levels of IFN-β approached those obtained with pSV529 HFIF1.[20] These results suggest the involvement of a putative polymerase II promoter located in the IFN-β 5'-end flanking region. Furthermore, they indicate that this region contains a cis-acting control element (which may or may not be different from the former) which responds to the poly rI:poly rC treatment. Formally, this postulated control sequence can be regarded as an element of negative control in that it constitutes the target for a repressor molecule which is inactivated by the induction procedure; alternatively, such an element may be looked upon as an element of positive control, e.g. by binding an activator molecule which is formed during the induction process, in a manner analogous to the action of cyclic AMP and CAP-protein (catabolite gene activator protein) which is well known[32] to be operative in positive regulation of many prokaryotic operons . Our recent data on the injection of the IFN-β gene into the nucleus of <u>Xenopus</u> oocytes provides some evidence for the latter.

The IFN-γ gene

Plasmids pHIIF-SV-γ1, γ5 and γ10 contain the IFN-γ cDNA inserted in the unique <u>Bam</u> HI site downstream from the SV40 late promoter. By restriction analysis pHIIF-SV-γ5 and γ10 were shown to be oriented in the nonsense orientation with respect to the SV40 promoter. Not surprisingly, these clones did not result in biological activity after transfection in monkey AP8 cells. Plasmid pHIIF-SV-γ1, however, directed the secretion into the medium of low but reproducible IFN-γ activity (equivalent to about 2 leukocyte international units per ml). Also, a derivative of pHIIF-SV-γ10 in which the <u>Bam</u> HI insert had been inverted, now produced the same levels of IFN-γ as did pHIIF-SVγ1. The IFN activity produced was neutralized by anti-human IFN-γ-serum but not by anti-IFN-α nor by anti IFN-β antiserum.

Next a <u>Sau</u> 3A fragment, internally located within the originally isolated <u>Bam</u> HI fragment and containing the complete coding region, was inserted into the <u>Bam</u> HI site of pSV529. Both the sense and the nonsense orientations with respect to the SV40 late promoter were obtained. These new plasmids did not contain the GC tails originally present at the 5'- and 3'-ends. Upon transfection of AP8 cells with the nonsense orientation no IFN-γ was detectable. But the plasmid with the sense orientation directed the synthesis and secretion into the medium of about 2×10^3 equivalents of international leukocyte units per ml. Remarkably, in this new pHIIF-SV-γ plasmid, elimination of the tails resulted in an increase of IFN-γ interferon synthesis of almost three orders of magnitude. In contrast, the presence of GC tails in IFN-β containing SV40 vectors hampered production of IFN-β by a factor of only four (see above). It should be noted that the precise lengths of the tails in the several plasmids have not yet been determined. Preliminary results, however, indicate that their sizes do not differ greatly (about 30 residues).

EXPRESSION IN PROKARYOTES

A number of parameters governing efficient expression of genes in a prokaryotic environment are well documented. Primarily, the level of expression of a particular gene is determined by the following three factors:
1) abundance of gene copy numbers;
2) the presence of a strong and preferably controllable promoter;
3) availability of a ribosome-binding site which governs efficient translation of downstream sequences.

Derivatives of plasmid Col El have been widely used as cloning vectors for propagation of foreign genes in the bacterium Escherichia coli. This replicon is present at about 30 copies per cell in normal growth conditions[33]. We have constructed a series of expression vectors employing the Col El replicon and have provided them with a segment of DNA containing the promoter-operator region of the leftward promoter, P_L, of bacteriophage λ[34]. The activity of this strong promoter is negatively controlled by a repressor protein - a product of the phage gene cI - and by the product of the cro-gene of the λ phage. We used mutant E. coli strains which harbour a defective λ prophage, devoid of cro-activity and which synthesised a temperature-sensitive cI-product. At low temperature (28°C), the activity of the P_L-promoter is shut off by an active repressor. Repression of P_L-activity is nearly complete due to the autoregulatory mode of synthesis of the cI-gene[35]. In contrast to the lac-operon system[36], for example, one single copy of the cI-gene of phage λ is apparently able to silence multiple copies of the P_L promoter present on a plasmid[34]. The ability to control promoter activity may be particularly advantageous in instances where a gene whose continuous expression would be lethal to the cell is cloned. For instance, Shimatake and Rosenberg[37] were unable to clone the cII gene of phage λ, unless its expression was put under P_L-control. We also showed that the lysis gene of the RNA bacteriophageMS2, which upon expression causes lysis of the cells, can be cloned into PL-expression vectors and be stably propagated at a low temperature[38]. The usefulness of the system was illustrated by placing the Salmonella typhimurium tryptophan synthetase A subunit under control of the P_L-promoter[34,39]. Under optimal conditions, the synthesis of this protein could be induced to account for about 40% of the total newly made protein. The protein then accumulated to over 10% of total cellular protein . Likewise, when the RNA bacteriophage MS2 replicase gene was put under P control, very soon after induction, the synthesis of the polypeptide[L] corresponded to 35% of total "de novo" protein synthesis[40]. In particular, the quick response of the replicase gene prompted us to explore its ribosome binding site and determine its effectiveness in directing high level translation. Although we were aware of the complexity of the factors that govern initiation frequency at a given ribosome binding site[41], we decided to adapt the region around the initiating ATG codon of the replicase polypeptide to make it suitable for in phase insertion of any given coding sequence. A plasmid was constructed that contained the P_L-promoter and downstream from it the ribosome-binding site and initiator ATG of the MS2 replicase. The sequence immediately after the ATG codon was engineered such that after cleavage with Sal I and treatment with nuclease S1, this ATG codon would be made available for blunt-end ligation to other blunt-ended fragments[42]. This plasmid served as an acceptor for a sequence coding for mature IFN-β interferon[4, 2, 43]. Bacterial cultures (strain K12ΔH1Δtrp[34]) containing the newly constructed plasmid (pPLc245HFIF25) were grown in rich medium to 4×10^8 cells per ml at $28\,^{\circ}C$ and then shifted to 42 C. Cells were harvested after 180 min after induction and assayed for IFN-β production. IFN-β activity was released from the bacterial pellets by boiling in 1% SDS; 1% β-mercaptoethanol; 5M urea; 30M NaCl; 50 mM HEPES buffer pH7.5[26]. (We were unable to obtain a reproducibly recoverable IFN-β activity by more gentle methods.) The activity found corresponded to 5×10^9 IU/l. By comparing intensities of Coomassie Brilliant Blue-stained protein bands after electrophoresis in polyacrylamide gels, we estimated the product to have accumulated to about 2% of total protein. The control culture which remained at $28\,^{\circ}C$ produced less than 3×10^4 IU/l; i.e. at least five orders of magnitude lower than the induced level. It is unclear whether this difference is caused by the sole action of repressor product. Indeed, uninduced basal levels of trp A gene for instance, are "only" 300-fold lower than the induced levels[34]. It should of course be noted that we have not determined actual transcription rates but rather rates of accumulation of protein. This value is possibly influenced by other factors such as differential initiation efficiency at both temperatures and/or of stability of the product. Preliminary pulse-chase experiments have shown that the IFN-β protein is not completely stable in the bacterial cell (unpublished work of this laboratory).

A remarkable effect seen in bacterial cultures containing pPLc245HFIF25 was the almost total cessation of growth after shift-up to 42 C. This is a rather specific effect not observed with either the vector itself or with a similar vector synthesizing large amounts of plasmid coded T_4 DNA-ligase . Inhibition of growth at $42\overset{o}{C}$ was likewise observed with other plasmid constructions containing the IFN-β gene downstream from the P_L-promoter, even though some of these produced only negligible amounts of IFN-β protein, detectable only in a mini-cell system[43]. It is not clear what causes the IFN-β protein to be "toxic" to the bacterial cell, even at very low levels. Perhaps the human IFN-β which is a highly hydrophobic protein, interacts with some essential and limiting bacterial component.

ACKNOWLEDGEMENT

These research projects were supported by grants from Biogen NV, the "Fonds voor Geneeskundig Wetenschappelijk Onderzoek" and the "Geconcerteerde Onderzoekakties" of the Belgian Ministry of Science. We thank S. Kaplan for taking care of the manuscript.

REFERENCES

1. Stewart, W.E. II The Interferon System (Springer, Berlin, 1979).
2. Derynck, R., Content, J., De Clercq, E., Volckaert, G., Tavernier, J., Devos, R. and Fiers, W. (1980) Nature 285, 542-547.
3. Volckaert, G., Tavernier, J., Derynck, R., Devos, R. and Fiers, W. (1981) Gene 15, 215-223.
4. Knight E., Jr., Hunkapiller, M.W., Korant, B.D., Hardy, R.W.F. and Hood, L.E. (1980) Science 207, 525-526.
5. Blobel, G., Walter, P., Chang, C.N., Goldman, B.M., Erickson, A.H. and Lingappa, R (1979) Soc. Exp. Biol. Symp. 33, 9-36.
6. Taniguchi, T., Ohno, S., Fujii-Kuriyama, Y. and Muramatsu, M. (1980) Gene 10, 11-15.
7. Goeddel, D.V., Shepard, H.M., Yelverton, E., Leung, D. and Crea, R. (1980) Nucleic Acids Res. 8, 4057-4074.
8. Crane, J.L., Glasgow, L.A., Kern, E.R. and Younger, J.S. (1978) J. Natl. Cancer. Inst. 61, 871-874.
9. Blalock, J.E., Georgiades, J.A., Langford, M.P. and Johnson, H.M. (1980) Cell. Immunol. 49, 390-394.
10. Gray, P.W., Leung, D.W., Pennica, D., Yelverton, E., Najarian, R., Simonsen, C.C., Derynck, R., Sherwood, P.J., Wallace, D.M., Berger, S.L., Levinson, A.D. and Goeddel, D.V. (1982) Nature 295, 503-508.
11. Devos, R, Cheroutre, H, Taya, Y. and Fiers, W. (1982), in press.
12. Devos, R., Cheroutre, H., Taya., Y., Degrave, W., Van Heuverswyn, H. and Fiers, W., (1982), submitted.
13. Mantei, N., Schwarzstein, M., Streuli, M., Panem, S., Nagata, S., and Weissmann, C. (1980) Gene 10, 1-10.
14. Goeddel, D.V., Leung, D.W., Dull, T.J., Gross, M., Lawn, R.M., McCandliss, R., Seeburg, P.H., Ullrich, A., Yelverton, E. and Gray, P.W. (1981) Nature 290, 20-26.

15. Goeddel, D.V., Yelverton, E., Ullrich, A., Heyneker, H.L., Miozzari, G., Holmes, W., Seeburg, P.H., Dull, T., May, L., Stebbing, N., Crea, R., Maeda, S., McCandliss, R., Sloma, A., Tabor, J.M., Gross, M., Familletti, P.C. and Pestka, S. (1980) Nature 287, 411-416.
16. Lawn, R.M., Fritsch, E.F., Parker, R.C., Blake, G. and Maniatis, T. (1978) Cell 15, 1157-1174.
17. Taya, Y et al, in preparation.
18. Nagata, S., Mantei, N. and Weissmann, C. (1980) Nature 287, 401-408.
19. Tavernier, J., Derynck. R. and Fiers, W. (1981) Nucleic Acids Res 9, 461-471.
20. Degrave, W., Derynck, R., Tavernier, J., Haegeman, G. and Fiers, W.(1981) Gene 14, 137-143.
21. Goldberg, M.L., 1979 Thesis, Stanford University
22. Benoist, C., O'Hare, K., Breathnach, R. and Chambon, P. (1980) Nucleic Acids Res. 8, 127-142.
23. Bell, G.I., Pictet, R.L., Rutter, W.J., Cordell, B., Tischer, E. and Goodman, H.M. (1980) Nature 284, 26-32.
24. Cochet, M., Gannon, F., Hen, R., Maruteaux, L., Perrin, F. and Chambon, P. (1979) Nature 282, 567-577.
25. Gheysen, D. and Fiers, W. (1982) accepted for publication, J. Mol. Appl. Genet.
26. Derynck R., Remaut, E., Saman, E., Stanssens, P., De Clercq, E., Content, J. and Fiers, W. (1980) Nature 287, 193-202.
27. Pirotta, V. (1976) Nucleic Acids Res. 3, 1747-1759.
28. Celmar, M.L., Dhar, R., Pan, J. and Weissmann, S.M. (1977) Nucleic Acids Res. 4, 2549-2559.
29. Haegeman, G. and Fiers, W. (1978) Nature, 273, 70-73.
30. Fitzgerald, M. and Shenk, T (1981) Cell 24, 251-260.
31. McCuchan, J.H. and Pagano, J.S. (1968) J. Natl. Cancer Inst. 41, 351-357
32. Pastan, I. and Perlman, R.L. (1968) Proc. Natl. Acad. Sci. U.S.A. 68, 1336-1342.
33. Clewell, D.B. and Helinski, D.R. (1972) J. Bacteriol. 110, 1135-1146.
34. Remaut, E., Stanssens, P. and Fiers, W. (1981) Gene 15, 81-93.
35. Ptashne, M., Backman, K., Kumayun, M.Z., Jeffrey, A., Maurer, R., Meyer, B. and Sauers, R.T. (1976) Science 194, 156-161.
36. Backman, K., Ptashne, M. and Gilbert, W. (1976) Proc. Natl. Acad. Sci. USA 73, 4174-4178.
37. Shimatake, H. and Rosenberg, M. (1981) Nature, 292, 128-132.
38. Kastelein, R., Remaut, E., Fiers, W. and Van Duin, J., (1982) Nature 295, 35-41.
39. Bernard, H.-U., Remaut, E., Hershfield, M.V., Das, H.K., Helinski, D.R., Yanofsky, C. and Franklin, N. (1979) Gene 5, 59-76.
40. Remaut, E., De Waele, P., Marmenout, A., Stanssens, P. and Fiers, W. J. (1982) The EMBO Journal, 1, 205-209.
41. Iserentant, D. and Fiers, W., (1980) Gene 9, 1-12.
42. Remaut, E. et al, in preparation
43. Stanssens, P. et al, in preparation
44. Remaut, E., Tsao, H. and Fiers, W. (1982), in preparation.

VAI RNA PLAYS A ROLE IN ADENOVIRUS LATE mRNA TRANSLATION.

CARY WEINBERGER, BAYAR THIMMAPPAYA,[+] ROBERT J. SCHNEIDER AND THOMAS SHENK
Department of Microbiology, Health Sciences Center, State University of New
York, Stony Brook, New York 11794

INTRODUCTION

The adenovirus (Ad) VA RNAs[1] are small RNAs (about 160 nucleotides)
encoded by two different genes and are designated VAI and VAII[2]. The genes
are located at about 30 map units on the Ad2 chromosome[2,3,4], and are
transcribed by RNA polymerase III[4,5]. These genes contain intragenic
transcriptional control regions; the VAI control region is located between
+10 and +69 relative to the initiation site for the VAI(G) species[6,7]. The
VA RNAs are made in remarkably large amounts late after infection, and are
found in both the nucleus and cytoplasm of infected cells[1,4]. The RNAs
exist as ribonucleoprotein particles in association with at least one
cellular protein antigen which is recognized by the anti-La class of lupus
sera[8,9]. VAI RNA is made in much larger amounts at late times after
infection than the VAII species (about 40:1).

We have constructed two Ad5 variants, each of which fails to synthesize
one of the VA RNAs. The mutant which fails to produce the minor VAII
species grows normally. The VAI⁻ virus, however, grows more poorly than
its wild-type parent. Its defect occurs at the level of translation of
late mRNAs.

MATERIALS AND METHODS

Mutant viruses defective in the synthesis of VA RNAs were constructed
utilizing segments of the viral genome derived from both recombinant
plasmids and viral DNA. The plasmid-derived segments included the VA RNA
genes and carried deletions within the VAI (pA2-dl4,-29bp) or the VAII
(pA2-dl8,-16bp) intragenic control regions[6]. The original plasmids
contained only small segments of the Ad genome (27-31.5 map units). To
reconstruct virus, small DNA segments carrying the VA-specific deletions
were first transferred to plasmids which contained much larger segments of

+ Present address: Department of Microbiology-Immunology, Northwestern
University School of Medicine, Chicago, IL 60611.

the viral chromosome (27-46 map units). Intact viral chromosomes were then rebuilt via a three-fragment ligation: 0-29 map unit fragment from dl313[10], 29-46 map unit fragment from the recombinant plasmid, and 46-100 map unit fragment from dl324 (which lacks the 79-85 map unit segment).

The 313 and 324 deletions were included in the construction for technical purposes. The 324 deletion is located within early region 3, a transcription unit known to be nonessential for growth of the virus in cell culture. The 313 deletion is located within early region 1. The 313 defect can be complemented by propagating the virus in 293 cells (a human embryonic kidney cell line which contains and expresses the Ad5 early region 1[11]). Thus the 313 and 324 deletions will not affect the phenotypes of the reconstructed viruses provided they are studied in 293 cells.

The variant carrying a deletion within its VAI RNA control region is designated dl330, and the mutant lacking a segment of the VAII RNA gene is dl328. dl331 is a derivative of dl330 which contains a wild-type early region 1. It was produced by marker rescue of dl330 for growth in HeLa cells.

RESULTS

Variants lacking one of the VA RNA species are viable. The variants which fail to produce one of the two VA RNA species are viable. Their relative growth abilities were quantitated by infecting 293 cells at a multiplicity of 3 plaque-forming units/cell and measuring virus yields after all cells were killed (Table 1). dl328 (VAI$^+$/VAII$^-$) generates a yield equivalent to its parental virus (dl324) and the wild-type virus (wt300). In contrast, dl330 (VAI$^-$/VAII$^+$) produces a 20-fold-reduced yield.

TABLE 1

GROWTH OF WILD-TYPE AND MUTANT VIRUSES

Virus	Yield (pfu/ml)[a]
wt 300	1.0×10^8
dl 324	1.2×10^8
dl 328	1.0×10^8
dl 330	5.9×10^6

[a]293 cells were infected at a multiplicity of 3 plaque-forming units cell, and the virus yield was measured by plaque assay on day 5 after infection.

The variants each fail to produce one of the VA RNA species in vivo.
The deletions within the VA RNA genes were originally shown to prevent
detectable transcription of the altered gene by analysis in a cell-free
polymerase III extract[6]. To test that the mutations also prevented
transcription in vivo, 293 cells were infected with the mutant viruses,
labeled with $^{32}PO_4^{2-}$, and their cytoplasmic RNAs analyzed (Figure 2).
dl330 (VAI⁻/VAII⁺) fails to produce the VAI species in vivo. It is also
clear that dl330-infected cells contain much larger than normal amounts of
the VAII RNA. This result is consistent with our earlier observation that
a functional VAI gene competively inhibits synthesis of the VAII species

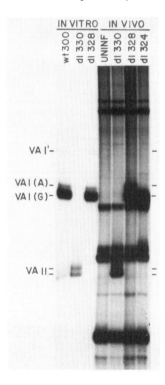

Fig. 1. Electrophoretic analysis of VA RNAs encoded by wild-type and mutant viral DNAs both in vitro and in vivo. In vitro transcription reactions were carried out as described by Fowlkes and Shenk[6] using whole viral DNAs as template. In vivo analysis was performed by infecting 293 cells with the indicated viruses at a multiplicity of 5 plaque-forming units/cell, labeling from 2 to 18 hr after infection with $^{32}PO_4^{2-}$ (200 µCi/ml), and extracting total cytoplasmic RNA. Electrophoresis was performed using a 6% polyacrylamide slab gel (0.6mm thick, 40 cm long, containing 8M urea in a Tris-borate buffer) for 12 hr at 500V.

in vitro[6]. It was not possible to identify the VAII RNA synthesized in
whole cells by the parental (dl324) virus. Analysis of RNAs synthesized in
isolated nuclei during a short labeling period comfirmed that dl324
produced VAII RNA while dl328 (VAI⁺/VAII⁻) did not (data not shown).

The dl330 growth defect is due to a lack of VAI RNA. It was conceivable that the poor growth of dl330 resulted from the disruption of the coding region for a previously unidentified polypeptide. To rule out this possibility, we tested the ability of a recombinant SV40 virus carrying and expressing the Ad5 VA RNA genes (SV-VA[12]) to complement the dl330 defect (Table 2). The experiment was performed in monkey kidney cells to permit optimal growth of SV40. As has been known for many years, wild-type Ad5 (wt300) grows very poorly in monkey kidney cells, and its poor growth is complemented by coinfection with wild-type SV40. dl331 (VAI$^-$/VAII$^+$, a derivative of dl330 described in Materials and Methods) also grows very poorly in monkey kidney cells. Wild-type SV40 only partially complements the dl331 growth defect, while coinfection with SV-VA generates a near wild-type yield. Since the only adenovirus-specific gene products encoded by SV-VA are the VA RNAs, we conclude that the dl330/331 growth defect results from its inability to synthesize VAI RNA.

TABLE 2
COMPLEMENTATION OF A VAI$^-$ MUTANT BY AN SV40 RECOMBINANT CARRYING THE VAI RNA GENE.

Viruses	Yield[a]
wt 300 alone	2.1×10^5
wt 300 + wt SV40	1.8×10^7
wt 300 + SV-VA	1.5×10^7
dl 331[c] alone	4.0×10^3
dl 331 + wt SV40	1.5×10^5
dl 331 + SV-VA	8.5×10^6

[a] CV-1P cells were infected with SV40 at a multiplicity of 5 plaque-forming units/cell; 40 hrs later the cells were superinfected with Ad. Yields of adenovirus were measured three days after infection with Ad by plaque assay on 293 cells.
[b] SV-VA is a recombinant SV40 which carries the Ad5 VAI RNA in its late region[12].
[c] dl331 is a VAI$^-$ derivative of dl330 in which the early region 1 deletion (dl313) carried by dl330 has been replaced by wild-type sequences.

Late mRNAs are inefficiently translated in dl330-infected cells. The defect in the dl330 (VAI$^-$/VAII$^+$) growth cycle responsible for its 20-fold

reduced yield was next identified. Analysis of early viral polypeptides
(Elb, 58K and E2a, 72K) by immunoprecipitation indicated they were
synthesized at normal levels in dl330-infected 293 cells (data not shown).
dl330 DNA replication was also normal (Figure 2). Further, late viral

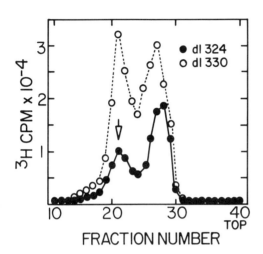

Fig. 2. Analysis by
equilibrium density
centrifugation of DNAs
synthesized in dl330 and
324-infected 293 cells.
Cells were infected at a
multiplicity of 5 plaque-
forming units per cell and
labeled with ^{3}H-thymidine
(10μCi/ml) from 9-12 hr
after infection. Total
cellular DNA was prepared
and analyzed by centrifu-
gation as described by
Jones and Shenk[10].

polyA$^+$ mRNAs were present in the cytoplasm of dl330-infected 293 cells at
wild-type levels (Figure 3a), they were properly processed (Figure 3b) and
capped (data not shown). However, synthesis of late polypeptides was
reduced about 7-fold in dl330 as compared to dl324 (parental virus)-
infected 293 cells (Figure 4). To confirm that dl330 infected cells
contained functional late mRNAs, total polyA$^+$ RNA was prepared from dl330-
and 324-infected cells at 20 hours after infection and equal amounts of the
isolated RNAs used to program cell-free translation in reticulocyte
lysates. As is evident in Figure 4, dl330-specific mRNAs were as efficiently translated as those encoded by its parent, dl324. The defect in the
dl330 growth cycle occurs at the level of translation of late viral mRNAs.

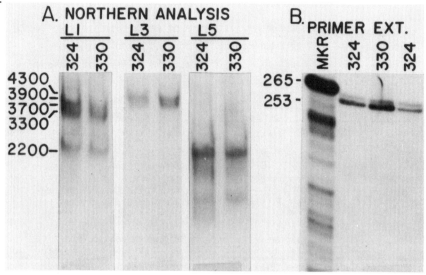

Fig. 3. Analysis of the late viral mRNA species produced in dl330-and 324-infected 293 cells. A. Northern-type analysis of late viral mRNAs. Cytoplasmic, polyA+ RNA was prepared from 293 cells 16 hr after infection with either dl330 or 324 at a multiplicity of 5 plaque-forming units/cell. RNAs were subjected to electrophoresis at 100V for 4 hr in 1% agarose gels containing 6% formaldehyde[13]. RNA was transferred to nitrocellulose, ^{32}P-labeled[14] plasmid DNA probes (2 X 10[8] cpm/ug) were hydridized to RNA on the paper, and the paper was washed using the procedure of Alwine et al.[15]. The plasmid DNA probes contained sequences specific for Ad2 late regions 1, 3 or 5. B. Primer extension analysis of hexon-specific (L3) mRNAs. A 46 nucleotide primer sequence (TaqI-AluI) specific for the 5' coding region of the hexon mRNA was annealed to total cytoplasmic polyA+ RNA prepared as described above. The primer was extended as described by Akusjarvi and Petterson[16] and the product analyzed by electrophoresis for 5 hr at 800 V in an 8% polyacrylamide gel containing 8M urea in Tris-borate buffer.

DISCUSSION

dl330-infected 293 cells contain normal levels of late viral mRNAs. These RNAs appear to be properly capped, spliced and polyadenylated, they are localized in the cytoplasm of infected cells and they are efficiently translated in vitro. Since dl330-infected 293 cells contain reduced levels of late polypeptides, we conclude that VAI RNA is required for efficient translation of late viral mRNAs. It is not yet clear whether VA RNA is absolutely required or merely enhances mRNA late translation since dl330-infected cells contain substantial amounts of the VAII species. We are presently attempting to construct a double mutant (VAI⁻/VAII⁻) to address this question.

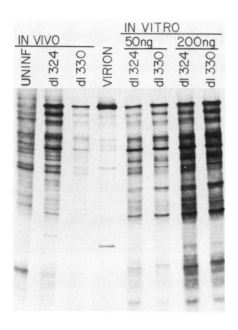

Fig. 4. Electrophoretic analysis of polypeptides synthesized at late times after infection by dl330 and 324 and polypeptides synthesized in rabbit reticulocyte extracts in response to late viral RNA. The in vivo experiment was performed by infecting 293 cells at a multiplicity of 5 plaque-forming units per cell with either dl330 or 324. Cells were labeled with ^{35}S-methionine (20 μCi/ml) from 16-20 hr after infection. For the in vitro analysis, cytoplasmic, polyA+ RNA was prepared from a second batch of infected cells and used to program mRNA-dependent reticulocyte lysates as described by Pelham and Jackson[17]. SDS-polyacrylamide gel electrophoresis was carried out using the system of Sarnow et al.[18].

How does VAI RNA function in translation? So far we have established that the rate of nascent polypeptide elongation during late translation in dl331 (VAI⁻/VAII⁺)-infected cells is identical to that in wild-type-infected cells (Schneider, Weinberger and Shenk, unpublished). This suggests a role for VAI RNA during initiation of late translation.

Do VA RNAs represent a specialized viral function or do cellular RNAs perform similar functions during translation in uninfected cells? The answer is not yet clear. However, there are a number of small, cytoplasmic RNAs present in a wide variety of uninfected animal cells. These include the 7S RNA (also termed ScL RNA)[19,20] which is partly complementary to human Alu-family sequences[21]. 7S RNA could play a role in protein synthesis since it has been found in association with polysomes [22,23]. Thus, it is an excellent candidate for a cellular RNA with a function similar to that of VA RNA. Perhaps there are a variety of small, cytoplasmic RNAs which facilitate translation of broad classes of mRNAs. Conceivably, these small RNAs play a role in certain types of translational control during development. A case in point is the surf clam embryo in which selective translation of RNA controls the pattern of protein synthesis during early development (ref. 24 and J.V. Ruderman, E.T. Rosenthal and J.R. Tansey, this volume).

ACKNOWLEDGMENTS

We acknowledge the competent technical assistance of Martha Marlow. This work was supported by a grant from the American Cancer Society (MV-45). R.J.S. is an NIH Postdoctoral Trainee (CA-09176). T.S. is an Established Investigator of the American Heart Association.

REFERENCES

1. Reich, P.R., Rose, J. Forget, B., and Weissman, S.M. (1966) Journal Molecular Biology 17, 428.
2. Mathews, M. (1975) Cell 6, 223.
3. Pettersson, U. and Philipson, L. (1975) Cell 6, 1.
4. Soderlund, H., Pettersson, U., Venstrom, B. Philipson, L., Mathews, M.B. (1976) Cell 7, 585.
5. Weinmann, R., Raskas, H.J. and Roeder, R.G. (1974) Proc. Nat. Acad. Sci. USA 71, 3426.
6. Fowlkes, D.M. and Shenk, T. (1980) Cell 22, 405.
7. Guilfoyle, R. and Weinmann, R. (1981) Proc. Nat. Acad. Sci. USA 78, 3378.
8. Lerner, M.R., Andrews, N.C., Miller, G., and Steitz, J.A. (1981) Proc. Nat. Acad. Sci. USA 78, 805.
9. Lerner, M.R., Boyle, J.A., Hardin, J.A., Steitz, J.A. (1981) Science 211, 400.
10. Jones, N. and Shenk, T. (1979) Cell 17, 683.
11. Graham, F.L., Smiley, J., Russell, W.C. and Naira, R. (1977) J. Gen. Virol. 36, 59.
12. Weinberger, C., Thimmappaya, B., Fowlkes, D.M., and Shenk, T. (1981). ICN-UCLA Sym. Mol. and Cell. Biol. 23, 509.
13. Rave , N., Crkenjakov, R. and Boedtker, H. (1979) Nucl. Acids Res. 6, 3559.
14. Rigby, P.W.J., Dieckmann, M., Rhodes, C. and Berg, P. (1977) Journal of Molecular Biology 113, 237.
15. Alwine, J.C., Kemp, D.J., Stark, G.R. (1977) Proc. Nat. Acad. Sci. USA 74, 5350.
16. Akusjarvi, G. and Petterson, U. (1978) Proc. Nat. Acad. Sci. 75, 5822.
17. Pelham, H.R.B. and Jackson, R.J. (1976) Eur. J. Biochem 67, 247.
18. Sarnow, P., Ho, Y.S., Williams, J. and Levine, A.J. (1982) Cell 28, 387.
19. Eliceiri, G. (1974) Cell 3, 11.
20. Zieve, G. and Penman, S. (1976) Cell 8, 19.
21. Weiner, A.M. (1980) Cell 22, 209.
22. Walker, T.A., Pace, N.R., Erikson, R.L., Erikson, E. and Behr, F.(1974) Proc. Nat. Acad. Sci. USA 71, 3390.
23. Gunning, P.W., Beguin, P., Shooter, E.M., Austin, L. and Jeffrey, P.L. (1981) J. Biol. Chem. 256, 6670.
24. Rosenthal, E.T., Hunt, T. and Ruderman, J.V. (1980) Cell 20, 487.

SOME ASPECTS OF METABOLIC REGULATION OF TRANSLATION IN CULTURED EUKARYOTIC
CELLS

KIVIE MOLDAVE, E. THALIA DAVID, JAMES S. HUTCHISON, STEWART A. LAIDLAW AND
ITZHAK FISCHER
Department of Biological Chemistry, University of California, College of
Medicine, Irvine, CA. USA

INTRODUCTION

It has been shown in a number of cell types in culture that, in contrast to
the exponential phase, the stationary phase is accompanied by a marked decline
in protein synthetic activity. Evidence has been reported suggesting that
important elements regulating protein synthesis, as a function of the growth
phase, include the activity or levels of mRNA[1-3], aminoacyl-tRNA synthetase[4,5],
initiation of protein synthesis[6-8], chain elongation[3,4,9], ribosome content[6,10],
protein degradation[11-13], etc. Although the rate of protein synthesis was
shown to decline in many cultured cell systems, as growth and cell division
decreased, the direct participation of a translational component in the regulation of protein synthesis, or whether the levels of mRNA can or cannot account
for the apparent changes in the rate of protein synthesis, remains controver-
sial.

Divergent conclusions regarding the mechanism that controls protein synthesis in cultured cell systems seem to indicate that this regulation is dependent
not only on the cell types used, but also on the conditions used to establish
various parameters of growth or cessation of growth. Also, in most reports,
the effects on protein synthesis were determined by measuring the incorporation
of amino acids into protein in intact cells, or by measuring the degree to
which polysomes disaggregated. Some studies used cell-free extracts that
carried out chain elongation only on endogenous templates, or with exogenous
synthetic polynucleotides that did not require or carry out the initiation or
termination sequence of reactions. This report describes the differences in
translational components, including mRNA, between exponentially growing and
stationary phase Chinese hamster ovary (CHO) cells; the studies made extensive
use of cell-free systems prepared from such cells, that were capable of carrying out actively and accurately all of the intermediary reactions involved in
polypeptide chain initiation, elongation and termination.

EFFECTS ON PROTEIN SYNTHESIS IN INTACT CELLS AND CELL-FREE SYSTEMS

When Chinese hamster ovary cells are cultured in media containing 2.5-5.0%
fetal calf serum, the cells grow at a constant exponential rate for several
days, then gradually cease to divide and grow as they enter stationary phase.
As shown previously[3], and in Table 1, incorporation of exogenous amino acid
into protein in these cultures was markedly lower in stationary phase cells
as compared to exponentially growing cells; in late log or early stationary
phase, incorporation was about 35% lower than in exponential phase, and after
24 hours in stationary phase incorporation was decreased over 80%.

TABLE 1

AMINO ACID INCORPORATION IN INTACT CULTURED CHO CELLS

Phase of culture	Specific activity[a]
Exponential	6.1
Late log	4.0
Stationary	1.6

[a]Specific activity = cpm (X 10^{-3}) incorporated per 10^6 cells. Aliquots containing approximately equivalent numbers of cells obtained from the midpoint of the
exponential phase, the beginning of the stationary or late log phase, and 24
hours into stationary phase, were incubated with [³H]leucine for 60 minutes at
34°C; the hot (90°C) trichloroacetic acid-insoluble fractions were then prepared and counted.

Studies reported previously[3], summarized in Table 2, also indicated that
translation of endogenous mRNA and of exogenous natural mRNA or synthetic
polynucleotide templates was markedly lower in cell-free systems prepared from
stationary phase cells, than in extracts prepared from exponentially growing
cells. Protein synthesis on endogenous mRNAs with stationary cell extracts
was only about 20% of that in similar preparations from exponential cells.
When mRNA-depleted stationary cell extracts were incubated with globin mRNA
plus radioactive leucine, or with poly(U) plus radioactive phenylalanine, amino
acid incorporation was 55-60% lower.

The finding that protein synthesis was lower in stationary than in exponential cell extracts when exogenous templates were translated, indicated either
that the amount or activity of a component or components of the translational
system was decreased, or the presence of a translational inhibitor. The much
greater differential in protein synthesis between the two culture phases when
endogenous mRNAs were translated *in vitro*, as compared to the differences

TABLE 2

TRANSLATION OF VARIOUS mRNA TEMPLATES IN CELL-FREE SYSTEMS FROM EXPONENTIALLY
GROWING AND STATIONARY PHASE CELLS

Template	pMol of [^{3}H]amino acid incorporated into protein	
	exponential	stationary
Endogenous mRNAs[a]	3.2	0.7
Globin mRNA[b]	2.4	1.0
Poly(U)[b]	21.5	9.5

[a]Postmitochondrial extracts from exponentially growing and stationary phase
cells, not treated with micrococcal nuclease, were incubated with radioactive
leucine, without added mRNA[3,14]; incubations were then analyzed for hot-acid
insoluble radioactivity.
[b]Similar extracts, depleted of mRNA, were incubated with radioactive leucine
and globin mRNA, or with radioactive phenylalanine and poly(U).

observed when exogenous templates were translated, suggested a decrease in the
amount of translatable polysome-associated mRNA, in addition to the direct
effect on the translation process.

EFFECTS ON mRNA

The total amount of poly(A)-containing mRNA in exponential and stationary
cells was measured by determining the amount of RNA that hybridizes with radioactive polyuridylic acid[15] (Table 3). Analyses for mRNA were carried out in

TABLE 3

DETERMINATION OF POLY(A)-CONTAINING mRNA, BY POLY(U) HYBRIDIZATION, IN PREPARA-
TIONS FROM EXPONENTIALLY-GROWING AND STATIONARY PHASE CELLS

Phase of culture	Cpm of [^{3}H]poly(U) hybridized/10^6 cells[a]	
	cytoplasm	cytoplasmic RNA
Exponential	12,500	11,000
Stationary	6,500	4,600

[a]Approximately 3 X 10^8 cells from each phase were lysed in isotonic buffered
salts solution containing 1% NP-40, and the nuclei removed by centrifugation
at 2000g. Hybridization was carried out with an excess of [^{3}H]poly(U) in the
presence of 0.5% SDS[16]. RNA was extracted from the cytoplasmic fraction with
phenol-chloroform[17,18] and hybridized to [^{3}H]poly(U) in excess[15]. The amount
of ribonuclease-resistant hybrids were determined by filtration through
Millipore membranes.

postnuclear lysates of CHO cells in order to minimize any changes that might occur during fractionation or isolation of mRNA, and in RNA extracted from the cytoplasmic fraction in order to minimize the possible effects of any inter-fering substances in crude lysates. With both preparations, the amount of mRNA in stationary cells was 50-60% lower than in exponential phase cells. Hybri-dization assays with tRNA and globin mRNA indicated that the reaction was specific for poly(A)$^+$ RNA and that the amount of nuclease-protected [^{3}H]poly(U) was proportional to the amount of poly(A)$^+$ mRNA used. The length of the poly-(A) segment was measured by digesting cytoplasmic RNA with pancreatic and T_1 ribonuclease, and subjecting the digest to polyacrylamide gel electrophoresis; the gel was sliced, individual fractions eluted, and hybridization with an excess of [^{3}H]poly(U) carried out[19]. The profiles of poly(A) lengths from growing and stationary cells were practically identical, indicating that the hybridization assays were an accurate measure of the amounts of mRNA in these preparations.

The distribution of mRNA among the various components in postmitochondrial extracts was also determined (Figure 1). The extracts were centrifuged through sucrose gradients and individual fractions were assayed for poly(A)$^+$ RNA by hybridization with radioactive poly(U). The mRNA pattern obtained with extracts from exponentially growing cells (A) revealed that approximately 25% of the poly(A)$^+$ mRNA sedimented in the region of the ribosomal subunits toward the top of the gradient, and that the rest was associated with polysomes, some

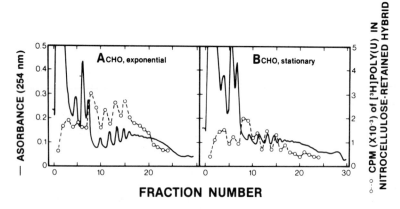

Fig. 1. Determination of poly(A)-containing mRNA, by poly(U) hybridization, in postmitochondrial extracts from exponential (A) and stationary (B) phase cells, resolved by gradient centrifugation.

containing up to a dozen ribosomes per polynucleotide. The pattern of hybridizable mRNA obtained with extracts from stationary cells (B) revealed a similar pattern; although the total amount of mRNA was considerably (about 50%) lower than that of exponential cells, approximately 30% of the mRNA was recovered in the non-polysomal region. There was no evidence from these analyses that there was an abnormal accumulation or sequestering of mRNA in stationary phase cells, such as in a messenger ribonucleoprotein complex.

Figure 2 presents results which indicated that the mRNA isolated from stationary cells was translated as efficiently as mRNA from exponentially growing cells. Estimates of the specific biological activity of the two RNA preparations, in terms of the pmol of amino acid incorporated into protein per μg of poly(A)$^+$ mRNA were within 25%, well within the experimental variations observed in such analyses. Thus, the mRNA in stationary phase cells, although lower in amount, seemed to be perfectly normal in its ability to act as template for protein synthesis.

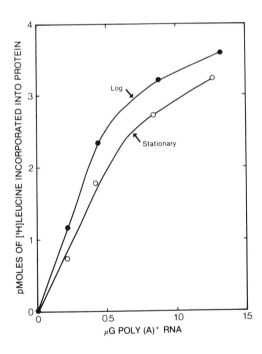

Fig. 2. The effects of cytoplasmic RNA from exponential (closed circles) and stationary (open circles) phase cells on protein synthesis, in a cell-free system from exponentially-growing CHO cells depleted of endogenous mRNA[3,14].

The nature of the polypeptides synthesized in exponential and stationary phase cells and those synthesized when mRNA from such cells were translated in cell-free systems, was also determined (Figure 3). In one set of experiments, cells obtained from the exponential (A) and the stationary (B) phase were in-

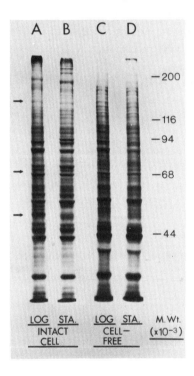

Fig. 3. SDS-gel electrophoretic analysis[14,20,21] of polypeptides synthesized in intact cells and in cell-free systems. Suspensions of exponentially growing (A) and stationary phase (B) cells labeled with radioactive methionine, and RNA from exponential (C) and stationary (D) cells translated in mRNA-depleted postmitochondrial extracts of CHO cells in the presence of radioactive methionine.

cubated with [^{35}S]methionine for 60 minutes, lysed, and the cytoplasms obtained by centrifugation were subjected to SDS-gel electrophoresis and autoradiography. In another set of experiments, the cytoplasmic poly(A)$^{+}$ mRNAs from exponential (C) and stationary (D) phase cells were translated in a messenger RNA-depleted cell-free system from exponentially growing CHO cells, in the presence of radioactive methionine, and the resulting polypeptide products were then analyzed by SDS-gel electrophoresis. In both analyses, using intact cells or cell-free translation of purified mRNAs, there were only minor differences in the products formed in stationary phase cells as compared to exponentially growing cells. For example, in intact cells, three proteins of molecular weights 55000, 60000 and 85000 appeared to be present in growing cells but were absent or present in very low amounts in resting cells; also, a protein of about 95,000 molecular weight seemed to be present in stationary cells in much higher concentrations than in exponential phase cells. Similar analyses, performed using the two dimensional gel electrophoretic system described by O'Farrell[22] did not reveal other major qualitative differences, although some quantitative variations were common. These observations are in agreement with others[23,24] with respect to changes in cellular proteins as a function of growth phases.

The data presented above suggested that the major changes between exponentially growing and stationary phase cells, with respect to mRNA, appeared to be quantitative rather than qualitative. Although less mRNA was present in resting cells, its quality as a template for translation, and the products for which they code were not significantly different from those in growing cells. One possible explanation of these findings was that the rate of synthesis of mRNA was lower in stationary than in exponential cells. In order to examine this possibility, cells obtained from the exponential and stationary phases were incubated with radioactive adenosine or radioactive uridine and the amount of radioactivity in the cold acid-insoluble fraction determined as a function of time of incubation. The results indicated that both the initial rate and the total extent of incorporation into RNA were at least 50% lower in resting as compared to growing cells. However, as shown in Table 4, transport of adenosine as revealed by the amount of radioactivity in the intracellular pool (TAS, total acid-soluble fraction), was also markedly lower in stationary phase cells and could account for the lower level of radioactivity in the RNA fraction. Even though adenosine transport was affected by the culture phase, a direct effect on other RNA biosynthetic reactions cannot be excluded; therefore, attempts to investigate RNA synthesis in cell-free preparations from CHO cells, in order to determine whether this process is directly affected by transition from the growing to the stationary phase, are in progress.

Also shown in Table 4 are the results of an experiment in which the uptake of radioactive leucine into the intracellular pool and into the protein fraction was examined. In contrast to the results obtained with adenosine, the transport of leucine into the cell was not appreciably different (within 20%) in resting cells as compared to exponential cells, while incorporation into protein was about 60% less in the stationary cells. Thus, transition from exponential to stationary phase did not affect amino acid transport, but did affect the amount of mRNA as described above, as well as the activity of the translational system as evidenced by the lower incorporation of amino acid in cell-free systems when globin mRNA and poly(U) were used as templates.

TABLE 4

INCORPORATION OF RADIOACTIVE ADENOSINE AND LEUCINE INTO THE INTRACELLULAR POOL AND INTO RNA AND PROTEIN

| Phase of culture | Cpm (X 10^{-3}) per 10^{5} cells[a] | | | |
| | $[^{3}H]$Adenosine | | $[^{3}H]$Leucine | |
	TAS	RNA	TAS	Protein
Exponential	326	9.3	6.8	2.3
Stationary	128	3.3	5.6	1.0

[a]Aliquots containing approximately equivalent number of cells from exponential or stationary phases were incubated with radioactive adenosine or leucine, for 30 minutes. The cells were then obtained by centrifugation, washed several times with buffered salts solution, and trichloroacetic acid (to 5%) added; the acid-soluble and acid insoluble fractions were separated by centrifugation and counted.

EFFECTS ON TRANSLATIONAL COMPONENTS

The change in activity of the translational system was localized to a decrease in the activity of elongation factor EF-1; other translational activities such as those involved in the initiation and termination of protein synthesis were not affected[3]. For example, as summarized in Table 5, the cell-free incorporation of radioactive methionine into the 40S-preinitiation complex containing 40S subunits and $[eIF\text{-}2 \cdot Met\text{-}tRNA_f \cdot GTP]$ ternary complex, and the formation of 80S initiation complex from the 40S-preinitiation intermediate were within 25% and 10%, respectively, in growing as compared to resting cells. These data indicated that the two preparations were similar in their ability to catalyze the formation of Met-tRNA$_f$ by methionyl-tRNA synthetase, the formation of ternary complex with eIF-2 and GTP, the interaction of ternary complex with 40S subunits, the binding of mRNA, and the joining of 60S ribosomal subunits. Also, analysis of the total amount of radioactive protein released from ribosomes when mRNA-depleted extracts from exponential and stationary phase cells were programmed with globin mRNA[3] suggested that termination of protein synthesis and release of the completed polypeptide was extensive (over 90% and not significantly different in the two preparations. The possibility

TABLE 5

FORMATION OF INTERMEDIARY INITIATION COMPLEXES WITH EXTRACTS FROM EXPONENTIALLY
GROWING AND STATIONARY PHASE CELLS[a]

Phase of culture	pMol of [^{35}S]Met in 40S preinitiation complex	% of [^{35}S]Met transferred to 80S initiation complex
Exponential	0.244	51
Stationary	0.185	48

[a]Postmitochondrial extracts depleted of mRNA were incubated with [^{35}S]methionine,
globin mRNA and cycloheximide, for 60 min. at 30°C, then centrifuged through a
linear 10-30% sucrose gradient, and analyzed for radioactive methionine-
containing 40S and 80S complexes as described previously[3].

existed in these experiments, however, that the release observed in stationary
phase extracts could reflect a premature and abnormal release of incomplete
peptides. Therefore, products of translation of globin mRNA, with exponential
and stationary cell extracts, were compared by SDS-gel electrophoresis (Figure
4). This figure shows that although the total amount of globin synthesized in
stationary cell extracts was less than in exponential cell preparations, both
were able to synthesize complete globin molecules; no evidence was obtained

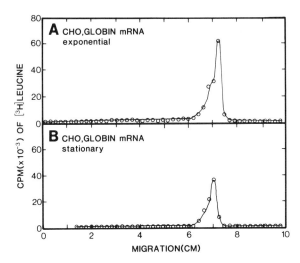

Fig. 4. SDS-gel electrophoretic analysis of reaction mixtures containing [^{3}H]-
leucine, globin mRNA and exponentially growing (A) or stationary phase (B)
postmitochondrial extracts, as described previously[3,20].

of low molecular weight polypeptides, migrating faster than globin, which would reflect premature or abortive release of incomplete products.

In contrast to the results obtained when the initiation or termination steps in protein synthesis were examined, which revealed no differences between growing and resting cells, chain elongation reactions were markedly different[3]. For example, the rate and total extent of translation of poly(U), which requires elongation factors EF-1 and EF-2, GTP, and ribosomes, was markedly lower with extracts from stationary phase cells than from exponentially growing cells (Figure 5). Further, varying concentrations of cytosols from the two

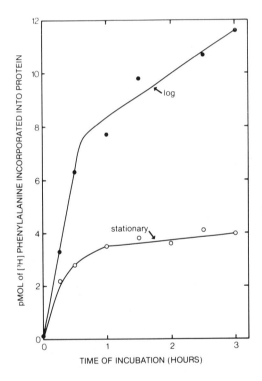

Fig. 5. Time-dependent synthesis of polyphenyl-alanine. Postmitochondrial extracts from exponential (closed circle) or stationary (open circle) cells, depleted of endogenous mRNA, were incubated with [³H]phenylalanine and poly(U) for varying periods of time, then analyzed for radioactive hot acid-insoluble protein.

cell preparations were assayed for poly(U)-dependent polyphenylalanine synthesis, in the presence of saturating amounts of EF-1 (to measure the activity of EF-2) or of EF-2 (to measure the activity of EF-1) as described[3,25]. Results obtained previously, summarized in Table 6, indicated that the activity of EF-1 was markedly lower in stationary cells but the activity of EF-2 was similar in the two extracts. Additional evidence that the component affected in the transition from exponential to stationary phase was indeed EF-1, and that there

TABLE 6

ACTIVITY OF CHAIN ELONGATION FACTORS IN EXPONENTIALLY GROWING AND STATIONARY
PHASE CELLS

Phase of culture	Specific activity of cytosol[a]	
	EF-1	EF-2
Exponential	0.51	0.36
Stationary	0.22	0.39

[a]Specific activity = pmol of [^{3}H]phenylalanine incorporated from [^{3}H]Phe-tRNA
into polyphenylalanine, per µg of cytosol protein, calculated from the linear
portion of the concentration curves. The cytosol fractions from exponential
and stationary phase cells, obtained by centrifugation at 100,000 g in 0.5 M
KCl solutions were incubated with rat liver derived 40S and 60S subunits[26],
[^{3}H]phenylalanyl-tRNA, poly(U), partially purified, resolved EF-2[27] to assay
for EF-1, or partially purified, resolved EF-1[27] to assay for EF-2. The assay
components and procedures have been described[25].

was no cytoplasmic inhibitor of the chain elongation reaction, is provided in
Table 7. When highly purified rat liver EF-1, completely resolved from EF-2,
was added to postmitochondrial extracts from stationary cells, the ability to
translate poly(U) was markedly increased, to within 20% of the activity
obtained with exponential cell extracts.

TABLE 7

THE EFFECT OF PURIFIED RAT LIVER EF-1 ON THE POLY(U)-DEPENDENT SYNTHESIS OF
POLYPHENYLALANINE WITH CELL-FREE EXTRACTS FROM STATIONARY PHASE CELLS[a]

Source of extract	Additions	pMol of [^{3}H]Phenylalanine incorporated
Exponential phase	none	6.20
Stationary phase	none	2.95
Stationary phase	EF-1	5.00

[a]Incubations contained nuclease-treated postmitochondrial extracts, [^{3}H]phenyl-
alanine, poly(U), other components required for cell-free protein synthesis[3,14],
and highly purified rat liver EF-1 where noted.

The decrease in the activity of EF-1 detected in stationary phase could be
the result of a modification of the protein factor such as phosphorylation,
methylation, acetylation, etc., which decreased its specific biological activity. If the changes involved a covalent modification, they could also affect
some of the properties of the factor such as its thermal lability or kinetic

behavior. EF-1 activity was therefore measured in cytosols from growing and resting cells, after preincubation for varying periods of time at 42°C. With cytosols from exponentially growing cells, 50% of the EF-1 activity was lost in about 10 minutes at 42°C; the results with cytosols from stationary phase cells, however, were inconsistent. With one such preparation the EF-1 appeared to be more stable, but others appeared to have the same half-life as that in exponential extracts. If the rates of inactivation are different, the results would suggest that the lower activity of EF-1 in stationary cells reflected a modification of the protein, which affected its thermolability and its specific biological activity. Such results would also argue against the total inactivation of some of the EF-1 molecules, while the rest remained unaffected and fully active. However, if the rates of inactivation of EF-1 are the same, the results would suggest that the lower activity of EF-1 in stationary cells reflected the complete inactivation of some of the EF-1 molecules, or a decrease in the concentration of EF-1. The data on the kinetic behavior of the factor, another sensitive measure of protein structure, as described below (Figures 6 and 7), are consistent with this latter interpretation.

Although the activity of EF-1 was lower in resting than in growing cells, the kinetic parameters of the reaction that it catalyzes did not appear to be affected. EF-1 activity was measured in cytosols by the extent to which they carried out the factor-dependent binding of radioactive Phe-tRNA to purified rat liver ribosomes in the presence of a guanine nucleotide. The ribosomes were preincubated with poly(U) and deacylated tRNA to eliminate the initial lag[28], the Phe-tRNA was chromatographed on a column of G-25 Sephadex to remove GTP in the preparation, and guanylyl-imidodiphosphate, a GTP analog, was used to prevent polymerization of phenylalanine in crude extracts which also contain EF-2. Thus, the assay measured the stoichiometric aspects of the reaction, since cyclic reutilization of EF-1 requires hydrolysis of the nucleotide. The kinetics of the EF-1 interaction with other components of the elongation reaction were analyzed in high salt cytosols from exponential and stationary cells, with varying concentrations of $[^3H]$Phe-tRNA (Figure 6) or of the ribosome-poly-(U)-tRNA complex (Figure 7), in short incubations in order to obtain initial rates. The figures show the concentration-dependent binding of Phe-tRNA (left panels), and the Lineweaver-Burk double reciprocal plots (right panels). The affinities for these two components, Phe-tRNA and ribosomes, were not appreciably different in the two cytosols. For example, the apparent Km values for Phe-tRNA and ribosomes, respectively, were 4.6 X 10^{-8} M and 8.4 X 10^{-8} M with

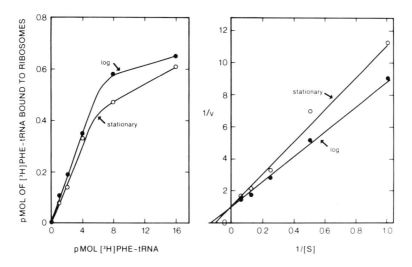

Fig. 6. The effect of increasing concentrations of Phe-tRNA on the aminoacyl-tRNA binding reaction catalyzed by EF-1. The cytosol fractions from exponential (closed circles) and stationary (open circles) phase cells were incubated for 3 minutes with rat liver 40S and 60S subunits, GTP and varying concentrations of [³H]Phe-tRNA. Ribosome-bound Phe-tRNA was analyzed by the Millipore filtration procedure[29]. Left panel, initial rates of the binding reaction; right panel, double reciprocal plot of velocity *versus* concentration.

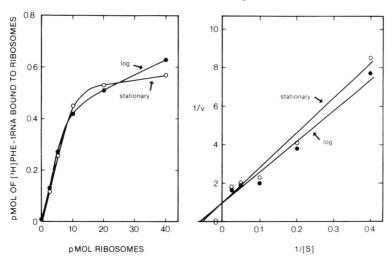

Fig. 7. The effect of increasing concentrations of ribosomes on the aminoacyl-tRNA binding reaction catalyzed by EF-1. Incubations with varying amounts of preformed ribosome-poly(U)-tRNA complex and analyses were carried out as described in the legend to Fig. 6.

exponential cytosols and 5.6×10^{-8} M and 10.5×10^{-8} M with stationary extracts.

The lower activity of EF-1 found in stationary cells could reflect a decrease in the amount of EF-1 protein; the finding that the kinetic parameters of the remaining EF-1 activity in stationary cells were unchanged is consistent with this possibility. The decrease could occur in the absence of protein synthesis if EF-1 has a particularly rapid turnover rate and was broken down to a greater extent than other translational components. In order to determine whether EF-1 had a particularly high turnover rate in CHO cells, and its steady state concentration was decreased as the result of rapid degradation under conditions where protein synthesis was blocked, chain elongation reactions were measured in extracts from cells where growth and protein synthesis were arrested by other than transition to stationary phase. A culture of a temperature-sensitive mutant of CHO cells altered in leucyl-tRNA synthetase[30] was grown exponentially at the permissive temperature (34°C), then divided into two portions; one of them was incubated at 34°C and the other at about 37.5°C for 24 hours. The 34°C-culture continued to grow, the number of cells doubled in about 25 hours, and the rate of amino acid incorporation per cell was unaffected. In the culture shifted to 37.5°C for 24 hours, there was no increase in the number of cells and protein synthesis was markedly depressed; in fact, amino acid incorporation at the elevated temperature was less than 30% of that at 34°C, even after only two hours. The poly(U)-dependent synthesis of polyphenylalanine in postmitochondrial extracts (Table 8), which measures EF-1, EF-2 and ribosomal activities, was essentially the same in preparations from growing and growth-arrested cells.

Experiments in which cell growth and protein synthesis were arrested by deprivation of glucose in the media led to similar conclusions. Exponentially growing cells were incubated for 24 hours with or without glucose. Whereas growth remained exponential, and amino acid incorporation into protein in intact cells was normal in the presence of glucose, growth did not occur in the absence of glucose and amino acid incorporation was markedly reduced even after 6 hours without glucose. The postmitochondrial extracts were prepared from each, depleted of endogenous mRNA, and assayed for chain elongation reactions with poly(U) and radioactive phenylalanine. The synthesis of polyphenylalanine was essentially the same with both preparations indicating that the activity of EF-1 was not affected after almost 24 hours without protein synthesis under these conditions.

TABLE 8

EFFECT OF ELEVATED CELL CULTURE TEMPERATURE ON THE POLY(U)-DEPENDENT SYNTHESIS
OF POLYPHENYLALANINE BY CELL-FREE EXTRACTS OF CHO tsHl

Culture temperature	Cpm of $[^3H]$phenylalanine incorporated[a]
34°C	185,600
34°C + 24 hrs. at 37.5°C	189,800

[a]Incubations contained CHO postmitochondrial extract depleted of endogenous
mRNA, from cells incubated for 24 hours at 34°C or at 37.5°C, $[^3H]$phenylala-
nine (6000 cpm/pmol), poly(U), and other components required for protein
synthesis[3,14].

These results, although indirect, suggested that the decrease in EF-1 was
not simply due to the degradation of existing factor, in the absence of *de novo*
protein synthesis, causing a decrease in the level of EF-1 protein. However,
the possibility must be considered that there may indeed be a decrease in the
amount of EF-1 protein, but that it may be related to events other than the
turnover rate of the factor.

Another possibility that must be mentioned is that the changes in EF-1
activity observed using the phenylalanine polymerization (polyuridylate
translation) assay may be due to an effect on an EF-Ts-like activity[31-35]
which catalyzes the exchange of GTP with guanosine diphosphate bound to EF-1,
and makes possible the cyclic reutilization of the factor.

In order to discriminate between the various possibilities that would explain the lower EF-1 activity in stationary cells, it was considered essential
to purify and compare the chemical and physical characteristics of EF-1s from
exponential and stationary phase cells. A preliminary step in this direction
is shown in Figure 8. Cytosols from growing and resting cells were centrifuged
through glycerol gradients and individual fractions were assayed for their
ability to carry out poly(U) translation in the presence of GTP and ribosomes
and EF-2 from rat liver. The results indicated that the high molecular weight
aggregates that characterize EF-1[31-38] did not appear to be appreciably different in the two preparations. In both cytosols, the complexes had molecular
weights of over 350,000. More definitive information on the structural and
biological differences in EF-1 in exponential and stationary cells must await
additional purification of the factor and the subunits that comprise this
complex protein.

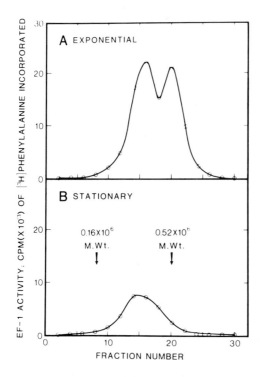

Fig. 8. Gradient sedimentation of EF-1 activity from cytosols. The high salt cytosol fractions from exponential (A) and stationary (B) phase cells were centrifuged through a linear 10-30% glycerol gradient, and individual fractions were assayed for EF-1 activity by the poly(U)-dependent incorporation of radioactive phenylalanine into polyphenylalanine with GTP, rat liver ribosomes, and an excess of rat liver EF-2.

EFFECTS OF SERUM

Another important problem with respect to the changes in EF-1 activity that accompany the transition from rapidly growing to the stationary phase concerns the nature of the causative agent(s) for this change. In these experiments, cells were not grown until they reached such a high concentration that cessation of growth would be a consequence of high cell density. Instead, the cells seemed to stop growing because some essential factor or factors in the serum appeared to be depleted. Consistent with this suggestion were the findings that when stationary phase cells were resuspended in fresh serum-containing media, or when fresh serum was added to the depleted media, growth was resumed and the level of activity of EF-1 returned to that in exponentially growing cells. Also, removal of serum from exponentially growing cells has been shown to cause a similar decrease in EF-1 activity[4]. These findings and the one described above with the temperature-sensitive mutant of CHO cells, or with cells incubated in the absence of glucose, in which EF-1 activity was not affected when growth in serum containing media was arrested by incubation at the nonpermissive temperature or in the absence of glucose, respectively, further suggested that (1) the lower EF-1 activity in stationary cells was related to the depletion of component(s) or growth factor(s), and (2) the decrease in the amount or activity of EF-1 appeared to be reversible.

In order to study more directly the effects of serum hormones and growth factors on growth and protein synthesis, it would be desirable to use a cultured cell type whose growth responds to well-characterized, pure components, in the absence of serum. Preliminary studies in collaboration with Dr. E. Rozengurt (Imperial Cancer Research Fund, London) indicate that both growth and amino acid incorporation into protein, in Swiss 3T3 cells, are markedly decreased as cells become quiescent, and that incorporation is stimulated within a short period of time after insulin, vasopressin and epidermal growth factor (or serum) are added. Examination of the effect of quiescence and of the subsequent stimulation with hormones and growth factors, on translation at a molecular level, is in progress.

SUMMARY

When Chinese hamster ovary cells in culture transit from the exponential to the stationary phase, growth, cell division and incorporation of amino acids into protein are markedly decreased. The amount of mRNA in stationary cells is also much lower than in rapidly growing cells, but its distribution among translational components such as polysomes, ribosomes and ribosomal subunits, and its ability to serve as a proper template for protein synthesis appear to be normal. Of all of the translational factors required for polypeptide chain initiation, elongation, and termination, only EF-1 is affected; its activity is significantly lower in stationary cells but its affinity for aminoacyl-tRNA and ribosomes remains unchanged. Whether the lower activity is due to a decrease in the amount of EF-1 protein, or to a chemical modification of the protein factor, remains to be clearly established; however, it does not appear to be due to turnover or degradation in the absence of protein synthesis. The changes in mRNA and EF-1, when growth is arrested under these conditions, appear to be reversible and due to the depletion of a component or components supplied with the serum in the culture media.

ACKNOWLEDGMENTS

This work was supported in part by Research Grants AM 15156 and AG 00538, and Training Grant CA 09054 (J.S.H.), from the National Institutes of Health. The valuable assistance of Eva Mack and Tom Ho is gratefully acknowledged.

REFERENCES

1. Johnson, L.F., Abelson, H.T., Green, H. and Penman, S. (1975) Cell, 1, 95-100.
2. Levis, R., McReynolds, L. and Penman, S. (1976) J. Cell Physiol. 90, 485-502.
3. Fischer, I. and Moldave, K. (1980) Biochemistry, 19, 1417-1425.
4. Engelhardt, D.L. and Sarnoski, J. (1975) J. Cell Physiol. 86, 15-30.
5. Conta, B.S. and Meisler, A.I. (1977) J. Biol. Chem. 252, 7640-7647.
6. Stanners, C.P. and Becker, H. (1971) J. Cell Physiol. 77, 31-42.
7. Rudland, P.S., Weil, S. and Hunter, A.R. (1975) J. Mol. Biol. 96, 745-766.
8. Meedel, T.H. and Levine, E.M. (1978) J. Cell Physiol. 94, 229-242.
9. Nielsen, P.J. and McConkey, E.H. (1980) J. Cell Physiol. 104, 269-281.
10. Weber, M.J. (1972) Nature, New Biol. 235, 58-61.
11. Tanaka, K. and Ichihara, A. (1976) Exp. Cell Res. 99, 1-6.
12. Tanaka, K. and Ichihara, A. (1977) J. Cell Physiol. 93, 407-416.
13. Warburton, M.J. and Poole, B. (1977) Proc. Natl. Acad. Sci. 74, 2427-2431.
14. Fischer, I. and Moldave, K. (1981) Anal. Biochem. 113, 13-26.
15. Bishop, J.O., Rosbash, M. and Evans, D. (1974) J. Mol. Biol. 85, 75-86.
16. Jagus, R. and Safer, B. (1979) J. Biol. Chem. 254, 6865-6868.
17. Singer, R.H. and Penman, S. (1973) J. Mol. Biol. 78, 321-329.
18. Aviv, H. and Leder, P. (1972) Proc. Natl. Acad. Sci. 69, 1408-1412.
19. Rosbash, M. and Ford, P.J. (1974) J. Mol. Biol. 85, 87-101.
20. Maizel, J.V., Jr. (1969) in Fundamental Techniques in Virology (Habel, K. and Salzman, N.P., eds.) Academic Press, N.Y. pp. 334-362.
21. Bonner, W.M. and Laskey, R.A. (1974) Eur. J. Biochem. 46, 83-88.
22. O'Farrell, P.H. (1975) J. Biol. Chem. 250, 4007-4021.
23. Lee, G.T. and Engelhardt, D.L. (1978) J. Cell Biol. 79, 85-96.
24. Garrels, J.I. (1979) J. Biol. Chem. 254, 7961-7977.
25. Moldave, K., Harris, J., Sabo, W. and Sadnik, I. (1979) Fed. Proc. Fed. Am. Soc. Exp. Biol. 38, 1979-1983.
26. Moldave, K. and Sadnik, I. (1979) Methods Enzymol. 59G, 402-410.
27. Moldave, K., Galasinski, W. and Rao, P. (1971) Methods Enzymol. 20C, 337-348.
28. Hardesty, B., Culp, W. and McKeehan, N. (1969) Cold Spring Harbor Symp. Quant. Biol. 34, 331-339.
29. Ibuki, F. and Moldave, K. (1968) J. Biol. Chem. 243, 791-798.
30. Thompson, L.H., Harkins, J.L. and Stanners, C.P. (1973) Proc. Natl. Acad. Sci. 70, 3093-3098.
31. Iwasaki, K., Motoyoshi, K., Nagata, S. and Kaziro, Y. (1976) J. Biol. Chem. 251, 1843-1845.
32. Slobin, L.I. and Moller, W. (1978) Eur. J. Biochem. 89, 69-77.
33. Grasmuk, H., Nolan, R.D. and Drews, J. (1978) Eur. J. Biochem. 92, 479-487.
34. Slobin, L.I. (1979) Eur. J. Biochem. 96, 287-296.
35. Hattori, S. and Iwasaki, K. (1980) J. Biochem. 88, 725-731.
36. Schneir, M. and Moldave, K. (1968) Biochim. Biophys. Acta, 166, 58-67.
37. Bollini, R., Soffientini, A.N., Bertani, A. and Lanzari, G.A. (1974) Biochemistry 13, 5421-5425.
38. Ejiri, S., Murakami, K. and Katsumata, T. (1977) FEBS, 82, 111-116.

FERTILIZATION TRIGGERS RAPID SEQUENCE-SPECIFIC CHANGES IN THE TRANSLATIONAL ACTIVITIES OF INDIVIDUAL MESSENGER RNAs.

JOAN V. RUDERMAN, ERIC T. ROSENTHAL and TERESE R. TANSEY.
Departments of Cell and Developmental Biology, and Anatomy
Harvard Medical School, Boston, Massachusetts, USA

INTRODUCTION

The mature oocytes of most organisms appear to contain all the components required for protein synthesis, including ribosomes, tRNA, amino acids, protein synthetic enzymes and mRNA, yet they generally exhibit very low rates of protein synthesis. Fertilization or, in some species the resumption of meiosis, sets off a massive cascade of metabolic and morphological changes that usually include either an increase in the rate of protein synthesis, a change in the pattern of protein synthesis or, in some cases, both.[1,2,3] In oocytes of the mollusc Spisula, fertilization causes a 2-4 fold increase in the overall rate of protein synthesis[4,5], and a dramatic shift in the kinds of proteins being made.[6] Both of these changes occur within 10 min of fertilization and are independent of new mRNA transcription. In fertilized sea urchin eggs, where the rate of protein synthesis increases 10-30 fold[7,8] but the pattern of protein synthesis remains essentially the same,[9] the rise in protein synthesis appears to be linked to an increase in the intracellular pH and, possibly, by a transient rise in intracellular calcium levels.[10,11] Similar changes have been reported for Spisula,[12,13] but a connection with protein synthesis has not been directly demonstrated. In both Spisula and urchins, these early changes in protein synthesis appear to be regulated entirely at the translational level.[6,14] The sudden and extreme (for eukaryotes) nature of these changes in Spisula oocytes make this a useful system for studying translational regulatory mechanisms as well as the function of selective mRNA utilization in early development.

In this paper, we briefly review the evidence that the early changes in the pattern of protein synthesis are regulated at the translational level. Previous studies of the switches in the translational activities of several specific individual mRNAs have been directly confirmed and extended using cloned cDNA probes complementary to those mRNAs. We have found that, except for changes in poly(A) tail lengths, fertilization does not result in any substantial change in mRNA size. These results show that, just as for histone mRNAs in sea urchins,[15,16] extensive processing of larger precursors cannot be responsible for the translational turn-on of specific mRNA sequences. We show that fertilization also causes the rapid polyadenylation of one set of poly(A)$^-$ oocyte mRNAs and the deadenylation of another set of maternal poly(A)$^+$ mRNAs, just at the time when the translational pattern is changing dramatically. However, several considerations indicate that these changes in adenylation are probably not directly related to concurrent changes in translational activities.

METHODS

Culture and Labeling of Oocytes and Embryos. Adult Spisula solidissima were collected locally by the Marine Resources Staff at the Marine Biological Laboratory, Woods Hole, Massachusetts. Gametes were prepared as described earlier.[6] Oocytes and embryos to be labeled with radioactive amino acids were pelleted gently and resuspended in 20 vols of Millipore-filtered sea water containing antibiotics. ^{35}S-methionine (New England Nuclear, 600 Ci/mmole) was added to a final concentration of 0.2 mCi/ml, and oocytes and embryos were labelled in vivo for 20 min. Samples were prepared for electrophoresis on polyacrylamide gels containing SDS.[17]

Preparation of 12K Supernatants. Oocytes or embryos were washed twice with cold calcium- and magnesium-free sea water and once with homogenization

buffer (0.3 M glycine, 70 mM potassium gluconate, 45 mM KCl, 2.3 mM MgCl$_2$,
1 mM DTT, 40 mM HEPES, pH 6.9).[18] Cells were resuspended in 3 vol of
homogenization buffer and broken in a Dounce homogenizer. Homogenates were
spun at 12,000 x g for 20 min. Aliquots of these "12K" supernatants were
frozen in liquid nitrogen and stored at -80°C until use.

Polysome gradients. 12K supernatants were diluted with 5 vol of gradient
buffer (0.5 M KCl, 6 mM MgCl$_2$, 0.5 mM DTT, 1 mM EDTA, 10 mM HEPES, pH 7.0).
Half of each sample was treated with 30 mM EDTA to release mRNA from
polysomes.[19] 1 ml of 12K supernatant was layered onto a 15-40% sucrose
(w/v) 11 ml gradient made up in gradient buffer, and gradients were
centrifuged in an SW41 rotor at 40,000 rpm for 70 min. Fractions were
collected and RNA was extracted as described previously.[6] RNA from each
gradient fraction was dissolved in an equal volume of water, usually 30 ul,
and stored at -80°C until use.

Cell-free translation. RNA samples were translated in a mRNA-dependent
rabbit reticulocyte lysate.[20] Total RNA was translated at a final
concentration of 600 ug/ml. When polysomal gradient RNA fractions were
translated, 2 ul of each RNA sample was added to 8 ul reticulocyte lysate.
When the translational specificities of unextracted Spisula oocyte or embryo
12K supernatant were assayed, 1 ul of unextracted 12K supernatant was diluted
with 1 ul of water and added to 8 ul of reticulocyte lysate. All incubations
were carried out for 1 hour at 30°C. Labeled protein samples were analysed
on SDS-polyacrylamide slab gels as above.

Isolation of cloned cDNAs complementary to translationally regulated
mRNAs. Double-stranded cDNAs complementary to poly(A)$^+$ RNA from 2-cell or
18 hr embryos were synthesized and inserted into the Pst I site of the plasmid
pBR322 by the GC-tailing method. Bacterial clones transformed with
recombinant plasmids were isolated and recombinant plasmid DNAs from selected

bacterial colonies were prepared from bacterial lysates.[22-24] The cDNA

sequences carried by individual clones were determined by selective

hybridization[25,26] of the complementary mRNA from total 2-cell or 18 hr

embryo RNA followed by translation in vitro and identification of the

translation products on 1-D gel electrophoresis. Details of this aspect of

the work are presented elsewhere.[27]

RNA blots. RNA samples were electrophoresed on 1% agarose gels containing

formaldehyde,[28] transferred to nitrocellulose and hybridized with [32]P-

labelled plasmid DNA probes.[29] Plasmid DNAs were labeled in vitro with

[32]P using a "nick-translation" reaction.[30] Total RNA was fractionated on

oligo(dT)-cellulose.[31] In some cases, RNA was chromatographed on

poly(U)-Sepharose (Pharmacia).

RNAse H Removal of poly(A) tails. Unfractionated 12K supernatant RNAs

were hybridized with oligo(dT) and the double-stranded portions of the hybrids

were digested with RNAse H (Bethesda Research Laboratories).[32] Nuclease-

resistant RNAs were then electrophoresed on an agarose gel, blotted to

nitrocellulose and hybridized to ^{32}P-labelled cloned cDNA probes as above.

RESULTS

The pattern of protein synthesis changes rapidly following fertilization.

Fertilization of the Spisula oocyte is followed by breakdown of the large

tetraploid oocyte nucleus, the germinal vesicle (10 min post-fertilization),

first meiosis and formation of the first polar body (40 min), second meiosis

and formation of the second polar body (50 min), union of the egg and sperm

haploid nuclei (55 min) and first cleavage (70 min). Subsequent asynchronous

cleavages occur rapidly without any significant change in the mass of the

embryo. About 5-6 hours after fertilization the young gastrula (200-300

cells) ciliates and becomes free-swimming. Several complex morphological

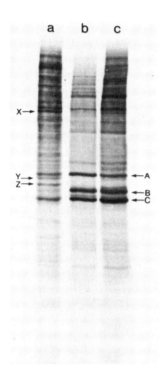

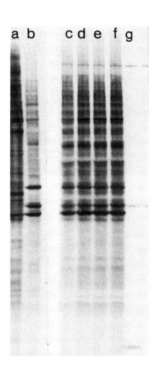

Figure 1 (left). Autoradiograms of proteins synthesized in vivo by oocytes and embryos. (a) oocytes labeled with ^{35}S-methionine for 20 min; (b) embryos labeled for 20 min after germinal vesicle breakdown; (c) embryos labeled for 20 min after first cleavage. From Rosenthal et al. (1980) Cell 20, 487. Reproduced with permission of MIT Press, Cambridge.

Figure 2 (right). Comparison of proteins synthesized in vivo by oocytes and embryos with those programmed in vitro by oocyte and embryo RNA. Proteins were labeled with ^{35}S-methionine. (a) Oocyte proteins synthesized in vivo; (b) 2-cell embryo proteins synthesized in vivo. Lanes c-f, in vitro translation products encoded by: (c) oocyte RNA; (d) embryo RNA, 15 min post-fertilization; (e) embryo RNA, 50 min post-fertilfization; (f) 2-cell embryo RNA. (g) endogenous incorporation (no RNA added). From Rosenthal et al. (1980) Cell 20, 487. Reproduced with permission of MIT Press, Cambridge.

changes occur over the next 18 hours, resulting in the formation of a
clam-shaped feeding larva about the same size as the oocyte.[33] During this
24 hr period, there are several changes in the overall pattern of protein
synthesis. The earliest of these occurs within 10 min of fertilization and is
mediated entirely at the translational level.[6] As development proceeds and
the maternal mRNA store is depleted, newly transcribed mRNAs make increasingly
prominent contributions to the pattern of protein synthesis.[34] In this
paper, we shall deal only with the early, translationally regulated changes in
protein synthesis.

Previous work has shown that fertilization results in a rapid change in
the pattern of protein synthesis[6] (Fig. 1). For example, the proteins
marked A, B and C are synthesized at very low levels, if at all, in the
oocyte, whereas they are prominently labeled in the embryo. Proteins X, Y and
Z are synthesized by the oocyte, whereas incorporation into these proteins is
not readily detectable in the embryo. Several other differences in the
patterns of proteins synthesis at the two stages are easily seen.

The change in the pattern of protein synthesis is controlled at the level
of translation. Comparison of the in vivo protein synthetic patterns of
oocytes and early embryos with the in vitro translation products encoded by
RNA isolated from these stages (Fig. 2) demonstrate that the switch in the
pattern of protein synthesis occurs in the absence of any significant changes
in mRNA concentrations or intrinsic translatabilities in the reticulocyte
lysate. This interpretation is further supported by the finding that
different subsets of the maternal mRNA pool are found on polysomes before and
after fertilization. 12K supernatants from oocytes and 2-cell embryos were
fractionated on high salt sucrose gradients. RNAs isolated from each gradient
fraction were dissolved in equal volumes of water and translated in vitro and
equal volumes of the translation products were analysed by gel electro-

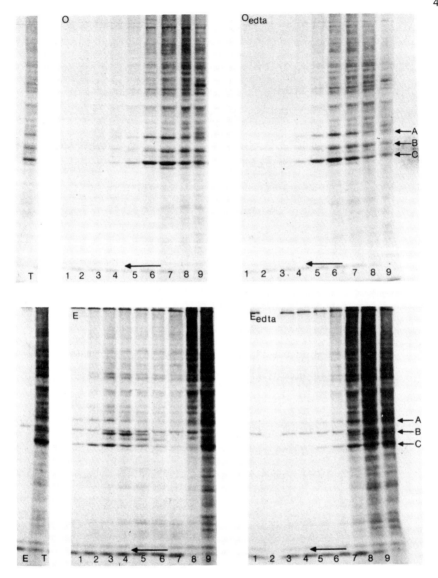

Figure 3. In vitro translation products, labeled with ^{35}S-methionine and
directed by RNA extracted from sucrose gradient fractions of oocyte and embryo
12K supernatants. (a) translation products encoded by total oocyte RNA [T];
(b) proteins encoded by total embryo RNA [T] or without added RNA [e];
Panels: translation products directed by RNAs isolated from sucrose gradient
fractionations of 12K supernatants from oocytes (O) and embryos (E), and of
EDTA-treated 12K supernatants from oocytes (O$_{EDTA}$) and embryos (E$_{EDTA}$).
Arrow indicates direction of sedimentation. 60S subunits are found in
fractions 7 and 8; 40S subunits are found in fractions 8 and 9.

phoresis. Figure 3 shows a typical result. In the exposure shown here, it can be seen that most mRNAs in the oocyte are not engaged on polysomes. Longer exposures show that there is a very small amount of mRNA in the polysome region that is released by EDTA treatment of oocyte 12K supernatant prior to gradient fractionation. In the 2-cell embryo, a considerable portion of the maternal mRNA is found loaded on polysomes. These and other comparisons of the polysomal vs. non-polysomal locations of specific mRNAs in the oocyte and embryo reveal that many mRNAs show stage-specific recruitment onto polysomes.[6,27]

Quantitative analyses of the translational status of individual mRNAs by this method are limited by two considerations. First, when the polypeptides encoded in vitro by two or more mRNAs co-migrate, it is impossible to conclude anything about the behavior of their mRNAs. Secondly, since the translation of Spisula maternal mRNA is quantitatively and qualitatively regulated in vivo and we do not understand the molecular mechanisms responsible for this selective translation, an in vitro translation assay of these mRNAs could yield misleading results. Therefore, we chose to use a second RNA assay method, hybridization to cloned cDNA probes, that does not depend on template activity. Recombinant DNA clones consisting of pBR322 and cDNAs complementary to 2-cell or 18 hr larval poly(A)$^+$ RNAs were isolated as described in the Methods section. Among the clones used in the following experiments were those carrying sequences for protein A (clone H2/1T55), protein C (clone D11/1T43), and a-tubulin (clone 10E10/3V4). Four other clones (6T21, 1T9, 6T23, and 29) that showed hybridization to RNA on Northern blots of 2-cell RNA, but did not selectively hybridize detectable amounts of template-active RNA, were also used in this study. The a-tubulin clone was isolated from the 18 hr library; all others came from the 2-cell library.

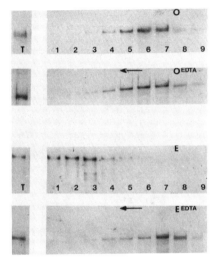
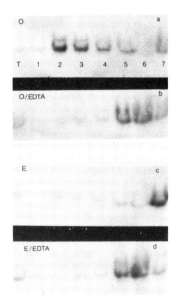

Figure 4 (left). Hybridization of ^{32}P-labelled H2/1T55 DNA to Northern blots of RNA from polysome gradient fractions. 12K supernatants from oocytes (O) and embryos (E) were fractionated on sucrose gradients as described in Methods. In panels O_{EDTA} and E_{EDTA}, the 12K supernatants were treated with EDTA prior to fractionation in order to release mRNA from polysomes. Arrow indicates direction of sedimentation. An equal portion of the corresponding 12K supernatant was run in each lane marked "T".

Figure 5 (right). Hybridization of ^{32}P-labelled clone 10E10/3V4 DNA to RNA isolated from sucrose gradient fractions of 12K supernatants and analysed as in Figure 4.

RNAs from 12K polysome gradient fractions were electrophoresed on agarose slab gels, blotted to nitrocellulose and hybridized with individual ^{32}P-labeled cloned DNAs. The hybridization pattern with clone 1T55 (protein A) shown in Figure 4 shows quite clearly that RNA for protein A is not present on polysomes in the oocyte and is almost completely loaded onto polysomes at the 2-cell stage. Similar results were obtained with clone 1T43 (protein C). RNAs complementary to clones 6T21 and 1T9 also show increased translation after fertilization (data not shown). In contrast, hybridization with clone 3V4 (Fig. 5) reveals that a considerable fraction of the RNA for a-tubulin is found on polysomes in the oocyte but is almost totally absent from polysomes in the 2-cell embryo.

As noted previously, the sedimentation properties of the presumed mRNPs in oocyte and 2-cell 12K supernatants are somewhat unusual.[6] Non-polysomal mRNAs appear to exist as rather large mRNPs, many of which sediment in the polysome region of the gradient even though they are not functionally associated with polysomes (as shown by the EDTA-release test). Also, both the in vitro translation and hybridization assays show that the sedimentation properties of many EDTA-released mRNPs change after fertilization and become lighter in the embryo. A third puzzling feature is exhibited in the case of a-tubulin mRNA: these mRNPs from EDTA-treated 12K supernatants appear to be heavier than their counterparts in untreated homogenates (Fig. 5). What this result reflects is not clear.

mRNA-specific adenylations and deadenylations occur at fertilization but do not appear to be coupled to translational changes. As part of another study of the features of maternal mRNA, we looked at the representation of mRNAs in poly(A)$^+$ and poly(A)$^-$ RNA in oocytes and 2-cell embryos. RNAs from both stages were fractionated on oligo(dT)-cellulose under standard conditions[31] and analysed by in vitro translation or blot hybridization. As

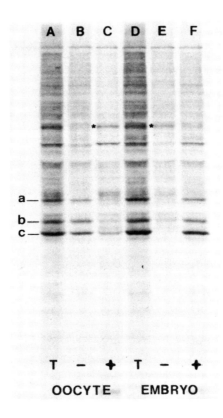

Figure 6. Translation products encoded by [T] total unfractionated RNA;
[-] poly(A)⁻ RNA; and [+] poly(A)⁺ RNA from oocytes and embryos.

shown in Figures 6 and 7, fertilization causes a large change in the adenylation of many maternal mRNAs as assayed by the ability to bind to oligo(dT)-cellulose. For example, mRNAs for proteins A, B, and C are predominantly poly(A)⁻ in the oocyte, whereas they are poly(A)⁺ in the embryo. mRNA complementary to clone 6T21 also undergoes this same switch. About one third of the 1T9 RNA sequences are adenylated in oocytes; after fertilization, essentially all of these sequences are found as poly(A)⁺ versions (data not shown). In contrast, a-tubulin mRNA is exclusively poly(A)⁺ in the oocyte, whereas in embryos it is found in both poly(A)⁺ and poly(A)⁻ RNA fractions. Almost all of these changes in the adenylation status occur between 2 min and 10 min of fertilization.[27]

Additional experiments indicate that the designation "poly(A)⁻" is not quite correct. All of the mRNAs examined so far, whether from oocytes or 2-cell embryos, bind to poly(U)-Sepharose under conditions where only 6-10 A residues are required for binding.[35] This finding suggests that these mRNAs have very short poly(A) tails or oligo(A) tracts, rather than being completely devoid of poly(A). Oocyte RNAs bearing short A tracts have been described previously in both sea urchin eggs and Xenopus oocytes.[36-39]

It is difficult to demonstrate that these conversions occur in the absence of new RNA synthesis because of uncertainties about the effectiveness of transcriptional inhibitors in very early Spisula embryos. However, two considerations strongly suggest that the changes in the proportions of poly(A)⁺ and poly(A)⁻ versions of these mRNAs occuring at fertilization are almost certainly due to adenylation and deadenylation of preexisting maternal mRNAs. First, these changes occur within 10 min of fertilization, during which time the chromosomes are undergoing condensation for meiosis. Secondly, even maximal transcriptional contributions from single-copy or low-copy genes to the large maternal mRNA pool are quantitatively insignificant during this interval.

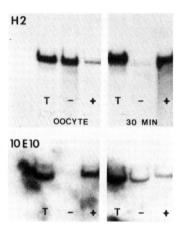

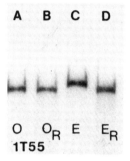

Figure 7 (left). Hybridization of ^{32}P-cloned DNA probes H2/1T55 (protein A) and 10E10/3V4 (-tubulin) to total RNA, poly(A)$^-$ RNA and poly(A)$^+$ RNA from oocytes and 30 min post-fertilization embryos. RNAs were electrophoresed on an agarose slab gel, blotted to nitrocellulose and hybridized with ^{32}P-labelled cloned DNAs.

Figure 8 (right). Comparison of the sizes of mRNA for protein A in oocytes and embryos. RNAs were isolated from oocytes (0) and 2-cell embryos (E). In the lanes O_R and E_R, the poly(A) tails were removed using RNAse prior to electrophoresis. Samples were electrophoresed on an agarose gel, blotted to nitrocellulose and hybridized with ^{32}P-labelled clone H2/1T55 DNA.

In some cases, the relationship between the adenylation status of a messenger RNA and its translational status in vivo appears intriguingly close. Several mRNAs, such as those encoding proteins A, B and C, are poly(A)⁻ at the stage when they are translationally inactive and these same mRNAs exist as polyadenylated versions at the stage when they are translationally active. However, other mRNAs do not fit this pattern. For example, mRNA complementary to clone 1T9 is found in both poly(A)⁺ and poly(A)⁻ RNA in oocytes, yet very little if any of this mRNA is detectable on polysomes in oocytes. a-tubulin mRNA is another case where polyadenylation and translation are not tightly coupled.

Large changes in the sizes of several maternal mRNAs do not occur at fertilization. A second finding to come out of these analyses is that the overall sizes of mRNAs do not change grossly at fertilization (Figs. 4,5,7). Closer comparisons of the sizes of several individual mRNAs in oocytes and 2-cell embryos (after their poly(A) tails had been removed by RNAse H)[32] have confirmed this result.[27] One example is shown in Fig. 8. These results suggest that, at least for the mRNAs examined, translational activation is not accompanied by any extreme changes in size that would be indicative of extensive processing of large precursors. More subtle alterations, such as changes in the 5' cap structures or the removal of just a few nucleotides, would not have been detected in these experiments.

Translational specifities are maintained in a mixed cell-free system. The results presented above show that oocytes and embryos contain the same sets of abundant mRNAs, that these mRNAs appear equally translatable in vitro when the standard mRNA-dependent reticulocyte lysate is used, and that no obvious changes in mRNA sizes (except adenylation or deadenylation) occur at fertilization. These findings suggest, but by no means prove, that major structural alterations in mRNA sequences are not responsible for the changes

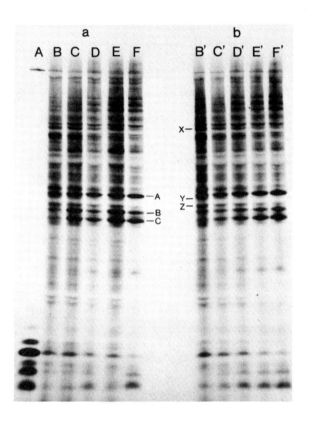

Figure 9. Results of mixing oocyte and embryo 12K supernatants before translation in reticulocyte lysate. (a) Proteins synthesized in the heterologous cell-free translation system when unextracted Spisula oocyte 12K supernatants are mixed with unextracted Spisula 2-cell embryo 12K supernatants and translated in reticulocyte lysate. Patterns of protein synthesis directed by reticulocyte lysate and: (A) endogenous (homogenization buffer added instead of 12K supernatant); (B) 100% oocyte 12K supernatant; (C) 25% embryo 12K supernatant: 75% oocyte 12K supernatant; (D) 50% embryo 12K supernatant: 50% oocyte 12K supernatant; (E) 75% embryo 12K supernatant: 25% oocyte 12K supernatant; (F) 100% embryo 12K supernatant. (b) For comparison with (a), the heterologous cell-free translation products shown in lanes B and F were mixed in comparable ratios after the translation reaction was terminated by addition of SDS gel sample buffer. Lanes B'-F' correspond to the same ratios as lanes B-F in (a). From Rosenthal et al. (1980) Cell 20, 487. Reproduced with permission of MIT Press, Cambridge.

in their translational activites following fertilization, and that some other
classes of molecules may mediate translational switching.

We have attempted to reproduce the stage-specific translation of
particular mRNAs in a cell-free system. Crude 12K supernatants were prepared
from Spisula oocytes or embryos homogenized in a buffer [0.3 M glycine, 70mM
potassium gluconate, 45 mM KCl, 2.3 mM $MgCl_2$, 1mM DTT, 40 mM HEPES, pH 6.9]
developed for a sea urchin egg cell-free system.[18] These 12K supernatants
do not initiate protein synthesis very efficiently. However, when the crude
12K supernatant is supplemented with mRNA-dependent reticulocyte lysate, this
mixed cell-free system now initiates and synthesizes protein rather
actively[6] (Fig. 9). Furthermore, the translation products encoded by the
oocyte-reticulocyte mix closely resemble those made by the oocyte in vivo
whereas the embryo-reticulocyte mix yields a pattern of protein synthesis that
is like the embryo pattern in vivo. Thus, the initiations that occur in these
mixed cell-free systems are proper, not promiscuous. Clearly, some
information that is present in the crude unextracted 12K supernatants, but
absent in the phenol-extracted RNA isolated from these 12K supernatants, is
responsible for this stage-specific, sequence-specific translational
discrimination. Very little is known about the nature of this "information."
When oocyte and embryo unextracted 12K supernatants are mixed in varying
proportions prior to addition to the reticulocyte lysate, the pattern of
protein synthesis simply resembles the sum of the oocyte and embryos
patterns (Fig. 9). This result suggests that there is no excess of either
a positive or negative diffusible regulator. One possible interpretation is
that these mRNAs in vivo are complexed with proteins as RNP particles in vivo
and that certain of these proteins can modulate translational activity in
different ways under different physiological circumstances. This notion is
consistent with evidence from other systems.[3] However, other explanations
are certainly not precluded by any of the experiments presented here.

DISCUSSION

Fertilization causes a rapid change in the pattern of protein synthesis that appears to be regulated entirely at the translational level. Within 10 min of fertilization, a specific subset of stored maternal mRNAs is recruited onto the polysomes and other mRNAs that had been engaged in protein synthesis in the oocyte are no longer found on polysomes. Recent experiments suggest that the signal for this change in Spisula, and in several other species as well, is meiotic activation rather than fertilization per se.[40] For example, artificial activation of Spisula oocytes by brief exposure to hypertonic salt results in the full complement of protein synthetic changes. Natural activation of starfish oocytes by 1-methyladenine similarly causes a one-step change in the pattern of protein synthesis; subsequent fertilization does not result in any further changes in the kinds of proteins being made at this early time.

The molecular basis of the translational switch is unknown. As is often the case, it has been easier to rule out mechanisms than to discover them. Extensive processing of translationally incompetent mRNA precursors cannot account for this switch, since RNAs purified from oocytes and early embryos are equally translatable in vitro in the message-dependent reticulocyte lysate. Also, except for various adenylations and deadenylations, mRNA sizes do not change significantly following fertilization. However, small changes in mRNA size or structure would not have been observed in the experiments reported here, and still remain a possibility.

Several translationally regulated mRNAs undergo a change in polyadenylation soon after fertilization. In some cases, such as the mRNAs for proteins A, B and C, translational activation of specific mRNAs occurs about the same time those sequences become polyadenylated - between 2 min and 10 min after fertilization. In other cases, such as a-tubulin mRNA, the loss

of translational activity in vivo is accompanied by deadenylation of a
substantial portion of the mRNA. However, several other mRNAs do not show any
obvious relationships between translational activity and adenylation.
Clearly, even if polyadenylation is involved in the modulation of
translational activity at fertilization, it cannot be the sole determinant.
One direct way of testing for a role of polyadenylation in the translational
activation of mRNAs following fertilization would be to fertilize oocytes in
the presence of a concentration of 3'-deoxyadenosine (cordycepin) sufficient
to block polyadenylation and ask if the switch in the pattern of protein
synthesis is affected. Unfortunately, Spisula oocytes are resistant to low
concentrations of cordycepin and become seriously deranged when cultured with
higher concentrations of the drug. Thus, we have found it impossible to carry
out this test. In sea urchins, where cordycepin does effectively inhibit the
2-fold increase in the poly(A) content at fertilization,[41-43] it has no
effect on the 30-fold increase in the rate of protein synthesis that occurs at
that same time.[44] However, since even under normal circumstances the
pattern of protein synthesis does not change when urchin eggs are
fertilized,[45] one cannot easily use cordycepin to test the effect of
adenylation of translational activation of specific mRNAs in this organism.
When the results from Spisula and sea urchin are considered together, it seems
unlikely that adenylation has any direct, primary role in translational
activation. Of course, more subtle roles may exist. Probably the most useful
results to come out of this brief study of polyadenylation is the realization
that many of the mRNAs which are preferentially translated in the oocyte are
probably not present in our cloned cDNA library of sequences complementary to
2-cell poly(A)$^+$ RNAs and might be preferentially represented in a library
made using oocyte poly(A)$^+$ RNA.

24. Grunstein, M. and Hogness, D.S. (1975). Proc. Natl. Acad. Sci. USA 72, 3961.

25. Ricciardi, R.P., Miller, J.S. and Roberts, B.E. (1979). Proc. Nat. Acad. Sci. USA 76, 4927.

26. Alexandraki, A. and Ruderman, J.V. (1981). Mol. Cell. Biol. 1, 1125.

27. Rosenthal, E.T., Tansey, T.R. and Ruderman, J.V. (1982). In preparation.

28. Rave, M., Crkvenjokov, R. and Boedtker, H. (1979). Nucleic Acids Res. 6, 3559.

29. Thomas, P.S. (1980). Proc. Nat. Acad. Sci USA 77, 5201.

30. Maniatis, R., Jeffrey, A. and van de Sande, H. (1975). Biochemistry 14, 3787.

31. Aviv, H. and Leder, P. (1972). Proc. Nat. Acad. Sci. USA 69, 1408.

32. Vournakis, J.H., Efstratiadis, A. and Kafatos, F.C. (1975). Proc. Nat. Acad. Sci. USA. 72, 2959.

33. Allen, R.D. (1953). Biol. Bull. 105, 213.

34. Tansey, T.R. and Ruderman, J.V. (1982). In preparation.

35. Palatnik, C.M., Storti, R.V. and Jacobson, A. (1979). J. Mol. Biol. 128, 371.

36. Duncan, R. and Humphreys, T. (1981). Dev. Biol. 88, 211.

37. Cabada, M.O., Darnbrough, C., Ford, P.J. and Turner, P.C. (1977). Dev. Biol. 57, 427.

38. Darnbrough, C. and Ford, P.J. (1979). Dev. Biol. 71, 323.

39. Sagata, N., Shiokawa, K. and Yamana, K. (1980). Dev. Biol. 77, 431.

40. Rosenthal, E.T., Ruderman, J.V., Brandhorst, B.T. (1982). Dev. Biol., in press.

41. Slater, D.W., Slater, I. and Gillespie, P. (1972). Nature 240, 333.

42. Wilt, F.H. (1973). Proc. Nat. Acad. Sci. USA 70, 2345.

43. Wilt, F.H. (1977). Cell 11, 673.

44. Mescher, A. and Humphreys, T. (1974). Nature 249, 138.

45. Brandhorst, B.T. (1976). Dev. Biol. 52, 310.

46 Spirin, A.S. (1966). Curr. Top. Dev. Biol. 1, 1.

47. Ehrenfeld, E. (1982). Cell 28, 435.

HEAT SHOCK IN *SACCHAROMYCES CEREVISIAE*: QUANTITATION OF TRANSCRIPTIONAL AND

TRANSLATIONAL EFFECTS

JUDITH PLESSET, JAMES J. FOY, LI-LI CHIA, AND CALVIN S. McLAUGHLIN
Department of Biological Chemistry, California College of Medicine, University
of California-Irvine, Irvine, California 92717

INTRODUCTION

A change in the pattern of protein synthesis after a sudden temperature elevation has been observed in numerous eukaryotic organisms including insects[1,2], mammalian cells[3], higher plants[4,5], and yeast[6,7,8,9]. A small set of polypeptides (heat shock proteins) is synthesized at a high rate in response to heat, whereas the synthesis of most cellular proteins is sharply curtailed. This response is termed the heat shock response and it is a valuable model system for studying the induction and regulation of protein synthesis in eukaryotes. This response has been most thoroughly studied in *Drosophila* where the effects of a heat shock are expressed at both the transcriptional and translational levels. Within minutes of a shift to a high temperature in this system, a small set of mRNAs is transcribed in abundance from specific puff sites on the chromosomes while the transcription of other mRNAs ceases[1]. In addition, these heat shock mRNAs appear to be translated preferentially at the expense of other pre-existing mRNAs, which are maintained in the cytoplasm but are not translated. The evidence for this is as follows: When polysomal RNA is isolated from heat shocked cells and translated in a heterologous cell-free system, only heat shock proteins are made. If, however, total RNA from heat shocked cells is used to program the cell-free system, many other "normal" cellular proteins are made, which are not made *in vivo* under heat shock conditions[2,10]. This suggests that the normal complement of mRNAs is somehow sequestered during the heat shock response and that the heat shock mRNAs are preferentially translated. This is supported by several additional findings. The total complement of mRNAs present at 23°C is also found in quantitatively similar amounts after a heat shock in *Drosophila*[11]. Upon recovery from a heat shock, normal protein synthesis can begin prior to new mRNA synthesis[11] or under conditions that prevent transcription of new mRNA (actinomycin D)[12].

In order to determine if the response observed in *Drosophila* represents the general response to heat shock in eukaryotic cells and to explore the details of the induction process both at the transcriptional and translational level, we have begun a quantitative examination of the heat shock response in

Saccharomyces cerevisiae. Previous reports have demonstrated a heat shock effect in *S. cerevisiae*[6,7,8,9]. Increased synthesis of several polypeptides is observed when cells are shifted from 23°C to 36°C although the synthesis of other proteins is not eliminated. When RNA is extracted from these cells and translated in a reticulocyte cell-free system, a characteristic heat shock pattern of protein synthesis is observed[8,9]. This indicates that the mRNA levels for these proteins are increased, i.e. a response to heat is occurring at the transcriptional level. To date, no evidence has been presented that suggests there is control at the level of translation in yeast.

Our experiments here confirm and quantitate the response at the transcriptional level. We have isolated RNA from heat shocked and control cells and translated it in a homologous cell-free protein synthesizing system prepared from cells that were not heat treated. The cell-free translation products were separated on O'Farrell two-dimensional gels and the relative synthesis of individual heat shock and non-heat shock polypeptides was determined. The relative abundance of heat shock mRNAs was determined by measuring *in vitro* translation levels of the heat shock proteins in both strain A364A and in ts 187, a temperature sensitive mutant derived from A364A that is defective in an initiation step of protein synthesis[13]. We found relative increases in the heat shock mRNAs in both strains. This indicated an induction at the transcriptional level in both strains.

In addition, we present evidence that regulation of the heat shock response can occur at the translational level in yeast and quantitate this response relative to the transcriptional effect. In ts 187 the rate of *in vivo* synthesis of heat shock proteins relative to other cellular proteins was greater than could be accounted for solely by the increased mRNA levels found. This implied, that upon a heat shock, there was an enhanced translation of heat shock mRNAs in ts 187 relative to other cellular mRNAs.

MATERIALS AND METHODS

Strains. A haploid strain of *Saccharomyces cerevisiae* A364A *(a, lys-2, tyr-1, his-7, gal-1, ade-1, ade-2, ura-1)* and a mutant derived from A364A, ts 187 *(prt-1)*, which has a temperature sensitive defect in initiation of protein synthesis[13], were used to examine the heat shock response. A diploid strain SKQ2n *(a/α, ade-1/+, +/ade-2, +/his-1)* was used to prepare the *in vitro* translation system.

Growth of cells. Strain SKQ2n was grown in YM-1 medium[14] as described previously[15]. Strain A364A and ts 187 were grown to early logarithmic phase

(50 Klett units) at 23°C on a rotary shaker in medium composed of: Difco yeast nitrogen base without amino acids and without ammonium sulfate, 1.7 g; ammonium sulfate, 5 g; each amino acid naturally occurring in proteins except methionine, cystiene, tyrosine, lysine, and histidine, 12 mg; tyrosine, 32 mg; lysine, 32 mg; histidine, 32 mg; adenine, 10 mg; uracil, 10 mg; succinic acid, 10 g; sodium hydroxide, 6 g; yeast extract, 200 mg; peptone, 200 mg; glucose, 20 g; and distilled water to 1 liter. The additional amounts of tyrosine, lysine, and histidine were added because of the auxotrophic requirements of A364A and ts 187.

Determination of mRNA half life. Half-lives of heat shock mRNAs were determined as described previously[16].

Isotopic labeling of cells. A364A or ts 187 cells grown to early logarithmic phase (50 Klett units) were shifted from 23°C to 36°C and radioactively labeled as follows: Each culture was divided into aliquots at time zero (t=0). At t=0 a 15 ml aliquot was removed and long term labeled for 3 h at 23°C with ^{3}H-yeast protein hydrolysate (10 μCi/ml) and L-[methyl-^{3}H]methionine (20 μCi/ml). Also at t=0, 3 ml aliquots were added to four separate 50 ml flasks. Three of these flasks had been pre-warmed to 36°C in a water bath while the fourth flask remained at 23°C. At t=5 min the 23°C aliquot and one aliquot of cells shifted to 36°C at t=0 were pulse-labeled with [^{35}S]methioine (70 μCi/ml) for 15 min. Similarly, the other two aliquots of cells shifted to 36°C at t=0 were pulse-labeled for 15 min starting at t=20 min and t=45 min, respectively. Growth and incorporation were halted by the addition of ice to the medium.

Heat shock of cells for RNA extraction. Concomitant with the pulse labeling described above, cells to be used for RNA extraction were heat treated. A total of 20 ml of cells were used for each extraction, but to ensure the same heat treatment as with labeled cells, these 20 ml were divided into four 5 ml aliquots in four separate 50 ml flasks. Again, as in the pulse labeling, one 20 ml culture was left at 23°C and three 20 ml cultures were shifted to 36°C at t=0 (into pre-warmed 50 ml flasks). At t=20 min, for the 23°C and one of the heat treated cultures, and 35 min and 60 min for the remaining 36°C cultures, growth was halted by the addition of ice.

Preparation of mRNA. Cells were broken by agitation on a Vortex mixer in the presence of glass beads as described previously[17]. The RNA was extracted according to the method of Gallis et al.[17]. This RNA was used to program the in vitro translation system.

Preparation of yeast cell-free translation system. A modified S-100 (S-100') of strain SKQ2n was prepared as described by Tuite et al.[17] with the following modifications. Zymolase (1 mg/10[5] Klett units) was used to prepare spheroplasts rather than glusulase. This required more careful washing of the growth medium from the cells prior to spheroplasting because zymolase is inhibited by some component of the growth medium (unpublished data). After harvesting, cells were washed once with water and once with 1 M sorbitol prior to spheroplasting.

In vitro mRNA-dependent protein synthesis. In vitro translation and precipitations prior to electrophoresis were done as described previously[17].

Sample preparation and two-dimensional gel electrophoresis; in vivo time course. For each time course an equal amount of pulse-labeled cells from each time point was used. Prior to gel electrophoresis ^{3}H-long term-labeled cells were added to each of the pulse-labeled samples to serve as an internal standard for recovery. Cells were suspended in cold distilled water and an equal amount of ^{3}H-labeled cells was added to each of the pulse-labeled samples such that the ^{3}H/^{35}S ratio was approximately 2.

Cell mixtures were centrifuged at 4°C in an Eppendorf 5414 centrifuge for 2 min. The supernatant was decanted and the cells were disrupted and prepared for electrophoresis as described previously[18]. Immediately before loading onto first-dimension isoelectric focusing (IEF) gels, solid urea was added to a concentration of 9.5M and then 20 μl of 2 X IEF sample buffer[19] was added.

Analysis of in vitro translation products. ^{3}H-long term-labeled extract combined with the [^{35}S]methionine-labeled in vitro translation products to serve as an internal standard for recovery. The ^{3}H-long term-labeled extract was prepared as in the previous section up to the point of urea saturation. At this point, the products of an individual in vitro translation (acetone precipitate) were suspended in 20 μl of 2 X IEF sample buffer and added to the ^{3}H-long term-labeled extract.

Electrophoresis was performed as described by O'Farrell[20] using modifications described previously[19].

Determination of relative rates of synthesis of heat shock and non-heat shock proteins. The amount of radioactivity in a polypeptide spot on the two-dimensional gel was determined as described previously[19].

The rates of protein synthesis were determined in A364A and ts 187 cells from the 23°C cultures and from cultures shifted to 36°C. The rate of protein synthesis for each polypeptide spot at each time point was measured as a ratio of the [35S]methionine incorporated during a 15 min pulse label to the 3H-amino acids incorporated into that spot during the 3 h long term label.

Since the rates of protein synthesis in A364A and ts 187 were very different at 36°C, the *in vivo* rate of synthesis of an individual heat shock or non-heat shock protein was expressed relative to the overall *in vivo* rate of synthesis at a given time and temperature. The overall rate of synthesis was determined as follows: The 27 major proteins (non-heat shock proteins) on the O'Farrell gels were cut and their radioactivity determined. The major proteins included the most abundant proteins, defined as the most heavily staining on gels and/or the most rapidly synthesized proteins, defined as those incorporating the most [35S]methionine during a 23°C pulse label. The ratio of total [35S]methionine incorporated into these 27 proteins to the 3H-amino acids incorporated was taken to be the average rate of protein synthesis under the conditions used. The ratio of [35S]methionine to the 3H-amino acids in an individual polypeptide relative to that of the average rate was taken to be the relative rate of synthesis of that individual polypeptide.

In order to quantitate the mRNA levels for a given protein after a heat shock, RNA was extracted from cells shifted to 36°C for the times previously indicated. This RNA was used to program a yeast cell-free protein synthesizing system prepared from non-heat shocked SKQ2n at levels where mRNA was limiting. This allowed a determination of the relative *in vitro* rate of synthesis of an individual heat shock or non-heat shock protein, relative to the overall *in vitro* rate of synthesis. As in the *in vivo* experiments, the 27 most abundant and/or most actively synthesized proteins were used to compute the average rate of synthesis. The ratio of [35S]methionine to 3H-amino acids in an individual polypeptide relative to that of the average rate of synthesis was taken to be the relative level of synthesis of that individual polypeptide. The amount of *in vitro* synthesis was dependent on the amount of mRNA present for that polypeptide. The relative levels of individual mRNAs were calculated for heat shock proteins present at the times during the heat shock at which RNA was extracted.

Chemicals and radioisotopes. [35S]methionine (1,160 Ci/mmol), 3H-yeast protein hydrolysate (1 mCi/ml), and L-[methyl-3H]methionine (18 Ci/mmol) were obtained from Schwarz/Mann. Zymolase 60,000 was obtained from Kirin Brewery Co., Ltd. All other fine chemicals were obtained from standard commercial sources.

RESULTS

Half-lives of heat shock mRNAs. The half-lives of heat shock mRNAs are shown in Table 1. The average mRNA half-life for yeast is 20-23 min at 36°C[16] and has been reported to be 16[21,22] and 22 min[23] at 23°C[21] depending on the method of determination.

TABLE 1

HEAT SHOCK mRNA HALF-LIVES.

Protein	$t^{\frac{1}{2}}$ of mRNA (min)
165	>1 hr
166	>1 hr
192	38
256a	19
256b	16
262	13
263	27
264	12

In vivo and in vitro protein synthetic patterns. Autoradiograms of O'Farrell gels are shown for A364A protein extracts pulse labeled in vivo with [35S]methionine for 15 min at 23°C and 36°C (Fig. 1A and 1B). Enhanced synthesis of heat shock proteins 165, 166; 256a,b; 258 and 264 was evident in the cells shifted to 36°C (Fig. 1B). Figure 2 shows the pattern of in vitro synthesis using RNA extracted from A364A cells grown at 23°C as mRNA. Actin (spot 116) and a major yeast protein (spot 146) are shown on these gels for reference. From the autoradiograms one can see the in vivo induction of the heat shock proteins after 15 min at 36°C and also the ability of the in vitro translation system to faithfully reflect the in vivo protein pattern.

Induction of the heat shock proteins was not maximal at 36°C. Much greater inductions can be seen at higher temperatures (unpublished data). However, 36°C was chosen as the heat shock temperature in order to compare A364A with the temperature sensitive mutant, ts 187, whose restrictive temperature is 36°C.

Rates of protein synthesis for non-heat shock proteins. The rates of protein synthesis in A364A and ts 187 were very different at 36°C (Fig. 3). In

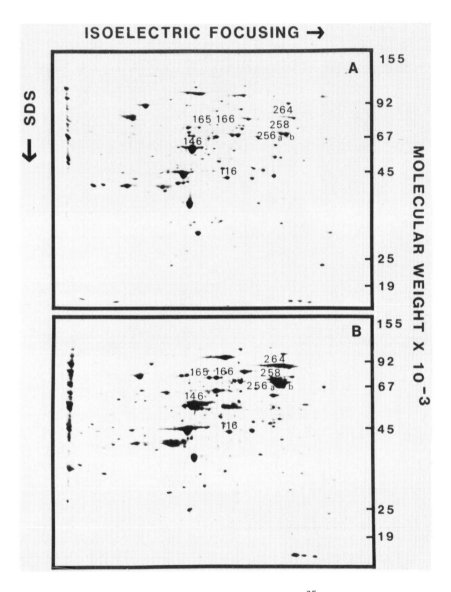

Fig. 1. (A) Autoradiogram of two-dimensional gel of [^{35}S]methionine proteins pulse labeled *in vivo* in A364A for 15 min at 23°C. (B) Autoradiogram of two-dimensional gel of [^{35}S]methionine proteins pulse labeled *in vivo* in A364A for 15 min at 36°C. Heat shock proteins 165, 166; 256a,b; 258, and 264 are indicated. Actin (116) and a major yeast protein (146) are shown for reference. Molecular weight marker proteins used were as described previously[18].

502

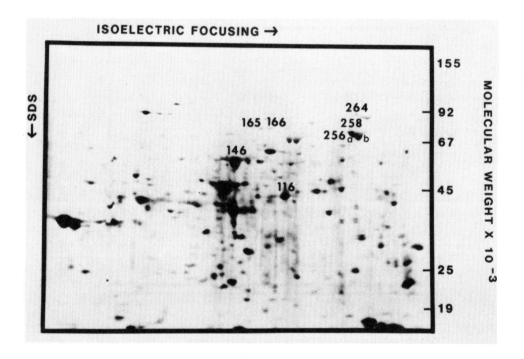

Fig. 2. Autoradiogram of two-dimensional gel of [^{35}S]methionine proteins synthesized *in vitro* using RNA extracted from A364A cells growing at 23°C as mRNA.

A364A, protein synthesis decreased slightly during the first 15 min after a
shift to 36°C and then rose to approximately twice the 23°C rate after 1 hr.
However, 36°C is the restrictive temperature for ts 187 and the rate of protein
synthesis decreased to less than 5% of the 23°C rate within minutes of a shift
to 36°C (Fig. 3).

Because of this difference in overall protein synthetic rate in ts 187 as
compared to A364A, it was necessary to express the rate of synthesis of individual polypeptides in each strain relative to the overall rate of protein synthesis in each strain. Figure 4 shows the relative synthetic rates for two
non-heat shock proteins in A364A and ts 187 after a shift from 23°C to 36°C.
Protein 116 has been identified as actin[18] and protein 146 has been identified
as a major yeast protein. When the synthetic rates were expressed in this

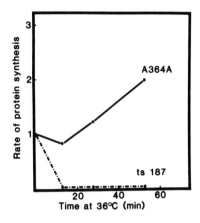

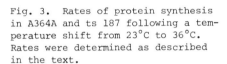

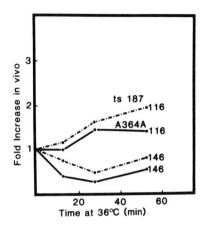

Fig. 3. Rates of protein synthesis in A364A and ts 187 following a temperature shift from 23°C to 36°C. Rates were determined as described in the text.

Fig. 4. Rates of synthesis of individual non-heat shock proteins in A364A and ts 187 following a temperature shift from 23°C to 36°C. Results are shown for an abundant yeast protein (146), and an identified protein, actin (116). Fold increase was determined as described in the text.

manner, it was clear that the relative synthetic rates of these proteins were similar in the mutant and in the parent strain. Also, the relative rate of synthesis of these individual proteins changed very little following a temperature shift of either strain. This occurred despite the fact that the overall rate of protein synthesis was quite different in the two strains (comparing Fig. 3 with Fig. 4). This was typical of the pattern of synthesis observed for the average non-heat shock proteins.

Rates of protein synthesis for heat shock proteins. When the synthetic rates of the heat shock proteins were expressed in the same manner, i.e. relative to the overall rate of synthesis of the non-heat shock proteins, a very different pattern was observed. Figures 5 and 6 show the rates of synthesis of the heat shock proteins following temperature shifts of A364A and ts 187, respectively. A relative increase in the rate of synthesis of the heat shock proteins of from 6 to 30 fold was observed in A364A (Fig. 5). This increase reached a maximum 15 to 30 min after the temperature shift. The pattern of protein synthesis began to return to the 23°C pattern after this time. This recovery was nearly complete by 1 hr. This transient induction, followed by a

504

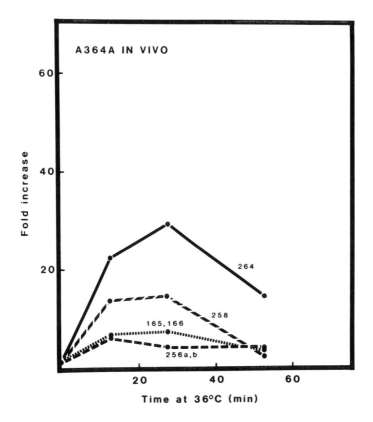

Fig. 5. *In vivo* heat shock protein synthesis in A364A. Fold increase was determined as described in the text.

recovery to the pre-shock pattern of protein synthesis, is characteristic of the heat shock response[6,7,24]. In ts 187 (Fig. 6), relative rates of synthesis of the heat shock proteins increased from 7 to 130 fold and exhibited a similar time course to that observed in A364A.

Quantitation of transcriptional and translational components of the heat shock response. In order to determine the relative contributions of transcription and translation to this increased synthesis of heat shock proteins, RNA was isolated from these two strains and translated *in vitro*. RNA was extracted from cells grown at 23°C and from cells that had been shifted to 36°C for various periods of time. This RNA was used to program a yeast cell-free protein synthesizing system prepared from non-heat shocked cells. Limiting concentra-

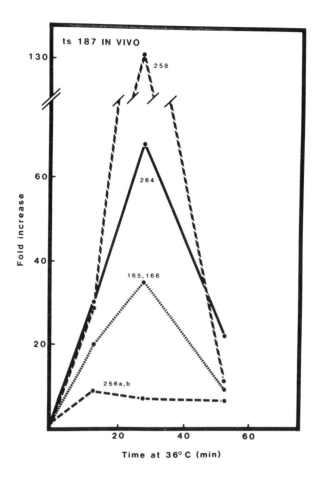

Fig. 6. *In vivo* heat shock protein synthesis in ts 187.

tions of mRNA were used so that the relative amount of an individual polypeptide synthesized *in vitro* was a reflection of the relative abundance of that individual mRNA in the RNA extract. ^{3}H-long term-labeled cells were disrupted, and mixed with [^{35}S]methionine-labeled cell-free products to serve as in internal standard.

The results of the *in vitro* incorporations are shown in Figs. 7 and 8 for A364A and ts 187, respectively. Both A364A and ts 187 exhibited relative increases in heat shock mRNAs after a shift to 36°C. The relative mRNA levels in A364A were increased up to 15 fold, while the levels in ts 187 were in-

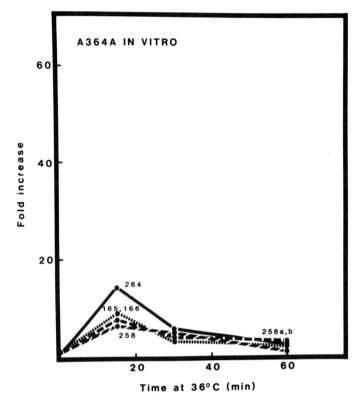

Fig. 7. *In vitro* synthetic levels for individual heat shock proteins in A364A. Fold increase was determined as described in the text.

creased up to 27 fold. A comparison of Figs. 7 and 8 shows that, in all cases but one (protein spot 256a,b), ts 187 showed a greater relative induction of heat shock mRNAs than was seen in A364A. These results provided a measure of the transcriptional response to a heat shock in these two strains. However, the response observed *in vivo* would consist of both the transcriptional component and any translational component, if any were present.

In order to demonstrate if the effect observed *in vivo* could be accounted for by the transcriptional response alone, the *in vivo* and *in vitro* data for ts 187 and A364A were directly compared in Figs. 9 and 10. Figure 9 shows the relative increase in mRNA abundance (from the *in vitro* incorporation levels) and the relative increase in protein synthetic rate (*in vivo*) for each heat shock protein examined in ts 187. In ts 187 there was a greater relative

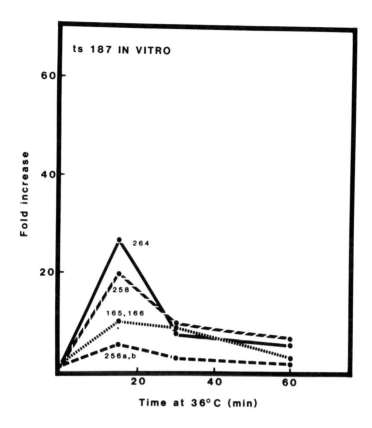

Fig. 8. *In vitro* synthetic levels for individual heat shock proteins in ts 187.

synthesis of each heat shock protein *in vivo* as compared with the relative
level of messenger RNA of each heat shock protein measured *in vitro*. This
indicated that, in addition to the transcriptional response to heat shock,
there was also a preferential translation of the heat shock mRNAs *in vivo*.
From this, we have calculated the relative magnitude of the transcriptional
and translational components of the overall heat shock response in ts 187.
For example, protein 264 showed a 27 fold increase in relative mRNA level and
a 68 fold increase in relative protein synthetic rate *in vivo*. This indicated
that there was an approximate 2.5 fold enhancement at the translational level
compared to the 27 fold increase observed at the transcriptional level. For
heat shock proteins 165, 166; 256a,b; and 258 the translational enhancements

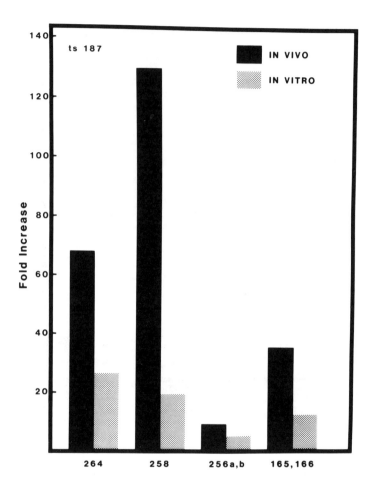

Fig. 9. Relative increases (fold increase) in mRNA abundance (*in vitro*) and protein synthetic rates (*in vivo*) for heat shock proteins in ts 187. An increase in the *in vivo* level over that of the *in vitro* level indicates the presence of a translational enhancement *in vivo*. The relative increase (fold increase) was determined as described in the text.

varied from 1.8 fold for protein 256a,b to 6.8 fold for protein 258. For each of the heat shock proteins examined the major effect was at the transcriptional level (Table 2).

When the same comparison was made for A364A (Fig. 10) the evidence for a translational effect was less evident. There was a 2 to 3 fold translational enhancement for proteins 258 and 264. For proteins 165, 166 and 256a,b no translational effect was observed.

Table 2

QUANTITATION OF TRANSCRIPTIONAL AND TRANSLATIONAL EFFECTS IN ts 187.

Heat shock proteins	Transcriptional effect (mRNA fold increase)	*In vivo* fold increase	Translational effect
165, 166	12	35	2.9
256a,b	5	9	1.8
258	19	130	6.8
264	27	68	2.5

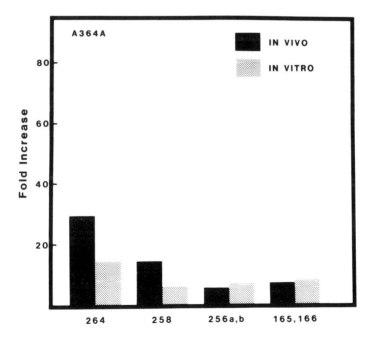

Fig. 10. Relative increases (fold increase) in mRNA abundance (*in vitro*) and protein synthetic rates (*in vivo*) for heat shock proteins in A364A.

510

DISCUSSION

Heat shock in yeast has been observed by us and others[7,8,9.25,26]. The results presented here indicate that in wild type *S. cerevisiae* (A364A) the enhanced rate of synthesis of heat shock proteins induced by a mild (36°C) heat shock was mediated by increasing the amount of heat shock mRNA available. That this enhanced mRNA level is due to an increased rate of transcription of heat shock mRNA is evident from the mRNA half-lives, which at the heat shock temperature[16,21,22,23] are not significantly different from the average mRNA half-life at either the permissive or restrictive temperature. The relative heat shock mRNA levels in the wild type strain (A364A) increased 7 to 15 fold within 15 min of the temperature shift. Each protein exhibited its own distinctive increase in its rate of synthesis and mRNA level in response to the heat shock. Under these conditions the rate of synthesis of each heat shock protein was controlled at the level of transcription. The mutant strain ts 187 (*prt-1*), which has a temperature sensitive lesion in protein synthesis[13], exhibited a very strong heat shock response at 36°C. The overall rate of protein synthesis was rapidly inhibited in the mutant by the shift from 23°C to 36°C in contrast to the wild type where the rate of protein synthesis increased upon a shift to 36°C. The transcriptional induction of heat shock proteins was seen in both mutant and wild type. This demonstrated that heat shock protein induction can occur under conditions that stimulate or inhibit the overall rate of protein synthesis.

The response of yeast cells to a heat shock is actually quite complex. The synthesis of some proteins, the heat shock proteins, is enhanced compared to the average proteins. At the same time the synthesis of another class of proteins, the heat stroke proteins, is strongly inhibited by the heat shock[6,27]. Within each class, heat shock, average, and heat stroke, each protein exhibits a specific individual pattern of induction or repression. The response as a whole represents something of a continuum separated into a major class of average proteins and two smaller classes, the heat shock proteins and the heat stroke proteins.

The function of most of the proteins whose synthesis is most affected by the heat treatment remains unknown. The major exception is the identification of the ribosomal proteins as a part of the heat stroke class[28]. These proteins undergo a transient inhibition in their rate of synthesis with a time course similar to that of the heat shock proteins. The response to heat is transient; the relative steady state rate of synthesis of the heat shock proteins at 36°C and at 23°C are comparable. This is also true for the heat stroke proteins[27]

The steady state rate of synthesis of the heat stroke proteins, including the ribosomal proteins, is high while the steady state rate of synthesis of the heat shock proteins is low. The outline of a reasonable hypothesis to explain the inhibition of ribosomal protein synthesis is suggested by the observation that the rate of protein synthesis immediately and transiently doubles during the shift from 23°C to 36°C[16,29,30]. The number of ribosomes per polysome remains constant under these conditions which indicates that the elongation rate doubles almost instantaneously during the temperature shift. Under these conditions the cell apparently senses this transient excess of ribosomes relative to the rest of the cells synthetic capabilities and quickly inhibits their rate of synthesis.

The control of the heat shock response in mutant ts 187 had both a transcriptional and a translational component. The control at the level of transcription was quantitatively the more important. For example, protein 264 showed a transcriptional response of 27 fold and a translational response of 2.5 fold. These results clearly indicated that yeast are capable of a translational response as well as a transcriptional response to heat shock.

The transcriptional induction of the heat shock proteins can occur at different rates of protein synthesis. However, the translational component may well depend on the precise conditions of the protein synthetic apparatus. We observed a much stronger translational effect in ts 187 at 36°C where protein synthesis was severely inhibited than in A364A at 36°C where protein synthesis was unhindered. Strong heat shocks inhibit the rate of protein synthesis in A364A. In A364A the optimal temperature for the induction of the heat shock proteins is higher than 36°C (unpublished data). As the temperature of the heat shock is increased above 36°C protein synthesis is inhibited at the level of initiation. Under the conditions of severely inhibited protein synthesis in the mutant at 36°C and the wild type at higher temperatures a transcriptional response alone may not ensure an adequate rate of synthesis of the heat shock proteins.

When one looks at the heat shock response in other oganisms the following correlation emerges. A moderate heat shock invokes a response at the transcriptional level. A more severe heat shock produces a stronger induction of the heat shock proteins and also causes an inhibition of protein synthesis at the level of initiation. In plants, moderate heat induces an effect upon transcription, whereas a more severe heat shock inhibits protein synthesis and elicits the maximum production of heat shock proteins. The optimal temperature for the induction of heat shock proteins in soybean cells is 39-40°C, a temperature at which overall protein synthesis is less than 50% of the 25°C level[4]. Polysomes are also reduced during the heat shock response in soybeans, implicating an inhibition of protein synthesis initiation at the optimal heat shock temperature[5].

Experiments in *Drosophila* have shown the following pattern of heat shock protein induction. At moderate heat shocks (up to 33°C[24] or 35°C[31]) the normal complement of cellular proteins is made but increased amounts of heat shock proteins are also synthesized. As the temperature of the heat shock is raised further, the heat shock proteins begin to predominate and by 41°C, where protein synthesis is inhibited. the heat shock mRNAs are clearly translated preferentially[24]. However, Lindquist[32] has reported that translational selectivity in *Drosophila* is not necessarily dependent upon a reduced initiation level, but in her experiments she does see a drop in polysomes from 80% to 50% during the temperature shift. This suggests either a limitation of initiation or a shift of ribosomes from polysomes to inactive monosomes.

A cell-free translation system has been prepared from *Drosophila* that maintains a translational selectivity[12,33,34]. Lysates prepared from control cells were shown to translate both control and heat shock mRNAs, whereas lysates prepared from heat shocked cells were shown to only translate heat shock mRNAs. The factor (or factors) responsible for this selectivity has not been characterized.

On the whole these experiments are consistent with the patterns observed in yeast, i.e. translational control is exerted when protein synthesis is inhibited. However, it is possible that several different methods of translational control may have evolved to ensure the synthesis of heat shock proteins under these more stressed conditions. Although further studies are required in this area to define the exact mechanisms of translational control involved, translational control may be a general response to a severe heat shock.

The models for a translational regulation are basically of two types. One invokes a change in level or specificity of an initiation factor which favors one mRNA over another. The second suggests that there is some structural feature of individual mRNAs that favor their preferential translation under some circumstances. There is evidence to suggest that each of these models operate in some systems. Lodish[35] has proposed that, under conditions where ribosomes are limiting, mRNAs with higher intrinsic affinities for ribosomes would be preferentially translated. He has presented evidence supporting this model using the translation of α and β hemoglobin mRNAs. Evidence suggesting that levels of initiation factors affect mRNA selectivity has also been presented.

DiSegni *et al.*[36] have shown that levels of eIF-2 can affect the competition between α and β globin synthesis and Lodish *et al.*[37] have shown that an unidentified factor required for chain initiation, which is not normally rate-limiting can, under some circumstances, affect competition for translation between two species of *Vesicular Stomatitis* virus mRNAs.

There are also other factors that can operate at a post-transcription level to affect translation - changes in mRNA stability or the sequestering of mRNAs. The precise mechanisms responsible for the translational control of the heat shock response are not yet clear. However, we do have some mechanistic information in yeast about the circumstances under which this translational control is observed. The lesion in ts 187 has been identified as being between the formation of the ternary complex (eIF-2:GTP:met-tRNA$_i$) and the binding and/or stabilization of the ternary complex to the 40S ribosomal subunit (K. Moldave; personal communication). This block would severely limit subsequent initiation steps, including mRNA binding. This may indicate that structural features in yeast heat shock mRNAs are responsible for their preferential translation under these conditions. This would be an elegant and simple method of ensuring their preferential translation.

A comparison of the transcriptional element of the heat shock response in the mutant and the wild type cell indicated that, at the same temperature, relative transcription was more elevated in the mutant than the wild type. This suggested that either the defect in the mutant was able to directly interact with the induction system or, more likely, the production of heat shock proteins was autoregulated. Some type of autoregulation would be consistent with the transient nature of the response in all organisms.

These studies suggest that yeast, with its complement of RNA and protein synthesis mutants, may be an ideal organism in which to study the mechanism of induction and function of the heat shock proteins.

ACKNOWLEDGMENTS

This work was supported by Research Grants AA 03506 from the National Institute of Alcohol Abuse and Alcoholism and AI 16252 from the National Institute of Allergy and Infectious Diseases. J.J.F. was supported in part by Public Health Service Grant CA 06504.

REFERENCES

1. Ashburner, M., Bonner, J.J. (1979) Cell 17, 241-254.
2. Mirault, M.E., Goldishmidt-Clermont, M., Moran, L., Arrigo, A.P., Tissie-res, A. (1977) Cold Spring Harbor Symposium on Quantitative Biology 42, 819-827.
3. Kelly, P.M., Schlesinger, M.J. (1978) Cell 15, 1277-1286.
4. Barnett, T., Altschuler, M., McDaniel, C.N., Mascareshas, J.P. (1980) Dev. Genetics 1, 3331-3340.
5. Key, J.L., Lin, C.Y., Chen, Y.M. (1981) Proc. Natl. Acad. Sci. USA 78, 3526-3530.
6. Miller, M.J., Xuong, N.-H., Geiduschek, E.P. (1979) Proc. Natl. Acad. Sci. USA 76, 5222-5225.
7. McAlister, L., Strausberg, S., Kulaga, A., Finkelstein, D.B. (1979) Current Genetics 1, 63-74.
8. McAlister, L., Finkelstein, D. (1980) J. Bacteriol. 143, 603-612.
9. Lindquist, S. (1981) Nature 293, 311-314.
10. Lindquist-McKenzie, S., Meselson, M. (1977) J. Mol. Biol. 117-279-283.
11. Peterson, N., Mitchell, H.K. (1980) Dev. Biol. 77, 463-479.
12. Sorti, R.V., Scott, M.P., Rich, A., Pardue, M.L. (1980) Cell 22, 825-834.
13. Hartwell, L.H., McLaughlin, C.S. (1969) Proc. Natl. Acad. Sci. USA 62, 468-474.
14. Hartwell, L.H. (1967) J. Bacteriol. 93, 1662-1670.
15. Gasior, E., Herrera, F., Sadnik, I., McLaughlin, C.S., Moldave, K. (1979) J. Biol. Chem. 254, 3965-3969.
16. Chia, L., McLaughlin, C.S. (1979) Molec. Gen. Genet. 170, 137-144.
17. Tuite, M.F., Plesset, J., Moldave, K., McLaughlin, C.S. (1980) J. Biol. Chem. 255, 8761-8766.
18. Ludwig II, J.R., Foy, J.J., Elliott, S.G., McLaughlin, C.S. (1982) Mol. Cell Biol. 2, 117-126.
19. Elliott, S.G., McLaughlin, C.S. (1978) Proc. Natl. Acad. Sci. USA 75, 4384-4388.
20. O'Farrell, P.H. (1975) J. Biol. Chem. 250, 4007-4021.
21. Hynes, N.E., Phillips, S.L. (1976) J. Bacteriol. 125, 595-600.
22. Peterson, N.S., McLaughlin, C.S., Nierlich, D.P. (1976) Nature 260, 7072.
23. Tonneson, T., Friesen, J.D. (1973) J. Bacteriol. 115, 889-896.
24. Peterson, N.S., Mitchell, H.K. (1981) Proc. Natl. Acad. Sci. USA 78, 1708-1711.
25. Plesset, J., Chia, L., Palm, C., McLaughlin, C.S. (1980) Abst. of Intl. Conf. of Yeast Genet. Mol. Biol. 365, 127.
26. Plesset, J., Foy, J.J., Chia, L., McLaughlin, C.S. (1982) Abst. Annu. Meet. Am. Soc. Microbiol. 486, 127.
27. Chia, L. (1980) Ph.D. Thesis, University of California, Irvine.
28. Gorenstein, C., Warner, J.R. (1976) Proc. Natl. Acad. Sci. USA 73, 1547-1551.
29. Hartwell, L.H., McLaughlin, C.S. (1968) J. Bacteriol. 96, 1664-1671.
30. Hartwell, L.H., McLaughlin, C.S., Warner, J.R. (1970) Molec. Gen. Genet. 109, 42-56.
31. Lindquist, S. (1980) Dev. Biol. 77, 463-479.
32. Lindquist, S. (1980) J. Mol. Biol. 137, 151-158.
33. Kruger, C., Benecke, B.J. (1981) Cell 23, 595-603.
34. Scott, M.P., Pardue, M.L. (1981) Proc. Natl. Acad. Sci. USA 78, 3353-3357.
35. Lodish, H.F. (1974) Nature 251, 385-388.
36. DiSegni, G., Rosen, H., Kaempfer, R. (1979) Biochem. 18, 2847-2854.
37. Lodish, H.F., Froshauer, S. (1977) J. Biol. Chem. 232, 8804-8811.

Index

DATE DUE